Clinical Antimicrobial Assays

Edited by

Professor D. S. Reeves, MD, FRCPath
Department of Medical Microbiology
Southmead Health Services NHS Trust
Bristol, United Kingdom

Professor R. Wise, MD, FRCPath
Department of Medical Microbiology
City Hospital NHS Trust
Birmingham, United Kingdom

Mrs J. M. Andrews, MSc
Department of Medical Microbiology
City Hospital NHS Trust
Birmingham, United Kingdom

Dr L. O. White, BSc, PhD
Department of Medical Microbiology
Southmead Health Services NHS Trust
Bristol, United Kingdom

Foreword by

Professor D. Speller
Emeritus Professor of Microbiology
Bristol University
Bristol, United Kingdom

OXFORD
UNIVERSITY PRESS

Oxford University Press, Great Clarendon Street, Oxford OX2 6DP, UK

Oxford   New York

*Athens   Auckland   Bangkok   Bogota   Buenos Aires   Calcutta
Cape Town   Chennai   Dar es Salaam   Delhi   Florence   Hong Kong   Istanbul
Karachi   Kuala Lumpur   Madrid   Melbourne   Mexico City   Mumbai
Nairobi   Paris   São Paulo   Singapore   Taipei   Tokyo   Toronto   Warsaw*

*and associated companies in
Berlin   Ibadan*

Oxford is a trade mark of Oxford University Press

*Published in the United States
by Oxford University Press Inc., New York*

A catalogue record for this book is available from the British Library

*Library of Congress Cataloging in Publication Data
(Data available)*

ISBN 0 19 922325 4

*This book is also available as a special hardback supplement to the
Journal of Antimicrobial Chemotherapy 1999
ISSN 0305-7453*

*Typeset by Footnote Graphics, Warminster, Wilts
Printed in Great Britain by Bell & Bain Ltd, Glasgow, UK*

Foreword

The original 'Blue Book', *Laboratory Methods in Antimicrobial Chemotherapy*, of which I have in my hands a battered and much annotated copy, was one of the first publications to arise from the activities of the British Society for Antimicrobial Chemotherapy. There was a perceived need in the 1970s to assemble, assess, organize and interpret diverse and scattered information about laboratory aspects of antibiotic therapy, and the first book included antibiotic sensitivity testing as well as assay. It was received gratefully and extensively used.

Now, twenty years on, we welcome—from the efforts of a team including two of the original editors—a greatly expanded and updated publication on antibiotic assay to greet a very different world of infection and chemotherapy. Whilst in the United Kingdom the methodology for sensitivity testing is undergoing a paroxysmal reassessment, the progress in assay of antimicrobials appears to have been steady. To be convinced of the advances we have only to consider: the comprehensive pharmacokinetic data expected and available on all new compounds; the improved technical quality of assays as evidenced by the results of quality assurance; or the ready service of assay and advice in the deployment of anti-biotics for which considerations of effectiveness or toxicity make this desirable. Nevertheless, as anyone who has worked in a reference laboratory or as an editor in the European forum will realize, there is still a pressing requirement for education and information in this field.

The new compilation begins critically with clinical interpretation, surely the commonest reason for consulting the text in a clinical laboratory setting. This chapter is followed by several general topics to promote understanding of processes *in vitro* and *in vivo*, before the individual classes of agent are reviewed. Recent concern about the threat to antibiotic treatment posed by multiple resistance among serious pathogens is forcing all prescribers of antibiotics to ensure that any course of treatment is focused optimally to the clinical need. This book will greatly assist in this endeavour.

David C.E. Speller M.A., B.M., B. Ch., F.R.C.P., F.R.C.Path.
Emeritus Professor of Clinical Bacteriology
University of Bristol
United Kingdom

Preface

In 1978 the forerunner of this book appeared as *Laboratory Methods in Antimicrobial Therapy* (editors David Reeves, Ian Phillips, David Williams and Richard Wise), a book which covered both susceptibility testing and assay of antimicrobials. In its Foreword Professor L. P. Garrod likened an antimicrobial prescriber who lacked laboratory assistance to a mariner without a chart or compass. It is certainly true that in antimicrobial chemotherapy there is an extremely close link between the laboratory and the prescriber, and this may be as true today as it was in 1978. Indeed as the choice of therapeutic agent becomes more limited by the greater frequency of antimicrobial resistance, it could be said that the link between the prescriber and the laboratory is becoming all the more imporant.

After publication of the original book attempts were made to produce an updated edition. These were overtaken by the decision of the British Society for Antimicrobial Chemotherapy (BSAC) to set up a Working Party on Susceptibility Testing, chaired initially by Ian Phillips and then Richard Wise. The advice offered and published by the Working Party covered the first half of the original *Laboratory Methods in Antimicrobial Therapy* so it was decided that new volume concentrating entirely on assays should be produced, a venture supported by the BSAC.

Assay methodologies in the clinical situation have changed considerably. Twenty years ago both chromatographic, including high performance liquid chromatography (HPLC), and immunological assays were in their infancy. The drugs themselves have changed also—the fluoroquinolones were as yet undeveloped and some compounds such as chloramphenicol were enjoying greater use than they are today. It seemed entirely appropriate that an updated book on assays should be published, and we as editors hope that members of the BSAC and the wider readership will find this a useful service.

In the preface to *Laboratory Methods in Antimicrobial Therapy* the editors commented that its production had proved to be 'a lengthy project'. Some things never change and we are grateful to the experts recruited to write chapters in this book in their patience both in waiting to see their labours in print and in updating their chapters when requested. Chemotherapy is a rapidly developing area and they are, unavoidably, a few *lacunae* in this book so as to not further delay publication. For example, there is nothing on the assay of streptogramins (where there is little published information on how to assay variable synergistic mixtures of compounds), on newer agents such as oxalidinones or everninomycins, or on older agents such as colistin being used in new ways. Assay methodology too is leaping into the next century. Enormous technological advance are being made and it is already possible to measure tissue levels of antimicrobials in intact live human subjects through the use of positron emission tomography.

The editors are pleased to express their thanks to all the expert contributors who have made this book possible. Since some areas, such as interpretation of drug levels in a clinical setting, are still to a greater or lesser extent a matter of individual judgment, the reader may note that not all chapter authors are in perfect harmony. This slight discord is both unavoidable and proper since it helps highlight areas were more research is needed.

D. S. R., R. W., J. M. A., L. O. W.

Contents

Contributors

J. M. Andrews
Department of Medical Microbiology
City Hospital NHS Trust
Birmingham B18 7QH
United Kingdom

N. M. Brown
Clinical Microbiology and Public Health Laboratory
Addenbrooke's Hospital
Cambridge CB2 2QW
United Kingdom

D. Felmingham
GR Micro Ltd
7–9 William Road
London NW1 3ER
United Kingdom

P. J. Jenner
Division of Mycobacterial Research
National Institute for Medical Research
London NW7 1AA
United Kingdom

E. M. Johnson
PHLS Mycology Reference Laboratory
Public Health Laboratory
Bristol BS2 8EL
United Kingdom

A. M. Lovering
Department of Medical Microbiology
Southmead Hospital
Bristol BS10 5NB
United Kingdom

A. P. MacGowan
Department of Medical Microbiology
Southmead Hospital
Bristol BS10 5NB
United Kingdom

D. A. Oliver
PHLS Mycology Reference Laboratory
Public Health Laboratory
Bristol BS2 8EL
United Kingdom

D. S. Reeves
Department of Medical Microbiology
Southmead Hospital
Bristol BS10 5NB
United Kingdom

G. Sweeney
Department of Bacteriology
Southern General Hospital NHS Trust
Glasgow G51 4TF
United Kingdom

C. M. Tobin
Department of Medical Microbiology
Southmead Hospital
Bristol BS10 5NB
United Kingdom

D. W. Warnock
PHLS Mycology Reference Laboratory
Public Health Laboratory
Bristol BS2 8EL
United Kingdom

L. O. White
Department of Medical Microbiology
Southmead Hospital
Bristol BS10 5NB
United Kingdom

R. Wise
Department of Medical Microbiology
City Hospital NHS Trust
Birmingham B18 7QH
United Kingdom

J. M. Woodcock
Department of Medical Microbiology
City Hospital NHS Trust
Birmingham B18 7QH
United Kingdom

Chapter 1

Clinical interpretation of antimicrobial assays

Alasdair MacGowan[a], **David Reeves**[a] **and Richard Wise**[b]

[a]*Department of Medical Microbiology, Southmead Hospital, Bristol BS10 5NB;* [b]*Department of Medical Microbiology, City Hospital NHS Trust, Dudley Road, Birmingham B18 7QH, UK*

Introduction

Although antimicrobial drugs (often colloquially termed antimicrobials) have been extensively used in the treatment of human disease for over 50 years, there is remarkably little information on the relationship of blood concentrations to either the outcome of treatment or the adverse effects of the drugs. The reasons for this are both understandable and readily apparent. Antimicrobials usually have a large therapeutic index and are thus traditionally given in large dosages, often exceeding those necessary for successful treatment, so in the unlikely event of blood concentrations being measured they are unlikely to relate to outcome. Furthermore, the body's own defence mechanisms frequently play a major or even predominant part in the eradication of infection such that cure would be possible in the absence of any antimicrobial therapy, which further obscures the effect of that therapy and the importance, or otherwise, of drug concentrations. Finally, it is difficult or impossible ethically to do placebo-controlled or dose-ranging studies in many acute infections, so information on the importance of concentrations is unlikely to emerge from clinical trials, although there has been recently interest in concentration-controlled trials (MacGowan, 1996). Much of the information on the relationship of antimicrobial concentrations to cure or adverse effects has been acquired opportunistically or by retrospective analysis of therapeutic drug monitoring data, hardly ideal modes of establishing a systematic and reliable knowledge base.

The tradition of happily giving supramaximal dosages of antimicrobials was severely perturbed by the advent of aminoglycoside therapy (other than streptomycin) in the 1970s since it was soon realized that these antimicrobials had a small therapeutic index and that concentrations of them in blood (or, for practical purposes, in serum) required monitoring if they were to be used effectively and safely. Since then, technical developments in assay methods have greatly facilitated the performance of speedy and accurate assays and the therapeutic drug monitoring of aminoglycosides and certain other antimicrobials has become an accepted part of managing therapy.

Antimicrobial assays are performed for many reasons. All new antimicrobials are assayed as part of a development programme to establish their pharmacokinetic properties, firstly in animals, then in healthy volunteers, and later often in patients to study drug or food interactions or tissue penetration. Serum concentration may subsequently be measured in clinical trials. Opportunities may be taken to study a drug's pharmacokinetics in the young, elderly and others in whom parameters may be modified, such as patients with renal or hepatic failure. The importance of pharmacodynamics in deciding optimal antimicrobial dosing and therapeutic drug monitoring has recently been reviewed (Dawson & Reeves, 1997; MacGowan & Bowker, 1997).

Therapeutic drug monitoring is performed for five main reasons:

- To ensure that serum concentrations are sufficiently high for efficacy but that toxicity is minimized for those agents with a narrow therapeutic index (e.g. aminoglycosides).
- To ensure that serum concentrations are sufficiently high for efficacy where recommended levels may vary with the pathogen and the drug is too expensive to justify giving all patients high doses (e.g. teicoplanin).
- To monitor patient compliance (e.g. anti-tuberculosis therapy).
- To confirm adequate bioavailability when changing from parenteral to oral therapy (e.g. quinolones, β-lactams).
- Given the increasing complexity of medical care in patients receiving therapies which will alter drug pharmacokinetics in an unpredictable way (for example, serious sepsis plus continuous veno-venous haemofiltration), monitoring of levels of antimicrobials can help to ensure that therapy is optimal.

Some new agents might, after their introduction into clinical practice, also be monitored in some patient groups where available data on likely concentrations are sparse and clinicians require assistance in deciding optimal dosing.

To be of value, therapeutic drug monitoring must be applied to those agents where there is a known or probable relationship between the measured concentration and the efficacy or toxicity. Furthermore, drug pharmacokinetics should be sufficiently unpredictable in the patient concerned that serum concentrations could not be modelled accurately based on patient factors such as age, sex, renal function and dosage alone.

Sampling blood for assays

The important elements of a therapeutic drug monitoring service are listed in Table I. Pre- and post-dose samples for assay will usually provide sufficient information.

Post-dose samples (often loosely and almost certainly incorrectly termed 'peak' samples) should be taken once the distribution phase is complete and thus early in the elimination phase. This time will vary between agents. For example, the appropriate time is 1 h post-dose for aminoglycosides but 2 h post-dose for vancomycin. Although accuracy in sampling should always be sought, in general it is better to take a post-dose sample slightly late than too early, so that it is in the elimination phase (when concentrations will usually be falling much less rapidly than in the distribution phase) and errors of timing will be minimized in terms of concentration changes. It is particularly important to sample in the elimination phase if an elimination-half life ($t_{1/2\beta}$) is to be estimated for the purposes of determing a dosing schedule. Sampling inadvertently in the distribution phase will cause the $t_{1/2\beta}$ to be underestimated. Sampling very early after a bolus intravenous dose is likely to reveal the transient and very high concentration of antimicrobial which is likely to have little or no relationship to efficacy or toxicity.

Pre-dose samples (often loosely and probably incorrectly termed 'trough' samples) are valuable in detecting drug accumulation, especially when measured at regular intervals, since they are much less subject to the vagaries of sample timing and drug administration than post-dose samples.

Only occasionally will an assay of a single blood sample taken at no particular time after a dose but with a knowledge of dose-sample timing (a so-called 'random' sample) may be useful, for example in patients with renal failure in whom the $t_{1/2\beta}$ of an antimicrobial is very prolonged.

It may seem an obvious precaution, but one sometimes honoured in the breach, that blood should never be taken down an intravenous line used for giving the antimicrobial, since the dosage form is in such high concentrations relative to that anticipated in the blood that even the slightest residual contamination by it will give a spurious highly elevated assay result.

Enquiry should be made by the clinician requesting the assay of the assaying laboratory as to the type of sample required for, e.g., plasma or serum (serum is almost always acceptable, plasma may not be), its volume and any transport conditions. Details of the dosage and its timing of the requested antimicrobial should always accompany a request, along with any other relevant patient details such as the existence of impairment of renal function.

In the rest of this chapter we will concentrate on the clinical data available to justify certain serum therapeutic ranges rather than on the delivery of a therapeutic drug monitoring service as a whole.

Aminoglycosides

There are data relating serum concentrations of aminoglycosides to clinical efficacy. Most were collected in the 1970/1980s and are related to regimens in which drugs were given by multiple daily doses. Since the introduction of once-daily dosing there have been few good studies to relate serum concentrations to clinical outcomes.

Table I. Elements of a successful therapeutic drug monitoring service for antimicrobials

1. Clear guidance as to when samples for assay should be taken, reinforced if possible by ward or computerized prescription review by clinical pharmacists. Where a computerized laboratory requesting system exists it should have appropriate algorithms for therapeutic drug monitoring.
2. Willingness to find out the dosing regimen when this information is not initially provided.
3. Availability of specimen analysis within a single dosing interval in most routine situations.
4. High quality analytical result based on:
   - internal and external quality control/assurance procedures
   - well trained, interested and committed staff of sufficient expertise and number
   - well maintained and appropriate equipment
   - range of assays offered on-site appropriate to clinical need and access to specialist assay services
   - clear Standard Operating Procedures
   - accreditation with Clinical Pathology Accreditation (CPA (UK)), or equivalent
   - appropriate funding and lines of managerial accountability
   - availability of ongoing postgraduate education, for example Continuing Professional Development and Continuing Medical Education (CPD/CME)
5. Availability of clinical advice on assay interpretation, with or without formal pharmacokinetic evaluation.
6. Availability of broad-based knowledge of infection management so alternative therapies can be offered if assayed agents are not most appropriate to clinical need.

In the early 1970s, post-dose serum gentamicin concentrations of >4 mg/L were associated with clearance of *Pseudomonas aeruginosa* bacteraemia (Jackson & Riff, 1971; Siber *et al*., 1975) but the first large studies to relate blood concentrations to outcomes were those of Noone and his colleagues (Noone *et al*., 1974a,b; Noone & Rogers, 1976). One study involved over 100 patients with infections caused by aerobic Gram-negative bacilli, mainly *Escherichia coli* or *P. aeruginosa*, and originating in urine, lung, skin and soft tissue and as primary bacteraemia. Favourable clinical responses were observed in those with skin infection, bacteraemia and urinary tract infection (UTI) if the post-dose concentration was >5 mg/L in the first 72 h. In patients with pneumonia a concentration of >8 mg/L at any time during the therapy was associated with cure. A further study of pneumonia alone caused by *P. aeruginosa* or Enterobacteriaceae confirmed the initial findings.

In the 1980s Moore and his colleagues studied the effect on outcome of (i) post-dose serum concentration and (ii) the ratio of post-dose concentration to the MIC (Moore *et al*., 1984a,b, 1987). Patients with pneumonia having a post-dose concentration of >7 mg/L had a higher response rate than those with concentrations of <6 mg/L, while in bacteraemia favourable outcomes were observed if the concentration was >5 mg/L. From a final study of a large group of patients with Gram-negative infection from various sites but mainly UTI, it was shown that the post-dose concentration:MIC ratio was related to outcome. There are at present no data to relate post-dose concentrations to outcome in patients dosed once-daily, although in laboratory models post-dose concentration:MIC ratios of >8–10 have been associated with optimal bacterial killing, avoidance of regrowth and reduced risk of emergence of resistance (Freeman *et al*., 1997; Kashuba *et al*., 1998).

While there is little doubt that aminoglycosides are potentially nephrotoxic, it is unclear whether initial high serum concentrations precede nephrotoxicity or are a result of it; high pre-dose concentrations occurring later in therapy are associated with both tissue accumulation and nephrotoxicity (Schentag *et al*., 1978; Matzke *et al*., 1983). Pre-dose concentrations (i.e. the minimum sustained gentamicin or tobramycin concentrations) of >2 mg/L have been related to raised serum creatinine concentrations (Goodman *et al*., 1975; Matzke *et al*., 1983; Cimino *et al*., 1987). Furthermore, some have related pre-dose concentrations of >2 mg/L and post-dose concentrations of >10 mg/L to nephrotoxicity (Li *et al*., 1989). While it is not possible to show that high pre-dose aminoglycoside concentrations are causally related to toxicity, aminoglycoside accumulation in the renal cortex is one of its main determinants.

Ototoxicity has been related to pre-dose serum concentrations of streptomycin (>3 mg/L; Line *et al*., 1970) and gentamicin (>3 mg/L, Nordstrom *et al*., 1973; Mawer *et al*., 1974). There have been suggestions that gentamicin post-dose concentrations of >12 mg/L are associated with ototoxicity. While this may be true in a general sense, there is no evidence for this or any other post-dose concentration being associated with an unexpected incidence of toxicity. In many studies both raised pre- and post-doses occurred in the same patients, making it difficult to attribute toxicity to one or the other. Furthermore, the recent use of once-daily therapy—which gives very high post-dose concentrations—has not given rise to increased ototoxicity, suggesting that previous studies associating post-dose concentrations with ototoxicity were erroneous. This matter has been reviewed by MacGowan & Reeves (1994) and Mattie *et al*. (1989). There have also been a number of studies which have failed to show any relationship between measured serum concentrations of aminoglycosides and ototoxicity.

While most toxicity data relate to multiple daily dosing, there is also evidence with once-daily therapy that serum levels are increased in patients with renal impairment, indicating accumulation and increased risk of toxicity (Blaser *et al*., 1994). When serum monitoring of aminoglycosides is performed, the goal is to tailor dosing to the individual in order to ensure that peak concentrations are high (to maximize antibacterial effect and reduce the risk of resistance developing) while at the same time avoiding drug accumulation (so minimizing the risk of toxicity). When aminoglycosides are given in multiple daily doses, pre- and post-dose concentrations are measured. With once-daily therapy, the most effective way of performing therapeutic drug monitoring has still not been addressed to any great extent. Several methods have been suggested: post-dose with midpoint concentration or trough concentration; midpoint or trough concentration alone; and area-under-the-curve determination (reviewed by Blaser & Konig (1995) and Barclay *et al*. (1995)).

Some investigators have developed nomograms to assist in initial dosing and monitoring, but recently these have been criticized on toxicity grounds (Nicolau *et al*., 1995; Singer *et al*., 1997). Others have not been able to relate trough concentration to nephrotoxicity in patients treated once-daily (Prins *et al*., 1996).

Apart from it being good clinical practice, because of the current medico-legal position we strongly recommend that all patients are monitored if they receive aminoglycoside therapy for more than 2 days. Table II summarizes the different interpretative criteria for aminoglycoside serum monitoring.

Glycopeptides

Retrospective reviews of teicoplanin in clinical trials have indicated that serum concentrations are related to clinical outcome. In a study of right-sided endocarditis mainly due to *Staphylococcus aureus*, when pre-dose concentrations were adjusted to 10–15 mg/L a post-dose concentration of ≥40 mg/L was associated with improved outcome (Leport

Table II. Recommendations for therapeutic drug monitoring of aminoglycosides

Antimicrobial(s)	Dosing regimen	Patient group	Peak/trough serum concentrations	Frequency of assay
Gentamicin, netilmicin and tobramycin	twice or three times a day	in normal renal function: assay at dose 2–4; assay earlier if there is renal impairment or other factors associated with increased toxicity	pre-dose: <2 mg/L; 1 h post-dose: >5 mg/L in most infections, >7 mg/L in pneumonia caused by aerobic Gram- negative bacilli	2–3 times per week
		infective endocarditis	pre-dose: <1 mg/L; 1 h post dose: 2–3 mg/L	2–3 times per week initially
	once-daily (i) any dose	normal renal function	pre-dose: <1 mg/L; 1 h post-dose: >10 mg/L, or pre dose <1 mg/L.	weekly; first assay within 48 h
	(ii) 4.5 mg/kg (iii) 7 mg/kg (iv) 5, 6 or 7 mg/kg/day		or 8 h: 1.5–6 mg/L. 6–14 h by nomogram. AUC: 72, 86 and 101 mg/L·h respectively[a].	
Amikacin	(i) any regimen, any dose (ii) twice-daily, any dose	normal renal function	C_{max}/MIC $\geq$ 10 within 48 h. pre-dose: <10 mg/L; 1 h post-dose: >20 mg/L.	2–3 times per week
	(iii) once-daily		pre-dose: <5 mg/L; 1 h post-dose: >40 mg/L.	
Streptomycin	see Table VI			

[a] Based on immediate post-dose concentration and 6–14 h mid point.

et al., 1989). In bone infection it has been suggested that a group of patients with a mean pre-dose concentration of >30 mg/L was more successfully treated than those with a mean concentration of about 10 mg/L (Greenberg, 1990). A review of over 50 cases published with sufficient data for pharmacodynamic and kinetic analysis to be possible showed a relationship between pre- or post-dose concentrations and outcome; again most of these patients had *S. aureus* infection (MacGowan *et al.*, 1996). Finally, a review of 92 patients with *S. aureus* bacteraemia using multivariate analysis showed that only patient age and trough concentration related to outcome (MacGowan *et al.*, 1997). A prospective study of teicoplanin to treat *S. aureus* infective endocarditis also showed a relationship between trough concentration of >20 mg/L and favourable outcome (Wilson *et al.*, 1994). Thrombocytopenia has been reported to occur with teicoplanin when trough concentrations exceed 60 mg/L (Wilson & Gruneberg, 1997).

Prospective clinical studies of vancomycin therapeutic drug monitoring could not relate trough concentration in the range 5–25 mg/L to clinical outcome (Rybak *et al.*, 1996, 1997). In contrast, a retrospective review of over 250 patients with proven Gram-positive infection was able to relate trough concentration of >10 mg/L to reduced number of fever days, improved WBC response but not length of stay or mortality (Zimmerman *et al.*, 1995).

Serum vancomycin concentrations have also been related to the incidence of relapse in CAPD infection (Mulhern *et al.*, 1995). Serum vancomycin levels cannot be related to ototoxicity and the data for nephrotoxicity would indicate that only trough concentrations of >30 mg/L may be important (Faber & Moellering, 1983). Therapeutic drug monitoring reduces the incidence of nephrotoxicity with vancomycin therapy, and vancomycin concentrations before the onset of nephrotoxicity were higher (23.4 ± 2.5 mg/L) in those who went on to develop a raised creatinine

than in those who did not (10.2 ± 3.8 mg/L) (Welty & Copa, 1994; Fernandez de Gatta *et al.*, 1996). Table III summarizes the present recommendations for glycopeptide monitoring (MacGowan, 1998).

Chloramphenicol

Systemic chloramphenicol was once widely used to treat acute meningitis or typhoid fever, but its clinical use is in decline, as other drugs are now used for these indications. However, in the future chloramphenicol may have a place in the treatment of susceptible vancomycin-resistant enterococci or methicillin-resistant *S. aureus*. It would seem prudent to monitor all patients, since (i) most prescribers have little experience of chloramphenicol use, (ii) the intravenous form has unpredictable pharmacokinetics in children, (iii) chloramphenicol has poor kinetic data in adults and (iv) it is toxic when overdosed. Serum concentrations of >25 mg/L are associated with reduced iron utilization by red cell precursors and morphological changes in erythroblasts and white blood cell and platelet precursors; such changes are different from those related to aplastic anaemia and occur more rapidly in those with hepatic disease (Scott *et al.*, 1965). 'Grey baby syndrome' may occur in all age groups and is now usually associated with accidental overdose: there is a high mortality. Serum concentrations are often very high, for example >100 mg/L (Thompson *et al.*, 1975), but grey baby syndrome has been reported with concentrations as low as 40 mg/L (Glazer *et al.*, 1980). Conversely, not all patients with high serum chloramphenicol concentrations will develop grey baby syndrome. Various therapeutic ranges have been recommended, mainly with reference to use in children: 15–25 mg/L is widely quoted (Mulhall *et al.*, 1983). When taking post-dose samples for chloramphenicol assay after iv administration, it is important to wait 1–2 h after the dose

Table III. Recommendations for therapeutic drug monitoring of vancomycin and teicoplanin

Antimicrobial	Patient group	Peak/trough serum concentrations	Frequency of assay
Vancomycin	all patients on treatment for >4–5 days; >2 days for those receiving other nephrotoxic therapy; or >2 days to be conservative	pre-dose only: range 5–15 mg/L; *or* pre-dose: 5–10 mg/L and post-dose: 20–40 mg/L at 1 h; *or* 18–26 mg/L at 2 h post-dose	1–2 times per week
Teicoplanin	severe non-staphylococcal infection, IVDA and renal impairment	pre-dose: >10 mg/L	1–2 times per week
	severe *S. aureus* infection	pre-dose: >10 mg/L (>20 mg/L may be better)	
	S. aureus IE	pre-dose: >20 mg/L	
	other IE	tentatively; pre-dose: 10–15 mg/L; post-dose: >40 mg/L	
	all groups	maintain pre-dose conc. <60 mg/L	

[a] IVDA, intravenous drug abuser; [b] IE, infective endocarditis.

to allow the inactive pro-drug, chloramphenicol succinate, to hydrolyse fully. More information relating to chloramphenicol can be found in *Chapter 12*.

Co-trimoxazole

Co-trimoxazole is occasionally monitored in the treatment of *Pneumocystis carinii* pneumonia (PCP), especially in patients with renal impairment. Monitoring may also have a place in treatment of patients <6 weeks of age. Routine monitoring has been proposed in those with HIV disease but does not seem to have a role (Joos *et al.*, 1995).

It has been suggested that trimethoprim concentrations maintained over 5 mg/L ensure efficacy in PCP and in small studies success was reported if concentrations were maintained over 5 mg/L (Lau & Young, 1976; Winston *et al.*, 1980; Sattler *et al.*, 1988). In addition it has been speculated that high concentrations of sulphamethoxazole or its metabolites may be responsible for side effects; leukopenia has been reported more often in patients with sulphamethoxazole concentrations of >200 mg/L (Fong, 1988). Higher concentrations of acetyl-sulphamethoxazole were reported as being associated with treatment-limiting toxicity in AIDS patients (Joos *et al.*, 1995). However, in-vitro data had indicated that side effects may be related to the hydroxyl amine metabolite (Rieder *et al.*, 1989). The correlation between serum concentrations and toxic effects for sulphonamides on bone marrow or renally is unclear but it has been suggested that concentrations are maintained below 100 mg/L. Co-trimoxazole monitoring is discussed further in *Chapter 19*.

Quinolones

Routine monitoring of quinolones is not indicated. It is, however, well understood from human studies with ciprofloxacin and grepafloxacin that the relationship between the area under the serum concentration–time curve and pathogen susceptibility (AUC:MIC ratio) is the main pharmacodynamic determinant of outcome (Forrest *et al.*, 1993, 1997). In a small group of patients with altered quinolone handling, the application of this knowledge in individualizing therapy may be of value as kinetics can be variable. More commonly serum concentrations are measured to ensure compliance or absorption in patients being treated with oral therapy for severe infection. Many factors are known to change quinolone absorption, including nasogastric feeds, sucralfate, bacteraemia, magnesium and aluminium salts, and tumour chemotherapy (reviewed by MacGowan & Bowker, 1998a).

β-Lactams

Benzylpenicillin is sometimes assayed in patients who have convulsions while receiving high doses or who are

Table IV. Rarely assayed β-lactams and potential indications

Agent	Potential indications
Flucloxacillin	IV oral switch therapy in severe infection, renal failure, liver failure and critical infection
Benzylpenicillin	renal failure or suspected neurotoxicity
Meropenem	renal failure; experimental dosing regimens, i.e. once-daily therapy or continuous infusion

receiving a high dosage in the the presence of renal failure. Brain tissue concentrations would appear to be the most relevant but for obvious reasons are rarely performed. CSF concentrations of around 10 mg/L or serum concentrations of >100 mg/L have been related to convulsion (see *Chapter 8*) but the human data are very limited (Lerner *et al.*, 1967).

Ceftazidime is also assayed when given by continuous infusion; this is to ensure that the concentration at steady state exceeds the pathogen MIC by a factor of 4–5 times. A 2 g bolus injection of ceftazidime followed by an infusion at a rate of 1 g each 4 h gives a steady-state concentration in the range 20–30 mg/L but there is wide variation between patients, so serum monitoring has a role (for review see MacGowan & Bowker, 1998b). A similar argument can be made for other β-lactams if they are dosed by continuous infusion, but the clinical and pharmacokinetic data to support their use is not as strong as with ceftazidime.

Other β-lactams are occasionally assayed to ensure compliance or in patients where the pharmacokinetics are likely to be unpredictable (Table IV).

Antifungal agents

Warnock and colleagues discuss the therapeutic drug monitoring of antifungal agents in *Chapter 17*. Flucytosine is the only agent which should be assayed in all patients; it is more useful to monitor renal function in patients receiving conventional amphotericin B than to assay amphotericin B levels. Table V summarizes some recent recommendations for the assay of antifungal agents (Working Party Report, 1991; Summers *et al.*, 1997).

Antimycobacterial agents

With the exception of aminoglycosides, macrolides and quinolones, which are covered in separate chapters, the drugs used for treatment of mycobacterial diseases are discussed in *Chapter 16*. Although all antimycobacterial drugs show unacceptable toxicity, only streptomycin and

Table V. Therapeutic monitoring of antifungals[a]

Agent	Patients requiring assay	Recommended concentration
Amphotericin B (any formulation)	assay seldom indicated	–
Flucytosine	all patients, especially in changing renal function, bone marrow suppression, those receiving amphotericin B, or suspected non-compliance	pre-dose: 30–40 mg/L 0.5 h after iv dose: 2 h after po dose: 70–80 mg/L
Fluconazole	suspected non-compliance, inadequate response to therapy, or potential drug interaction	unclear
Itraconazole	life-threatening infection, malabsorption, acid-modifying agents, non-compliance, inadequate response, or potential interactions	4 h post-dose: ≥0.25 mg/L at steady state (after >7 days of therapy); may assay earlier if needed
Ketoconazole	malabsorption, acid-modifying agents, suspected non-compliance, or inadequate response	detectable

[a]Working Party (1991); Summers *et al.* (1997).

Table VI. Suggested serum levels for antimycobacterial agents and reasons for their assay

Agent	Normal dose (mg)	Serum concentration (mg/L)		Patient category requiring assay	Frequency of assay
		pre-dose	post-dose		
Isoniazid[a]	300	<1	3–5	patients on multi-drug HIV therapy, or significant liver disease	c
Ofloxacin	600	0.5–1	4–8	poor response to therapy	d
Ciprofloxacin	750	1–2	4–6	poor response to therapy	d
Cycloserine	500	10–20	20–35	all patients receiving the drug	b
Ethambutol	15–25 mg/kg	<1	4–6	patients with severe renal impairment	bd
Rifabutin	300	<0.3	0.5–1.0	patients on multi-drug HIV therapy, or receiving macrolides	c
Rifampicin	300–600	<0.5	4–10	patients on multi-drug HIV therapy, or significant liver disease	c
Streptomycin	750–1000	<5	15–40	all patients receiving the drug	b

[a] Instances where there is a need to assay isoniazid are rare; data for isoniazid from Peloquin (1997).
[b] After fifth dose and then every 2–4 weeks or if there is a change in renal function.
[c] After at least the seventh dose and subsequently only if there is an alteration in hepatic function or a change in medications that may affect cytochrome P450.
[d] After at least the fifth dose and subsequently only if there are alterations in medications that may affect cytochrome P450. Ethambutol not recommended for patients on renal dialysis.

cycloserine are routinely monitored to minimize toxicity. The other main reason for assaying antimycobacterials is to monitor compliance or poor oral absorption. Since multiple therapy is the norm, monitoring rifampicin is probably the simplest way to do this. Although simple urine assays are described in *Chapter 16*, these may give misleading results if the urine has been collected in a rather random fashion, since renal excretion of rifampicin occurs only in the first few hours following a dose (Acocella, 1983). Measurement of post-dose serum concentrations is the most reliable way to confirm oral absorption, although in the authors' experience the C_{max} (peak) may be occur 1–4 h post-dose and may be missed if only a single blood sample is taken. Monitoring may have a role in AIDS patients (Peloquin, 1996, 1997). Table VI summarizes reasons for assaying some drugs used to treat mycobacterial infections, with recommended concentrations.

Acknowledgement

The authors would like to thank Dr Andrew Lovering for his help in compiling Table VI.

References

Acocella, G. (1983). Pharmacokinetics and metabolism of rifampin in humans. *Reviews of Infectious Diseases* **5**, Suppl., S428–32.

Barclay, M. L., Duffull, S. B., Begg, E. J. R. & Buttimore, R. C. (1995). Experience of once-daily aminoglycoside dosing using a target area under the concentration–time curve. *Australian and New Zealand Journal of Medicine* **25**, 230–5.

Blaser, J. & Konig, C. (1995). Once-daily dosing of amino-glycosides. *European Journal of Clinical Microbiology and Infectious Diseases* **14**, 1029–38.

Blaser, J., Konig, C., Simmen, H. P. & Thurnheer, U. (1994). Monitoring serum concentrations for once-daily netilmicin dosing regimens. *Journal of Antimicrobial Chemotherapy* **33**, 341–8.

Cimino, M. A., Rotstein, C., Slaughter, R. C. & Emrich, L. J. (1987). Relationship of serum antibiotic concentrations to nephrotoxicity in cancer patients receiving concurrent aminoglycoside and vancomycin therapy. *American Journal of Medicine* **83**, 1091–7.

Dawson, S. J. & Reeves, D. S. (1997). Therapeutic monitoring, the concentration–effect relationship and impact on clinical efficacy of antibiotic agents. *Journal of Chemotherapy* **9**, Suppl. 1, 84–92.

Farber, B. F. & Moellering, R. C. (1983). Retrospective study of the toxicity of preparations of vancomycin from 1974 to 1981. *Antimicrobial Agents and Chemotherapy* **23**, 138–41.

Fernandez de Gatta, M. D., Calvo, M. V., Hernandez, J. M., Caballero, D., San Miguel, J. F. & Dominguez-Gil, A. (1996). Cost-effectiveness analysis of serum vancomycin concentration monitoring in patients with hematologic malignancies. *Clinical Pharmacology and Therapeutics* **60**, 332–40.

Fong, I. W. (1988). Correlation of the side effects of trimethoprim–sulphamethoxazole with blood levels in AIDS patients treated for *Pneumocystis carinii* pneumonia. In *Program and Abstracts of the Twenty-Eighth Interscience Conference on Antimicrobial Agents and Chemotherapy, Los Angeles, October 23–26*, Abstract 1226, p. 328. American Society for Microbiology, Washington, DC.

Forrest, A., Nix, D. E., Ballow, C. H., Goss, T. F., Birmingham, M. C. & Schentag, J. J. (1993). Pharmacodynamics of intravenous ciprofloxacin in seriously ill patients. *Antimicrobial Agents and Chemotherapy* **37**, 1073–81.

Forrest, A., Chodosh, S., Amantea, M. A., Collins, D. A. & Schentag, J. J. (1997). Pharmacokinetics and pharmacodynamics of oral grepafloxacin in patients with acute bacterial exacerbations of chronic bronchitis. *Journal of Antimicrobial Chemotherapy* **40**, Suppl. A, 45–57.

Freeman, C. D., Nicolau, D. P., Belliveau, P. P. & Nightingale, C. H. (1997). Once-daily dosing of aminoglycosides: review and recommendations for clinical practice. *Journal of Antimicrobial Chemotherapy* **39**, 677–86.

Glazer, J. P., Danish, M. A., Plotkin, S. A. & Yaffe, S. J. (1980). Disposition of chloramphenicol in low birth weight infants. *Pediatrics* **66**, 573–8.

Goodman, E. L., Van Gelder, J., Holmes, R., Hull, A. R. & Sanford, J. P. (1975). Prospective comparative study of variable dosage and variable frequency regimens for administration of gentamicin. *Antimicrobial Agents and Chemotherapy* **8**, 434–8.

Greenberg, R. N. (1990). Treatment of bone, joint and vascular-access-associated Gram-positive bacterial infection with teicoplanin. *Antimicrobial Agents and Chemotherapy* **35**, 2392–7.

Jackson, G. G. & Riff, L. J. (1971). *Pseudomonas* bacteremia: pharmacologic and other bases for failure of treatment with gentamicin. *Journal of Infectious Diseases* **124**, Suppl. 124, 185–91.

Joos, B., Blaser, J., Opravil, M., Chave, J.-P. & Luthy, R. (1995). Monitoring of co-trimoxazole concentrations in serum during treatment of *Pneumocystis carinii* pneumonia. *Antimicrobial Agents and Chemotherapy* **39**, 2661–6.

Kashuba, A. D. M., Bertino, J. S. & Nafziger, A. N. (1998). Dosing of aminoglycosides to rapidly attain pharmacodynamic goals and hasten therapeutic response using individualized pharmaco-kinetic monitoring of patients with pneumonia caused by Gram-negative organisms. *Antimicrobial Agents and Chemotherapy* **42**, 1842–4.

Lau, W. K. & Young, L. S. (1976). Trimethoprim–sulfamethoxazole treatment of *Pneumocystis carinii* pneumonia in adults. *New England Journal of Medicine* **295**, 716–8.

Leport, C., Perronne, C., Massip, P., Canton, P., Leclercq, P., Bernard, E. *et al.* (1989). Evaluation of teicoplanin for treatment of endocarditis caused by Gram-positive cocci in 20 patients. *Antimicrobial Agents and Chemotherapy* **33**, 871–6.

Lerner, P. I., Smith, H. & Weinstein, L. (1967). Penicillin neuro-toxicity. *Annals of the New York Academy of Sciences* **145**, 310–8.

Li, S. C., Ioannides-Demos, L. L., Spicer, W. J., Berbatis C., Spelman, D. W., Tong N. *et al.* (1989). Prospective audit of aminoglycoside usage in a general hospital with assessments of clinical processes and adverse clinical outcomes. *Medical Journal of Australia* **151**, 224–32.

Line, D. H., Poole, G. W. & Waterworth, P. M. (1970). Serum streptomycin levels and dizziness. *Tubercle* **51**, 76–81.

MacGowan, A. (1996). Concentration controlled and concentration defined clinical trials: do they offer any advantages for anti-microbial chemotherapy? *Journal of Antimicrobial Chemotherapy* **37**, 1–5.

MacGowan, A. P. (1998). Pharmacodynamics, pharmacokinetics and therapeutic drug monitoring of glycopeptides. *Therapeutic Drug Monitoring*, in press.

MacGowan, A. P. & Bowker, K. E. (1997). Pharmacodynamics of antimicrobial agents and rationale for their dosing. *Journal of Chemotherapy* **9**, Suppl. 1, 64–73.

MacGowan, A. P. & Bowker, K. E. (1998a). Sequential antimicrobial therapy: pharmacokinetic and pharmacodynamic considerations in sequential therapy. *Journal of Infection*, in press.

MacGowan, A. P. & Bowker, K. E. (1998b). Continuous infusion of β-lactam antibiotics: a consensus. *Clinical Pharmacology*, in press.

MacGowan, A. & Reeves, D. (1994). Serum aminoglycoside concentrations: the case for routine monitoring. *Journal of Antimicrobial Chemotherapy* **34**, 829–37.

MacGowan, A., White, L., Reeves, D. & Harding, I. (1996). Retrospective review of serum teicoplanin concentrations in clinical trials and their relationship to clinical outcome. *Journal of Infection and Chemotherapy* **2**, 197–208.

MacGowan, A. P., White, L. O., Reeves, D. S., Reed, V. & Harding, I. (1997). Teicoplanin in *Staphyloccus aureus* septicaemia; relationship between trough serum levels and outcome. In *Program and Abstracts of the Thirty-Seventh Interscience Conference on Antimicrobial Agents and Chemotherapy, Toronto, September 28–October 1, 1997*, Abstract A-45, p. 9. American Society for Microbiology, Washington, DC.

Mattie, H., Craig, W. A. & Pechère, J. C. (1989). Determinants of efficacy and toxicity of aminoglycosides. *Journal of Antimicrobial Chemotherapy* **24**, 281–93.

Matzke, G. R., Lucarotti, R. L. & Shapiro, H. S. (1983). Controlled comparison of gentamicin and tobramycin nephrotoxicity. *American Journal of Nephrology* **3**, 11–7.

Mawer, G. E., Ahmad, R., Dobbs, S. M., McGough, J. G., Lucas, C. B. & Tooth, J. A. (1974). Prescribing aids for gentamicin. *British Journal of Clinical Pharmacology* **1**, 45–50.

Moore, R. D., Smith, C. R. & Lietman, P. S. (1984a). Association of aminoglycoside plasma levels with therapeutic outcome in Gram-negative pneumonia. *American Journal of Medicine* **77**, 657–62.

Moore, R. D., Smith, C. R. & Leitman, P. S. (1984b). The association of plasma levels and mortality in patients with Gram-negative bacteraemia. *Journal of Infectious Diseases* **149**, 443–8.

Moore, R. D., Lietman, P. S. & Smith, C. R. (1987). Clinical response to aminoglycoside therapy: importance of the ratio of peak concentration to minimal inhibitory concentration. *Journal of Infectious Diseases* **155**, 93–9.

Mulhall, A., de Louvois, J. & Hurley, R. (1983). The pharmacokinetics of chloramphenicol in the neonate and young infant. *Journal of Antimicrobial Chemotherapy* **12**, 629–39.

Mulhern, J. G., Braden, G. L., O'Shea, M. H., Madden, R. L., Lipkowitz, G. S. & Germain, M. J. (1995). Trough serum vancomycin levels predict the relapse of Gram-positive peritonitis in peritoneal dialysis patients. *American Journal of Kidney Diseases* **25**, 611–5.

Nicolau, D. P., Freeman, C. D., Belliveau, P. P., Nightingale, C. H., Ross, J. W. & Quintiliani, R. (1995). Experience with a once-daily aminoglycoside program administered to 2,184 adult patients. *Antimicrobial Agents and Chemotherapy* **39**, 650–5.

Noone, P. & Rogers, B. T. (1976). Pneumonia caused by coliforms and *Pseudomonas aeruginosa*. *Journal of Clinical Pathology* **29**, 652–6.

Noone, P., Parsons, T. M. C., Pattison, J. R., Slack, R. C. B., Garfield-Davies, D. & Hughes, K. (1974a). Experience in monitoring gentamicin therapy during treatment of serious Gram-negative sepsis. *British Medical Journal* i, 477–81.

Noone, P., Pattison, J. R. & Davies, D. G. (1974b). The effective use of gentamicin in life-threatening sepsis. *Postgraduate Medical Journal* **50**, *Suppl. 7*, 9–16.

Nordstrom, L., Banck, G., Belfrage, S., Juhlin, I., Tjernstrom, O. & Toremalm, N. G. (1973). Prospective study of the ototoxicity of gentamicin. *Acta Pathologica et Microbiologica Scandinavica–Section B, Microbiology and Immunology, Suppl. 241*, 58–61.

Peloquin, C. A. (1996). Therapeutic drug monitoring of the antimycobacterial drugs. *Clinics in Laboratory Medicine* **16**, 717–29.

Peloquin, C. A. (1997). Using therapeutic drug monitoring to dose the antimycobacterial drugs. *Clinics in Chest Medicine* **18**, 79–87.

Prins, J. M., Weverling, G. J., de Blok, K., van Ketel, R. J. & Speelman, P. (1996). Validation and nephrotoxicity of a simplified once-daily aminoglycoside dosing schedule and guidelines for monitoring therapy. *Antimicrobial Agents and Chemotherapy* **40**, 2494–9.

Rieder, M. J., Uetrecht, J., Shear, N. H., Cannon, M., Miller, M. & Spielberg, S. P. (1989). Diagnosis of sulfonamide hypersensitivity reactions by in-vitro 'rechallenge' with hydroxylamine metabolites. *Annals of Internal Medicine* **110**, 286–9.

Rybak, M. J., Cappelletty, D. M., Kang, S. L., Levine D. P. & Levison, M. E. (1996). Pharmacodynamic evaluation of teicoplanin versus vancomycin in the treatment of Gram-positive

bacteraemia and endocarditis. In *Program and Abstracts of the Thirty-Sixth Interscience Conference on Antimicrobial Agents and Chemotherapy*, Abstract A-36. American Society for Microbiology, Washington, DC.

Rybak, M. J., Cappelletty, D. M., Rulting, R. C., Mercier, R. C., Houlihan, H. H., Kepser, M. E. *et al.* (1997). Influence of vancomycin serum concentration on the outcome of patients being treated for Gram-positive infections. In *Program and Abstracts of Thirty-Seventh Interscience Conference on Antimicrobial Agents and Chemotherapy Toronto, September 28–October 1, 1997*, A-46. American Society for Microbiology, Washington, DC.

Sattler, F. R., Cowan, R., Nielsen, D. M. & Ruskin, J. (1988). Trimethoprim–sulfamethoxazole compared with pentamidine for the treatment of *Pneumocystis carinii* pneumonia in the acquired immunodeficiency syndrome. A prospective, noncrossover study. *Annals of Internal Medicine* **109**, 280–7.

Schentag, J. J., Cumbo, T. J., Jusko, W. J. & Plaut, M. E. (1978). Gentamicin tissue accumulation and nephrotoxic reactions. *Journal of the American Medical Association* **240**, 2067–9.

Scott, J. L., Finegold, S. M., Belkin, G. A. & Lawerence, J. S. (1965). A controlled double blind study of the haematologic toxicity of chloramphenicol. *New England Journal of Medicine* **272**, 1137–43.

Siber, G. R., Echeverria, P., Smith, A. L., Paisley, J. W. & Smith, D. H. (1975). Pharmacokinetics of gentamicin in children and adults. *Journal of Infectious Diseases* **132**, 637–51.

Singer, C., Smith, C. & Krieff, D. (1996). Once-daily aminoglycoside therapy: potential ototoxicity. *Antimicrobial Agents and Chemotherapy* **40**, 2209–11.

Summers, K. K., Hardin, T. C., Gore, S. J. & Graybill, J. R. (1997). Therapeutic drug monitoring of systemic antifungal therapy. *Journal of Antimicrobial Chemotherapy* **40**, 753–64.

Thompson, W. L., Anderson, S. E., Lipsky, J. J. & Lietman, P. S. (1975). Overdoses of chloramphenicol. *Journal of the American Medical Association* **234**, 149–50.

Welty, T. E. & Copa, A. K. (1994). Impact of vancomycin therapeutic drug monitoring on patient care. *Annals of Pharmacotherapy* **28**, 1335–9.

Wilson, A. P. R. & Gruneberg, R. N. (1997). *Teicoplanin, the First Decade*, pp. 137–44. The Medicine Group, Abingdon.

Wilson, A. P. R., Gruneberg, R. N. & Neu, H. (1994). A critical review of the dosage of teicoplanin in Europe and the USA. *International Journal of Antimicrobial Agents* **4**, *Suppl. 1*, S1–30.

Winston, D. J., Lau, W. K., Gale, R. P. & Young, L. S. (1980). Trimethoprim–sulfamethoxazole for the treatment of *Pneumocystis carinii* pneumonia. *Annals of Internal Medicine* **92**, 762–9.

Working Party of the British Society for Antimicrobial Chemotherapy. (1991). Laboratory monitoring of antifungal chemotherapy. *Lancet* **337**, 1577–80.

Zimmermann, A. E., Katona, B. G. & Plaisance, K. I. (1995). Association of vancomycin serum concentrations with outcomes in patients with Gram-positive bacteremia. *Pharmacotherapy* **15**, 85–91.

9

Chapter 2

Pharmacology and pharmacokinetics

Richard Wise

Department of Medical Microbiology, City Hospital NHS Trust, Birmingham B18 7QH, UK

Introduction

The term 'pharmacokinetics' is often used erroneously when 'pharmacology' is meant. Pharmacology is the broad study of drugs in treatment whereas pharmacokinetics confines itself to the mathematical study of the rate processes involved in absorption, distribution, metabolism and elimination. The term 'pharmacodynamics' is being increasingly employed; this refers to the time-course of drug effects and other interactions between the antimicrobial and the bacterium. These would include the MIC, MBC, both sub-MIC and supra-MIC effects, post-antibiotic effects and any interactions between the immune system of the host and the antimicrobial (Figure 2.1).

Pharmacology, unlike pharmacokinetics, is not an exact science. To those interested in the treatment of infectious diseases, pharmacology is concerned with the efficacious and safe use of antimicrobial agents. There are areas of uncertainty and controversy even in the use of many of the well-established drugs described in this book. There is still little rationale for therapy, and what constitutes an adequate dose, a reasonable dosing interval and what is a toxic concentration, are often debatable. For example, for historical reasons drugs with short elimination half-lives (such as some cephalosporins) are given with the same frequency as agents with long half-lives such as fusidic acid and metronidazole. It is, however, usually possible from the knowledge of the concentration of drug necessary to inhibit an organism *in vitro*, information obtained from animal experiments and a knowledge of tissue penetration and human experience, to arrive at a reasoned and safe dosage schedule (Schentag, 1991).

In this chapter I will discuss the basic principles of pharmacokinetics and illustrate them with examples drawn from the antimicrobials. Absorption, distribution, metabolism and excretion of antimicrobials are interrelated processes and to study them separately is somewhat artificial. Therefore, each will be examined and then the pharmacokinetics, which interrelate these processes, will be reviewed.

Absorption

Oral administration

Bioavailability is a measure of absorption and refers to the relative amount of the drug from the administered form which enters the systemic circulation. For an orally administered drug, the amount reaching the systemic circulation is usually less than that administered (as a result of poor formulation, enzymatic destruction in the gut and enteric elimination). The absolute bioavailability can, therefore, be determined experimentally by comparing the area under the time-versus-serum-concentration curve (AUC) of the oral formulation, with that of the systemically administered drug. Table I gives the bioavailability of a number of commonly used oral antibacterials.

Absorption of drugs can occur at any site in the gastrointestinal tract. The majority of antimicrobials are absorbed from the upper small intestine although some are absorbed from the stomach, lower intestines or even the colon. The site of absorbtion depends mainly upon the physicochemical properties of the molecule and the physiology of the gastrointestinal tract. It is useful to consider the lining

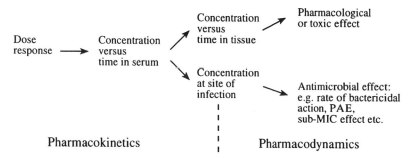

Figure 2.1 The interrelationship between pharmacokinetics and pharmacodynamics.

Table I. Bioavailability (% of oral dose administered) of commonly used oral
antibacterial agents

Antibacterial agent	Bioavailability (%)	Antibacterial agent	Bioavailability (%)
Ampicillin	40	Ofloxacin	90
Amoxycillin	60	Norfloxacin	40
Cephalexin	90	Pefloxacin	90
Cefaclor	60	Metronidazole	75
Cefixime	45	Roxithromycin	40
Cefuroxime axetil	50	Azithromycin	40
Ciprofloxacin	70	Clarithromycin	55

of the gastrointestinal tract as a lipid membrane and only drugs which are lipid-soluble can readily pass through this membrane. Most drugs are either weak acids or weak bases and the ratio of that portion which is ionized (water-soluble) to that which is un-ionized (lipid-soluble) is dependent on the drug's pK_a and the pH of that region of the lumen of the intestinal tract. For example, for a weakly acidic drug, HA, at a certain pH, the relative concentrations of the ionized and un-ionized moieties of the drug are governed by the Henderson–Hasselbalch equation:

$$K_a = ([H^+][A^-])/[HA]$$

where K_a is the dissociation constant and where the acid HA can dissociate reversibly to give H^+ and A^- ions.

At pH 7, benzylpenicillin is 99.9% ionized and one would not expect it to be absorbed. Despite this, the small un-ionized portion of the compound is absorbed from the small intestine and, as this occurs, more un-ionized fraction is formed in the gut lumen (i.e. the law of mass action is obeyed). Lipid solubility depends not only on the degree of ionization, but also on the intrinsic lipid solubility of the drug as measured by a somewhat artificial system such as the n-octanol/buffer or chloroform/buffer partition. The clinical relevance of such partition data is difficult to ascertain. Experience with the quinolones shows that lipid solubility, as measured by the n-octanol/water partition, possibly does not explain their absorption, which might be more related to their zwitterionic nature (Ashby et al., 1985). Highly water-soluble drugs, such as gentamicin, will not pass through the intact gastrointestinal membrane, and therefore are hardly absorbed at all when given by the oral route. Once admitted to the membrane the majority of antimicrobials are transported across by simple diffusion, and the rate of the process will depend upon the concentration gradient across the membrane, but some oral cephalosporins are thought to have an active transport mechanism.

A few antimicrobials are partially absorbed from the stomach. These are low molecular weight drugs which exist as weak acids, such as the sulphonamides (Antonioli et al., 1971). The amount of drug absorbed from this site often tends to be rather erratic and dependent upon factors such as the presence of food, the rate of gastric emptying, the physical properties of the tablet or capsule, and the presence of other drugs which may alter the pH of the gastric secretions. For example, the concomitant administration of an alkali or an H_2 antagonist will tend to raise the pH of the stomach and increase the ionization and hence reduce absorption of sulphonamides. Ketoconazole, which is poorly soluble in water at pH 7, is readily soluble at pH 2 or less and is, in the most part, absorbed from the stomach; again, though, cimetidine or achlorhydria will lead to poor ketoconazole absorption (Van Tyle, 1984).

Food, by nature of its buffering action, can delay the absorption of drugs from the stomach but, as discussed below, it can also increase absorption. It is difficult to judge how important the stomach is as a site of drug absorption, but it is reasonable to suppose that, in comparison with the small intestine, its role is relatively insignificant.

Most antimicrobials are absorbed from the small intestine. The microvilli and folds of Kerckring present a very large surface area and motility ensures that the contents come in contact with the villi. Absorption is usually fairly rapid, with most oral quinolones and β-lactams being essentially fully absorbed by 1.5–2.5 h post dose (Schentag et al., 1990). The amount of absorption of even closely structurally related drugs can vary considerably. For example, cephradine is almost completely orally absorbed (Zaki et al., 1974) while only 30–50% of an oral dose of ampicillin is absorbed (Neu, 1974).

Certain antimicrobials are administered in the form of an inactive precursor or prodrug designed to facilitate oral absorption both by protecting the drug from acid hydrolysis in the stomach and by increasing its lipid solubility. Examples are chloramphenicol (chloramphenicol palmitate), ampicillin (the piroloyloxymethyl ester, pivampicillin, and the phthalidyl ester, talampicillin), mecillinam (pivmecillinam), cetetamet (cefetamet pivoxil), carbenicillin (carfecillin and carindacillin), cefuroxime (cefuroxime axetil) and cefpodoxime (the isopropyloxy-carbonyloxyethyl ester, cefpodoxime proxetil). The ubiquitous esterases of the gastrointestinal mucosa and portal system rapidly hydrolyse these drugs to their parent compounds during absorption. In one instance, two compounds, namely ampicillin and the β-lactamase inhibitor sulbactam, are chemically combined via an ester link as the mutual prodrug sultamicillin (Wise, 1981), each facilitating the absorption of the other. Although not strictly considered a prodrug, dirithromycin after absorption is hydrolysed almost completely to erythromycylamine, both compounds being microbiologically active. Similarly, a considerable amount of pefloxacin is desmethylated to norfloxacin and a mixture of the two compounds circulates; whether pefloxacin is a prodrug of norfloxacin or norfloxacin is a metabolite of pefloxacin is a matter for debate.

Factors other than lipid solubility must affect the absorption of certain similar antimicrobials. A good example is the more rapid and complete absorption of amoxycillin when compared with ampicillin. Since amoxycillin is less lipid-soluble than ampicillin, this may be related to active transport of the former. The physical formulation of an antimicrobial, i.e. the form in which it is presented to the small bowel, is crucial in determining the extent of absorption. A good example is the enhanced bioavailability of the micro-encapsulated form of erythromycin base compared with other formulations (McDonald et al., 1977). Similarly, microcrystalline formations of nitrofurantoin show improved absorption.

Absorption may be hindered by many mechanisms. The pH of the stomach may assist in the hydrolysis and, hence, inactivation of a number of agents. The macrolides are a good example (Kirst & Sides, 1989). The poor bioavailability of erythromycin results from internal molecular rearrangement, first to a hemi-ketal form and then to a spiro-ketal form, neither of which has any antimicrobial activity. To overcome this problem, the C-9 position has been modified to produce new macrolides with improved absorption such as erythromycylamine (the hydrolysis product of dirithromycin) and clarithromcyin; ring expansion at this site has yielded azithromycin; replacement of the cladinose moiety by a keto function has led to the development of the ketolides.

The buffering effect of food has already been mentioned but food can also dilute the drug and so hinder its access to the mucosa. Food may stimulate peristalsis but will increase gastric retention. In some instances food can enhance absorption, for example of cefuroxime axetil (Williams & Harding, 1984) and cefpodoxime proxetil (Hughes et al., 1989); although the mechanism is uncertain it is possibly related to an effect on the non-specific esterases in the gut. The absorption of cefetamet pivoxil increases from 41% to 51% when given after a standard meal (Blouin et al., 1989). Other drugs can interfere with the absorption process. For example, lincomycin absorption is slowed down by kaolin in the form of 'Kaopectate' (Wagner, 1971). Tetracycline absorption may be reduced as a result of chelation with the calcium ions found in the capsule itself or milk (Bolger & Gavin, 1959). The quinolones are similarly chelated by divalent cations and

the peak serum concentrations achieved are considerably lowered by those antacids which contain them (Hoffken et al., 1985).

Disease states affecting the bowel can alter absorption. Parsons & Paddock (1975) found in coeliac disease that the absorption of cephalexin and co-trimoxazole was increased, but others have shown that sulphonamide absorption is decreased in villous atrophy (Matilla et al., 1973). It might be thought that diarrhoea, by reducing transit time, would decrease drug absorption, but patients in Calcutta with possible dysentery and treated with norfloxacin or ciprofloxacin had blood concentrations of the agents similar to those found in healthy volunteers (Bhattacharya et al., 1992). Although little work has been conducted on the absorption of antimicrobials in general disease states, it is probable that shock and other factors would affect the process. Interestingly, however, the absorption of cephalexin by patients with very severe underlying disease was not very different from that seen in a matched convalescent group (Dean et al., 1979). Drug absorption in cystic fibrotics is often studied and generally this disease has little effect on the serum concentrations attained. In patients treated with ofloxacin before and after cancer chemotherapy the peak serum concentrations achieved were greater before chemotherapy than afterwards (Brown et al., 1993). Similarly, six patients with newly diagnosed haematological malignancy who received ciprofloxacin prophylactically demonstrated reduced absorption after anticancer chemotherapy (Johnson et al., 1990).

In the UK and USA, few antimicrobials are given by the rectal route, and the factors affecting this mode of administration are less well understood. It would appear that lipid-soluble agents can diffuse across the rectal mucosa. Metronidazole is highly lipid-soluble and good blood concentrations are obtained after rectal suppository or enema administration (Giamarellou et al., 1977). Rectal absorption is usually less efficient than oral absorption, but more sustained levels may be attained.

Parenteral administration

The rate of drug absorption from intramuscular or subcutaneous injection sites is often extremely rapid. Penetration of the capillary walls by diffusion, the rate of which is related to the lipid solubility of the compound, is one means of absorption. Lipid-insoluble molecules pass through aqueous pores in the capillary endothelium. The process can be slowed by administering these drugs in a poorly soluble form, for example, procaine penicillin. The site of the intramuscular injection may also affect the serum concentrations of drug that are attained. Reeves et al. (1974) found higher serum concentrations of cefacetrile after thigh rather than buttock injections. After bolus iv administration of a drug, peak serum concentrations are attained almost immediately. If one is investigating the pharmacokinetics of a drug, this is the ideal mode of administration, inasmuch as if the drug is given by any other route there is the addition of a further compartment to the mathematical model (see below).

It may be necessary to prepare more water-soluble derivatives for parenteral administration. Some of the newer fluoroquinolones, for example tosufloxacin and trovafloxacin, have been modified for this purpose.

Distribution

The differing types of tissues within the body, such as the brain, bone or subcutaneous tissue, can be considered as theoretical volumes, into which a drug might penetrate, surrounded by theoretical membranes (these may be actual membranes, in the case of the meninges, or the cellular membranes of the tissue itself), thus forming so-called compartments as referred to in pharmacokinetic modelling (see below). The concentration of a drug in a particular compartment depends on the physico-chemical properties of the drug (for example the pK_a, lipid solubility and protein binding) and the physiology of its compartment (for example the pH in the tissue concerned). The concentration of any drug will vary from compartment to compartment throughout the body. This is best explained by an example, the penetration of an antimicrobial into the prostate of the dog (Reeves et al., 1973) (Figure 2.2). The un-ionized portion of a weak base, such as clindamycin, diffuses into the prostatic secretions in the lumen of prostate. The lower pH of the prostatic fluid causes a greater proportion of the drug to be ionized and consequently it cannot diffuse back, 'trapping' it in the ionized form. Thus tissue concentrations in excess of those in the serum may occur. The reverse occurs with a weak acid such as cephalexin. Another example of this phenomenon is the so-called 'proton pump'. Figure 2.3 shows the fate of a weak base, B, being concentrated at an intracellular focus. In Table II, the pH values of different body fluids are shown.

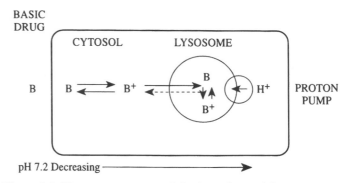

BASIC DRUG

pH 7.2 Decreasing ⟶

Figure 2.3 The proton pump and the fate of a weak base, B.

The examples quoted are an over-simplification of the true mechanisms involved in drug distribution and do not include the concept of the chloroform/buffer lipid solubility partition and other factors. Lipid solubility is important in determining the apparent volume of distribution (V_d, see below) of a drug. A highly water-soluble drug, such as gentamicin, will have a low V_d of 12–15 L, while a highly lipid-soluble drug, such as griseofulvin, will distribute itself throughout most of the tissues of the body and the V_d may be well in excess of 40 L. The quinolones, such as ciprofloxacin, ofloxacin and enoxacin, have extremely large apparent V_ds, in excess of 150 L (Wise et al., 1986; Schentag et al., 1990) which is a reflection of the high concentration of these drugs found at many sites in the body (see below).

It is also possible that a metabolic acidosis, which may occur in severe infection, may alter the pH sufficiently to increase or decrease the ionized portion of an antimicrobial and hence affect tissue distribution.

The foregoing, of course, suggest that drugs are not evenly distributed throughout the body tissues. This is clearly shown by the quinolones where concentrations in macrophages may be very high (about 5- to 10-fold greater than the serum concentration) yet concentrations in CSF may be relatively low (c.10% of the serum concentration).

The effect of regional blood flow upon the tissue concentrations of antimicrobials is considerable. The low

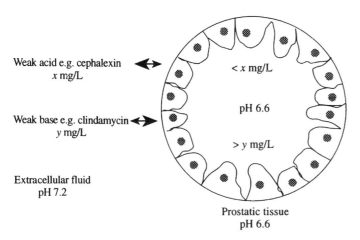

Figure 2.2 The penetration of antibiotics into an acid environment; x and y are the relative concentrations of cephalexin and clindamycin, respectively.

Table II. The pH of various body fluids

Body fluid	Usual pH	pH range
Blood (arterial)	7.4	7.35–7.45
CSF	7.34	7.32–7.37
Saliva	6.4	5.8–7.1
Sweat	4.5	4.0–6.8
Breast milk	6.6	6.6–7.0
Stomach	2.2	1.0–3.5
Duodenum and ileum		6.5–7.5
Colon	7.9	5.8–8.4

antimicrobial penetration of a poorly vascularized tissue such as cortical bone is well known. Inflammation increases blood flow and alters the permeability of the capillary endothelium, and antimicrobial concentrations are greater in such circumstances. The greater penetration of penicillins into the CSF through inflamed meninges is a good example (Thrupp *et al.*, 1965). Generalizing, inflammation of the meninges may double or treble the degree of penetration into the CSF compared with the uninflamed state. For example, the penetration of the poorly lipid-soluble benzylpenicillin is increased from 5–10% to about 20%, and the more lipid-soluble chloramphenicol from about 20% to about 50%.

One of the most important factors determining the tissue concentration of an agent is the concentration in serum: the higher the serum concentration, the higher that in tissue. Therefore, it is preferable to give an antimicrobial by iv bolus injection, rather than orally, if maximal tissue concentrations are desired.

The measurement of tissue concentrations of drugs is a developing area and several methods of investigation have been tried, such as the earlier use of animal tissue cages (Chisholm *et al.*, 1973), skin windows (Raeburn, 1971) and blistering agents (Simon *et al.*, 1973). There has been a considerable amount of work using the last model (Wise *et al.*, 1980a) and two examples are given in Figure 2.4. It can be seen that higher blister fluid concentrations were found in the case of amoxycillin (*c*.20% protein binding) compared with flucloxacillin (*c*.90% protein binding). The penetration into peritoneal fluid of a number of β-lactams and quinolones has also been studied (Wise *et al.*, 1983a,b; Kavi *et al.*, 1989) and in general all penetrate rapidly and achieve tissue concentrations close to those in serum, except when the serum protein binding is high (possibly >80%).

Serum protein binding can affect the renal elimination of some antimicrobials. For example, there is a good relationship between the degree of serum protein binding and renal clearance of tetracyclines (Kunin *et al.*, 1973), but not the β-lactams, this possibly being related to the high active renal tubular elimination of the latter agents. Some cephalosporins are not eliminated by tubular secretion and if their protein binding is high (95–98% for

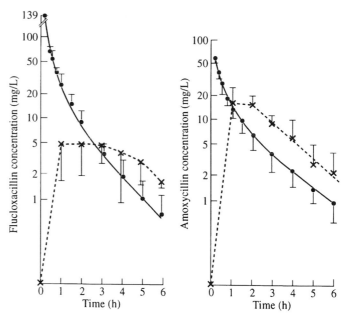

Figure 2.4 Concentrations in serum (●) and inflammatory fluid (×) after flucloxacillin or amoxycillin 1 g iv.

ceftriaxone, for example) then the elimination half-life is prolonged.

Advances in endoscopic procedures have led to a greater ability to perform biopsies at many different sites. For example, in the respiratory tract, as well as measuring drug concentrations in sputum it is now possible to assay concentrations in the bronchial mucosa, alveolar epithelial lining fluid and alveolar macrophages (Baldwin *et al.*, 1992). Table III summarizes some information on the penetration of antimicrobials into the respiratory tree.

What constitutes an effective tissue concentration of an antimicrobial is poorly understood and is dependent on tissue and cell penetration and the site of the pathogen. For example, gentamicin is active against *Legionella pneumophila in vitro*, yet ineffective *in vivo*. What is required to treat a patient infected with such a pathogen is an agent which will penetrate into the infected cell, such as a macrolide or quinolone. Although it is generally accepted that concentrations at least equal to the MIC for the infecting pathogen should be attained, it is not known

Table III. Approximate percentage penetration (compared with serum) of some different classes of antimicrobial into the respiratory tree

Antimicrobial agent	sputum	Penetration (%) into		
		bronchial mucosa	epithelial lining fluid	alveolar macrophages
β-Lactams	5–15	40	25	<10
Quinolones	70–90	150–200	200–300	≥900
Azithromycin	5–50	200–1000	≥1000	≥2000

Table IV. Approximate percentage penetration (compared with serum) of quinolone antimicrobials into various tissues

Site	Penetration (%)
Sputum	70
Bronchial mucosa	150
Lung epithelial lining fluid	200–300
Macrophages	>1000
Subcutaneous tissue	70
Muscle	70
CSF	
uninflamed	10
inflamed	≤60
Anterior chamber of eye	30–60
Bone	30
Saliva	80
Endometrium	200
Prostate	150

what multiple of the MIC should be the aim of therapy. Little is known of the effect of sub-inhibitory concentrations of drugs at sites of infection. Similarly the role of the post-antibiotic effect (PAE), although well described *in vitro*, is little understood in the clinical setting, although the pronounced PAE of aminoglycosides possibly has relevance to the efficacy of once-daily therapy with these agents.

The quinolones appear to have a propensity to distribute themselves into mucosal sites and concentrate approximately two-fold in bronchial mucosa (Honeybourne *et al.*, 1988) and gastric mucosa (Stone *et al.*, 1988). As many infections occur at mucosal sites, this has important therapeutic implications. The efficacy, albeit controversial, of quinolones in pneumococcal disease may be ascribed to the fact that bronchial mucosal concentrations exceed the MIC of the pneumococcus, although serum concentrations may be below this MIC. Table IV summarizes the approximate degree of tissue penetration of quinolones. The macrolides also appear to penetrate tissues well, in particular azithromycin, where concentrations in lung tissue, gastric mucosa, prostate, tonsil and bone were >1 mg/L when the concurrent serum concentration was <0.1 mg/L (Foulds *et al.*, 1990).

Metabolism

Many antimicrobials, for example the aminoglycosides, are not metabolized but are excreted unchanged in the urine. Some other drugs, by nature of their physico-chemical properties, being highly lipid-soluble are partially reabsorbed by the renal tubules and renal excretion is slower. The metabolism of these highly lipid-soluble drugs

produces metabolites with increased water solubility and enhances excretion by reducing tubular reabsorption. Drug metabolism can be considered in two phases. In Phase 1, reactions including oxidation, reduction and hydrolysis may occur, usually in the smooth endoplasmic reticulum (the liver being the most important site). These Phase 1 reactions (often catalysed by the multifunctional cytochrome P450 enzymes) reveal or add reactive groups which are further altered in Phase 2. Phase 2, which again mainly takes place in the liver, consists most commonly of glucuronidation, sulphation, methylation or acetylation, and is dependent on adequate functional hepatic tissue. Chloramphenicol, a very lipid-soluble drug, is excreted in the urine, but <10% is in the form of the active unmetabolized drug; the majority of the remainder is the microbiologically inactive glucuronide (Lindberg *et al.*, 1966). Certain cephalosporins containing a 3-acetoxy-methyl side chain (such as cephalothin, cephacetrile and cefotaxime) undergo desacetylation. This is of importance in the case of cefotaxime as the metabolite is a potent antimicrobial in its own right, but has a narrower spectrum than cefotaxime (Wise *et al.*, 1980b) and has different pharmacokinetic properties. Wise *et al.* (1981) observed that even mild impairment of liver function (as seen in, for example, mild congestive cardiac failure) is accompanied by diminished formation of desacetyl cefotaxime. This is surprising, as it is usually considered that the liver has a considerable reserve capacity. Imipenem is enzymically hydrolysed in the renal tubular cells by dehydropeptidase I, hence an inhibitor of this enzyme, cilastatin, is co-administered. However, the newer carbapenems, such as meropenem and biapenem, are stable to hydrolysis and hence do not require an enzyme inhibitor. Many sulphonamides are excreted both unchanged and as the acetyl metabolite. Some of those sulphonamides (which are termed 'long acting' because they are eliminated very slowly), such as sulphametopyrazine, owe this property, in part, to the fact that they are acetylated more slowly and hence tubular reabsorption of the active drug can take place. A high serum protein binding may also contribute to this slow elimination of certain sulphonamides.

In the case of ceftriaxone, a highly protein-bound cephalosporin, the kinetics can be dosage-dependent as the protein binding can become saturated and a further dosage increase can lead to an increase in the free fraction of drug.

Some antimicrobials are excreted by the liver directly into the bile. It is often difficult with these drugs to account fully for their excretion via this route, and many drugs, such as erythromycin and sodium fusidate, have complex routes of excretion, mainly biliary, some urinary and some undetermined. Antimicrobials can also affect the metabolism of other drugs. Rifampicin can act as an enzyme inducer for its own metabolism and can also increase the metabolism of steroids found in the oral contraceptive agents and reduce their biological activity. Chloram-

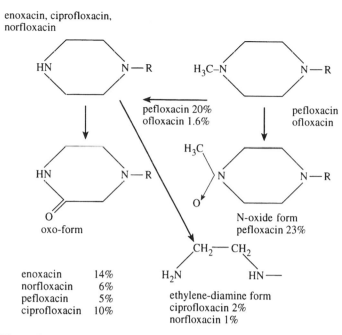

enoxacin, ciprofloxacin, norfloxacin

pefloxacin 20%
ofloxacin 1.6%

pefloxacin
ofloxacin

oxo-form

N-oxide form
pefloxacin 23%

enoxacin	14%
norfloxacin	6%
pefloxacin	5%
ciprofloxacin	10%

ethylene-diamine form
ciprofloxacin 2%
norfloxacin 1%

Figure 2.5 The urinary metabolites of the piperazine ring of fluoroquinolones accounting for >1% of the excreted drug. R is the body of the fluoroquinolone molecule.

phenicol would appear to reduce the metabolism of certain drugs such as tolbutamide and elevated blood concentrations may occur (Christensen & Skovsted, 1969). Some, but not all, quinolones can affect the metabolism of xanthine compounds such as theophylline and caffeine. This is most marked in the case of enoxacin when co-administration with theophylline led to neurotoxic serum concentrations of theophylline (Wijnands et al., 1986).

The metabolism of the piperazine ring of the fluoroquinolone agents is exceedingly complex (Figure 2.5; see also *Chapter 11*) (White, 1987). For example, norfloxacin is converted to an oxo-metabolite and an ethylenedamine metabolite and up to four other metabolites (Osaki et al., 1981) The oxo-metabolite of enoxacin has modest antimicrobial activity and accounts for 12% of the excreted drug found in the urine (Wise et al., 1986). Some quinolones are also conjugated before excretion. In the case of pefloxacin, the position is further complicated since its desmethyl metabolite is chemically identical to norfloxacin (Roquet et al., 1983).

The azole antifungals can undergo complex and extensive metabolism. In ketoconazole, the piperazine and imidazole rings are first split followed by further metabolism. Very little renal elimination occurs.

Various factors, including age, genetic make-up and disease state, can influence drug metabolism. The 'grey baby' syndrome associated with the failure to conjugate and excrete chloramphenicol in the immature infant is well known (Weiss et al., 1960). Alteration of metabolism in the elderly is poorly understood in the antimicrobial field. Genetic influences, such as the variability of isoniazid

acetylation and hence excretion, have practical implications (Jessamine et al., 1963). The concomitant administration of other drugs which influence enzyme induction can also affect drug metabolism. Phenobarbitone and phenytoin, both enzyme inducers, can lead to lower serum concentrations of doxycycline when given at the same time (Neuvonen & Pentilla, 1974; Pentilla et al., 1974). The effect of certain quinolones upon theophylline metabolism and the effect of H_2 antagonists upon quinolone pharmacology can be ascribed to effects upon the cytochrome P450 enzymes (Rubinstein & Seqev, 1986).

Excretion

Many antimicrobials are excreted unchanged in the urine; this has already been discussed above. Elimination via this route and via the bile are by far the most important methods of excretion of antimicrobials. Other routes of elimination are, however, also of importance. The excretion of tetracycline into maternal milk and the possible effect this may have on the infant's teeth are well known (Takyi, 1970). It is possible that a portion of clavulanic acid not excreted in the urine is broken down completely in the body and exhaled as CO_2.

Renal excretion

The renal excretion of antimicrobials is the overall result of the processes of filtration and of tubular secretion, reabsorption and metabolism. The more lipid-soluble drugs can be reabsorbed through the tubular cells and the pH of the urine can influence the extent of this process. Weak acids, such as the majority of sulphonamides, have an increased excretion in an alkaline urine, hence the importance of urinary alkalization when the early, poorly soluble sulphonamides were given. Other antimicrobials, such as the aminoglycosides, which are essentially lipid-insoluble, will not be reabsorbed at all from the renal tubules. In this case the rate of excretion is dependent on glomerular filtration (Jao & Jackson, 1964). Tubular secretion occurs with certain antimicrobials, notably the penicillins, some cephalosporins and certain quinolones, such as norfloxacin (Shimada et al., 1983), cinoxacin (Israel et al., 1978) and ciprofloxacin. Coadministered probenecid competes in this process and can decrease the tubular secretion of the other drug, leading to elevated blood concentrations. The total effect of probenecid is more complex as this drug also appears to affect the transport of penicillin across the choroid plexus into the CSF (Dacey & Sande, 1974). A further unexplained effect of probenecid was noted by Welling et al. (1979) when studying the pharmacology of oral cephalosporins. Peak cephalosporin concentrations, greater than could be explained by the tubular action of probenecid, were noted. It is possible that probenecid can decrease the V_d of such agents.

An interesting observation is the interaction between

piperacillin and the β-lactamase inhibitor tazobactam where the former appears to inhibit the elimination of the latter (Wise *et al.*, 1991). It is not known if this results from competitive inhibition in the renal tubule.

Impairment of renal function reduces the excretion of those antimicrobials largely excreted via the kidneys and this is reflected in elevated blood concentrations. Because the elimination rate of drugs such as the aminoglycosides declines with falling glomerular filtration rate, it is possible to predict approximately the effect of a known degree of renal failure upon the serum concentrations of the drug. Various biochemical indices are used to assess renal function. Blood urea concentration was used by Gingell & Waterworth (1968) to predict gentamicin dosage and a more sophisticated variation of this, use of the serum creatinine concentration, was introduced by Mawer *et al.* (1974) to assist the determination of the dosage of this and related drugs. Whilst such methods are commonly employed to determine initial dosage regimens in patients with impaired renal function, regular assays of serum aminoglycoside concentrations are an essential part of the therapeutic monitoring of these agents.

Whether and to what degree drug dosage should be adjusted in renal failure depends on the metabolism and excretion of that drug and its potential for toxicity. Some drugs handled almost totally by glomerular filtration have to be carefully monitored, the above-mentioned aminoglycosides being the best example. Other drugs, such as erythromycin, lincomycin, sodium fusidate, ketoconazole, itraconazole and the rifamycins, do not require dosage reduction in renal failure, accumulation being relatively rare since they undergo extensive hepatic clearance. There is an intermediate group in which it is preferable to reduce somewhat the dose or frequency of dosing, although it is less critical than in the case of the aminoglycosides. The penicillins, cephalosporins and some quinolones come into this category. Other drugs to be avoided in renal failure are those which are either nephrotoxic themselves, or produce toxic metabolites. For example, chloramphenicol glucuronide can accumulate and this may be toxic to the haematopoietic tissues (Weiss *et al.*, 1960). Care should be exercised in giving certain tetracyclines to patients in renal failure as these drugs can cause a further deterioration in renal function, probably as a result of an anti-anabolic effect (Edwards *et al.*, 1970).

Biliary excretion

As has already been discussed, some antimicrobials, such as erythromycin, sodium fusidate and the rifamycins, are excreted mainly by the bile. Other drugs, such as ampicillin, certain tetracyclines and cephalosporins (for example, cefoperazone and ceftriaxone), are also excreted by this route. They may be concentrated in the gall bladder, and are used clinically for biliary infections. However, it is uncertain if biliary concentrations of antimicrobials are the essential determinant of outcome in biliary infections, or if good serum/tissue concentrations are more important. Little is known of the mechanisms of biliary excretion of antimicrobials and the subject is complex. Sodium fusidate, for instance, is excreted as seven inactive metabolites as well as active fusidate (Godtfredsen, 1967). The pharmacology of these drugs can be further complicated by the systemic reabsorption of the excreted drug from the gastrointestinal tract (enterohepatic recycling). Such recycling also possibly occurs with some quinolones (Saito *et al.*, 1988).

If liver impairment is superimposed on renal failure, the serum concentrations of some antimicrobials can be highly unpredictable. For example, with cefotaxime the serum half-life can increase from 1.0 to 5.6 h if there is modest impairment of liver function as well as severe renal failure (glomerular filtration rate < 10 mL/min) (Wright & Wise, 1981). Both metabolism and excretion of those drugs handled by the liver depend upon the presence of adequate hepatic tissue. If there is evidence of moderate to severe liver damage these drugs should be avoided.

An interesting and well known consequence of high biliary elimination is the development of biliary pseudolithiasis seen with ceftriaxone (Schaad *et al.*, 1990). It is reversible but can be associated with symptoms.

Other excretory mechanisms

Some quinolones may be excreted via the gastrointestinal tract using a transmucosal route from the blood to the gut via the mucosal cells (transintestinal elimination). Stone *et al.* (1988) showed that the concentrations of ciprofloxacin in the gastric mucosa exceed those found in blood. The clinical significance of this is that concentration in the gastrointestinal mucosa may explain efficacy in infections at this site. In addition, transmucosal elimination of ciprofloxacin (Rohwedder *et al.*, 1990) contributes to the less stringent need to reduce dosage in patients with renal failure (MacGowan *et al.*, 1994).

Pharmacokinetics

This is the mathematical study of drugs in the body, and measures the rate processes of drug absorption, distribution, metabolism and elimination. The concept requires the body to be considered as one or more compartments, separated by membranes which are passively permeable to the drug. The information obtained from a pharmacokinetic study is employed to determine or improve the way in which the drug being investigated is used clinically.

One-compartment model

In the simplest case we can consider the rapid iv injection of a drug into a single compartment, the blood, central or

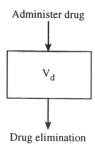

Administer drug

V_d

Drug elimination

Figure 2.6 Single compartment pharmacokinetic model, V_d = volume of distribution.

vascular compartment (Figure 2.6) of volume V_d (where V_d represents the apparent volume of distribution of the drug in the body). In this model, if C is the blood concentration of the drug at a certain time, t, after administration, the rate of change of the concentration (dC) with time (dt), i.e. the excretion rate, is proportional to the amount of drug available for this process (first order kinetics). Therefore $dC/dt = -K_{el}C$, where K_{el} is a constant referred to as the elimination rate constant.

If the logarithm of the drug concentration in the blood is plotted against time there is a linear relationship, where the slope is K_{el}.

Another way of expressing this model is to say that there is a simple exponential fall in the serum drug concentration with time and the elimination half-life ($T_{1/2}$, the time interval over which the blood concentration is halved) is related to K_{el} thus:

$$T_{1/2} = (\log_e 2)/K_{el}$$

where $\log_e 2 = 0.693$.

From assaying serum concentrations at various times after administration we can derive useful information about a drug. In Figure 2.7 the log serum concentrations of a drug determined 1, 2, 3 and 4 h after iv administration are

plotted against time. Assuming a one-compartment model, as described above, if we extrapolate the concentration-versus-time curve back to the concentration axis this will give an estimate of the concentration at time 0, C_0. The fictive initial concentration, C_0, can bc uscd to calculate the apparent volume of distribution V_d from the relationship $C_0 = D/V_d$, where D is the amount of drug given as a single dose. From this the total clearance (Cl_{tot}) of the drug from the body can be calculated as $V_d K_{el}$ or as $0.693 V_d/T_{1/2}$.

Two-compartment model

In practice, for most drugs the simple linear relationship shown in Figure 2.7 will probably not be seen; instead it is more likely that a biphasic curve will be seen, as shown in Figure 2.8. In this case we say that the pharmacokinetics of this drug cannot be described by a simple single-compartment model, but during the first hour the drug is being distributed between compartments (the distribution phase) and the elimination rate, K_{el}, from the blood (the vascular compartment) is more rapid than seen later. This distribution phase continues for much longer than is apparent from the curve, but the degree after the first hour or two is probably insignificant for most well-absorbed antimicrobials. The pharmacokinetics of such a drug are said to conform to a two-compartment open model, distribution from the first to the second compartment occurring during the first phase. The first compartment (the central or vascular compartment) is best considered as the blood and the well perfused organs of the body, and the second compartment (the peripheral or tissue compartment)as the more peripheral tissues. In Figure 2.8, C'_1 and C'_2 are the actual serum drug concentrations measured

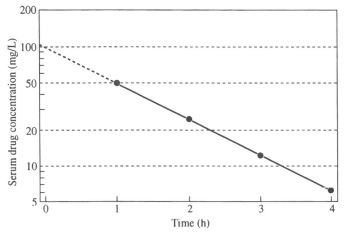

Figure 2.7 Typical log concentration versus time curve for a one-compartment model. The slope of the line is equal to the elimination rate constant (K_{el}). The elimination half-life = $0.693/K_{el}$.

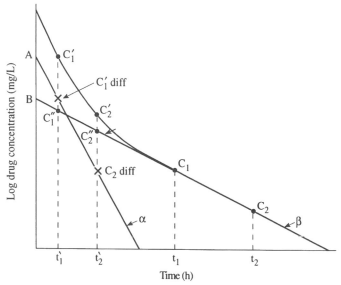

Figure 2.8 Typical log concentration versus time curve for a two-compartment model. α is the slope of the early part of the curve (the distribution or α phase), and β is the slope of the latter part of the curve (the elimination or β phase).

at times t'_1 and t'_2. C''_1 and C''_2 are the hypothetical drug concentrations obtained from the back-extrapolation of the later part of the curve through points C_1 and C_2 (the so-called β or elimination phase). C_1diff and C_2diff are the concentrations determined from $(C'_1 - C''_1)$ and $(C'_2 - C''_2)$. Back-extrapolation to the concentration axis of the line plotted through C_1diff and C_2diff with slope α (the so-called α or distribution phase) gives the concentration A at time 0. This is distinct from B, the concentration at time 0 obtained by back-extrapolation to the concentration axis of the line plotted through C_1 and C_2 with slope β.

If K_{12} is the distribution rate constant for transfer from the central to the peripheral compartment and K_{21} is the distribution rate constant for transfer from the peripheral to the central compartment, then various pharmacokinetic values can be calculated:

$$\beta = (\log_e (C_1 - C_2))/(t_2 - t_1)$$

$$\alpha = (\log_e (C_1\text{diff} - C_2\text{diff}))/(t'_2 - t'_1)$$

$$T_{1/2}\beta \text{ (elimination half-life)} = 0.693/\beta$$

$$T_{1/2}\alpha \text{ (distribution half-life)} = 0.693/\alpha$$

$$C_0 = A + B$$

$$K_{12} = (AB(\beta - \alpha)^2)/(C_0(AB + \beta\alpha))$$

$$K_{21} = ((AB) + (\beta\alpha))/C_0$$

The volume of distribution (V_d) was mentioned previously in relation to the one-compartment model. In two-compartment models estimates of the volume of distribution into the central compartment (V_{dC}) are similarly calculated from the equation $V_{dC} = D/C_0$. The total volume of distribution when the drug has finally equilibrated is known as the volume of distribution at steady state (V_{dss}) and can be calculated from:

$$V_{dss} = ((K_{21} + K_{12})V_{dc})/K_{21}$$

when the drug is given intravenously.

Multiple dosing

The discussion so far has related to a single dose of a drug. Repeat dosing may well lead to drug accumulation and pharmacokinetic analysis can be a guide to the rational choice of an appropriate dosing interval.

Steady state is achieved when no further accumulation occurs for a given dosing regime. This is when the rate of drug elimination equals the rate of administration. The time taken to reach steady state depends on the dosing interval (τ) and the elimination half-life ($T_{1/2}$) and is expressed as ε (when $\varepsilon = \tau/T_{1/2}$). The number of doses, n, to reach 95% of steady state is calculated from the equation: $0.95 = 1 - 2^{-n\varepsilon}$. Therefore if $\tau = T_{1/2}$ (i.e. the drug is administered every half-life) then $n = 4.4$ doses, if $\tau = 2T_{1/2}$ then $n = 2.2$ doses, and if $\tau = 0.5T_{1/2}$ then $n = 8.7$ doses.

The accumulation factor, R, following multiple doses depends upon ε; for example, if $\varepsilon = 0.5$ then $R = 3.4$, if $\varepsilon = 1$ then $R = 2.0$, and if $\varepsilon = 4$ then $R = 1.1$. This implies that virtually no accumulation occurs if a drug is given every four half-lives, but if it is administered twice in each $T_{1/2}$ then an accumulation of 3.4 occurs. These data can then be used to suggest a loading dose. As an accumulation factor of 2 occurs if the drug is given every $T_{1/2}$, a loading dose twice the maintenance dose will achieve steady-state levels immediately.

Intra-muscular administration

Important clinical data can also be obtained by studying the intramuscular administration of a drug, but now a further compartment, the site of the injection, is introduced. The pharmacokinetic analysis now becomes very complex and of little practical benefit to microbiologists and clinicians. Those interested in this subject are advised to read the reviews by Janku (1971), Ritschel (1976) or Gibaldi & Perrier (1982).

Non-compartmental analysis

Increasingly the use of models not requiring a compartmental analysis are being used to estimate clearance, volumes of distribution, bioavailability, elimination rates, etc. The methodology, based on statistical moment theory, has many advantages (Gibaldi & Perrier, 1982).

Pharmacokinetic software

There are a large number of computer programs available for those interested in the subject of pharmacokinetics. Two commercial packages which are used extensively are Siphar (Simed S.A., 9–11 rue G. Enesco, 94008 Créteil, France) and WinNonlin (Scientific Consulting Inc., 8509 Burnside Drive, Apex, NC 27502, USA). Public-domain software and shareware are also widely available.

Influence of protein binding on pharmacokinetics

Antimicrobials, like most drugs, are reversibly bound to a variable extent to plasma protein, which is usually (but not always) albumin. It is only the free, microbiologically active component (D_F) which diffuses into the extra-vascular compartment. It should be remembered that assays measure total antimicrobial in the plasma, D_B (bound drug) + D_F (Figure 2.9), unless specially designed to measure free drug only.

It is well known that, in vitro, a highly protein-bound antimicrobial will exhibit reduced antimicrobial activity in the presence of serum when compared with the activity in serum-free broth. From a knowledge of the broth MIC and the percentage binding one can quite accurately predict

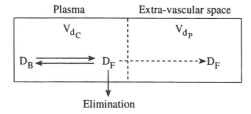

Figure 2.9 The protein binding of drug D. D_B and D_F are the concentrations of protein-bound and free drug, respectively, V_{dC} is the volume of blood plasma (central compartment) and V_{dP} the volume of the extravascular space (peripheral compartment) into which the drug distributes.

the MIC in serum (Kunin, 1969). When the in-vivo situation is considered the problem is far less clear. Pharmacokinetic analysis may help (Keen, 1971).

Assuming the drug is protein-bound only in plasma (i.e. there is negligible protein in the extravascular tissues), then the total amount of drug in the body is equal to $V_{dC}D_T + V_{dP}D_F$ where D_T is the total concentration in plasma (i.e. $D_B + D_F$) and V_{dC} and V_{dP} are the volumes of the central and peripheral compartments respectively.

These data can be used to calculate curves (Figure 2.10) showing the effect of the degree of protein binding on the free concentration of the drug in the tissue (D_F). The two curves shown in Figure 2.10 are for theoretical and widely different drugs. Curve A is derived for a lipid-insoluble drug which distributes itself into a blood volume V_{dC}, of 3 L and, as it cannot enter cells, to an extracellular volume V_{dP} of (say) 9 L. Curve B, on the other hand, is derived for a highly lipid-soluble drug which can penetrate cells and has an extravascular distribution volume V_{dP} of 39 L.

Protein binding only causes a dramatic reduction in the concentration of free drug in tissue when it exceeds 80% and the effect is greater with lipid-insoluble drugs.

Wise *et al.* (1980a), using a cantharides blister technique, elucidated further the effect of protein binding on tissue concentrations of antimicrobials. In Figure 2.11 the effect of serum protein binding upon the amount of antibiotic penetrating the blister fluid is shown. The degree of penetration was quantified by using the area under the blister concentration versus time graph. It can be seen that only when protein binding is greater than about 80% does the total amount of antimicrobial penetrating into the tissue fluid decline substantially. In Figure 2.12 the influence of protein binding in tissue fluid upon the amount of free antimicrobial penetrating into tissue fluid is shown. A linear relationship was observed, indicating the importance of protein binding upon the penetration of microbiologically active free drug into tissue fluid.

A point to be stressed is that the protein binding of a drug should not be considered in isolation. The V_d of the drug (determined additionally by such parameters as lipid solubility) and serum elimination half-life will also influence the tissue concentrations of a drug.

It is reasonable to conclude that protein binding will probably affect a drug's activity *in vivo* and *in vitro*, its distribution and, in the case of certain antimicrobials, its elimination. If the protein binding is high it is reasonable to assume that a drug may be less efficacious than an equal dose of an equally active drug of lower protein binding.

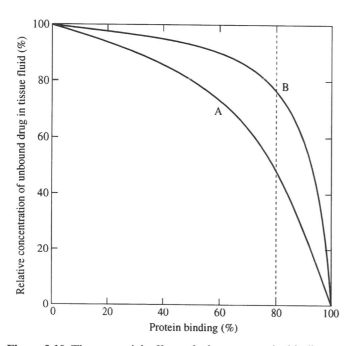

Figure 2.10 The potential effect of plasma protein binding on tissue penetration (from Keen, 1971).

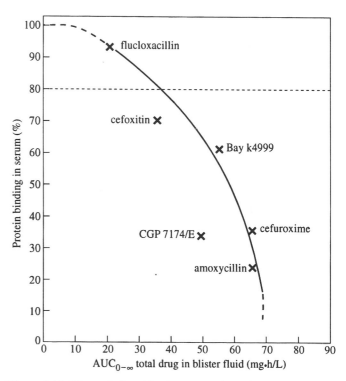

Figure 2.11 The relationship between the percentage protein binding in serum with the AUC of total drug in blister fluid (Wise *et al.*, 1980).

21

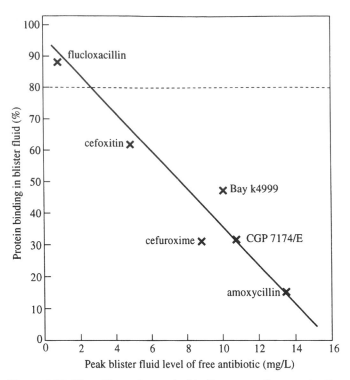

Figure 2.12 The effect of protein binding upon the penetration of 'free' antibiotic (as measured by peak concentration) into tissue fluid of six β-lactams administered to humans (Wise *et al.*, 1980).

Animal studies using a series of penicillins (Merrikin *et al.*, 1983) have shown this to be the case. However, such clear-cut data will be almost impossible to obtain in human studies. It is important to remember that when assessing pharmacological data in animals there may be gross differences in the way animals handle a drug compared with humans (O'Grady, 1976). Considerable inter-species variation in the degree of protein binding can also occur (Zak, O., personal communication, 1976). The small amount of information from human studies has been reviewed (Wise, 1985).

Clinical relevance of pharmacokinetics

As already stated, the value of pharmacokinetics is to assist the clinician in the choice of the most efficacious dosing regime. Many publications on pharmacokinetics can confuse rather than help, often by indulging in irrelevant mathematical detail. Those treating infections need understand only a few of the parameters. The elimination half-life ($T_{1/2}\beta$) is one important factor in choosing a dosing schedule; the peak serum concentration (C_{max}) and the time taken to achieve it (T_{max}) will give a measure of rapidity of absorption and whether enough is absorbed to inhibit the infecting pathogens; the volume of distribution (V_d) will give an indication of how well a compound penetrates the tissues of the body.

In recent years, data—at first from animals but now

Table V. Pharmacodynamics and outcome: ciprofloxacin in nosocomial pneumonia

	Patients cured (%)	
Parameter	clinical cure	microbiological cure
MIC (mg/L)		
<0.125	82	79
0.5	54	79
1	33	44
AUIC		
0–125	32	42
125–250	81	88
250–1000	79	71
Dosage regimen		
200 mg bd	78	67
300 mg bd	25	38
400 mg bd	70	84

Schentag *et al.* (1993)

accumulating from patients—that combine pharmacodynamics with pharmacokinetics are giving more precise information on how we should use antimicrobials. Drusano (1988) has shown that aminoglycosides are more efficacious on a once-daily basis. As they kill in a concentration-dependent fashion, a high C_{max} is important. In contrast, β-lactams only show marginal concentration-dependent bacterial killing and constant infusion is probably the most efficient way to employ these agents. In the case of the quinolones it is not quite so clear, but neutropenic animal models of infection have shown that the ratio of C_{max} to MIC was linked to survival (Drusano *et al.*, 1993). Further work by Schentag *et al.* (1993) and Craig (1995) has clarified the subject. The former authors have shown that the most useful predictor of outcome in patients treated with ciprofloxacin for nosocomial pneumonia is the ratio of AUC to MIC (AUMIC) (Table V) and further generalization can be made (Table VI) in order to maximize the efficacy of antimicrobials.

Conclusions

The pharmacology of many antimicrobials is well understood and a basic knowledge and understanding of pharmacokinetics should enable us to use these drugs in a rational and predictable way. There are still, however, many 'grey areas'. The effect of the patient's disease upon the way in which a drug is handled is only beginning to be investigated. The relevance of the many in-vitro measurements, such as MIC, MBC and post-antibiotic effect, to the in-vivo situation is often, at best, only an educated guess. After more than 50 years of antibiotic use we still are uncertain of the importance of a number of basic points such as the relevance of serum concentrations and tissue

Table VI. Pharmacodynamic factors and in-vitro correlations

Antimicrobial agent	Pharmacodynamic characteristic	Goal of regimen	Correlating parameter
Aminoglycosides, fluoroquinolones, metronidazole	Concentration-dependent killing, post-antibiotic effect (PAE)	Maximum peak concentrations	C_{max}/MIC, 24 h AUC/MIC
Penicillins, cephalosporins	Time-dependent killing, little or no PAE	Maximum exposure time	Time for which the serum concentration is above the MIC or MBC
Carbapenems, vancomycin, macrolides, clindamycin	Time-dependent killing, PAE	Maximum exposure time but concentration is above the concentrations may fall below MIC	Time for which the serum MIC or MBC

Craig (1995).

concentrations to clinical outcome (Schentag, 1989). Studies in models of meningitis suggest that the concentration of an antimicrobial in CSF should exceed the MIC of the infecting organism by 8- to 10-fold to optimize therapy (McCracken *et al.*, 1982; Decazes *et al.*, 1983). As mentioned above, there is now evidence from some patients that aminoglycosides are preferably given once- or twice-daily in order to obtain high peak concentrations whereas β-lactams are better used as continuous infusions (Drusano, 1989). As mentioned above, there is fairly strong evidence from new quinolones and macrolides that tissue concentrations, rather than serum concentrations, are the important parameter in determining the outcome of infection. Almost 40 years ago, in the treatment of chest infections, May (1955) showed that sputum drug concentrations greater than those required to inhibit the pathogens were required and he later showed that higher concentrations were found in purulent rather than mucoid sputum (May & Delves, 1965). Much work still needs to be done on antimicrobial pharmacology before we can claim to be thoroughly conversant with these widely used drugs.

References

Antonioli, J. A., Schelling, J. L., Steininger, E. & Borel, G. A. (1971). The effect of gastrectomy and an anticholinergic drug on the gastrointestinal absorption of a sulphonamide in man. *Zeitschrift für Klinische Pharmakologie, Therapie und Toxikologie* **5**, 212–5.

Ashby, J., Piddock, L. J. V. & Wise, R. (1985). An investigation of the hydrophobicity of the quinolones. *Journal of Antimicrobial Chemotherapy* **16**, 805–8.

Baldwin, D. R., Honeybourne, D. & Wise, R. (1992). Pulmonary disposition of antimicrobial agents: methodological considerations. *Antimicrobial Agents and Chemotherapy* **36**, 1171–5.

Bhattacharya, M. K., Chakrabati, M. K., Siuha, A. K., Nashipurd, J. N. & Bhattacharya, S. K. (1992). Absorption and toxicological study of fluoroquinolones in humans. *Indian Journal of Physiology and Allied Science* **46**, 82–8.

Blouin, R. A., Kneer, J. & Stoeckel, K. (1989). Pharmacokinetics of intravenous cefetamet (Ro 15-8074) and oral cefetamet pivoxil (Ro 15-8075) in young and elderly subjects. *Antimicrobial Agents and Chemotherapy* **33**, 291–6.

Bolger, W. P. & Gavin, J. J. (1959). An evaluation of tetracycline preparations. *New England Journal of Medicine* **261**, 827–32.

Brown, N. M., White, L. O., Blundell, E. L., Chowrn, S. R., Slade, R. R., MacGowan, A. P. *et al.* (1993). Absorption of oral ofloxacin after cytotoxic chemotherapy for haematological malignancy. *Journal of Antimicrobial Chemotherapy* **32**, 117–22.

Chisholm, G. D., Waterworth, P. M., Calnan, J. S. & Garrod, L. P. (1973). Concentration of antimicrobial agents in interstitial fluid. *British Medical Journal i*, 569–73.

Christenson, L. K. & Skovsted, L. (1969). Inhibition of drug metabolism by chloramphenicol. *Lancet ii*, 1397–9.

Craig, W. A. (1995). Kinetics of antibiotics in relation to effective and convenient outpatient therapy. *International Journal of Antimicrobial Agents* **5**, 19–22.

Dacey, R. G. & Sande, M. A. (1974). Effect of probenecid on the cerebrospinal fluid concentrations of penicillin and cephalosporin derivatives. *Antimicrobial Agents and Chemotherapy* **6**, 437–41.

Dean, S., Harding, L. K., Wise, R. & Wright, N. (1979). Absorption and excretion of cephalexin in health and acute illness. *European Journal of Clinical Pharmacology* **16**, 73–4.

Decazes, J. M., Ernst, J. D. & Snade, M. A. (1983). Correlation of *in vitro* time–kill curves and kinetics of bacterial killing in cerebrospinal fluid during ceftriaxone therapy of experimental *Escherichia coli* meningitis. *Antimicrobial Agents and Chemotherapy* **24**, 463–7.

Drusano, G. L. (1988). Role of pharmacokinetics in the outcome of infection. *Antimicrobial Agents and Chemotherapy* **32**, 289–97.

Drusano, G. L., Johnson, D. E., Rosen, M. & Standiford, H. C. (1993). Pharmacodynamics of a fluoroquinolone antimicrobial agent in a neutropenic rate model of pseudomonas sepsis. *Antimicrobial Agents and Chemotherapy* **37**, 483–90.

Edwards, O. M., Huskisson, E. C. & Taylor, R. T. (1970). Azotaemia aggravated by tetracycline. *British Medical Journal i*, 26–7.

Foulds, G., Shepard, R. M. & Johnson, R. B. (1990). The pharmacokinetics of azithromycin in human serum and tissues. *Journal of Antimicrobial Chemotherapy* **25**, Suppl. A, 73–82.

Giamarellou, H., Kanellakopoloulou, K., Pragastis, D., Tagaris, N. & Daikos, G. K. (1977). Treatment with metronidazole of 48 patients with serious anaerobic infections. *Journal of Antimicrobial Chemotherapy* **3**, 347–53.

Gibaldi, M. & Perrier, D. (1982). In *Pharmacokinetics*, 2nd edn. Marcel Dekker, New York.

Gingell, J. C. & Waterworth, P. M. (1968). Dose of gentamicin in patients with normal renal function and renal impairment. *British Medical Journal ii*, 19–22.

Godtfredsen, W. O. (1967). *Fusidic Acid and Some Related Antibiotics*, pp. 66–71. Leo Laboratories, Copenhagen.

Höffken, G., Borner, K., Glatzel, P. D., Koeppe, P. & Lode, H. (1985). Reduced external absorption of ciprofloxacin in the presence of antacids. *European Journal of Clinical Microbiology* **4**, 345.

Honeybourne, D., Andrews, J. M., Ashby, J. P., Lodwick, R. & Wise, R. (1988). Evaluation of the penetration of ciprofloxacin and amoxycillin into bronchial mucosa. *Thorax* **43**, 715–9.

Hughes, G. S., Heald, D. L., Barker, K. B., Patel, R. K., Spillers, C. R., Watts, K. C. *et al.* (1989). The effects of gastric pH and food on the pharmacokinetics of a new oral cephalosporin, cefpodoxime proxetil. *Clinical Pharmacology and Therapeutics* **46**, 674–85.

Israel, K. S., Black, H. R., Nelson, R. L., Brunson, M. K., Nash, J. F., Brier, G. L. *et al.* (1978). Cinoxacin: pharmacokinetics and the effect of probenecid. *Journal of Clinical Pharmacology* **18**, 491–9.

Janku, I. (1971). *Fundamentals of Biochemical Pharmacology* (Bacq, A. M., Ed.), p. 203. Pergamon Press, Oxford.

Jao, R. L. & Jackson, G. G. (1964). Gentamicin sulphate: a new antibiotic against Gram-negative bacilli. *Journal of the American Medical Association* **189**, 817–22.

Jessamine, A. G., Hamilton, E. J. & Eidus, L. (1963). A clinical study of isoniazid inactivation. *Canadian Medical Association Journal* **89**, 1214–7.

Johnson, E. J., MacGowan, A. P., Potter, M. N., Stockley, R. J., White, L. O., Slade, R. R. *et al.* (1990). Reduced absorption of oral ciprofloxacin after chemotherapy for haematological malignancy. *Journal of Antimicrobial Chemotherapy* **25**, 837–42.

Kavi, J., Ashby, J. P., Wise, R. & Donovan, I. A. (1989). Intraperitoneal penetration of cefpirome. *European Journal of Clinical Microbiology and Infectious Diseases* **8**, 556–8.

Keen, P. (1971). *Handbook of Experimental Pharmacology* (Brodie, B. B. & Gillette, J. R., Eds), p. 224. Springer-Verlag, Berlin.

Kirst, H. A. & Sides, G. D. (1989). New directions for macrolide antibiotics. *Antimicrobial Agents and Chemotherapy* **33**, 1413–8.

Kunin, C. M. (1969). Clinical significance of protein binding of the penicillins. *Annals of the New York Academy of Science* **145**, 282–90.

Kunin, C. M., Craig, W. A., Korngust, M. & Manson, R. (1973). Influence of protein binding on the pharmicologic activity of antibiotics. *Annals of the New York Academy of Science* **226**, 214–24.

Lindberg, A. A., Nilsson, L. H., Bucht, H. & Kallings, L. O. (1966). Concentration of chloramphenicol in the urine and blood in relation to renal function. *British Medical Journal ii*, 724–8.

MacGowan, A. P., White, L. O., Brown, N. M., Lovering, A. M., McMullin, C. M. & Reeves, D. S. (1994). Serum ciprofloxacin concentrations in patients with severe sepsis being treated with ciprofloxacin 200 mg iv bs irrespective of renal function. *Journal of Antimicrobial Chemotherapy* **33**, 1051–4.

Mattila, M. J., Jussila, J. & Takki, S. (1973). Drug absorption in patients with intestinal villous atrophy. *Arzneimittel Forschung (Drug Research)* **32**, 583–5.

Mawer, G. E., Ahamd, R., Dobbs, S. M. & Tooth, J. A. (1974). Experience with a gentamicin monogram. *Postgraduate Medical Journal* **50**, *Suppl. 7*, 31–2.

May, J. R. (1955). Laboratory background to the use of penicillin in chronic bronchitis and bronchiectasis. *British Journal of Tuberculosis* **49**, 166–73.

May, J. R. & Delves, D. M. (1965). Treatment of chronic bronchitis with ampicillin. *Lancet i*, 929–33.

McCracken, G. H., Nelson, J. D. & Grimm, L. (1982). Pharmacokinetics and bacteriological efficacy of cefoperazone, cefuroxime, ceftriaxone and moxalactam in experimental *Streptococcus pneumoniae* and *Haemophilus influenzae* meningitis. *Antimicrobial Agents and Chemotherapy* **21**, 262–7.

McDonald, P. J., Mather, L. E. & Story, M. J. (1977). Studies on absorption of a newly developed enteric-coated erythromycin base. *Journal of Clinical Pharmacology* **17**, 601–6.

Merrikin, D. J., Briant, J. & Rolinson, G. N. (1983). Effect of protein binding on antibiotic activity *in vivo*. *Journal of Antimicrobial Chemotherapy* **11**, 233–8.

Neu, H. C. (1974). Antimicrobial activity and human pharmacology of amoxycillin. *Journal of Infectious Diseases* **129**, *Suppl.*, S123–31.

Neuvonen, P. J. & Pentilla, O. (1974). Interaction between doxycycline and barbiturates. *British Medical Journal i*, 535–46.

O'Grady, F. (1976). Animal models in the assessment of antimicrobial agents. *Journal of Antimicrobial Chemotherapy* **2**, 1–3.

Osaki, T., Uchida, H. & Irikura, T. (1981). Studies on metabolism of AM-715 in humans by HPLC. *Chemotherapy (Tokyo)* **92**, *Suppl. 4*, 128–35.

Parsons, R. L. & Paddock, G. M. (1975). Absorption of two antibacterial drugs, cephalexin and co-trimoxazole, in malabsorption syndromes. *Journal of Antimicrobial Chemotherapy* **1**, *Suppl. 3*, S59–67.

Pentilla, O., Neuvonen, P. J., Aho, F. K. & Lehtovaara, R. (1974). Interaction between doxycycline and some antiepileptic drugs. *British Medical Journal ii*, 470–2.

Raeburn, J. A. (1971). A method of studying antibiotic concentrations in inflammatory exudate. *Journal of Clinical Pathology* **24**, 633–45.

Reeves, D. S., Rowe, R. C. G., Snell, M. E. & Thomas, A. B. W. (1973). Further studies on the secretion of antibiotics in the prostatic fluid of the dog. In *Urinary Tract Infection* (Brumfitt, W. & Asscher, A. W., Eds), pp. 197–205. Oxford University Press, London.

Reeves, D. S., Bywater, M. J., Wise, R. & Whitmarsh, V. B. (1974). Availability of three antibiotics after intramuscular injection into thigh and buttock. *Lancet ii*, 1421–2.

Ritschel, W. A. (1976). *Handbook of Basic Pharmacokinetics*. Drug Intelligence Publication, Hamilton, IL.

Rohwedder, R., Bergan, T., Thorsteinsson, B. & Scholl, H. (1990). Transintestinal elimination of ciprofloxacin. *Chemotherapy* **36**, 77–84.

Roquet, F., Montay, G. & Goueffon, Y. (1983). Pefloxacin main metabolites in mouse, dog, rabbit and man. In *Proceedings of the Thirteenth International Congress of Chemotherapy, Vienna, 1983*. Abstract S4–6/4–3.

Rubenstein, E. & Segev, S. (1986). Drug interactions of ciprofloxacin with other non-antibiotic agents. In *Ciprofloxacin Workshop, Fort Lauderdale, 1986*. Abstract 35.

Saito, A., Nagata, O., Takahara, Y., Okezaki, E. Yamada, T. & Ito, Y. (1988). Enterohepatic circulation of NY-198, a new difluorinated quinolone, in rats. *Antimicrobial Agents and Chemotherapy* **32**, 156–7.

Schaad, U. B., Sutter, S., Gianella-Borradori, A., Pfenninger, J., Auckenthaler, R., Bernath, O. *et al.* (1990). A comparison of ceftriaxone and cefixime for the treatment of bacterial meningitis in children. *New England Journal of Medicine* **322**, 141–7.

Schentag, J. J. (1989). Clinical significance of antibiotic tissue penetration. *Clinical Pharmacokinetics* **16**, Suppl. 1, 23–31.

Schentag, J. J. (1991). Correlation of pharmacokinetic parameters to efficacy of antibiotics: relationships between serum concentrations, MIC values and bacterial eradication in patients with Gram-negative pneumonia. *Scandinavian Journal of Infectious Diseases*, Suppl. 74, 218–34.

Schentag, J. J., Wise, R. & Nix, D. E. (1990). In *Pharmacokinetics and Tissue Penetration of Quinolones* (Siporin, C., Heifetz, C. L., Domagala, J. M., Eds), pp. 189–222. Marcel Dekker, New York.

Schentag, J. J., Nix, D. E. & Forrest, A. (1993). Pharmacodynamics of fluoroquinolones. In *Quinolone Antimicrobial Agents* (Hooper, D. C. & Wolfson, J. S., Eds), pp. 259–71. American Society for Microbiology, Washington, DC.

Shimada, J., Yamaji, T., Ueda, Y., Uchida, H., Kusajima, H. & Irikura, T. (1983). Mechanisms of renal excretion of AM-715, a new quinolone carboxylic acid derivative, in rabbits, dogs and humans. *Antimicrobial Agents and Chemotherapy* **23**, 1–7.

Simon, C., Malerczyk, V., Brahmstaedt, E. & Toefler, W. (1973). Cefazolin, ein neues breitspektrum-antibiotikum. *Deutsche Medizinische Wochenschrift* **98**, 2448–50.

Stone, J. W., Wise, R., Donovan, I. A. & Gearty, J. (1988). Failure of ciprofloxacin to eradicate *Campylobacter pylorii* from the stomach. *Journal of Antimicrobial Chemotherapy* **22**, 92–3.

Takyi, B. E. (1970). Excretion of drugs in human milk. *Journal of Hospital Pharmacy* **28**, 317–26.

Thrupp, L. D., Leedom, J. M., Ivler, D., Wehrle, P. F., Portnoy, B. & Mathies, A. A. W. (1965). Ampicillin levels in the cerebrospinal fluid during treatment of bacterial meningitis. *Antimicrobial Agents and Chemotherapy* **5**, 206–13.

Van Tyle, J. H. (1984). Ketoconazole: mechanism of action, spectrum of activity, pharmacodynamics, drug interactions, adverse reactions and therapeutic use. *Pharmacotherapy* **4**, 343–73.

Wagner, J. G. (Editor) (1971). In *Biopharmaceutics and Relevant Pharmacokinetics*, p. 238. Drug Intelligence Publications, Hamilton, IL.

Weiss, C. F., Glazko, A. J. & Weston, J. K. (1960). Chloramphenicol in the newborn infant. *New England Journal of Medicine* **262**, 787–94.

Welling, P. G., Dean, S., Selen, A., Kendall, M. J. & Wise, R. (1979). Probenecid: an unexplained effect on cephalosporin pharmacology. *British Journal of Clinical Pharmacology* **8**, 491–5.

White, L. (1987). Metabolism of 4-quinolones. *Quinolones Bulletin* **3**, 1–3.

Wijnands, W. J. A., Vree, T. B. & van Herwaarden, C. L. A. (1986). The influence of quinolone derivatives on theophylline clearance. *British Journal of Clinical Pharmacology* **22**, 677–83.

Williams, P. E. O. & Harding, S. M. (1984). The absolute bioavailability of oral cefuroxime axetil in male and female volunteers after fasting and after food. *Journal of Antimicrobial Chemotherapy* **13**, 191–6.

Wise, R. (1981). After pro-drugs—mutual pro-drugs. *Journal of Antimicrobial Chemotherapy* **7**, 503–4.

Wise, R. (1985). The relevance of pharmacokinetics to in-vitro models: protein binding—does it matter? *Journal of Antimicrobial Chemotherapy* **15**, Suppl. A, 77–83.

Wise, R., Wills, P. J., Andrews, J. M. & Bedford, K. A. (1980b). Activity of the cefotaxine (HR 756) desacetyl metabolite compared with those of cefotaxime and other cephalosporins. *Antimicrobial Agents and Chemotherapy* **17**, 84–6.

Wise, R., Gillett, A. P., Cadge, B., Durham, S. R. & Baker, S. (1980a). The influence of protein binding upon the tissue fluid levels of 6 β-lactam antibiotics. *Journal of Infectious Diseases* **142**, 77–82.

Wise, R., Wright, N. & Wills, P. J. (1981). Pharmacology of cefotaxime and its desacetyl metabolite in renal and hepatic disease. *Antimicrobial Agents and Chemotherapy* **19**, 526–31.

Wise, R., Donovan, I. A., Drumm, J., Dent, J. & Bennett, S. A. (1983a). Intraperitoneal concentration of cefotetan. *Antimicrobial Agents and Chemotherapy* **24**, 279–81.

Wise, R., Donovan, I. A., Drumm, J., Dyas, A. & Cross, C. (1983b). The intraperitoneal penetration of temocillin. *Journal of Antimicrobial Chemotherapy* **12**, 93–6.

Wise, R., Lister, D., McNulty, C. A. M., Griggs, D. & Andrews, J. M. (1986). The comparative pharmacokinetics and tissue penetration of four quinolones including intravenously administered enoxacin. *Infection* **14**, Suppl. 3, 196–202.

Wise, R., Logan, M., Cooper, M. L. & Andrews, J. M. (1991). Pharmacokinetics and tissue penetration of tazobactam administered alone and with piperacillin. *Antimicrobial Agents and Chemotherapy* **35**, 1081–4.

Wright, N. & Wise, R. (1981). Ceftriaxone pharmacokinetics in subjects with varying degrees of renal dysfunction. *Current Chemotherapy and Immunotherapy* **1**, 464–5.

Zaki, A., Schreiber, E. C., Weliky, I., Knill, J. R. & Hubscher, J. A. (1974). Clinical pharmacology of oral cephradine. *Journal of Clinical Pharmacology* **14**, 118–26.

Chapter 3

Antimicrobial pharmacokinetics—expected concentrations in blood and urine

Richard Wise

Department of Medical Microbiology, City Hospital NHS Trust, Birmingham B18 7QH, UK

Introduction

In order to undertake an assay of any antimicrobial in a body fluid or tissue, it is necessary to have an appreciation of the concentrations likely to be encountered in order that the appropriate calibrators can be chosen, or so that the specimen can be diluted appropriately. The following Table gives the expected concentrations in blood and urine, and describes some other properties of the antimicrobials which may be important to the clinical laboratory. For example, if an agent is highly protein bound, human serum might be considered an appropriate diluent if MIC determinations or agar diffusion tests are to be performed. The route of elimination and the influence of renal failure can have a dramatic effect on serum concentrations attained. The term 'renal failure' is generally meant to imply a severe degree of impairment, i.e. a glomerular filtration rate of <15 mL/min.

It is not possible to mention every antimicrobial available, but it is hoped that the most commonly employed agents are included. For any drugs not included, the reader is advised to contact the relevant manufacturer and/or search the scientific literature. Dosages quoted are those used in particular studies, the reader should also consult the most recent data sheet for current dosage recommendations.

Drug	Dosage	Protein binding	C_{max} (mg/L)	T_{max} (h)	Half-life (h) normal	Half-life (h) renal failure	Route of elimination	Urine concentration and excretion	Remarks
Amikacin	500 mg	<20%	40–70	0 (iv)	2.3	>48	renal over 6 h	500–1000 mg/L	
Amoxycillin	125–250 mg tds po or iv	<20%	4–6	2 (po)	1.2	8	mainly renal	300–>1000 mg/L over 6 h after 250 mg	
Amphotericin B	0.3–1.0 mg/kg/day (or 50 mg) or alternate day 2–6 h infusion	c.10%[a]	1–2	end[b]	15 days	>24	partially renal	1–5 mg/L over 24 h	Can be given as bladder instillation (15 mg in 100 mL sterile water). May be detected in urine 4 weeks after parenteral administration
Ampicillin	250–500 mg qds po im or iv	<20%	2–3 after 250 mg	2 (po)	1.2	8	mainly renal	200–>1000 mg/L over 6 h after 500 mg po	
Azithromycin	250–500 mg/day po	12–50%[c]	0.8 3 h after 500 mg po	3	14–35[d]	e	mainly biliary	5% of dose excreted in urine	High tissue concentrations
Azlocillin	2–5 g tds iv	40%	100 after 5 g iv	2[f]	1.5	6	mainly renal	>1000 mg/L over 6 h after 5 g iv	
Aztreonam	500–1000 mg tds iv or im	50%	60–80	1	1.7	6	70% renal	>300 mg/L over 8 h after 1000 mg	
Bacampicillin[g]	400 mg tds po	<20%	7–10	1	1.2	8	mainly renal	200–>1000 mg/L over 6 h	Hydrolysed to ampicillin in vivo
Capreomycin	1000 mg/L im	NA	15–40	1.5	2–4		mainly renal	500–1000 mg/L over 6 h	
Carbenicillin	1 g im or 2–5 g up to qds	45%	20–30	1	1.3	10	mainly renal	500–>1000 mg/L after 1 g im over 6 h	
Cefaclor	250–500 mg tds or qds po	25%	8–10 after 500 mg	1	0.6	3	mainly renal	>500 mg/L for 6 h after 500 mg po	In vivo degradation occurs (54% 12 h urinary recovery)
Cefadroxil	500 mg bd tds or po	<20%	16	1.5	1.5	25	renal	>500 mg/L	
Cefamandole	1–2 g im or iv 4–6 times a day	74%	80 after 1 g iv	0.25	0.8		mainly renal	600–>1000 mg/L over 4 h after 1 g iv	
Cefepime	1–2 g bd or tds im or iv	<20%	50 at 1 h after 2 g iv	–	2	up to 15	renal	c.25 mg/L 12–24 h after 1 g iv	
Cefetamet	500 or 1000 mg bd	'low'	4 after 500 mg po	4	2.3	15–18	renal	50% bioavailability	Administered as pivoxil ester
Cefixime	200–400 mg/day po	65%	4.8 after 400 mg	3.8	3.4	11	mainly renal	20% of dose recovered in urine	
Cefodizime	1 g iv bd	75–90%	215 after 1 g	-	2	7–10	mainly renal	>1000 mg/L over 4 h after 1 g	
Cefoperazone	1–2 g tds or qds iv or im	94%	375 after 2 g	0	1.7	42	mainly biliary	2000 mg/L for 1 h after 1 g iv	c.25% eliminated by kidneys

Expected concentrations in blood and urine

Drug	Dose	%	Blood conc (mg/L)				Elimination	Urine conc	Notes
Cefotaxime	1–2 g bd im or iv	40%	200 after 2 g iv	0	0.9	8	mainly renal	>1000 mg/L for 6 h after 2 g iv	40% desacetylation *in vivo*
Cefotetan	0.5–2 g bd	87%	250 after 1 g iv	1.6	3		mainly renal	>1000 mg/L for 6 h after 1 g iv	
Cefotiam	0.5–2 g qds iv	62%	100	h	1.2		80% renal	>1000 mg/L over 6 h after 1 g	
Cefoxitin	0.5–2 g tds im or iv	75%	100 after 2 g iv	0.25	0.9	13–20	renal	>1000 mg/L after 2 g iv	
Cefpirome	1–2 g bd iv	<10%	23 after 1 g iv	–	2		renal		
Cefpodoxime	200 mg bd po	–	2.6	2.5	2.2	25	mainly renal	40% of dose recovered in urine	Administered as a pro-drug
Cefprozil	500 mg bd	–	10	2	1.5	–	mainly renal	>8 mg/L 8 h after dose	
Cefroxadine	500 mg tds or qds po	10%	15	1–2	0.9	–	mainly renal	>500 mg/L for 6 h after 500 mg po	
Cefsulodin	0.5–2 g bd or tds im or iv	34%	100 after 1 g im	0	1.5	10	mainly renal	>400 mg/L for 6 h after 0.5 g im	
Ceftazidime	1–2 g bd iv or im	17%	45 after 1 g iv	1	1.8	8	mainly renal	>800 mg/L for 6 h after 1 g iv	
Ceftibuten	200 mg bd	60%	10	2	2.5	13	mainly renal	70% of dose	60% excreted in urine by 48 h
Ceftizoxime	1 g iv 2–4 times a day	31%	100	1[i]	1.4	8	mainly renal	600–>1000 mg/L over 6 h	
Ceftriaxone	500 mg–2 g iv once a day	95%	110 after 500 mg iv	0	8.5	15–50	mainly renal	>100 mg/L over 12 h	
Cefuroxime	750 mg iv or im bd or tds	35%	28 after 750 mg im	1.2	1.2	18	renal	>500 mg/L for 6 h after 1 g iv	
Cephacetrile	0.5–1 g bd or tds iv	35%	60 after 1 g iv	0.3	0.6	–	mainly renal	>500–>1000 mg/L over 6 h after 1 g	20% desacetylation *in vivo*
Cephalexin	0.5–1 g tds or qds	<20%	10–14 after 500 mg	1	0.9	c. 20	renal	>300 mg/L for 6 h	
Cephalothin	1–4 g iv up to six times a day	72%	300 after 4 g iv	0.5	0.6	4–6	renal	>1500 mg/L for 8 h	30–40% desacetylation *in vivo*
Cephapirin	1–2 g im tds or qds	48%	20 after 1 g	1	0.6	3–5	mainly renal	>800 mg/L for 6 h after 4 g iv	40% desacetylation *in vivo*
Cephazolin	0.5–1 g tds im or iv	80%	140 after 1 g iv	0.3	1.8	15	mainly renal	700–1000 mg/L over 6 h after 0.5 g	
Cephradine	0.5–1 g tds po im or iv	<20%	10 after 1 g im	2	0.9	10	renal	>300 mg/L for 6 h after 1 g im	
Chloramphenicol	500 mg qds po or iv	50%	15–25	2–3[j]	2.5	4	metabolizes 90% to glucuronide	5–10% of dose recovered in urine	
Cinoxacin	500 mg bd	62%	15	1.5	1.5		mainly renal	200 mg/L 4–6 h after dose	97% eliminated in urine
Ciprofloxacin	500–750 mg bd po	20–30%	2 after 500 mg	1.5	4	6–8	mainly renal	>100 mg/L 0–4 h after 100 mg iv	only 30% of oral absorbed
	200–400 mg iv bd		0.8–1.0 1 h after 200 mg iv						A number of metabolites (some active) formed

Drug	Dosage	Protein binding	C_{max} (mg/L)	T_{max} (h)	Half-life (h) normal	Half-life (h) renal failure	Route of elimination	Urine concentration and excretion	Remarks
Clarithromycin	250 mg bd po	42–70%[k]	1 mg/L	1	35	>30	18% in urine		Up to one-third of dose metabolized to a bioactive 14-hydroxy compound
Clavulanic acid	125–250 mg tds po or iv[l]	30%	3–6	1	1	5	mainly renal	100–500 mg/L 0–6 h after 250 mg	c.50% excreted in urine
Clinafloxacin	200 mg bd	2–7%	1.3	2	5.6	–	mainly renal	>3 mg/L 24 h after dose	75% excreted in urine
Clindamycin	150–300 mg po im or iv	94%	2–4	1 h[m]	2–3	c.5	mainly biliary	25–50 mg/L 6 h after 150 mg	c.10% excreted in urine
Cloxacillin	500 mg po or im qds	94%			0.75	6	mainly renal		
Co-trimoxazole	see trimethoprim and sulphamethoxazole								
Colistin sulphamethate	2 megaunits tds		6–12 after 50 mg dose		c.4	–	mainly renal		Pharmacology very complex as is broken down into at least five compounds
Cycloserine	250–500 mg bd or tds po	<20%	6–10[n]	3–4	10	–	mainly renal	100 mg/L over 6 h after 500 mg po	
Dicloxacillin	see cloxacillin (pharmacology is similar)								
Doxycycline	200 mg first day then 100 mg/day po	30%	1–2 after 100 mg	2–3	c.20	c.25	renal and hepatic	20–40 mg/L 4–8 h after dose	
Enoxacin	600 mg po bd	c.20%	3.7	2	6.2	10–12	renal	300 mg/L 4–8 h after dose	At least two urinary metabolites accounting for 12% of administered dose
Erythromycin	500–1000 mg po or infusion	90%	3.1 after 500 mg po	3	2–4	4–6	mainly hepatic	urine levels low	
Fleroxacin	200–400 mg/day po	32%	3 after 200 mg	2	10	c.30	mainly renal	55% urinary recovery	
Fluconazole	100–400 mg po or iv	12%	20 after 400 mg po	3.3[o]	27–37	≤98	mainly renal	11% found as metabolite in urine	
Flucytosine	50 mg/kg infusion	low	50–80	1	5	50	mainly renal	c.128 mg/L in urine 36 h after dose	
Fosfomycin[p]	3 g po			3	4				35–40% absorption
Gatifloxacin	400 mg	20%	3.5	2	7–8		mainly renal		80–90% unchanged plus piperazine ring metabolite
Gentamicin	80–160 mg iv or im up to tds	20%	8 after 120 mg iv/im	1 after im	2–2.5	>48	renal	30–100 mg/L over 6–8 h	
Grepafloxacin	400 mg daily	10%	1	2	12		6% excreted in urine	>15 mg/L 8–24 h after dose	Five metabolites in urine

Expected concentrations in blood and urine

Drug	Dose	Oral absorption	Peak serum concn	Time to peak (h)	Half-life (h)		Excretion	Expected concentration in urine	Notes
Griseofulvin	500–1000 mg/day po		1 after 500 mg po	4	24	4	metabolite excreted in urine	<1% griseofulvin found in urine	50% metabolized to desmethyl form
Imipenem	500–1000 mg bd or tds 20 min infusion	20%	35	end[b]	1	4	renal	>100 mg/L 8 h after 500 mg iv	Metabolized in renal tubule, hence co-administration of cilastatin
Isoniazid	150 mg bd po		c.1	2	1–4	4–8	mainly renal with metabolite		
Itraconazole	100–200 mg once or twice a day	99.8%	40–140	3–4	17–24	?	hepatic metabolism		Increased absorption with food
Kanamycin	500 mg bd or tds im	c.20%	10–30	1	2.5	>48	renal	500–1000 mg/L over 6 h	
Ketoconazole	200 mg/day po	99%		2–3	2–3		faecal (60%)	only 2–4% unmetabolized in urine	Multiple metabolites
Latamoxef	1–2 g bd, im or iv	50%	60 after 1 g	end[b]	2–3	c.20	mainly renal	>400 mg/L over 6 h after 1 g	S and R bioactive isomers present in serum and urine
Levofloxacin	250–500 mg po/day	20%	6.5 after 500 mg	1–2	6–7	>10	renal, slight metabolism	>100 mg/L 6 h after 500 mg po	l-isomer of ofloxacin
Lincomycin	500 mg tds po or 300 mg bd im or iv	72%	8–18 after im	0.5	4–6		mainly biliary	low except after iv injection	
Lomefloxacin	400 mg once daily	10%	3–5	1–2	6–9	≤48	renal (80%)	100 mg/L 12–24 h after 400 mg po	Small % of five metabolites found in urine
Mecillinam	5–5 mg/kg qds im or iv	c.10%	4	1 h[q]	1	5	mainly renal	>100 mg/L 0–4 h after 200 mg	Relatively unstable in serum
Meropenem	500–1000 mg bd or tds 30 min infusion	5%	55	end[b]	1	4	70% renal recovery	>100 mg/L 8 h after 500 mg	Not metabolized (unlike imipenem)
Metronidazole	200–600 mg tds qds po, 1 g tds per rectum, 500 mg tds iv	45%	2.5–13 after 200 mg po	1–2	7	10	15% in faeces 70% in urine	50–200 mg/L over 6 h	50% metabolized to microbiologically active compounds.[r]
Mezlocillin	2–5 mg iv tds or qds	55%	100 2 h after 5 g iv	2	1.3	≥3	renal, biliary	>1000 mg/L for 6 h after 5 g iv	
Miconazole	250 mg po qds, 600 mg iv tds	90%	6 after 600 mg iv	2	2 (t_{1/2α}) 24 (t_{1/2β})	25	hepatic, faecal	<1% unmetabolized	
Minocycline	100 mg bd po	85%	1–3	1	1.5	25–32	renal, biliary		
Moxifloxacin	200–400 mg daily	26%	1 after 200 mg	2	12–14		c.20% urinary	2–8 mg/L over 8 h	Sulphate and glucuronide metabolites in urine
Nalidixic acid	1 g qds po	90%		2	4		renal[s]	50–500 mg/L, very variable	
Netilmicin	80–160 mg iv or im up to tds	<20%	8 after 120 mg iv/im 1 h after im	2.5	>48		renal	30–100 mg/L over 6–8 h	

Drug	Dosage	Protein binding	C_{max} (mg/L)	T_{max} (h)	Half-life (h) normal	Half-life (h) renal failure	Route of elimination	Urine concentration and excretion	Remarks
Norfloxacin	400 mg po bd	c.20%	1.5	1.5	3.3	7–8	mainly renal	>5 mg/L 12–24 h after 400 mg po	Poor absorption, 27% oral dose recovered in urine, several metabolites
Ofloxacin	600 mg po bd	c.20%	10	1.2	10–11	20–25	renal	150 mg/L 0–4 h	
Pefloxacin	400 mg po or iv	c.25%	5 0.25 h after 400 mg iv	–		16	mainly renal	>5 mg/L 12 h after 400 mg po/iv	Metabolized mainly to norfloxacin and oxo-pefloxacin
Penicillin G (benzylpenicillin)	0.5–5 megaunits im or iv 4–5 times a day	c.50%	8–15 units/mL after 1 megaunit im	0.5	0.3	3	mainly renal	'high'	c.40% urinary recovery
Piperacillin	2–5 g iv tds or qds	16%	30–50 1 h after 2 g iv	0	1	3	renal	>1000 mg/L 0–4 h after 5 g iv	
Pivampicillin	see bacampicillin								
Pivmecillinam	200–400 mg bd po	c.10%	5 after 400 mg	1.5	1	5	renal	>100 mg/L 0–6 h after dose	
Polymixin	see colistin								
Pyrazinamide	20–35 mg/kg/day po	17%	33 mg/L after 1.5 g	2	9–10		mainly metabolized	similar to serum	
Rifampicin	450–600 mg/day po	60–80%	6 after 600 mg po	2	3		faecal/renal metabolism	20–100 mg/L 0–8 h after 600 mg	60% of dose recovered in faeces
Roxithromycin	150 mg bd	85–90%	25	3	11–14	18	mainly faecal	urine levels low (5 mg over 24 h)	At least four metabolites
Rufloxacin	400 mg/day po followed by 200 mg	60%	4	1–2	30	c.40	mainly renal	85 mg/L at 24 h	Levels of 30 mg/L reported 4 days after end of therapy
Sparfloxacin	400 mg once daily	40%	1.6	2.7	17	–	mainly renal	5–20 mg/L 12–24 h after dose	Only 8% urinary elimination over 48 h
Spectinomycin	2–4 g im	'low'	100 after 2 g	2	2		renal	up to 1000 mg/L after 2 g	
Sulbactam	500 mg iv tds, qds	38%	30–50 after 500 mg iv	–	0.8	13.4	renal	>900 mg/L after 500 mg	Usually given with ampicillin[a]
Sulphamethoxazole	800 mg bd (in co-trimoxazole)	60%	40 after 400 mg po	2–4	14	≥15	renal/metabolized		20% excreted in urine unchanged
Sultamicin	see sulbactam								
Talampicillin	see bacampicillin								
Tazobactam	500 mg qds iv	25%	20 0.5 h after iv injection	–	1.1	–	renal	>300 mg/L after 500 mg iv over 6 h	
Teicoplanin	200–800 mg/day iv or im	>90%	20 at 1 h after 200 mg	–	30–40	>100	urine; possibly other routes	16–156 mg/L after 200 mg iv	50% urinary elimination[a]

Expected concentrations in blood and urine

Temocillin	1–2 g bd	85%	100 after 1 g	1	4.5	17	mainly renal	>100 mg/L up to 6 h after 1 g	Use should be avoided in renal failure[w]
Tetracycline[v]	250–500 mg po qds	25%	3 after 250 mg po	3	5–8	30–>100	renal/faecal	up to 500 mg/L 0–6 h after 250 mg	
Ticarcillin	2–5 g up to qds	<20%	20–30		1.2	11	urine	500–1000 mg/L after 1 g iv over 6 h	
Tinidazole	1–2 g /day		50 after 2 g	2	12		faecal/urinary and metabolism		At least four metabolites
Tobramycin	see gentamicin								
Trimethoprim	160 mg bd po[x] or 200 mg bd	30–60%	2–3	2	9	30	mainly renal	500 mg 0–8 h after 160 mg	Inactive metabolites in urine
Trovafloxacin	200 mg once-daily	87%	2	1	12–14		mainly faecal		Glucuronide is the major metabolite
Vancomycin	0.5–1 g iv infusion bd to qds	55%	25 after 1 g	2	6	>100	renal	100–400 mg/L 0–12 h after 1 g	

[a] Also bound to lipid membranes. [b] At end of infusion. [c] Twelve percent with 0.5 mg/L: 50% at 0.05 mg/L. [d] Not log linear. [e] Little effect on excretion. [f] Two hours after iv infusion. [g] Ampicillin pro-drug. [h] At end of 0.5 h infusion. [i] After 2 g iv. [j] After 500 mg. [k] Concentration-dependent. [l] Together with amoxycillin or ticarcillin (iv). [m] After palmitate po. [n] After 500 mg. [o] After oral administration. [p] Trometamol salt. [q] After 200 mg iv. [r] Major metabolite is hydroxy-metronidazole. [s] Including active and inactive metabolite. [t] Also available as a mutual pro-drug with ampicillin in some countries (sultamicillin). [u] Consists of a number of closely related structures. [v] Including oxytetracycline and chlortetracycline. [w] Serum levels of oxytetracycline and chlortetracycline tend to be somewhat lower. [x] As co-trimoxazole.

Chapter 4

Microbiological assays

Jenny M. Andrews

Department of Medical Microbiology, City Hospital NHS Trust, Birmingham B18 7QH, UK

Introduction

Microbiological assays have been used in clinical laboratories since the introduction of antibiotics and other antimicrobials. Many of the original assay methods were initially devised by the pharmaceutical industry to control commercial preparations, but methods were also developed to study the pharmacokinetics of new antimicrobials. With the introduction of potentially toxic agents, such as the aminoglycosides, microbiological assays were undertaken by diagnostic laboratories to monitor the therapy of patients. Although microbiological assays are relatively simple to perform, many aspects have to be considered in their development, to ensure that they are sufficiently precise and accurate. This is particularly important if data from different studies are to be compared with confidence, in the knowledge that any observed differences are not attributable to differences in assay methodology. In a clinical setting it is important to remember that the development of an assay method requires time and that results obtained from 'one-off' assays using methods which have not been validated, may be highly misleading.

The aim of this chapter is to suggest a reference method for measuring concentrations of an antimicrobial in serum or plasma, which can then be adapted to measure concentrations of the same antimicrobial at other potential sites of infection. As a consequence, although turbidimetric assays were used in the past results from UK National External Quality Assessment Scheme (UK NEQAS) distributions revealed that results obtained with aminoglycoside assays were highly misleading (M. J. Bywater, personal communication, 1988). A description of this method will therefore not be included in this chapter and only agar plate diffusion methods will be discussed.

Development of an agar diffusion assay method for measuring the concentration of an antimicrobial in plasma or serum

Agar diffusion assays are relative rather than absolute; they determine the concentration of antimicrobial present in a sample by comparison with a calibration curve (sometimes called a standard curve) produced by the effect of varying concentrations of drug diffusing into agar on or within which a bacterial population of a chosen bacterial strain is growing (production of zones of inhibition). A zone of inhibition is formed when that concentration of drug which is capable of inhibiting growth (the critical concentration) reaches a population of bacterial cells with a cell density too large for it to inhibit. Diffusion tests are therefore a 'race' between the diffusion of the drug from a reservoir into the agar and the growth of the seeded microbial population microorganism. The size of zone produced can be greatly affected by environmental conditions and the microorganism's growth rate. Once inoculated on to nutrient medium, a microorganism begins to grow, passing from a lag phase to log phase, during which multiplication is rapid. At this stage growth of the microorganism may proceed more rapidly than the diffusion of the drug through the agar; this helps to explain why drugs that diffuse poorly into agar produce small zones of inhibition. Eventually, microbial growth becomes visible to the naked eye, showing distinctly the edge of the zone of inhibition (clearly, the organism will not grow where the drug concentration is equal to or higher than the critical concentration). It is possible to adjust the final diameter of the zones of inhibition by altering either environmental conditions such as medium composition and incubation temperature or by increasing or decreasing the concentration of microorganisms inoculated on to or into the agar, or by using an inoculum which is already in the logarithmic growth phase. A comprehensive explanation of zone formation can be found in Hewitt & Vincent (1989).

Microbiological assays can be classified as one-, two- or three-dimensional (Edberg, 1986). In one-dimensional assays, seeded agar is placed in a test-tube or capillary tube and then the sample is layered on the top. Antimicrobial diffusion in one dimension (along the tube) produces the zone of inhibition. In two-dimensional assays, antimicrobial diffuses out from a well cut into seeded agar or from an antimicrobial-impregnated disc placed on the surface of the agar seeded, on the surface only, with the indicator microorganism. Diffusion is two-dimensional from the source and equilibration occurs within the agar. In a three-dimensional system the antimicrobial in the sample diffuses from the agar surface (the sample being placed within cylinders, fish spines or discs) both downward and outward into the agar which has been seeded throughout with indicator microorganism. The concentration of antimicrobial may therefore not be the same in all directions throughout the agar. This method is greatly influenced by the depth of agar: if the agar is too deep a 'double-zone' effect may be seen, making it difficult to measure the zone diameter accurately.

Although one-dimensional plate assays have been used by Japanese workers to assay antimicrobials in paediatric samples, two- and three-dimensional plate assays are the methods most often used.

The variables which have to be considered when developing a microbiological assay method for the measurement of an antimicrobial in plasma or serum are as follows: medium; microorganism (sometimes called the indicator strain or assay strain); diluent for calibrator and tests; preparation of calibrator material; application of samples; incubation temperature; reading and calculating results. These variables are discussed below.

Medium

A variety of media have been used for the assay of antimicrobials (Grove & Randal, 1955). It is important that the pH of the chosen medium is optimum for the activity of the drug, thus ensuring maximum sensitivity, and that the medium shows good batch-to-batch consistency of composition and pH. It must not be assumed that variation in the composition of commercial media does not occur (Andrews et al., 1990). For this reason, 'in-house' quality control of media, if not mandatory, is strongly advised. Commercial media are usually prepared following the manufacturer's instructions and cooled to 50°C before pouring on to the assay plate. This prevents excessive evaporation which may alter the gelling properties of the medium and the diffusion of the antimicrobial, and also prevents any thermal distortion of plastic assay plates. Assay plates should be levelled with the aid of a spirit level and a levelling stand in order to obtain a uniform thickness of agar. The depth of agar needs to be carefully controlled to ensure that it is not too thin, which can result in areas of the plate drying out during incubation, and not too thick, which can result in double-zoning (particularly applicable to three-dimensional assays) (Edberg, 1986). In order to overcome the problems associated with the depth of media, methods have been developed which use two layers of agar, the lower (base) layer being sterile and the top layer being seeded with the assay organism.

Microorganism (indicator strain, assay strain)

When choosing an assay organism (indicator strain), the sensitivity (the lowest concentration of antimicrobial which can be confidently measured), specificity (certainty that the result reflects the measurement of only one agent) and availability to other laboratories should be considered. Ideally, organisms should be selected from the American Type Culture Collection (ATCC) or the UK National Collection of Type Cultures (NCTC). If no published data or pharmaceutical company information are available on the microbiological assay of a particular antimicrobial, then the following information is needed to assist in the the selection of a suitable indicator strain: (i) the spectrum of activity of the drug and typical MICs; (ii) whether, in vivo, the drug is metabolized to a microbiologically active metabolite or metabolites; (iii) the pH at which the drug exhibits maximum antibacterial activity; (iv) percentage binding to human serum protein (protein binding, see Chapters 2 and 3); (v) approximate concentrations likely to be found in the samples to be assayed (see Chapter 3).

For example, consider a new antimicrobial, fluoroquinolone, 'X'. Its MIC ranges are as follows: Escherichia coli, 0.008–0.06 mg/L; staphylococci, 0.03–16 mg/L; Klebsiella spp., 0.03–16 mg/L; Proteus mirabilis, 0.06–0.5 mg/L. No active metabolites have been reported. X has optimum activity at pH 7. Typical concentrations achieved are 0.2–2.0 mg/L in serum and 0.04 mg/kg in tissues; 10% is protein-bound (human serum). Given these data, a strain of E. coli with an MIC of c.0.008 mg/L growing in a medium of pH 7.2 would be appropriate for preliminary experiments.

Long-term storage of strains of microorganisms for use in microbiological assays is usually best achieved using conventional freeze-dried cultures, suspensions stored below –40°C, or as gelatin discs stored at 4°C. Repeated serial subculture is not recommended for daily use because variants may be inadvertently selected after multiple subcultures. A preferred method is to subculture several slopes from the original pure culture. After overnight incubation these slopes can then be placed at 4°C and used subsequently to seed the plates. It is possible, with many strains, to make direct suspensions from these stored slopes. A written record of passage number and date should be kept on all assay strains. It is also possible to prepare working suspensions which can be added directly to the assay medium without the necessity of subculture; these should be stored below –40°C. However, the recovery of organisms after storage should be determined for every strain. To ensure that inocula are standardized, organism suspensions are usually prepared to a given optical density at a known wavelength (usually 640 nm). With regard to the aerobic spore-bearing indicator strains used for assay, these organisms are usually maintained as spore suspensions which are stored at 4°C ready for use.

Microbiological assay may not always seem appropriate for the determination of some antimicrobials; for example if the drug is metabolized in vivo to a microbiologically active metabolite or if samples are likely to contain additional antimicrobials. To overcome the problems associated with mixtures of antimicrobials, either selectively-resistant

Table I. Methods of removing or inactivating antimicrobials in microbiological assays

Antibiotic to be inactivated	Sample pretreatment	Inactivating agent added to agar
Aminoglycosides	Phosphocellulose (powder or paper disc)	2% sodium lauryl sulphate, 0.6–1% sodium polyanethol sulphonate
Amphotericin B	Heat at 100°C for 45 min, ultrafilter through collodion membrane	
Cephalosporins	β-lactamase[a]	β-lactamase[a]
Flucytosine		10 mg/L cytosine
Cycloserine		20 mg/L D-alanine
Penicillins	β-lactamase[a]	β-lactamase[a]
Polymixin		0.6–1% sodium polyanethol sulphonate
Sulphonamides		0.1% para-aminobenzoic acid, 5 mg/L thymidine
Trimethoprim		5 mg/L thymidine
Tetracyclines		50 mg/L magnesium sulphate

[a]Effective β-lactamase concentrations should be determined experimentally.

assay indicator strains have been used or antimicrobials have been selectively removed or inactivated (Reeves et al., 1978). Appropriate controls containing the agent(s) to be removed or inactivated should always be prepared and treated in the same way as the samples to ensure that the removal/inactivation procedure is effective. Some inactivation/removal procedures are listed in Table I.

Antimicrobial-free serum or plasma may also have an inhibitory effect on the indicator strain (Andrews et al., 1987); this should be considered when defining the lower limits of assay detection.

The indicator strain can be incorporated into the agar (agar incorporation) or flooded on to the surface (surface inoculation). If vegetative organisms are to be incorporated into agar it is important to cool the medium to 50°C before adding the suspension of organisms. This is less important when incorporating spores since these will usually be more heat-resistant. This cooling process is also necessary if antimicrobials are to be added to the agar, as is the case with the assay of β-lactamase inhibitors.

Diluent for calibrators and tests

Antimicrobials bind not only to serum albumin but also to other serum components and this can have a major effect on microbiological assay (Andrews et al., 1987) Interspecies variations in protein binding have been observed for some β-lactams (Andrews & Wise, 1987) and macrolides (Andrews et al., 1987). It may not be prudent, therefore, to substitute animal serum for human serum in the preparation of calibrators or for diluting clinical samples. It is also suggested that new batches of human serum (which should have always been screened and

shown to be negative for hepatitis B and C and for HIV) are checked for protein-binding properties and also for any inhibitory effect on assay organisms before use. It should be noted that treatment of human plasma with certain antiviral preparations may make it inhibitory to bacterial growth (Dalhoff, A., Stabb, H. & Andrews, J. M., unpublished data).

It may be necessary to dilute test samples to bring them within the calibrator range. Great care should be taken with this procedure as major inaccuracies can be caused by the use of an inappropriate diluent; in general the diluent should closely resemble the sample matrix.

Preparation of calibrators

It is essential to use the appropriate form of the antimicrobial in the preparation of calibrator solutions. Where possible, analytical reference material (powder or solution) should be obtained from the supplier, manufacturer or other appropriate institution (see Chapter 7). This material should be supplied with details of batch number, potency (usually expressed as 'as is' potency in μg/mg), solubility, stability (as powder and in solution) and expiry date. It is important to ensure that the composition of the calibrator material is, as nearly as possible, identical to the tests. This is not always possible with antimicrobials which are not pure chemical substances but fermentation products. For example, gentamicin is a mixture of four chemically related aminoglycosides (the gentamicin C components). Ideally the component compostion of the calibrators and tests should be similar (Andrews & Wise, 1984). Teicoplanin is another example of an antimicrobial that is not a pure compound.

Antimicrobial powders are usually protected from light, as photodegradation of drugs is well recognized (Bhadresa & Sugden, 1981), but they should also be protected from heat and moisture. This is usually best done by storing in the dark, at 4°C over a desiccant such as silica gel. Powders stored below room temperature should be allowed to warm to room temperature before the container is opened to avoid condensation of water on to the surface of cold powder.

Some substances may be poorly soluble in water and any initial solution should be prepared using a prodedure recommended by the supplier or manufacturer. This may prove to be initial solution in a small volume of acid, alkali or organic solvent such as methanol or dimethylsulphoxide (DMSO).

All assays will require internal control solutions as well as calibrator solutions. Internal controls are available commercially for some commonly assayed analytes, such as aminoglycosides, but it must be remembered that commercial internal controls often contain multiple analytes and preservatives, making them unsuitable for use with microbiological assay methods. Consequently controls will probably need to be prepared in-house.

Analytical balances, volumetric glassware and pipettes should be maintained to a high standard. If variable-volume pipettes are used they should be checked regularly to see that they deliver the correct volume by weighing known volumes of distilled water. Pipettes performing outside the acceptable limits should be returned to the supplier for recalibration or recalibrated in-house if a trained person is available.

Application of sample

Samples can brought into contact with the inoculated agar by a variety of methods, including the use of fish spines, steel cylinders, wells cut into the agar or blotting paper discs (Edberg, 1986; Hewitt & Vincent, 1989). Wells may be cut either manually (using a cork borer) or semi-automatically (using multiple hole punches attached to a vacuum pump). Commercially available 6 mm blotting paper discs are usually used but in some studies 13 mm discs have been employed. In an attempt to increase sensitivity, some workers have pipetted excess fluid on to blotting paper discs, which have been subsequently dried in air before being placed on the agar surface (personal communication, Rhône–Poulenc Rorer). The volume of sample chosen to be applied to the plate will depend on the required sensitivity of the assay method or the diffusibility of the antimicrobial.

Choice of plate size

Although assay methods have been described that use multiple small Petri dishes (Edberg, 1986), I have shown that highly misleading results can be obtained when using multiple plates (unpublished data). For this reason application of calibrators, internal controls and test samples on the same large plate is recommended highly.

Reusable 30 cm square glass plates, which may be autoclaved after use, are available from Mast Laboratories (Bootle, UK) and disposable 25 cm square polystyrene plates are available from Life Sciences (Paisley, UK).

Samples and calibrators should be applied to each plate in triplicate (at least) and the positioning should expose the replicates of calibrator, internal control and test randomly to any variability (e.g. agar thickness, temperature) which might be observed over the entire plate. Samples and calibrators are therefore best distributed over the assay plate in a predetermined random pattern. Some examples of random patterns suitable for use with 25 or 30 cm plates are shown in Figure 4.1. Hewitt & Vincent (1989) give more patterns.

(a) Thirty wells or discs per plate

| Well contents | Well number | | | | | |
	code 1			code 2		
Calibrator A	1	13	25	5	19	26
Calibrator B	7	20	21	9	12	29
Calibrator C	5	12	23	3	16	27
Calibrator D	4	15	22	7	14	22
Calibrator E	3	19	27	1	18	24
Test 1	6	17	28	10	11	23
Test 2	2	14	26	4	17	30
Test 3	8	11	30	8	20	21
Test 4	9	16	24	2	13	25
Test 5	10	18	29	6	15	28

(b) Forty-five wells or discs per plate

| Well contents | Well number | | | | | |
	code 1			code 2		
Calibrator A	3	22	33	11	24	35
Calibrator B	13	27	40	5	28	34
Calibrator C	11	24	34	1	23	32
Calibrator D	9	17	44	4	20	36
Calibrator E	2	23	42	7	19	43
Test 1	5	28	37	2	29	38
Test 2	1	19	43	14	16	42
Test 3	10	26	39	15	17	39
Test 4	7	25	41	3	27	40
Test 5	12	30	38	9	21	33
Test 6	4	16	32	6	30	31
Test 7	14	21	35	13	26	44
Test 8	6	29	45	10	18	41
Test 9	15	18	36	12	25	37
Test 10	8	20	31	8	22	45

Calibrator A is the lowest concentration and calibrator E the highest.

Figure 4.1 Examples of random layout patterns for microbiological assay plates

Well filling

When filling wells in agar, the well should be filled to the very top, the fluid thus producing a slightly convex meniscus when viewed from the side. Standardized, reproducible, well filling requires careful training, constant practice, care and attention, and a certain degree of technical aptitude. Underfilling of wells may result in inaccuracies, as may use of pipettes that deliver a fixed volume rather than well filling (unpublished observations). However, slight overfilling of wells does not seem to have a major effect on assay performance (Edberg, 1986) and is therefore preferable to underfilling. If so much excess fluid is put into the well that it breaks away and flows across the plate, though, zones may become distorted.

Incubation temperature

Assay plates are generally incubated immediately after the application of the sample. However, the lower limit of detection of some assays has been increased when plates have been pre-incubated at room temperature or 4°C for several hours before final incubation. Incubation temperatures between 35°C and 37°C are usually used, but the sensitivity of some assays has been increased by incubation at 30°C. Stacking the plates in the incubator should be avoided as this may result in uneven heating. Ideally, incubators should be fitted with a temperature-recording system. The duration of incubation depends on the method used. Rapidly growing strains of Enterobacteriaceae have been used for gentamicin assays which give visible results within 4–6 h (Reeves *et al.*, 1978), but most assay plates are incubated overnight.

Reading and calculating results

Rulers, callipers, zone-readers and image-analysers have been used to measure zones of inhibition. Usually zone diameters or, occasionally, zone areas have been used to construct calibration curves. Viewers that magnify the zones of inhibition (magnification ≈ 8×) may be used to measure zones of inhibition more precisely. Unless the zone sizes are entered directly into a computer, it is important to have a worksheet for recording zone sizes. Some typical worksheets for 'derandomizing' 30 and 45 hole codes are shown in Figure 4.2.

With the aid of semi-logarithmic graph paper (or by determining the log of each calibrator concentration), zone diameter can be plotted against the log of calibrator concentration to construct a calibration curve. Because the true line-of-best-fit obtained with such a plot is often sigmoidal, a straight line should not be forced through the points; the calibration curve should be contructed by joining individual points or, ideally, a calculation determining the line-of-best-fit should be used since a mathematical method will be more reliable and reproducible

than a visual method of curve fitting (Bennett *et al.*, 1966; Perkins, 1978), Table II and Figure 4.3 demonstrate curve-fitting using the method of Bennett *et al.* (1966), Table 4.3 illustrates Perkins method of curve fitting. From the calibration curve the assayed concentrations for the internal controls and tests are calculated. Results from a plate are only accepted if the internal controls give appropriate results. For most clinical microbiological assays internal controls should be within 10–15% of their assigned concentration for the test results to be acceptable (see *Chapter 7*).

Preliminary studies to be done when setting up a microbiological assay

The following section gives some advice as to the considerations to be made and the recommended preliminary studies to be carried out when developing a microbiological assay for an antimicrobial.

Literature search

Find out as much about the compound as possible. Search for existing published assay methods (pharmacokinetic papers may provide this information). Also research MIC data, pharmacokinetic data, chemical data, etc., as all of these will help in the design of your assay. Obtain as much information as possible from the relevant pharmaceutical company. Study closely any existing assay methods, but try to reproduce them only if you feel they will be appropriate for the use to which you wish to put your assay. Be sure that microbiological assay is appropriate for your study. If it is not, see *Chapter 5*.

Selection of medium

Choose a medium for which the pH is optimum for the activity of the antimicrobial to be assayed. pH 7 is recommended for fluoroquinolones, pH 8 for macrolides, and pH 6.6 for β-lactams.

Selection of assay indicator strain

Look at the in-vitro spectrum of activity of the antimicrobial. Confirm that the compound is not metabolized to a microbiologically active form *in vivo*, or that more than one antimicrobial will be present in the samples. Survey the data on expected blood and tissue concentrations and ensure that the strains you test will give you the required degree of sensitivity. If possible select an organism from the NCTC or ATCC collections rather than a local clinical isolate as this will make your assay easier for other centres to reproduce.

Calibrators

Prepare a range of calibrators covering at least two doubling dilutions below the lowest expected concentration and two

(a)

Date:		Antimicrobial:	Study:		Medium:
Organism:		Diluent:	Well/disc size:		Temperature:
Plate no:					

Well contents	Well numbers (enter zone size alongside relevant well number)			Mean
1.0 mg/L calibrator	7 …	19 …	43 …	…
0.5 mg/L calibrator	4 …	20 …	36 …	…
0.25 mg/L calibrator	1 …	23 …	32 …	…
0.125 mg/L calibrator	5 …	28 …	34 …	…
0.062 mg/L calibrator	11 …	24 …	35 …	…
Test 1: IC 0.8 mg/L	2 …	29 …	38 …	…
Test 2: IC 0.1 mg/L	14 …	16 …	42 …	…
Test 3	15 …	17 …	39 …	…
Test 4	3 …	27 …	40 …	…
Test 5	9 …	21 …	33 …	…
Test 6	6 …	30 …	31 …	…
Test 7	13 …	26 …	44 …	…
Test 8	10 …	18 …	41 …	…
Test 9	12 …	25 …	37 …	…
Test 10	8 …	22 …	45 …	…

(b)

Date:		Antimicrobial:	Study:		Medium:
Organism:		Diluent:	Well/disc size:		Temperature:
Plate no:					

Well contents	Well numbers (enter zone size alongside relevant well number)			Mean
1.0 mg/L calibrator	3 …	19 …	27 …	…
0.5 mg/L calibrator	4 …	15 …	22 …	…
0.25 mg/L calibrator	5 …	12 …	23 …	…
0.125 mg/L calibrator	7 …	20 …	21 …	…
0.062 mg/L calibrator	1 …	13 …	25 …	…
Test 1 IC 0.8 mg/L	6 …	17 …	28 …	…
Test 2 IC 0.1 mg/L	2 …	14 …	26 …	…
Test 3	8 …	11 …	30 …	…
Test 4	9 …	16 …	24 …	…
Test 5	10 …	18 …	29 …	…

Figure 4.2 Example worksheets for (a) 45-hole and (b) 30-hole plates.

doubling dilutions above the highest expected concentration. Also include an antimicrobial-free blank sample. Use an appropriate matrix for your calibrators.

Trial runs

Carry out the following trial runs: (i) evaluate a selection of strains for use in the assay; (ii) compare the well and disc methods; (iii) incubate plates at 35–37°C and at 30°C. The data obtained from these studies will indicate the environmental conditions giving best all-round performance and should indicate which assay conditions should be used in the internal and external validation of the assay. Assay validation procedures are discussed further in *Chapter 7*.

A reference method for assaying an antimicrobial in human serum or plasma

Preparation of media

Prepare the chosen medium following the manufacturer's instructions. It is preferable to use media that are designed for use in microbiological assays and supplied by reputable

Table II. An example of Bennett's calculation to obtain a line of best fit with five calibrators

Calibrator concentration (mg/L)	Coded concentration x	Mean zone size			
		y*	x^2	xy	x^2y
0.062	1	129	1	129	129
0.125	2	169	4	338	676
0.25	3	197	9	591	1773
0.5	4	226	16	904	3616
1	5	250	25	1250	6250
		$\Sigma y = 971 = e$		$\Sigma xy = 3212 = f$	$\Sigma x^2y = 12444 = g$

Substitute the above values in the following formulae:

a = $4.6e + 0.5g - 3.3f$
  = $4466.6 + 6222 - 10599.6 = 89$

b = $(18.7f - 3g - 23.1e)/7$
  = $(60064.4 - 37332 - 22430.1)/7 = 43.18$

c = $g + 7e - 6f$
  = $12444 + 6797 - 19272 = -2.21$

Line of best fit (predicted zone size for each calibrator) is calculated from:

$a + bx + cx^2 = y$
89 + 43.18 – 2.21 = 130.0 (Calibrator code 1)
89 + 86.18 – 8.84 = 166.5 (Calibrator code 2)
89 + 129.54 – 19.89 = 198.6 (Calibrator code 3)
89 + 172.72 – 35.36 = 226.3 (Calibrator code 4)
89 + 215.9 – 55.35 = 249.6 (Calibrator code 5)

*y in mm or arbitrary units are typical zone size values for illustration purposes.

manufacturers. Although commercial media are prepared to defined specifications, batch-to-batch variations in media performance have been observed (Andrews et al., 1990). It may be prudent, therefore, to purchase sufficient supplies of a single batch to complete specific studies and, for routine methods, to test new batches of media by comparing zone diameters obtained with the new batch alongside those obtained with the currently used batch. Dispense the medium into bottles containing the exact volume of agar required for a single plate.

Preparation of suspensions of the assay indicator strain

Details are given below for spore and non-spore preparations.

Preparation of spore suspensions
The following items are required:

- the reference strain obtained from a suitable source (ATCC or NCTC); prepare a pure culture on blood agar
- 0.1% glucose broth in 20 mL bottles
- spore production medium comprising, per litre: poly-peptone 5 g, beef extract 3 g, NaCl 8 g, agar 15 g, $MnCl_2$

2 mg (or $MnCl_2\cdot4H_2O$ 2.72 mg), $NaNO_3$ 2 mg, in distilled water. Sterilize at 121°C for 20 min, check the pH is 7.3 (adjust if necessary) and pour 250 mL into a Roux flask. Allow to set on a slant to obtain a large surface area.

- a 70°C water bath
- a spectrophotometer reading percentage transmission at 540 nm

The method of spore production is as follows:

1. Subculture the organism into 20 mL of 0.1% glucose broth and incubate overnight at 37°C.

2. Add 3 mL of overnight broth culture to the Roux flask containing spore production medium. Ensure that the whole surface of the agar is flooded. Incubate at 37°C for 1 week for *Bacillus cereus* or *Bacillus subtilis*.

3. Harvest the growth from each flask into 50 mL of sterile distilled water.

4. Heat the suspension in a 70°C water bath for 30 min to kill remaining vegetative cells. This stock suspension remains viable at 4°C for up to 10 years.

5. To use the spore suspension, dilute it in sterile distilled water until a reading of 80% light transmission is recorded at 540 nm.

Table III. Fitting a curve using Perkins' method. (For use with four to seven calibrators with any fixed dilution factor between them)
Actual calibrator concentrations (e.g. 1, 2, 4, 8, 16, etc mg/L) are changed to codes values (x_1, x_2, x_3, x_4, x_5, etc) using the Table below. Appropriate values for the constants d, e, f are also chosen

No. of calibrators (N)	x_1	x_2	x_3	x_4	x_5	x_6	x_7	d	e	f
4	−3	−1	+1	+3				164	20	256
5	−4	−2	0	+2	+4			544	40	1120
6	−5	−3	−1	+1	+3	+5		1414	70	3584
7	−6	−4	−2	0	+2	+4	+6	3136	112	9408

Suppose then that the mean zone sizes for the N calibrators are $y_1, y_2, \ldots y_n$, then:

$$\Sigma y = y_1 + y_2 \ldots y_n$$
$$\Sigma xy = x_1 y_1 + x_2 y_2 + \ldots x_n y_n$$
$$\Sigma x^2 y = x_1^2 y_1 + x_2^2 y_2 \ldots x_n^2 y_n$$

If the mean size of zone for Test $t = y_t$ then the Code for this test, x_t, is calculated from:

$$x_t = \text{square root } [((b^2 - 4ac + 4cy_t) - b)/2c]$$

where

$$a = (d(\Sigma y) - e(\Sigma x^2 y))/f$$
$$b = (\Sigma xy)/e$$
$$c = (\Sigma y) - (Na))/e$$

To convert this coded result for Test t to a concentration it is necessary to know the following:

$$d = \text{dilution factor of calibrators expressed as } \text{Log}_{10}$$
$$v = \text{concentration of lowest calibrator expressed as } \text{Log}_{10}$$

Hence result for Test t in mg/L:

$$t = \text{Antilog } [v + (d/2) (x_t + (N - 1))$$

Preparation of vegetative cell suspensions

1. Obtain the reference strain from a suitable source (e.g. ATCC or NCTC) and prepare pure cultures on agar slopes. These may be stored at 4°C for up to a month.

2. Before using, check purity by subculture

3. After the purity check, subculture into digest or infusion broth and incubate overnight at 35–37°C. Cultures of some organisms, for example *Micrococcus lutea*, may require shaking to obtain homogeneous growth.

4. Dilute the culture before adding to the plate. Dilution of *M. lutea* in sterile water to an optical density of 0.2 at 630 nm is appropriate for most purposes; 2 mL of this is added to 100 mL of cooled, molten agar. Dilution of coliforms to an optical density of 0.04 at 640 nm is suitable if the suspension is to be used to flood the plate.

Preparation of calibrators

Usually five calibrators should be chosen at doubling dilution intervals and used to construct a calibration curve with the line of best fit calculated using the method of Bennett *et al.* (1966) (see Table II). For other numbers of calibrators, the method of Perkins (1978) may be used for curve fitting (Table III).

An example of the preparation of calibrators is outlined below. The declared potency P_A of the drug powder is 998 μg/mg. To prepare a solution containing 1000 mg of active antimicrobial per litre in a final volume V mL, then the weight of powder (W) in mg is calculated from:

$$W = (1000/P_A) \times V_f$$

where V_f is the final volume. Thus if the final volume is 25 mL, then $W = 1000/998 \times 25 = 25.05$ mg. Therefore

25.05 mg of drug powder is weighed accurately and made up to 25 mL in the volumetric flask with appropriate diluent. This gives a 1 g/L (1000 mg/L) solution.

Further stock antimicrobial solutions may then be prepared in appropriate volumetric flasks. Thus, for example, by taking 100 μL of the 1000 mg/L solution and making it up to 10 mL with diluent, a 10 mg/L solution is formed. By taking 1 mL of this 10 mg/L solution and making it up to 10 mL with diluent, a 1 mg/L solution is formed.

These stock solutions can then be diluted using appropriate volumes of human serum in order to make calibrators of, say, 1.0, 0.5, 0.25, 0.125 and 0.062 mg/L.

The initial diluent depends on the antimicrobial being assayed but is usually water. All solutions should be made using Grade A volumetric glassware.

Preparation of internal controls

Internal controls should usually be prepared as described above from (solubility permitting) a 1 g/L solution obtained by weighing known potency powder (preferably from a different batch from that used for making the calibrators), by a laboratory worker other than the one who prepared the calibrators. These might be, for example, of 0.75 mg/L (made by taking 375 μL of the 10 mg/L stock solution and making it up to 5 mL with human serum) or 0.1 mg/L (made by taking 500 μL of the 1 mg/L stock solution and making it up to 5 mL with human serum).

Agar incorporation assay procedure

In this procedure the organisms are incorporated into the agar.

1. Cool the melted medium to 50°C. (In some assays spores are added to medium that is hotter than this in order to stimulate growth by 'heat shock'. However, care must be taken when handling hot media. Another disadvantage of this procedure is that moisture may be lost, which could alter the gelling properties of the agar).

2. Add spore suspension or organism suspension adjusted to the appropriate optical density.

3. Place 100 mL of agar in a disposable plate (25 cm × 25 cm; Life Technologies) or place 200 mL into a reusable plate (30 cm × 30 cm; Mast). Ensure that the plate is level before pouring the agar.

4. Allow the agar to set and then dry the surface of the agar in a drying cabinet (approximately 10 min) or a 37°C incubator (approximately 30 min).

5. Store the plates at 4°C until needed (the maximum storage time will depend on the assay organism).

Surface inoculation assay procedure

In this procedure the organisms are flooded on to the surface of the agar, as follows.

1. Ensure that the plate is level before pouring the molten agar.

2. Pour cooled agar (50°C) into the assay plate (100 mL into a disposable plate (25 cm × 25 cm), or 150 mL into a reusable plate (30 cm × 30 cm)).

3. Allow the agar to set and then dry the surface in a drying cabinet (approximately 10 min) or a 37°C incubator (approximately 30 min).

4. Adjust an overnight broth culture in sterile distilled water to the required optical density.

5. Flood the surface of the agar with approximately 30 mL of diluted organism.

6. Drain excess water from the plate then re-dry the surface of the agar (the drying procedure should be performed as quickly as possible).

7. Store plates at 4°C until needed (the maximum allowable storage time will depend on the assay organism).

Subsequent processing of agar incorporation or surface inoculation assays

At all times protect calibrators and tests from heat and light (keep in a refrigerator). First prepare the calibrators and internal controls. Then, if necessary dilute test any samples in an appropriate matrix (human serum). Apply calibrators, internal controls and test samples using an appropriate random pattern (examples of random patterns are shown in Figure 4.1), by one of the following methods.

Method A

1. Using a well-sharpened cork borer, cut holes in the agar using an appropriate template to ensure even spacing. Thirty or 45 wells can be accommodated on a 25 cm × 25 cm plate.

2. Using a scalpel, remove the agar plugs and discard into a suitable disinfectant.

3. Fill the wells to give a convex meniscus. Do not dose wells with a fixed volume as this may reduce the reproducibility.

4. Incubate overnight at the temperature optimum for the assay.

Method B

1. Dip blotting paper discs into the sample.

2. Drain excess fluid from the disc by placing on to blotting paper.

3. Place disc on to the surface of the agar. (Note that if spores are incorporated in the medium it may be necessary to kill the surface growth by exposing the plate to a UV sterilizing lamp for 5 min since surface growth may mask the true zone edges. However, UV irradiation may have a harmful effect on some antimicrobials

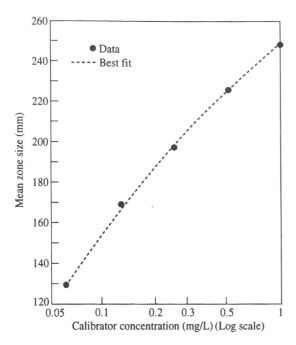

Figure 4.3 Typical microbiological assay calibration curve with curve fitted using Bennett's method (data from Table II).

(especially those which absorb UV radiation and are photolabile) and therefore, in the case of uncertainty, the possible effects of this procedure on the antimicrobial should be investigated.)

4. Incubate overnight at the temperature optimum for the assay.

For both methods A and B, interpret the results as follows. (i) Measure the zones or zone areas using a device which magnifies at least 8×. Rulers and callipers are less precise than zone readers and so not recommended. (ii) Derandomize your zones, calculate mean zone sizes (see Figure 4.2) and then construct a calibration curve a line of best fit. An example of how to calculate a line of best fit using the calculation of Bennett *et al.* (1966) and typical data are shown in Table II and Figure 4.3. This process of

derandomization and calibration curve construction can be programmed into a computer. Commercial spreadsheets such as Lotus 123 or Excel can be 'programmed' to derandomize, construct the calibration curve, determine curve statistics and calculate the results of the tests and internal controls.

References

Andrews, J. M. & Wise, R. (1984). A comparison of the homogenous enzyme immunoassay and polarization fluoroimmunoassay of gentamicin. *Journal of Antimicrobial Chemotherapy* **14**, 509–20.

Andrews, J. M. & Wise, R. (1987). Practical problems in the assay of new antimicrobials. In *Proceedings of the Fifth International Symposium on Rapid Methods and Automation in Microbiology and Immunology, Florence* 1987. Abstract p. 269. Brixia Academic Press, Brescia.

Andrews, J. M., Ashby, J. P. & Wise, R. (1987). Factors affecting the in-vitro activity of roxithromycin. *Journal of Antimicrobial Chemotherapy* **20**, *Suppl. B*, 31–7.

Andrews, J. M., Ashby, J. P. & Wise, R. (1990). Problems with IsoSensitest agar. *Journal of Antimicrobial Chemotherapy* **26**, 596–7.

Bennett, J. V., Brodie, J. L., Benner, E. J. & Kirby, W. M. (1966). Simplified accurate method for antibiotic assay of clinical specimens. *Applied Microbiology* **14**, 170–7.

Bhadresa, B. & Sugden, J. K. (1981). Light transmittance through amber glass medicine bottles. *Pharmaceutica Acta Helvetiae* **56**, 122–4.

Edberg, S. C. (1986). The measurement of antibiotics in human body fluids: techniques and significance. In *Antibiotics in Laboratory Medicine*, 2nd edn (Lorian, V., Ed.), pp. 381–476. Williams & Wilkins, Baltimore, MD.

Grove, D. C. & Randall, W. A. (1975). *Assay Methods of Antibiotics, A Laboratory Manual.* Medical Encyclopedia Inc, New York, NY.

Hewitt, W. & Vincent, S. (1989). *Theory and Application of Microbiological Assay.* Academic Press, London.

Perkins, A. (1978). Statistics of plate assays. In *Laboratory Methods in Antimicrobial Chemotherapy* (Reeves, D. S., Phillips, I., Williams, J. D. & Wise, R., Eds), pp. 157–63. Churchill Livingstone, Edinburgh.

Reeves, D. S., Phillips, I., Williams, J. D. & Wise, R. (1978). *Laboratory Methods in Antimicrobial Chemotherapy.* Churchill Livingstone, Edinburgh.

Chapter 5

Non-microbiological assays

Les O. White and Andrew M. Lovering

Department of Medical Microbiology, Southmead Hospital, Bristol BS10 5NB, UK

The need for non-microbiological assays

Microbiological assays have, for two reasons, traditionally been used to assay antimicrobials. Firstly, the important property of any antimicrobial is its ability selectively to inhibit the growth and sometimes kill a susceptible microorganism; if this property is used to form the basis for the assay then inactive impurities or degradation products will not interfere. Secondly, many early antimicrobials were relatively impure fermentation products (gentamicin is a good example) and their antimicrobial potency could not be determined precisely chemically.

In the pharmaceutical industry microbiological assays are performed to a high level of accuracy and reproducibility for quality control purposes. Official regulations laid down by, for example, the *British Pharmacopoeia* or US Code of Federal Regulations may demand that a microbiological assay technique is used (see Hewitt & Vincent, 1988). In contrast, the clinical laboratory has different needs and constraints and microbiological assays suffer three major disadvantages: long turnaround time, poor specificity, and poor performance in the field.

Long turnaround time

This is a problem because most antibiotics have short pharmacokinetic elimination half-lives and are therefore administered two, three or four times a day. If the results of a microbiological assay cannot reach the clinician until 24 h after the blood was taken the patient may have received one or more additional doses. To overcome this, rapid microbiological assays for aminoglycosides were developed but these still suffered from the other disadvantages described below.

Poor specificity

Multiple antibiotic therapy will cause problems of poor specificity if the indicator organism used in the microbiological assay is susceptible to drugs in the sample other than the one to be assayed. Not only antibacterials cause this problem but also anticancer drugs, barbiturates, beta-blockers, diuretics and non-steroidal antiinflammatory agents (see *Chapter 7*). To re-establish acceptable assay specificity, the laboratory needs to be notified of the presence of other medications and these will require removal or inactivation unless a suitably resistant indicator organism is used. Newer agents such as β-lactamase-resistant cephalosporins or fluoroquinolones present the most difficult problems.

Poor performance in the field

That microbiological assays for aminoglycosides performed in clinical laboratories suffered from poor performance is clearly shown in the early external quality assessment (EQA) returns from the UK NEQAS for Antibiotic Assays. On only two occasions between 1973 and 1985 did more than 50% of participating laboratories using a microbiological assay for gentamicin achieve a satisfactory standard of performance (White, 1986). Good performance of microbiological assays requires proper training, particularly careful attention to technique and a considerable degree of technical expertise and practice. Many clinical laboratories had difficulty acquiring and maintaining these skills. It was these serious quality issues with microbiological assays in the clinical setting which created interest in the application of non-microbiological methods.

Non-microbiological assay techniques

Non-microbiological techniques applicable to antimicrobial assay fall into three broad classes:

- *Enzyme-based assays*. Here the antimicrobial acts as a substrate for a specific enzyme. Assay specificity equates with enzyme specificity. Methods based on this principle are usually rapid, sensitive and precise.
- *Immunoassays*. For this type of assay, specific antibodies against the drugs to be assayed are required. Assay specificity is high because it equates with antibody specificity. These assays can be very rapid, very sensitive and very reproducible.
- *Physico-chemical assays (including chromatography)*. These assays usually rely on some relatively non-specific physical or chemical property such as UV

absorbance, fluorescence, lipid solubility, chemical activity, presence of particular chemical groups, etc. It is difficult to generalize about such techniques except to say they may show poor specificity (especially when assaying body fluids) unless some form of sample clean-up procedure is included. This might be some form of solvent extraction, chromatographic separation, or both. Overall specificity is then the sum of the specificity of the separation procedure plus the specificity of the method of detection and quantification. These methods are usually rapid compared with microbiological assays. Some individual physico-chemical assays are described where appropriate in the chapters devoted to specific classes of antimicrobial. In this chapter only chromatographic assays will be discussed since they have universal application.

Enzyme-based assays

The antimicrobial resistance mechanisms found in bacteria include specific enzymes which inactivate the drug. Examples are β-lactamases, chloramphenicol acetyltransferase (CAT) and the many different aminoglycoside-modifying enzymes which include aminoglycoside acetyltransferases (AACs), aminoglycoside nucleotidyl transferases (ANTs, sometimes referred to as adenylyltransferases or AADs) and aminoglycoside phosphotransferases (APHs). CAT and several aminoglycoside-modifying enzymes have been used to assay the substrate antimicrobial.

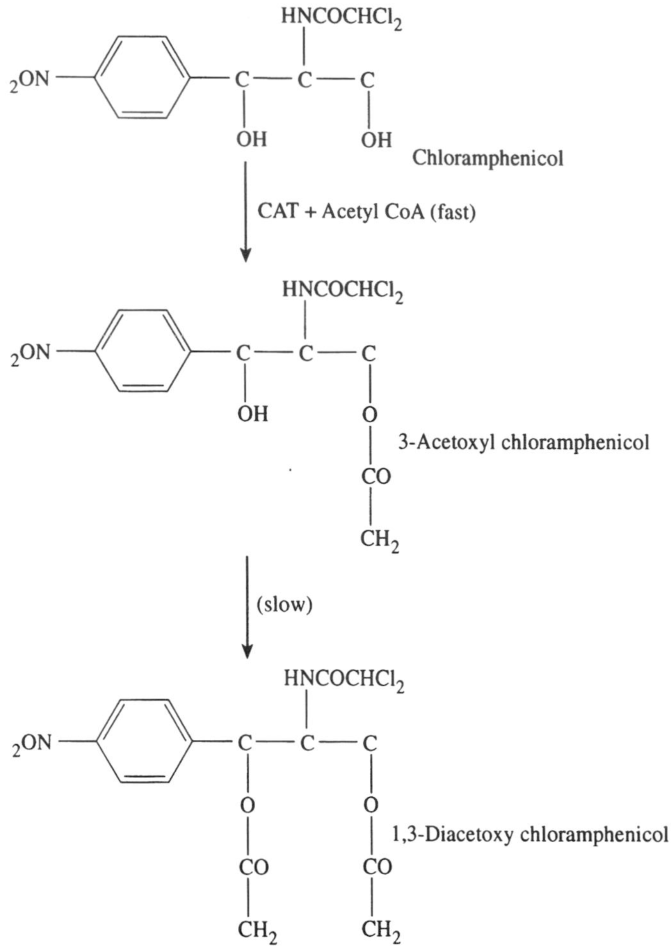

Figure 5.1 Acetylation of chloramphenicol.

Chloramphenicol acetyltransferase

CAT acetylates chloramphenicol at the 3-position (Figure 5.1) employing the co-factor acetyl coenzyme A. Following this there is a slow, pH-dependent, migration of the acetyl group to the 1-position with additional slow acetylation at the 3-position to give 1,3-diacetoxy chloramphenicol. CAT used to assay chloramphenicol is not susceptible to interference from the pro-drug chloramphenicol succinate.

Aminoglycoside-modifying enzymes

In the case of the aminoglycosides the various enzymes may act at one of several susceptible groups present in susceptible compounds. The number of the carbon atom at which the acetylation/nucleotidylation/phosphorylation occurs together with the prefix AAC, ANT or APH is used in the classification and nomenclature of these enzymes (see van de Klundert & Vliegenthart, 1993; Rather et al., 1993). Figure 5.2 illustrates the site of action of the

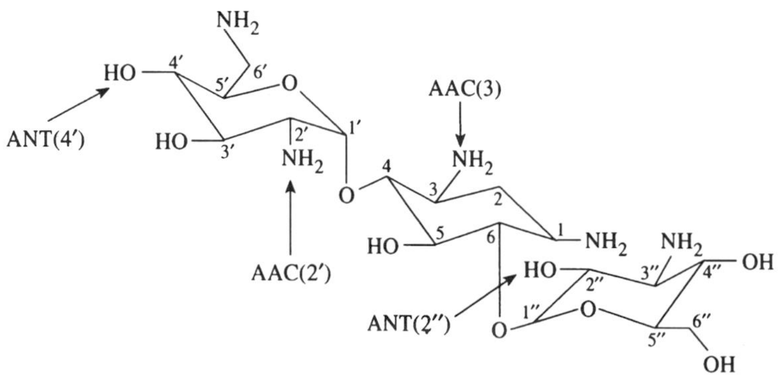

Figure 5.2 Site of action of some aminoglycoside-modifying enzymes.

Table I. Some aminoglycoside antibiotics and enzymes suitable for use in assays

	Chemical group on the corresponding carbon atom									Suitable assay enzymes	
Antimicrobial	3	6	9	2'	3'	4'	6'	2"	3"	AAC	ANT
Amikacin	NH_2	–	–	OH	OH	OH	NH_2	OH	NH_2	6'	4'
Apramycin	NH_2	OH	–	NH_2	H	H	OH	–	–	1, 3	
Butirosin B	NH_2	OH	–	NH_2	OH	OH	NH_2	OH	OH	2', 6'	4'
Gentamicin C1	NH_2	–	–	NH_2	H	H	ma[b]	OH	ma[b]	2', 3	2"
Gentamicin C1a	NH_2	–	–	NH_2	H	H	NH_2	OH	ma[b]	2', 6', 3	2"
Gentamicin C2	NH_2	–	–	NH_2	H	H	NH_2	OH	ma[b]	2', 6', 3	2"
Gentamicin C2a	NH_2	–	–	NH_2	H	H	NH_2	OH	ma[b]	2', 6', 3	2"
Kanamycin A	NH_2	–	–	OH	OH	OH	NH_2	OH	NH_2	6', 3	4', 2"
Kanamycin B	NH_2	–	–	NH_2	OH	OH	NH_2	OH	NH_2	2', 6', 3	4', 2"
Lividomycin	NH_2	OH	–	NH_2	H	OH	OH	OH	–	2', 3	4'
Neomycin B	NH_2	OH	–	NH_2	OH	OH	NH_2	OH	–	2', 6', 3	4'
Paromomycin	NH_2	OH	–	NH_2	OH	OH	OH	OH	–	2', 3	4'
Ribostamycin	NH_2	OH	–	NH_2	OH	OH	NH_2	OH	OH	2', 6', 3	4'
Spectinomycin[a]	–	–	OH	–	–	–	–	–	OH	–	3", 9
Streptomycin	–	OH	–	–	OH	CH_3	–	ma[b]	OH	–	3"
Tobramycin	NH_2	–	–	NH_2	H	OH	NH_2	OH	NH_2	2', 6', 3	4', 2"

[a]Spectinomycin is an aminocyclitol; [b]ma = $NHCH_3$.

enzymes AAC(2'), AAC(3), AAC(6'), ANT(4') and ANT(4"). The aminoglycoside requires an NH_2 group on the target carbon atom to be susceptible to AAC enzymes or an OH group to be susceptible to ANT or APH enzymes. Possession of the appropriate group does not, however, guarantee enzyme susceptibility since other factors are involved. For example, although amikacin has an NH_2 group at the 3-position and an OH group at the 2"-position, it is not a substrate for AAC(3) or ANT(2"). Each type of enzyme has a number of different isoenzymes differing in various properties including substrate specificity. Isoenzymes are classified by the use of roman numerals, for example AAC(3)-I to AAC(3)-IV, but this classification is still somewhat controversial (van de Klundert & Vliegenthart, 1993; Rather et al., 1993).

Both AAC and ANT enzymes have been used in assays (Table I). Such enzyme-based assays are relatively rapid, require only a small sample volume and show very good reproducibility. Specificity is very high and non-aminoglycosides do not interfere. Gentamicin is best not assayed with AAC(6') since only three of its four major components are substrates for this enzyme and it would be impossible in practice to ensure that the assay calibrators had identical component composition to the gentamicin in the patient's sample (White et al., 1983a).

Since most assay protocols use radiolabelled cofactor, the cost of this, and the need for access to expensive scintillation counting equipment, are important cost factors. In comparative studies the results of enzyme-based assays correlated well with results obtained with other methods (Ratcliff et al., 1981; White et al., 1980,

1981). For practical details of these types of assays, see *Chapter 9*.

Immunoassays

In an immunoassay the drug being assayed acts as an antigen or hapten and binds with specific anti-drug antibody. Most drug immunoassays are of the competitive binding type shown in Figure 5.3. A fixed concentration of tracer-labelled drug reagent (a solution of a derivative of the drug to be assayed which has been chemically labelled with some easily traced group such as a radioactive atom, a fluorophore or an enzyme) competes with any drug present in the sample being assayed for a limited amount of specific anti-drug antibody reagent. The greater the concentration of drug in the sample, the smaller the amount of tracer-labelled drug which can bind to antibody. The reagents used in an immunoassay are formulated by the manufacturer such that, over a chosen range of concentrations of drug in the sample, there is a graded concentration–response relationship between drug concentration and percentage binding of tracer-labelled drug to the antibody (Figure 5.4). Thus any immunoassay has three essential requirements: suitable antibody, tracer-labelled drug, and sample to be assayed.

Antibody reagent

Antiserum prepared for use in an immunoassay must have suitable properties with regard to specificity (the measure of the degree of cross-reactivity with other closely-related

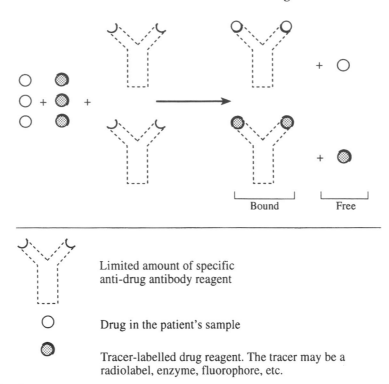

Limited amount of specific
anti-drug antibody reagent

Drug in the patient's sample

Tracer-labelled drug reagent. The tracer may be a
radiolabel, enzyme, fluorophore, etc.

Figure 5.3 Competitive binding immunoassay.

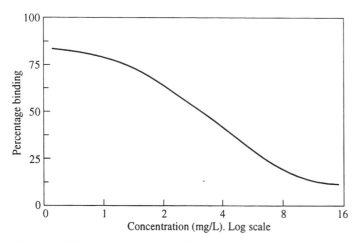

Figure 5.4 Theoretical relationship between concentration of drug and percentage binding of tracer-labelled drug to specific anti-drug antibody.

antigens), avidity (the measure of the energy of the antigen–antibody binding) and titre (the measure of the specific antibody content). Antisera for immunoassay were traditionally produced in animals. By chemically coupling the drug to a carrier protein such as bovine serum albumin or human thyroglobulin, this conjugate was used to evoke an antibody response in rabbits, goats or sheep (see Marks, 1981). However, during recent years, monoclonal antibodies have, in some kits, replaced these animal antisera. Once a suitable antibody-producing cell line has been found, it is possible to synthesize unlimited quantities of antibody with carefully controlled and extremely repro-

ducible specificity, avidity and titre. In addition, in the long term, monoclonal antibodies are cheaper and easier to make in bulk and eliminate the increasingly controversial need to use animals.

Almost all immunoassays for antimicrobials are commercial kits and the manufacturers are responsible for formulating the antibody reagent. There are some differences between the antibodies used by different manufacturers. For example, antibody in the Abbott TDX vancomycin assay reagents cross-reacts with certain vancomycin degradation products whereas those used in the Syva EMIT vancomycin assay and the Abbott Axsym assay do not.

Tracer reagent

The prototype immunoassay was used in 1959 to measure hormones in blood but the first immunoassay for a drug (digoxin) was not published until 1968 (Oliver et al., 1968). The tracer was a radiolabel and the antibody-bound tracer was determined by separating it from the unbound tracer. This assay was what is usually called a separation (or heterogeneous) immunoassay. Several separation radioimmunoassays for aminoglycosides have been described using either [3]H- or ^{125}I-labelled aminoglycoside as the tracer (see White & Reeves, 1981; Edberg & Sabath, 1988). However, radioimmunoassays for antibiotics are now all but obsolete in the clinical setting thanks to the development of non-radioactive tracers. These non-radioactive tracers brought two improvements: they obviated the need to work with potentially hazardous radioactive reagents and they allowed the development of procedures whereby antibody-bound tracer could be measured without the need to separate it from the bound tracer. These assays are referred to as homogeneous (or non-separation) and are technically simpler than the early separation immunoassays. Many have the additional advantage of improved speed and reproducibility.

Assay protocols

Early heterogeneous assays involved a time-consuming separation step which often introduced imprecision. Methods used to separate the free from the bound tracer (Marks, 1981) included:

- antibody absorption to a solid phase (e.g. charcoal)
- antibody precipitation using a protein precipitant
- a second (precipitating) anti-antibody
- solid-phase antibody.

In contrast, the homogeneous techniques involve simply mixing the sample and the reagents and then determining the degree of binding in some way, either kinetically (see EMIT below) or after a fixed incubation period.

All immunoassays require only a small volume of sample, and are therefore particularly suited to neonatal patients, and are usually calibrated with five or six calibrators covering the required concentration range and including a drug-free blank (sometimes called the zero calibrator). With the early heterogeneous techniques a calibration curve would be run with each batch of assays. Many of the new homogeneous methods, however, have very stable calibration curves and they are often only run occasionally. Like all assays every batch of assay results should be validated by the use of appropriate internal controls (see *Chapter 7*).

Quantification (calibration)

With separation radioimmunoassays, the calibration curve—a plot of percentage tracer binding against the log calibrator drug concentration—normally gives a sigmoid curve (Figure 5.4) which can be mathematically 'straightened' by a curve-fitting procedure such as a log–logit transformation. Kits would usually include suitable graph paper for plotting this curve. Similarly, homogeneous techniques produce complex calibration curves which require appropriate curve fitting to ensure accurate quantification. Today the analysers used to perform the immunoassays are also programmed to calculate and store these curves. As stated above, the various homogeneous immunoassay calibration curves generated and stored by the analysers are likely to be very stable and calibration with every analytical run is now very much the exception rather than the rule.

Performance expectations

All immunoassays for antimicrobials show very high specificity but they are only available for a few, frequently assayed, drugs (aminoglycosides, chloramphenicol, teicoplanin, vancomycin). Since others are assayed only infrequently it is not commercially viable to develop immunoassays.

There is partial cross-reactivity between some aminoglycoside immunoassays. For example, most gentamicin assay kits will cross-react with netilmicin and vice versa. If the aminoglycoside is incorrectly identified on the report form, for example if netilmicin is assayed in the belief that it is gentamicin, an inaccurate result will be obtained. However, this cross-reactivity can be exploited. Some gentamicin assay kits can be adapted for netilmicin assay by using appropriate netilmicin calibrators (White *et al.*, 1983b).

Homogeneous assay kits produce results in under 15 min, use small volumes of sample, are very reproducible, and the results obtained correlate well with results obtained using other methods (O'Leary *et al.*, 1980; Standiford *et al.*, 1981; Delaney *et al.*, 1982; Stobberingh *et al.*, 1982; Acton *et al.*, 1982; Masterton *et al.*, 1983; Witebsky *et al.*, 1983; Nilsson, 1984; Glew & Pavuk, 1985; Redmond *et al.*, 1985).

Costs

Once the capital equipment has been purchased, the cost of performing an immunoassay is essentially the cost of the kit (ignoring staff, overheads, etc.). Some manufacturers operate reagent-rental agreements where the analyser is supplied and serviced as a part of the price paid for each kit. The cost of each assay depends on the frequency of calibration (since calibration uses reagent) and the frequency with which internal controls (and EQA samples) are run. The cost of a single assay will therefore *never* be less than the cost of the kit divided by the number of assays in the kit and will, in practice, be somewhat more than this figure. Immunoassay kits are expensive and so the more assays which can be obtained from one kit the better. Attempts to cut costs by skimping on calibration, internal controls or EQA are unacceptable since quality is compromised, and unreliable or inaccurate results may go unnoticed.

Despite high reagent costs there has been a huge swing in clinical laboratories away from the use of microbiological assays to immunoassays for routine aminoglycoside assays because they produce reliable high quality results (White, 1986). Now that immunoassay kits are firmly established as the standard methods for routine clinical antimicrobial assays the current emphasis is on controlling and reducing costs in the laboratory. Rationalization of assays is one way to cut costs. It is now common practice in many laboratories to assay only the antibiotics most commonly prescribed in their hospital in-house, but refer the other assays to a reference laboratory. Once rationalization has been taken as far as is practicable, the cost of the kit again becomes the most important factor. Increasing competition in the area of immunoassay kits is slowly resulting in a reduction in kit costs without any reduction on quality.

Commercial immunoassays

There are many commercial immunoassay kits on the market for the assay of antimicrobials. Most but not all are non-isotopic, non-separation immunoassays. Some of the newer or more popular kits are described below.

Opus immunoassay system. The Opus analyser (Behring) is a dedicated instrument which can process two types of innovative immunoassay: (i) a fluorigenic ELISA test designed to measure large molecular weight molecules and (ii) a dry, thin-film, multilayer immunoassay for various drugs including gentamicin, tobramycin and vancomycin. This immunoassay, unlike the other assays described below, is a separation fluoroimmunoassay using immobilized antibody. The separation step does not, however, require operator intervention and is achieved by natural diffusion.

Reagents for the assay come as dried thin films built into a plastic moulding (the test module). The operator enters the test requirements and any identifying codes into the analyser via an interactive touch screen. Batch, random access, panel and stat modes of operation are user-selectable. After data entry the analyser automatically generates the worklist. Next the appropriate test modules and samples are loaded by the operator into the analyser together with disposable pipette tips. A computer-controlled pipette on a pipettor arm picks up a pipette tip and then dispenses 10 μL of sample into the test module. A new tip is used for every sample to eliminate carry-over. A 20 position rotor then transports the test modules within a rotating enclosure at 37°C.

Once the sample has been dispensed into the test module, it passes into the spreader grid (100 μm thick) which evenly distributes it over the surface of the top coat film (Figure 5.5). The sample then naturally diffuses through the top coat (10 μm thick) which contains buffer and surfactants (which eliminate any protein binding) and which also retains large particles and large molecules. The analyte molecules continue to diffuse through the iron oxide layer (10 μm thick) into the signal layer (1 μm thick) which contains immobilized antibody and rhodamine-labelled analyte. The analyte from the sample and the rhodamine-labelled analyte then compete for the antibody-binding sites. Non-bound rhodamine-labelled analyte then naturally diffuses back through the iron oxide layer. The base of the test module is then illuminated by a tungsten–halogen lamp; the light first passing through a 550 nm excitation filter and the fluorescence emitted

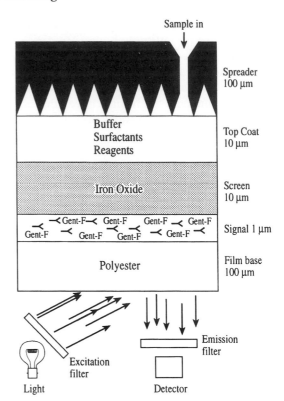

Figure 5.5 Dry thin-film multilayer fluoroimmunoassay: diagrammatic cross-section of the Opus test module.

by any antibody-bound rhodamine-labelled analyte is measured by a sensitive silicon photodiode detector after passing through a 580 nm emission filter. The amount of fluorescence generated is inversely related to the amount of analyte in the sample. A single sample takes 6–12 min to process depending on the analyte but the machine has a throughput of up to 75 tests an hour.

Each assay is calibrated by processing six calibrators in duplicate, the analyser automatically calculating and storing the calibration curve and printing a graphical plot of the curve together with curve statistics (Figure 5.6). Standard curve stability of 2–6 weeks is claimed.

EMIT assays. EMIT is an enzyme immunoassay developed by Syva (USA) for the assay of many therapeutic drugs and drugs of abuse in body fluids. There are currently three Syva EMIT formulations: the original traditional EMIT formulation, which comes freeze-dried and is suitable for use with a number of different analysers; EMIT 2000, formulated for use on a variety of clinical chemistry autoanalysers such as the Cobas Mira; and a liquid EMIT formulation for use on the Solaris analyser. EMIT antimicrobial assays include amikacin, gentamicin, netilmicin, tobramycin (all as 100-test kits), vancomycin and, uniquely, chloramphenicol (both 50-test kits) but not all of these are available in all three formulations. All EMIT assays work on the same principle. An EMIT kit comprises Reagent A and Reagent B. Reagent A con-

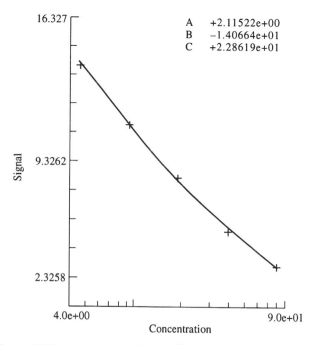

A +2.11522e+00
B −1.40664e+01
C +2.28619e+01

Figure 5.6 Print-out of the Opus calibration curve.

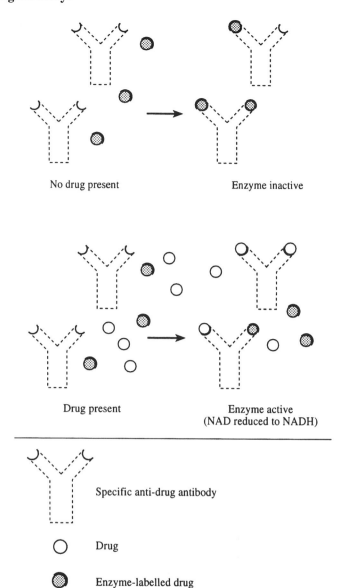

Figure 5.7 Principle of EMIT immunoassays.

tains specific anti-drug antibody, nicotinamide adenine dinucleotide (NAD) and substrate for the enzyme glucose-6-phosphate dehydrogenase. Reagent B contains a drug derivative chemically labelled with glucose-6-phosphate dehydrogenase. (Note: some EMIT assays have these reagents reversed.) Traditional EMIT reagents come freeze-dried and require reconstitution with distilled or deionized water. The use of tap water or other low quality water for reconstitution may cause reagent problems and is to be avoided. For reliable performance reconstituted reagents must be kept at 20–25°C for at least 1 h, or 2–8°C overnight, before use. The reagents should also always be allowed to reach room temperature before use. The liquid EMIT formulations avoid problems associated with reconstitution. Kits using the EMIT principle are packaged by some other companies for use on particular auto-analysers. One example is the IL Gentamicin Test (International Laboratory Company, Lexington, KY, USA) for use on the Monarch analyser; another are the gentamicin, tobramycin and vancomycin reagents for use with the ACA discrete clinical analyser.

All EMIT formulations work on the same principle (Figure 5.7). When reagents and sample are mixed, the specific antibody binds to any specific drug in the sample and the enzyme-labelled drug competes for this antibody. When enzyme-labelled drug becomes bound then the enzyme activity is reduced. Unbound enzyme remains active and this activity is directly related to the concentration of drug in the sample. During enzyme activity NAD is reduced to NADH and this results in a change in absorbance at 340 nm (Figure 5.7). By comparing the absorbance change for the sample with that seen with the calibrators, the concentration of drug is determined.

EMIT antimicrobial assays performed on the Solaris are calibrated with six calibrators (including a drug-free blank). The instrument computes a calibration curve from the calibrator concentrations and the appropriate enzyme reaction rates (the rate of change of absorbance as the NAD is reduced to NADH) using an appropriate mathematical model. Various statistical quality checks are applied to the curve and, if it is acceptable, it is stored in memory. The operator is given three options when assaying a batch of samples which may be programmed to proceed with or without operator intervention:

1. If there is a standard curve stored in the instrument this may be used and the instrument will automatically assay the samples.

2. If no curve is in the memory (or if the operator chooses to recalibrate), all six calibrators are run. The zero calibrator is run in duplicate and if the rates are ≤0.008

units apart the mean value is stored; the other calibrators are assayed in singlicate. The instrument checks that there was an increase in rate with each increase in calibrator concentration and if this was so makes the appropriate mathematical curve fit of the data. If none of the calibrator rates are >0.008 units from the curve fit the curve is accepted and the samples are assayed.

3. Alternatively the operator can choose to validate a stored curve by running a control sample. If duplicate assays of the control are ≤0.008 units apart the mean is used to determine the concentration and this is compared with the programmed acceptable range. If it is outside this range the zero calibrator is assayed in duplicate and if the rates are ≤0.008 units apart the instrument automatically updates the stored curve by making an appropriate parallel shift up or down the rate axis, and recalculates the control concentration. If it is now acceptable the assays proceed; if not a full recalibration is performed.

Standard curve stability is a minimum of 8 h but may be much longer (several days possibly).

Abbott TDX fluorescence polarization immunoassay (FPIA). The Abbott TDX, FLX and Axsym analysers are instruments designed to perform a number of investigations, including antimicrobial assays, based on the principle of fluorescence polarization. The TDX is the original instrumentation, the FLX is a somewhat refined design which essentially performs like the TDX and the Axsym is the most recent analyser designed for these assays. Kits are available for amikacin, gentamicin, netilmicin, tobramycin and vancomycin (all 100-test kits). Uniquely a teicoplanin assay kit is available (Oxis International, Portland, OR, USA) although this is not produced by Abbott. The reagents and calibrators come in liquid form and do not require reconstitution; they are normally stored in a refrigerator and do not need to be brought to room temperature before use.

The reagents pack is placed directly into the analyser. Each Abbott pack comprises three bottles labelled S, T and P. One contains the antibody, one the fluorophore-labelled drug reagent and one a pretreatment solution. The pack carries a bar code which the analyser reads automatically to determine which assay is being performed. The FLX bar code is more complex than the TDX bar code and is unique to every pack, whereas the TDX barcode is only unique to each analyte. The FLX bar code will not allow more than 101 assays per pack to be performed. Samples (up to 20) are placed in disposable plastic boats which fit into a carousel which also carries a bar code. A robot arm in the analyser performs all the pipettings and dilutions. When the reagents and sample are mixed, the fluorophore-labelled drug competes for specific antibody with any drug in the patient sample. A

glass cuvette in the instrument contains the final reaction mixture and is illuminated with light from a tungsten–halogen lamp. The light initially passes through an interference filter allowing only blue light (481–489 nm) to pass and a liquid crystal polarizer which polarizes this blue light. This polarized light beam excites the fluorophore, causing it to fluoresce at 525–550 nm. When the fluorophore is antibody-bound the fluorescence retains the polarization because the large molecular complex only undergoes a small amount of Brownian movement between the excitation and emission (Weber, 1953). Conversely, any free fluorophore emits light that is no longer polarized (Figure 5.8). The degree of polarization of the emitted fluorescence is therefore inversely related to the concentration of drug in the sample.

The assay is calibrated with six calibrators including a drug-free blank. These should be processed in duplicate but the instrument can be programmed to accept only a single series of calibrators. A typical print-out after a

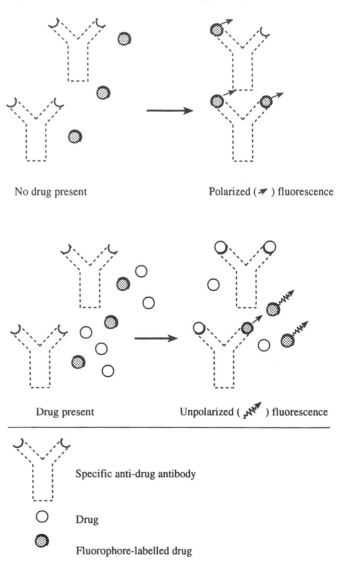

No drug present   Polarized (↗) fluorescence

Drug present   Unpolarized (⤨) fluorescence

⌐Y⌐ Specific anti-drug antibody

○ Drug

◉ Fluorophore-labelled drug

Figure 5.8 Principle of fluorescence polarization immunoassay (FPIA).

DATE: *(date and time of calibration)*
TIME:

SERIAL #: *(instrument serial number)*
LOCK = 1OP ID #: *(optional operator I.D.)*
RGT LOT #: *(optional reagent I.D.)*
ASSAY: *(drug name)*

CALIBRATION

VOL = 2.00 *(sample volume used for calibration)*
REPS = 2 *(duplicate calibrators required)*
GAIN = 40

CONC = MG/L *(concentration units in use)*

I.D.	NET P	NET I	BLANK I	
1A	242.39	7105.84	460.6	*(P = polarization,*
2A	244.11	7488.64	452.9	*I = intensity)*
3B	217.79	7276.94	472.1	
4B	217.11	7276.94	472.1	
5C	196.02	7551.34	419.3	
6C	196.99	7323.04	422.9	
7D	169.24	7572.54	433.4	
8D	167.98	7654.14	431.1	
9E	135.77	7871.24	440.7	
10E	135.06	7776.14	448.0	
11F	101.86	8023.64	444.8	
12F	102.31	8096.44	452.1	

I.D.	CONC	AVGP	FITP	PERR
A	0.00	243.25	243.25	0.00
B	2.50	217.45	216.80	0.65
C	5.00	196.51	197.09	−0.58
D	10.00	168.61	168.80	−0.19
E	20.00	135.42	134.95	0.47
F	40.00	102.09	102.27	−0.18

RMSE = 0.36

- -

AVGP = average actual polarization reading
FITP = expected polarization reading from the fitted curve
PERR = polarization error (AVGP - FITP).

Figure 5.9 Print-out of successful calibration of an Abbott TDX analyser.

calibration is shown in Figure 5.9. Only if the polarization error (PERR), root-mean-square error of the curve (RMSE) and other preset values are within acceptable limits will the curve be accepted. If a calibration is not accepted the whole procedure must be repeated since assays cannot be performed until an accepted calibration curve is stored. A useful tip is to run the zero calibrator as a test in position 13 after the top calibrator. This eliminates the possibility of carry-over from the top calibrator (position 12) to the zero calibrator (position 1).

TDX calibration curves are very stable and may remain valid when a new kit with the same batch number as the kit used for calibration is brought into use. For aminoglycoside assays curves may remain valid for several months.

An interesting feature of the TDX or FLX is the dilution protocol. If a sample concentration is above the top calibrator the analyser will print out 'HI' as the result. The assay parameter for sample volume can be simply edited from the key pad and a smaller volume (e.g. half or a quarter of the normal volume) entered. If the sample is now reassayed the analyser only picks up the reduced sample volume, calculates the concentration and multiplies this value by the appropriate dilution factor before printing it. This features enables concentrations above the top calibrator to be assayed without the need for manual dilution of the sample. After a dilution protocol assay is finished the instrument automatically resets the sample volume to the default value (this automatic reset can be over-ridden if required). The dilution protocol should not be used if the sample volume is reduced to < 5 μL.

The dilution protocol may also be used in reverse to assay accurately very low concentrations of aminoglycosides as might be encountered in trough samples from patients on once-daily therapy (White *et al.*, 1994, also see *Chapter 9*). The sample volume may be increased five-fold.

Because of the ease of operation, the performance characteristics and the standard curve stability, the Abbott TDX or FLX has become the instrument of choice for most laboratories performing aminoglycoside assays. Other companies, such as Sigma (Poole, UK) and Oxis International, sell FPIA kits for use on the TDX or FLX. A laboratory owning a TDX might choose to use these non-Abbott reagents. Abbott does not sanction the use of any non-Abbott reagents with the TDX or FLX analysers and therefore laboratories working to Good Laboratory Practice or other some standards may find it mandatory to use approved reagents.

Ciba–Corning chemiluminescent immunoassay. This is a separation immunoassay designed to be run on the Ciba–Corning ACS:180 analyser. Patient sample is mixed with antibody reagent (solid-phase on paramagnetic particles) and tracer reagent (drug labelled with dimethylacridium ester), and after 7.5 min incubation the solid-phase antibody is separated and washed. A chemiluminescent reaction is then initiated and the amount of light emitted is inversely related to the drug concentration. Assays are available for gentamicin and vancomycin.

Bayer (Technicon) latex agglutination. This assay is designed to be run on Bayer autoanalysers such as the RA1000. Patient sample is mixed with drug–Ficoll conjugate (the agglutinator reagent) and monoclonal antibody attached to engineered microparticles. Agglutination of the antibody-coated particles causes an increase in absorbance at 600 nm and this increase is reduced in proportion to the concentration of drug in the sample. The assay is available for gentamicin.

High-performance liquid chromatography

High-performance liquid chromatography (HPLC) is the most likely alternative to microbiological assay for those antimicrobials where enzyme assays or immunoassays are not available. HPLC is a form of liquid chromatographic assay in which the drug of interest must be chromatographically separated from any other compounds in the sample being assayed and then quantified using some relatively non-specific property such as UV absorption. Because detection and quantification are non-specific, chromatographic assays can be developed for almost any antimicrobial. Other forms of chromatography including thin-layer chromatography (TLC) and gas–liquid chromatography (GLC) have been used to assay some antimicrobials, but the vast majority of assays have been by HPLC and consequently this will be discussed in some detail. The theory and principles discussed for HPLC apply to all forms of chromatographic assay.

Theory of column chromatography

In column chromatography a *mobile phase*, which in GLC will be a gas and in HPLC a liquid, is pumped through a column packed with a *stationary phase* on which the chromatographic separation takes place. The sample is introduced into the flowing mobile phase by means of an *injector*. After eluting from the stationary phase the mobile phase containing the separated solutes passes through a *detector* where each solute produces a response by which it can be quantified. The output from the detector plotted against time is called a *chromatogram.*

Retention. As a solute in the mobile phase passes over the stationary phase then, according to its relative affinity for the mobile phase compared with the stationary phase, its passage will be slowed to a greater or lesser extent. Solutes with no affinity for the stationary phase will pass straight through. Such solutes elute in the *void volume* (V_0), defined as the volume of mobile phase between the injector and the detector. Any solutes which are retarded as a result of their interaction with the stationary phase will elute after a greater volume of mobile phase has passed through the column. For example Compound A in the chromatogram shown in Figure 5.10 elutes after a volume, V_a, has been pumped through and Compound B elutes only after a greater volume, V_b, has passed. If, as is usually the case, the flow rate of the mobile phase remains constant, then these elution volumes can be replaced by measures of time, the *retention times.* The retention time is the most commonly quoted chromatographic property of a compound and is easily determined from a chromatogram. For any given set of chromatographic conditions the retention time of a particular compound will remain constant as long as the flow rate remains constant. Since retention time is dependent on flow rate and column dimensions then it is meaningless to quote a retention time unless the flow rate and column dimensions are also specified. Nevertheless, it is surprising how many peer-reviewed articles in scientific journals quote retention

times without quoting the flow rate used. If comparing retention times for two columns of different dimensions then the equivalent flow-rate for the second column (F_2) can be determined from the formula:

$$F_2 = F_1(L_2/L_1)(R_2/R_1)^2$$

where F_1 is the flow rate used with the first column of length L_1 and radius R_1, and the second column is of length L_2 and radius R_2.

A more fundamental measure of retention is the *capacity factor* (K'). K' is independent of flow rate and for Compound A (Figure 5.10) is given by the formula:

$$K'_a = (V_a - V_0)/V_0$$

K' is used in the formula for calculating the degree of separation of two peaks (see below).

Separation. A successful set of chromatographic conditions must separate the solute of interest (say Compound B, Figure 5.10) from any interfering compound(s) in the sample (say Compound A, Figure 5.10). The quantitative measure of the ability of any chromatographic system to separate two compounds is the *selectivity*, α, which is determined from:

$$\alpha = K'_b/K'_a = (V_b - V_0)/(V_a - V_0)$$

Efficiency. As each band of solute passes over the stationary phase, various factors such as diffusion, flow irregularities, large void volumes of mobile phase, etc., will cause the band to broaden. Column efficiency is a measure of how little the band of solute spreads as it passes through the system. High efficiency is a very important aspect of HPLC as it makes rapid separation of closely related compounds possible.

The theory behind the equations for calculating efficiency derive from distillation theory and involve a concept known as a *theoretical plate*. The more theoretical plates in a column, the greater its efficiency. The number

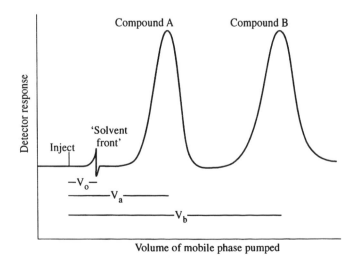

Figure 5.10 Parameters used in HPLC.

of theoretical plates (N) in a column may be simply determined from any chromatogram. There are several ways in which N can be calculated; the simplest to use in the laboratory is:

$$N = 5.54(Rt/W_{hh})^2$$

Where Rt is the retention time of the peak being measured and W_{hh} is the width of the peak at half-height. Rt and W_{hh} can be measured from a chomatogram and may be in any units, such as millimetres or minutes, but note, both measurements must be in the same units.

Another version of this equation uses peak width at 4.4% peak height. In this case the formula is:

$$N = 25(Rt/W_{4.4})^2$$

The first equation usually gives the best result for N, whereas the latter usually gives a lower value because of peak asymmetry due to 'fronting' or 'tailing'. Column suppliers usually quote values of N for their stationary phases (often quoted as plates/metre rather than plates/column since the latter figure allows comparisons between columns of different lengths to be made). It is important to know which formula has been used to calculate N when comparing published claims for column efficiency. Regular in-house determinations of N will give an early warning of any deterioration in a column's performance.

Resolution. All the terms discussed above go to define the *resolution* (R) of an HPLC method for two peaks A and B from the equation:

$$R = 0.25((\alpha - 1)/\alpha)(\sqrt{N})(K'_B/(K'_B + 1))$$

Where K'_B is the capacity factor of the longest-retained peak. When $R = 1.5$ then baseline separation of two compounds has been achieved. Clearly α must be greater than 1 and, given this, increments in α, K' or N will improve resolution. However two points are worth remembering. Firstly, as K'_B gets larger so the effect on resolution gets progressively smaller. Secondly, since the square root of N is used, a four-fold increase in efficiency will only produce a doubling of the resolution. In HPLC where columns are usually very efficient and retention times are relatively short, the most powerful tool for improving resolution is by making changes to the selectivity, usually by making slight modifications to mobile phase composition, or, if this fails to produce the required results, changing the stationary phase.

HPLC equipment

A working HPLC instrument will comprise a pump, injector, detector (including an output device such as a chart recorder or computing integrator), stationary phase in an analytical column and mobile phase. In addition it may have a guard column fitted in front of the analytical column (Figure 5.11). These components are connected by

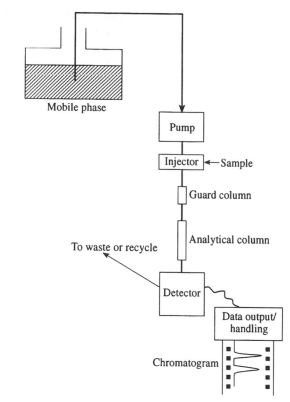

Figure 5.11 Diagram of HPLC equipment.

narrow-bore stainless steel or polymer tubing. The most commonly used polymer tubing is polyetheretherketone (PEEK) which is inert to most HPLC solvents with the exception of methylene chloride, tetrahydrofuran and concentrated nitric or sulphuric acids. Between the pump and the injector, relatively wide-bore (c.1 mm internal diameter) tubing may be used but in the rest of the system narrow-bore (c.0.2–0.5 mm internal diameter) tubing is essential to minimize the dead volume of the equipment and thus maximize efficiency. All the individual components must function correctly and efficiently. A problem with, or breakdown of, any component will render the instrument unusable until the faulty part can be repaired or replaced.

Pump. An HPLC pump must pump the mobile phase through the stationary phase at a constant flow rate. HPLC pumps are usually single- or dual-piston reciprocating devices with some form of pulse dampening incorporated to produce a steady, pulse-free, flow. The high pressures generated in HPLC are due to the resistance of the stationary phase to the solvent flow and are a by-product of the solvent flow, not an essential feature of HPLC. Solvent flow rates will usually be less than 2.5 mL/min although most HPLC pumps can be set to pump at flow rates up to 10 mL/min, usually in 0.1 mL/min increments. The higher flow rates are used when priming the system and changing mobile phases rather than during an analytical run. High back-pressures will be encountered during an analytical run and the pump should be fitted with a gauge to monitor this back-pressure. Although the SI unit of pressure is the pascal (Pa), HPLC back-pressures are usually quoted in pounds per square inch (psi). Pressure may also sometimes be quoted as bar or atmospheres; conversion factors are as follows:

1000 psi = 6900 kPa
1 bar = 100 kPa
1 atmosphere = 101.3 kPa

It is important, therefore, to know which units have been used to calibrate your pump pressure gauge. When the pump is working correctly, the back-pressure will remain constant and the solvent will flow at the correct rate. Most pumps will cut out if the back-pressure rises above some pre-set limit (often 6000 psi). Erratic flow (as shown by a fluctuating pressure reading) may be due to leaks, air in the system, or worn or sticking valves. Incorrect flow rate may be due to air in one chamber of a dual piston pump. Irregular flow may lead to assay imprecision. Before starting an analytical run it is worth collecting mobile phase in a measuring cylinder for 1–2 min to ensure the pump is pumping at the correct flow rate. High back-pressure can develop if the system gets blocked at any point. The point of blockage can be determined by systematically disconnecting the system begining at the furthest point from the pump. The most usual causes of high back-pressure is a blocked frit or filter at the head of the column.

Injector. The injector is used to introduce the sample into the flowing mobile phase. Injectors may be simple manual devices or sophisticated programmable autoinjectors capable of dealing with large numbers of samples unattended. Reproducible injection volumes are essential for good assay reproducibility unless an *internal standard* is used (see below). In analytical HPLC, injection volumes are typically 5–200 µL. Most injectors in use today work on the *sample-loop* principle, the loop comprising a narrow-bore stainless steel (or PEEK) tube which can be isolated from the mobile phase flow by a system of valves. The sample is introduced into the loop with a micro-syringe. Variable volumes may be injected or the loop itself may have a finite volume (say 20 µL) and it is deliberately overfilled. In the latter case the injection volume is fixed at 20 µL no matter what volume >20 µL is introduced into the loop. When the valves are actuated the loop is introduced into the mobile phase flow path and the sample is washed on to the stationary phase. When working with biological samples, sample stability and sample size may be a problem and this must be taken into consideration when choosing an autoinjector. For labile samples an auto-injector with a cooled sample tray may be necessary. If sample volume is likely to be limited then care must be taken over the choice of injector. Some of the less expensive injectors work on the basis of picking up an excess of sample and using a fixed-volume loop (see above), thus a minimum of 200 µL may be required for an actual injection volume of 20 µL. Other autoinjectors (e.g. Waters WISP) pick up exactly the volume required from a sample containing very little more than the volume required for injection.

Detectors. HPLC detectors measuring UV absorbance, fluorescence, refractive index and electrochemical activity are available commercially. All of these except refractive index detectors (which have very low sensitivity) have application to the measurement of antimicrobials in body fluids. The essential property of any detector is the *signal-to-noise ratio* since this will set the level of sensitivity of your assay. Figure 5.12 illustrates the difference between two hypothetical detectors. Detector 1 appears more sensitive since the peak is larger, but its poor signal-to-noise ratio makes it inferior to detector 2. Manufacturers make claims for the performance of their detectors but it is essential to have a demonstration of a detector before purchase. Detector output is usually directed to a chart recorder or a computing integrator. The latter device can be programmed to determine retention times, peak heights and peak areas and, if required, results from calibration curves. However, with biological samples, especially if low concentrations of antimicrobial are being assayed, it is often difficult to program an integrator reliably. The results obtained from an integrator should *never* be accepted without a visual inspection of the chromatogram.

Detector 1. Good signal peaks but unacceptable noise on baseline

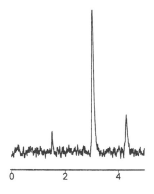

Detector 2. Weaker signals but acceptable baseline noise

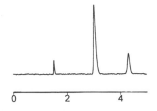

Figure 5.12 Importance of signal-to-noise ratio in HPLC detectors.

UV absorbance detectors. Most compounds absorb UV light with a characteristic absorption spectrum and consequently the UV absorbance detector has very wide applicability. Certain groups confer UV absorbance properties; these include double or triple bonds, benzene (204 and 256 nm), quinoline (228, 270 and 315 nm), imidazole (207 nm), isoxazole (211 nm) and thiazole (240 nm) groups. UV detectors may be of fixed (selectable) wavelength or variable wavelength. The former uses a lamp which has a maximum emission at a particular wavelength (usually 254 or 214 nm). A filter between the lamp and sample selects this wavelength. The latter uses a prismatic monochromator which allows any particular wavelength to be selected by a simple dial. In theory compounds could be detected with greatest sensitivity at the wavelength of highest absorption (λ_{max}); in practice, however, using a fixed wavelength detector at 214 or 254 nm is often acceptable. With both types of detector, after passing through the sample, the radiation falls on to a photomultiplier which accurately measures its intensity. Output from the photomultiplier is usually available in two forms. Firstly, an unattenuated 1–10 V output which can be connected to a computing integrator and, secondly, an attenuated 5–10 mV output for a chart recorder. The required attenuation setting to ensure peaks remain on-scale is selected by buttons or a dial. Typical detectors will have attenuation settings measured in absorbance units full-scale (AUFS) of 1.0 to 0.005 AUFS. Choosing a setting of 0.02 AUFS means a change in UV absorption of 0.02 absorption units (AU) will produce a full-scale deflection on the chart. Whether to choose a fixed or variable wavelength detector depends on several factors. Fixed wavelength detectors are relatively inexpensive and robust and ideal if adaptability is not important. Variable wavelength detectors cost more but give maximum flexibility if assaying a variety of different compounds with differing UV absorption properties. Fixed-wavelength detectors usually have a better signal-to-noise ratio than variable-wavelength detectors but the difference is often marginal. Also available are multiwavelength or diode array detectors which are able to determine UV spectra of peaks as they elute. These are specialist research instruments which would not normally be found in a routine laboratory.

Fluorescence detectors. Some compounds, when illuminated with radiation of a particular wavelength (*excitation wavelength*), absorb the energy and then emit some of it as radiation of a longer wavelength (*emission wavelength*). Such compounds are said to be fluorescent. Fluorescence detection is important in the field of antimicrobial assay because many fluoroquinolones are naturally fluorescent. Fluorescence has two advantages over UV absorbance: it is often more sensitive and it offers improved specificity since few compounds fluoresce and only fluorescent compounds will interfere. Sometimes assay sensitivity and specificity can be improved by derivatizing drugs to produce fluorescent compounds. This approach has been used for aminoglycosides, some β-lactams and teicoplanin (see relevant chapters). Such derivatization may be done before the sample is analysed (*pre-column derivatization*) or using a special device which sits between the column and the detector (*post-column derivatization*).

Fluorescence detectors may have filters or monochromators to set the excitation and emission wavelengths or a combination of both. It is very important when choosing a fluorimeter to know what excitation and emission wavelengths, and what degree of sensitivity, will be required. Degrees of sensitivity vary enormously between fluorescence detectors and it may be impossible to achieve the level of sensitivity claimed by a published procedure if a different model of detector is used. The Perkin–Elmer LS1 has filters to set the excitation wavelength and a monochromator to set the emission wavelength. It is perfectly adequate for measuring serum concentrations of fluorescent fluoroquinolones such as ciprofloxacin or ofloxacin. However, to measure tissue levels or metabolite levels, which may be below 0.01 mg/L, a detector with a higher specification (improved signal-to-noise ratio) will probably be needed.

Electrochemical detection. Electrochemical detectors may be used to detect molecules that do not fluoresce or absorb UV but that can be easily oxidised or reduced in solution. Electrochemical detectors are operationally more complex than other detectors. The detector contains a flowthrough electrolysis cell (FEC) which contains a working electrode and a control module (potentiostat). Eluant from the column passes through the FEC over the surface of the electrode which is tuned to a potential (usually between +1.3 and –1.2 V) to suit the analyte being assayed. As the analyte passes the electrode oxidation or reduction causes an electrical current to flow and this is measured as a function of time. The strength of the current depends on the concentration and ease of oxidation or reduction of analyte. The most commonly used electrode is composed of glassy carbon (an impermeable material made by heat treatment of phenoformaldehyde resin) (Jenkins & Kawamura, 1971).

Unlike UV or fluorescence detectors, electrochemical detectors are not passive, the analyte being oxidized or reduced during transit. There are two types of electrochemical detector on the market. In an *amperometric detector* only 1–5% of the analyte is oxidized or reduced; in a *coulometric detector* this figure is up to 100%. Most detectors are amperometric but some (for example the 5100A Coulochem detector (Environmental Sciences Associates, Bedford, MA, USA)) may be run in either mode.

When using electrochemical detection it is important to use the highest purity solvents and chemicals and to filter the mobile phase through a 0.2 μm filter before use (otherwise electrode clogging and high background 'noise' may result).

The particular advantage of these types of detectors is their high sensitivity and selectivity. Coulometric detectors may also show better reliability and the electrodes may need replacing less frequently. The disadvantage is that few antimicrobials can be detected this way. The most common application in the antimicrobial field is to the assay of macrolides (*Chapter 13*).

Stationary phases

In analytical HPLC the stationary phase is usually packed into a stainless steel column of approximately 4 mm internal diameter and 100–300 mm in length. For rapid high-efficiency separations the particles of the stationary phase should be no more than 10 μm in diameter; many commercial packings have 10, 5 or 3 μm particle versions available. When complex biological mixtures are being analysed it is common practice to fit a short *guard column* before the analytical column. The function of the guard column is to protect the analytical column from potentially harmful solutes and thereby prolong its useful life. The guard column should be packed with a stationary phase with a lower efficiency but similar chromatographic properties to the stationary phase in the analytical column. Guard column packings should be replaced regularly.

In HPLC any mode of chromatography is possible (e.g. ion exchange, partition, gel permeation, etc.) and an extensive and possibly bewildering range of commercial stationary phases is available. However, the majority of antimicrobial assays are performed using the technique of reversed-phase chromatography with a bonded C_{18} packing material.

Reversed-phase chromatography. Traditional chromatographic techniques used a polar stationary phase such as paper or silica and an organic, non-polar, mobile phase (usually a mixture of organic solvents which were immiscible with water). This form of chromatography, where the stationary phase is polar and the mobile phase is non-polar, is referred to as *normal-phase chromatography* (NPC). If the situation is reversed and the stationary phase is non-polar and the mobile phase is polar (watery) then this is referred to as *reversed-phase chromatography* (RPC). RPC has several advantages over NPC not the least of which is the use of a watery mobile phase allowing many aqueous biological fluids to be injected directly, making sample preparation relatively straightforward.

Bonded stationary phases. Chemical reaction of silanol groups (–SiOH) in silica with various organosilanes allows a variety of different bonded stationary phases to be produced commercially. By choosing an appropriate organosilane, packings ranging from normal-phase through those suitable for both NPC and RPC to hydrophobic reversed-

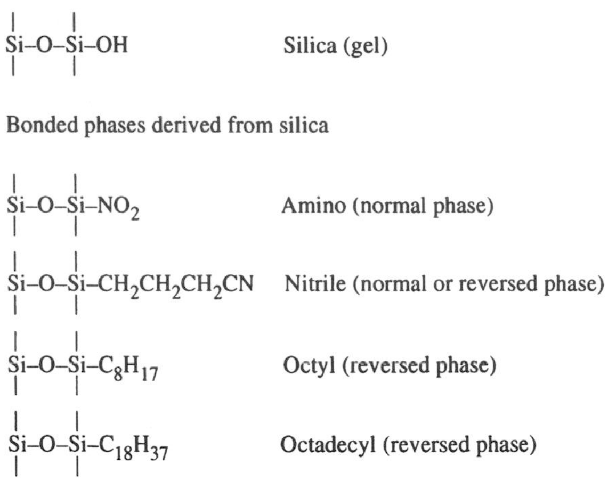

Figure 5.13 Some bonded packing materials for normal- and reversed-phase HPLC.

phase packings have been synthesized (Figure 5.13). The octadecyl packing (also referred to as C_{18} or ODS) is one of the most versatile reversed-phase packings and a very large number of different C_{18} packings are available commercially. These are all subtly different in their properties and a separation which works for one brand of C_{18} column may not work with a different brand. Properties to consider are particle size, particle shape, pore size, carbon load and end capping, since each of these will affect the chromatographic performance of the material. *End capping* is a process where residual unreacted silanol groups in the material are reacted with a small silanizing reagent. If not end-capped these residual groups could interact with basic or polar solutes and cause peak tailing. End capping aims to eliminate this problem. The greater the carbon load then, in general, the more retentive the material. Some relevant properties of some commercially available C_{18} packings are described in Table II. Most manufacturers and suppliers will advise alternative packings with identical chromatographic properties to those quoted in the Table. There is some evidence that some brands have changed their properties over the years. Consequently a method published in 1983 using C_{18} column 'Brand X' may not work when you attempt to reproduce it over 10 years later using a new 'Brand X' column. This should always be borne in mind when trying to reproduce a published procedure.

Bonded-phase silica columns are ideal for acid mobile phases since they are stable at low pH. However, above pH 7.5 silica begins to dissolve. Saturating the mobile phase with silica before it reaches the column can minimize column deterioration at alkaline pH.

RPC is ideal for separating acidic drugs since their ionization is suppressed at pH values below the pK_a (ions are not retained by RPC packings). The problem of assaying basic drugs without the use of alkaline mobile phase or at the same time as an acidic drug in the same sample (e.g. sulphamethoxazole and trimethoprim, see

Table II. Some commercially available C_{18} stationary phases

Name	Manufacturer/supplier	Particle size (μm)	Particle shape	End capped?	Carbon load (%)	Pore size (Å)
Microbondapak C_{18}	Waters	10	irregular	yes	10	125
Novapak C_{18}	Waters	4	spherical	yes	7	60
Hypersil ODS	Shandon	3, 5, 10	spherical	yes	10	120
Spherisorb ODS-I	Phase Separations	3, 5, 10	spherical	partially	7	80
Spherisorb ODS-II	Phase Separations	3, 5, 10	spherical	yes	12	80
Zorbax ODS	DuPont	3, 5, 7	spherical	yes	20	70
Nucleosil 100 C_{18}	Macherey-Nagel	3, 5, 10	spherical	yes	14	100
Nucleosil 100 C_{18} AB	Macherey-Nagel	5	spherical	no	25	100
Nucleosil 120 C_{18}	Macherey-Nagel	3, 5, 7, 10	spherical	yes	11	120
Partisil ODS (1)	Whatman	10	irregular	no	5	85
Partisil ODS (2)	Whatman	10	irregular	no	16	85
Partisil ODS (3)	Whatman	5, 10	irregular	yes	10.5	85

Chapter 19) is overcome by the use of ion-pair chromatography (see mobile phase section below).

A novel approach to RPC is *internal-surface reversed-phase chromatography* (ISRP) (Hagestam & Pinkerton, 1985). Here the stationary phase is a 5 μm spherical porous silica particle with only the internal surface of the particle covered by a bonded hydrophobic group. The outer surface of the silica spheres remains hydrophilic and non-absorptive. Small molecules, such as most drugs, enter the pores and are separated by RPC. Large molecules such as serum proteins cannot penetrate the pores and pass straight through the column, thus becoming separated from the small molecules as in conventional gel-permeation chromatography. In theory such a packing would allow the direct injection of serum samples provided the proteins remained soluble in the mobile phase.

Mobile phase

For any chosen stationary phase such as an ODS packing, the desired selectivity and capacity are determined by optimizing the mobile phase composition. In RPC the mobile phase is usually a mixture of two or three solvents, one of which is water, usually with the addition of some form of buffering agent (*modifier*) to maintain a constant pH (see below). Water is considered a *weak solvent* which means it is relatively poor at causing solutes to be eluted from the stationary phase. In contrast organic solvents such as methanol or acetonitrile are *strong solvents* in as much as they have a greater capacity to elute solutes from the stationary phase. In a mobile phase which is a mixture of a strong and weak solvent (say for example methanol and water) increasing the proportion of strong solvent (methanol) will cause solutes to be retained less (i.e. retention times will become shorter). Conversely, increasing the proportion of weak solvent (water) will cause greater retention (retention times will increase). If a mobile phase

composition is given as (say) A:B (50:50, v/v) then it is important to measure out and mix equal volumes of A and B rather than fill a measuring cylinder or volumetric flask to the half-way mark with A and then fill to volume with B. This is because mixtures of solvents may have somewhat different densities from the individual components. For example mixing 500 mL of water with 500 mL of methanol produces a final volume of approximately 980 mL. The buffering agent is also important. In RPC the pH should also be 1–2 pH units below the pK_a of weakly acidic drugs to ensure they are not significantly ionized in solution.

Mobile phase selectivity. The selectivity of the mobile phase is finely tuned by the modifiers such as a buffer, paired-ion reagent, an additional solvent, etc., and a large part of method development involves progressively modifying mobile phase composition until the necessary separation is secured. When the baseline separation of two solutes is complete the resolution R is >1.5.

Paired-ion chromatography. Sometimes attaining the required selectivity will require the use of a *paired-ion reagent*. Basic antimicrobials such as aminoglycosides or trimethoprim are best separated by *paired-ion chromatography*. In paired-ion chromatography a hydrophobic *counter-ion* (the paired-ion reagent) is added to the mobile phase. This counter-ion has affinity for the stationary phase by virtue of a large hydrophobic group. Counter-ions can be acids or bases. Acidic counter-ion reagents include: pentane-sulphonic acid, hexane-sulphonic acid, heptane-sulphonic acid and octane-sulphonic acid. These are used in the separation of basic molecules. Basic counter-ion reagents include: tetramethylammonium chloride, tetraethylammonium chloride, tetrabutyl-ammonium hydroxide and hexadecyltrimethylammonium bromide. These are sometimes used in the separation of acidic drugs.

The mobile phase composition is adjusted so that both the solute of interest and the counter-ion are extensively ionized. Concentration-dependent equilibria are then formed as below:

(i) For a basic drug:

$$\text{drug}^+ + \text{counter-ion}^- \rightarrow \text{counter-ion}^{-+}\text{drug}$$

(positive ion)　(negative ion)　(neutral ion pair)

(ii) For an acidic drug:

$$\text{drug}^- + \text{counter-ion}^+ \rightarrow \text{counter-ion}^{+-}\text{drug}$$

(negative ion)　(positive ion)　(neutral ion pair)

Thus the ionized drug which would normally show no retention on an RPC column is retained because the neutral ion pair is retained by virtue of the affinity of the counter-ion portion of the ion pair. Since the whole process is dynamic, precise control of retention is possible using paired-ion chromatography. For example, if a mobile phase containing pentane sulphonate gave a retention time of x min for a given solute and the same mobile phase with the pentane sulphonate replaced by the same concentration of heptane sulphonate ion gave a longer retention time of y min, then by mixing the two counter-ions in appropriate proportions any retention time between x and y min could be produced.

Sample preparation

Using RPC, many biological samples can be analysed with minimal sample preparation. Any particulate solid matter *must* be removed by centrifugation or filtration, since this this would otherwise clog the column inlet eventually. Most published methods also require biological samples to be deproteinated since proteins may precipitate in mobile phase containing >30% methanol, or bind irreversibly to, and denature, the stationary phase.

The simplest method of removing proteins is by precipitation with either organic solvents (acetonitrile, methanol), organic acids (perchloric acid, trichloroacetic acid), or, less frequently, ultrafiltration or solid-phase extraction.

Acetonitrile mixed 50:50 with serum produces a rapidly formed, easily sedimented precipitate, and is probably the simplest most effective form of sample preparation. Kees (1986) preferred acetonitrile to methanol for β-lactams indicating that β-lactam esters may form in methanolic solution. Unfortunately, acetonitrile precipitation will not be suitable for all analytes and will sometimes impair the chromatography or cause analyte loss or degradation.

Solid-phase extraction procedures involve the initial use of a short, disposable colums such as Bond-Elut (Varian) or Sep-Pak (Waters) and work on the same principle as HPLC. The sample is absorbed on to the extraction column and then eluted with a suitable eluant. Solid-phase sample preparation procedures can be used both to deproteinate and to concentrate a sample.

The most suitable sample preparation procedure is determined during method development. If large numbers of samples are to be assayed in batches then it is essential to confirm that the analyte is stable for a suitable time period after sample preparation. Organic acids should therefore be avoided when dealing with acid-sensitive drugs. Not withstanding this, however, if samples are being processed rapidly, using an organic acid it may be possible to obtain acceptable results even with with an acid-sensitive drug. Performance-related aspects of sample procedures are discussed in *Chapter 7*.

Quantification

Like all the assays discussed in this book, HPLC assays require calibration using appropriate calibrator solutions and appropriate curve fitting procedures. Ideally a series of drug solutions prepared to calibrate a validated HPLC assay will produce a linear standard curve with an intercept of zero and a doubling of concentration producing a doubling of peak height and area (concentration plotted against peak height or peak area). This may not always be exactly the case in practice, and the subject of HPLC assay calibration is discussed in detail by Philips *et al.* (1990). Simple linear regression with the curve forced through zero usually produces an acceptable curve fit in most situations. This type of calibration can be performed manually using graph paper, or most spreadsheet software packages, or automatically using a computing integrator. Nonetheless, as pointed out by Philips *et al.* (1990), whilst forcing the regression curve through zero avoids problems with low concentrations, such as the apparent association of a low concentration result (or even a negative concentration) with a peak height of zero, it may produce systematic inaccuracies and a distorted limit of quantification.

Once acceptable linearity is obtained, it may be possible in many situations (for example clinical assays) to obtain results of acceptable accuracy using a single calibrator of an appropriate concentration (*single point calibration*). Then, making the assumption that the curve goes through zero on both axes (i.e. no peak = no drug), the concentration (c) in the sample is:

$$c = C(R_{\text{sample}}/R_{\text{cal}})$$

Where C is the concentration of the calibrator and R_{sample} and R_{cal} are the responses (peak height or peak area) obtained in the sample and calibrator chromatograms respectively.

As pointed out in *Chapter 7*, where body fluids are being assayed, recovery of drug from the sample must be determined during assay validation. Where recovery is substantially below 100% then calibrators must be prepared in a similar matrix to the samples and treated in an identical way.

In some instances the use of an internal standard may be

essential to give acceptable results or may produce improved accuracy and reproducibility.

Internal standards. Many published methods use internal standards. Internal standardization is *not* an alternative to the calibration procedures described above but is a means of improving the accuracy and reproducibility of a method *assuming proper calibration.* An internal standard is a chemical substance added to the samples and calibrators at some stage in the sample preparation procedure. It must have the following properties:

- It must be chemically stable.
- It must be detected using the same detector used for the analyte and must elute with an acceptable peak shape.
- It must be completely separated from all other compounds in the sample but should have an appropriate, convenient, retention time.
- It must be reproducibly recovered from any sample preparation procedure.

When an internal standard is used then quantification is based not on the height or area of the analyte peak obtained from a particular sample but on the ratio of the analyte peak height or area to the internal standard peak height or area in the same chromatogram. Using this procedure any potential errors due to unpredictable dilution factors occurring during sample preparation, or any irreproducibility in injection volume are eliminated since, while they will affect absolute values for peak height or area measurements, they will not change the ratio of the analyte peak to the internal standard peak. Using an internal standard the peak height or area ratio is plotted against concentration in the calibration curve. If single-point calibration is used the formula remains:

$$c = C(R_{sample}/R_{cal})$$

where C is the concentration of the calibrator and R_{sample} and R_{cal} are now the ratio of peak height or peak area of the analyte peak to the peak height or area of the internal standard peak, in the sample and calibrator chromatograms respectively.

With biological samples it should not be assumed that better results will always be obtained if an internal standard is used. Solid-phase extraction procedures probably need an internal standard because of the changes in sample volume which are involved, but simpler protein precipitation methods may not require the use of an internal standard. It must be established at assay validation that no other substances are likely to interfere (co-elute) with the internal standard as well as the analyte peak. The addition of an internal standard therefore potentially reduces assay specificity.

If assay validation has shown that acceptable performance can be obtained without an internal standard then it might prove counter-productive to add one. It does not follow that a published method using an internal standard will give better results than a method not using one. On the other hand, a published method that involves a long complex extraction procedure without an internal standard should be viewed with suspicion.

Performance expectations

HPLC assay methods are either developed in-house or are attempted reproductions of published methods. Often it is easier to develop a method from scratch, using published methods as a general guide, than reproduce exactly a published method. Whatever approach is used the method should be thoroughly validated and characterized before it is commissioned (see *Chapter 7*). Problems in attempting to reproduce published methods may arise if:

- The published method was not very robust but the authors failed to mention this. This may be particularly true if specialized 'home-made' equipment was involved.
- An essential piece of information was omitted from the published method.
- There was a typographical error in the published method.
- The published method gave inadequate detail. For example, it merely stated that a C_{18} column was used.
- The method was published some time ago and the stated stationary phase now has subtly different performance characteristics.
- A different make of column from that originally stated is (by necessity) being used.
- The equipment in the laboratory has inferior performance to that used in the published method. This is of particular importance with fluorescence and electrochemical detectors.
- The chemicals being used are not of a sufficiently high grade of purity.
- The internal standard used is not commercially available.
- The types of samples assayed in the originally published method are different from the types of sample to be assayed. For example, a published method devised to assay a drug in serum from healthy volunteers will not necessarily be suitable for urine or serum from patients since other interfering compounds (for example, other drugs) may be encountered. As a general rule of thumb the increasing likelihood of encountering specificity problems: is normal serum < urine < uraemic serum < patient's serum < uraemic patient's serum < seriously ill patient's serum.

The laboratory worker should not be afraid to make changes to a published method if they appear to be necessary for the separation to be achieved sucessfully, bearing in mind that the amended procedure must be fully validated when it is finalized. However, sometimes it may be necessary to reproduce exactly an official method; in which case amendments and alterations are not allowed. In some instances attempts to reproduce exactly a defined procedure have failed due to apparently trivial differences between instruments which turned out to have an un-expectedly important effect on assay performance. When attempting to reproduce an established assay procedure exactly, it may be necessary to invest in some appropriate new equipment.

Cost and performance issues

Since samples must be assayed sequentially by HPLC, a rapid turnaround can only be obtained if the work load is modest. This is of little concern for pharmacokinetic studies with chemically stable drugs, where samples can be stored and assayed as batches. Where assays are being used for therapeutic drug monitoring, however, this presents a problem since most antimicrobial assay results are needed within 24 h or less if they are to be of any use in the management of the patient. This requirement for rapid turnaround is one of the reasons for the almost universal use of immunoassays rather than HPLC for therapeutic monitoring of aminoglycosides and glycopeptides.

Most HPLC methods use only 10–100 μL of sample and could therefore be used with samples from neonates or for other small samples. Although it is theoretically possible to assay almost any antimicrobial on a single C_{18} column using an appropriate mobile phase, changing from one mobile phase to another can be time-consuming, making HPLC unsuitable for rapid changes of method at short notice. Also packing materials may respond unfavourably to frequent changes in mobile phase composition. For example, it is usually said that a column that has been used for paired-ion chromatography will behave differently when subsequently used for a non-ion-pair method. This limitation can only be overcome by reserving individual columns for individual drugs; this requires investment in additional columns.

Once the capital equipment has been purchased, the major costs are new columns and maintenance/repair. Mobile phase constituents are relatively inexpensive and here costs can be minimized by recycling mobile phase from the detector outlet back to the original reservoir several times rather than to waste. This practice may improve the performance of some assays. For example, if the mobile phase contains a minor impurity which slowly binds to and inactivates the stationary phase, recycling will counteract this since all the impurity be removed during the first cycle.

If a single HPLC column analyses only 100 samples before it requires replacement, the costs still compare favourably with the cost of immunoassays. Hovever, despite high reagent costs, the vast majority of laboratories will continue to use immunoassays (where they are com-

mercially available) for those assays where an analysis is available, the number of samples is high and a rapid turnaround time is essential.

References

Acton, W. J., van Duyn, O. M., Alled, L. V. & Flournoy, D. J. (1982). Evaluation of four assay methods for determination of tobramycin in human serum. *Clinical Chemistry* **28**, 177–80.

Delaney, C. J., Opheim, K. E., Smith A. L. & Plorde, J. J. (1982). Performance characteristics of bioassay, radioenzymic assay, homogeneous enzyme immunoassay, and high-performance liquid chromatographic determination of serum gentamicin. *Antimicrobial Agents and Chemotherapy* **21**, 19–25.

Edberg, S. C. & Sabath, L. D. (1980). Determination of antibiotics levels in body fluids: techniques and significance. In *Antibiotics in Laboratory Medicine* (Lorian, V., Ed.), pp. 206–64. Williams & Wilkins, Baltimore, MD.

Glew, R. H. & Pavuk, R. A. (1985). Comparison of the Beckman Auto ICS and the Syva Autolab 6000 for determination of gentamicin levels in serum. *Journal of Clinical Microbiology* **21**, 8–11.

Hagestam, I. H. & Pinkerton, T. C. (1985). Internal surface reversed-phase silica supports for liquid chromatography. *Analytical Chemistry* **57**, 1757–63.

Hewitt, W. & Vincent, S. (1988). *Theory and Application of Microbiological Assay*. Academic Press, London.

Jenkins, G. M. & Kawamura, K. (1971). Structure of glassy carbon. *Nature* **231**, 175–6.

Kees, F. (1986). Preparation for HPLC of samples in biological matrices, and problems of degradation. In *High Performance Liquid Chromatography in Medical Microbiology* (Reeves, D. S. & Ullmann, U., Eds), pp. 7–19. Gustav Fischer Verlag, Stuttgart.

Marks, V. (1981). Immunoassays for drugs. In *Therapeutic Drug Monitoring* (Richens, A. & Marks, V, Eds), pp. 155–82. Churchill Livingstone, Edinburgh.

Masterton, R. G., Tettmar, R. E., Strike, P. W. & Williams, S. (1983). An evaluation of gentamicin EMIT—its performance on the Kem-O-Mat and its role in the small laboratory. *Journal of Clinical Pathology* **36**, 1241–5.

Nilsson, L. (1984). Correlation of bioluminescent assay of gentamicin in serum with agar diffusion assay, latex agglutination inhibition card test, enzyme immunoassay, and fluorescence immunoassay. *Journal of Clinical Microbiology* **20**, 296–9.

O'Leary, T. D., Ratcliff, R. M. & Geary, T. D. (1980). Evaluation of an enzyme immunoassay for serum gentamicin. *Antimicrobial Agents and Chemotherapy* **17**, 776–8.

Oliver, G. C., Parker, B. M., Brasfield, D. L. & Parker, C. W. (1968). The measurement of digoxin in human serum by radio-immunoassay. *Journal of Clinical Investigation* **47**, 1035–42.

Phillips, L. J., Alexander, J. & Hill, H. M. (1990) Quantitative characterization of analytical methods. In *Analysis for Drugs and Metabolites, Including Anti-infective Agents* (Reid, E. & Wilson, I. D., Eds), pp. 23–36. Royal Society of Chemistry, Cambridge.

Ratcliff, R. M., Mirelli, C., Moran, E., O'Leary, D. & White, R. (1981). Comparison of five methods for the assay of serum gentamicin. *Antimicrobial Agents and Chemotherapy* **19**, 508–12.

Rather, P. N., Shaw, K. J., Hare, R. S. & Miller, G. H. (1993). Nomenclature of aminoglycoside resistance genes. *Antimicrobial Agents and Chemotherapy* **37**, 928.

Redmond, L., Chang, A. & Lynch, M. (1985). Gentamicin and tobramycin EMIT assays in the CentrifiChem 500. *Clinical Chemistry* **31**, 1408–9.

Standiford, H. C., Bernstein, D., Nipper, H. C., Caplan, E., Tatem, B., Hall, J. S. *et al.* (1981). Latex agglutination inhibition card test for gentamicin assay: clinical evaluation and comparison with radioimmunoassay and bioassay. *Antimicrobial Agents and Chemotherapy* **19**, 620–4.

Stobberingh, E. E., Houben, A. W. & van Boven, C. P. (1982). Comparison of different tobramycin assays. *Journal of Clinical Microbiology* **15**, 795–801.

van de Klundert, J. A. M. & Vliegenthart, J. S. (1993). Nomenclature of aminoglycoside resistance genes: a comment. *Antimicrobial Agents and Chemotherapy* **37**, 927–8.

Weber, G. (1953). Rotational Brownian motion and polarization of the fluorescence of solutions. *Advances in Protein Chemistry* **8**, 415–59.

White, L. O. (1986). The development of microbiological and immunological assays for antibiotics. *Trends in Analytical Chemistry* **5**, 29–31.

White, L. O. & Reeves, D. S. (1981). Antibiotics: analytical techniques. In *Therapeutic Drug Monitoring* (Richens, A. & Marks, V., Eds), pp. 457–70. Churchill Livingstone, Edinburgh.

White, L. O., Scammell, L. M. & Reeves, D. S. (1980). An evaluation of a gentamicin fluoroimmunoassay kit: correlation with radioimmunoassay, acetyl transferase and microbiological assay. *Journal of Antimicrobial Chemotherapy* **6**, 267–73.

White, L. O., Scammell, L. M. & Reeves, D. S. (1981). Serum aminoglycoside assay and enzyme-mediated immunoassay (EMIT); correlation with radioimmunoassay, fluoroimmunoassay, acetyl transferase and microbiological assay. *Antimicrobial Agents and Chemotherapy* **19**, 1064–6.

White, L. O., Lovering, A. M. & Reeves, D. S. (1983a). Variation in gentamicin C1, C1a, C2 and C2a content of some preparations of gentamicin sulphate used clinically as determined by high-performance liquid chromatography. *Therapeutic Drug Monitoring* **5**, 123 6.

White, L. O., Bywater, M. J. & Reeves, D. S. (1983b). Assay of netilmicin in serum by substrate labelled fluoroimmunoassay. *Journal of Antimicrobial Chemotherapy* **12**, 403–6.

White, L. O., MacGowan, A. P., Lovering, A. M., Holt, H. A., Reeves, D. S. & Ryder, D. (1994). Assay of trough serum gentamicin concentrations by fluorescence polarization immunoassay. *Journal of Antimicrobial Chemotherapy* **33**, 1068–70.

Witebsky, F. G., Silva, C. A., Selepak, S. T., Ruddel, M. E., MacLowry, J. D., Johnson, E. E. *et al.* (1983). Evaluation of four gentamicin and tobramycin assay procedures for clinical laboratories. *Journal of Clinical Microbiology* **18**, 890–4.

Chapter 6

The assay of antimicrobials in tissues and fluids

Jenny M. Andrews

Department of Medical Microbiology, City Hospital, Birmingham B18 7QH, UK

Introduction

It is the general belief that the measurement of anti-microbial concentrations at potential sites of infection will be an indication of the efficacy of particular drugs in particular infections. Data on site concentration are often used in conjunction with MICs to predict the outcome of antimicrobial therapy and also as guidelines for selecting in-vitro breakpoint concentrations which are necessary for routine susceptibility testing. There are, however, few data correlating site concentrations, in-vitro activity and clinical efficacy. In fact, although some studies have shown apparent accumulation of active drug at the site of infection, clinical response has been poor. It has been suggested by Schentag (1989) and Nix *et al.* (1991a) that inappropriate study design may be responsible for this, including (i) assuming even distribution of antimicrobial throughout tissue; (ii) interpreting outcome using whole tissue homogenates rather than concentration in extra-cellular fluid (ECF) (the possible site of infection); and (iii) sampling when an equilibrium does not exist between blood and ECF.

It is my intention in this chapter to highlight the problems associated with the assay of antimicrobials in tissues and fluids and also to suggest methods which will minimize them.

The sequences of two approaches to the handling of tissue samples are shown in the Figure. In Approach A, a weighed amount of tissue is homogenized in a known volume of fluid and the concentration of antimicrobial is usually expressed as mg/kg of tissue. Using this approach the result assumes even distribution of antimicrobial throughout the tissue (which may be true for quinolones and macrolides but not for β-lactams or aminoglycosides) and does not differentiate between intracellular and ECF antimicrobial concentrations. β-Lactam antibiotics do not penetrate eukaryotic cells and are located almost exclusively in the extracellular space, so this method of calculation will underestimate the concentration of β-lactam at the site of infection (extracellular bacteria located in the ECF, rather than evenly distributed in tissue), because the ECF, which contains the anti-microbial, is diluted by the cell mass, which does not. Since approximately 10–20% (by weight or volume) of most

tissue is composed of ECF (Barza, 1990), a more meaningful result would be obtained after a correction had been made for this dilution of site concentration (ECF) by tissue mass.

In Approach B an attempt is made to measure concentrations of antimicrobial in unmanipulated tissue (Cars, 1981). Whole tissue pieces (10–30 g) are placed in a 4 mm well cut in an agar plate. The well is subsequently filled with molten agar to ensure complete contact between the tissue and the agar in the assay plate. Using this approach there is no homogenization of the tissue and therefore the antimicrobial is prevented from coming into contact with tissue or cellular structures or contents with which it would not normally be associated *in vivo*, avoiding errors due to possible inactivation or non-specific binding. This approach also circumvents the problems associated with loss of antimicrobial activity during any extraction or

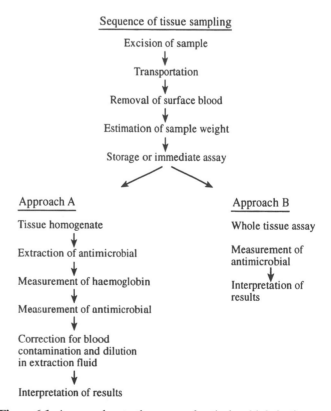

Figure 6.1 Approaches to the assay of antimicrobials in tissues.

homogenization process and may be suitable for very labile compounds.

A third approach has used freeze drying to remove the moisture which makes up approximately 20% of the dry weight of tissue samples. After drying samples are crushed to a powder, resuspended in buffer and then assayed (Thurmann-Nielson *et al.*, 1990).

Problems encountered in the sequence of tissue handling

Excision of sample

Ideally the chosen sample should only represent one type of tissue and be free from contamination with other tissues or body fluids. In the study of Philpott *et al.* (1991), histological examination of frozen sections of prostatic tissue was used to separate stroma and epithelium before assay. Other investigators have trimmed away connective tissue (Kinzig *et al.*, 1992). Attempts have also been made to identify the composition of bronchial biopsy tissue used in pharmacokinetic studies (Baldwin *et al.*, 1992). In order to reduce the amount of capillary blood within the tissue before excision, ligatures have been applied to blood supplying vessels to allow blood to flow out of the prospective sample area (Bouvet *et al.*, 1992). Removal of superficial blood from the surface of tissue is usually achieved by blotting, or immediate rinsing followed by blotting (Kinzig *et al.*, 1992; Shyu *et al.*, 1993) Another source of contamination is the presence of urine in prostatic tissue. Contamination with urine can be avoided if samples are collected at transurethral resection using continuous intraoperative irrigation with 1.5% glycine, simultaneous suction and low intravesical pressure (Hoogkamp-Korstanje *et al.*, 1984). The presence of bile within gall bladder tissue is another source of contamination.

Transportation

As soon as tissue is removed, water is lost through evaporation. With smaller pieces of tissue as much as 20% of the original mass can be lost in the first 15–20 min, resulting in an overestimation of the true antimicrobial concentration within the tissue (Cars & Ogren, 1985). Humidity chambers have been used to prevent this loss of moisture (Honeybourne *et al.*, 1988). It may also be necessary to 'snap freeze' samples in liquid nitrogen if particularly labile compounds are to be assayed (Brouwer *et al.*, 1984).

Removal of surface blood

If it is not possible to remove superficial blood as soon as the tissue is taken in theatre, then it should be removed as soon as possible on receipt in the laboratory. Methods for removal are detailed above. Residual blood contamin-

ation can be determined by measuring haemoglobin concentration in the tissue homogenate. If tissue samples weigh less than 500 mg then it is probable that there will be insufficient homogenate for both assay and determination of blood contamination by measurement of haemoglobin content. In this situation heavily blood-contaminated samples may have to be discarded.

Sample weight

The weight of tissue samples may be simply determined by collecting them in pre-weighed bottles and then re-weighing the bottles. Surface contamination (blood, urine, bile, etc.) should of course be removed before the sample is weighed.

Sample storage

As mentioned above, desiccation of tissue samples before weighing will lead to overestimation of the true antimicrobial concentrations. The thermal instability of β-lactams and the possibility that enzymic breakdown of antimicrobial may occur have resulted in it becoming common practice to store tissue samples at −20°C or below. However, tissue samples will continue to desiccate at low temperatures and therefore it is essential either that weights are recorded before storage or that samples are stored in containers with small air spaces. Alternatively samples may be suspended in known volumes of liquid before storage (Davey *et al.*, 1991). It must be remembered that some antimicrobials may continue to be metabolized in tissues during storage. This may result in erroneous measurements of parent:metabolite ratios.

Optimally, where feasible, tissue should not be stored and assays should be performed as soon as possible after the tissue has been excised.

Sample processing

There are many methods of dealing with tissue samples before assay. Tissues have been combined with aqueous solutions or other matrices to form homogeneous suspensions which are suitable for microbiological assay or for further sample preparation in the case of non-microbiological assays. It is the intention of all processes used to recover as much of the antimicrobial as possible. Initial sample treatment usually involves combining known weights of tissue with known volumes of fluid followed by homogenization (Brouwer *et al.*, 1984; Benoni *et al.*, 1987; Thurman-Nielsen *et al.*, 1990). Various fluids have been used, including sterile distilled water (Muller-Serieys *et al.*, 1992) and buffer (Brouwer *et al.*, 1984; Benoni *et al.*, 1987; Thurman-Nielsen *et al.*, 1990). However, for compounds that are only partially recovered from tissues by conventional methods, for example the aminoglycosides, it has been necessary to use either sodium hydroxide to solubilize protein or trichloroacetic acid to precipitate protein, to ensure complete recovery (Fox, 1989). Usually all manipulations of the tissue sample are performed on ice to prevent or minimize thermal degradation; this is particularly important if very labile antimicrobials are under investigation. The devices usually used to disrupt tissue are either ultrasonicators or mechanical homogenizers. A microhomogenizer has been developed which minimizes the loss of homogenate during centrifugation since the whole unit is centrifuged (Hearse, 1984). UK suppliers of such devices include Life Sciences Ltd (Sedgewick Road, Luton, UK) and Merck BDH (Lutterworth, UK). Following homogenization a clear supernatant for assay is usually obtained by centrifugation (Philpott et al., 1991; Bouvet et al., 1992; Kinzig et al., 1992). For more resilient tissues, enzymes have also been used to disrupt tissue (Waldren et al., 1986). Drug inactivation during the tissue processing procedure is a real possibility and therefore controls of tissue, spiked with antimicrobial, should also be examined during method validation. Such spiked controls should also be used to determine the recovery of antimicrobial from the tissue homogenate.

Estimation of blood contamination

Blood present in the capilliaries of the tissue and the superficial blood contaminating the surface of the tissue will also be incorporated in the homogenate. The presence of blood in the tissue sample may affect the result obtained, depending on the antimicrobial. If an antimicrobial is equally distributed between tissues, cells, plasma and red blood cells, then the presence of blood will have little effect on the result. However, if an antimicrobial is concentrated within erythrocytes or phagocytic cells, as is the case with sulphonamides and azalides respectively, then the presence of blood could falsely elevate the tissue concentration. Conversely, with antimicrobials that do not penetrate cells, the presence of blood could result in an underestimation of the true tissue concentration.

If it is not possible to measure directly the amount of blood present in a tissue sample, it may be possible to infer blood volumes from assays performed on other individuals under similar conditions (Cars & Ögren, 1985). A simple indirect method for measuring blood volume has been employed also. Here the initial tissue concentration following an iv injection, before distribution of drug has occurred, is compared with that of the simultaneous serum concentration (Cars & Ögren, 1981). The percentage of blood in the tissue sample (assuming no distribution from blood to tissue) is $(C_{tissue}/C_{blood}) \times 100$, where C_{tissue} is the tissue concentration and C_{blood} the concentration in blood.

The amount of blood present in tissues has also been measured by estimation of haemoglobin content, radio-labelled albumin, and number of erythrocytes; the most commonly used marker for determining blood content is haemoglobin (Lowry & Hastings, 1942; Cars & Ögren, 1985) (see below). The amount of blood present in most tissue masses is estimated to be between 3 and 6% (Nix et al., 1991a).

Measurement of antimicrobial concentration

The major problems that have to be addressed when assaying concentrations of antimicrobials in tissue (apart from drug stability and correction for blood content, which have already been dealt with) are the smallest amount of drug which can successfully be measured (lower limit of detection or lower limit of quantification) and the medium that should be used for the preparation of the assay calibrators. With regard to the sensitivity of the assay, it is essential to consider the lower limit of detection (LLD) of the assay and the theoretical penetration into tissue (whole tissue homogenates) when compared with serum. Using arbitrary serum levels, the LLD and expected tissue weights, it is possible to predict whether an assay will have an adequate detection limit and thus be suitable for use. This is especially important when assaying small biopsy samples which weigh <100 mg. An example of how to predict the smallest amount of drug that can be measured for varying weights of tissue is given below.

The formula of Honeybourne et al. (1988) derives the concentration (C, in mg/kg) of antimicrobial in a small biopsy sample of known weight (W, in mg) homogenized in a known volume (V_B, in μL) of buffer:

$$C = C_A \times (1000/W) \times ((V_B + W)/1000)$$

where C_A is the assayed concentration in mg/L. Then if C_A = LLD = 0.05 mg/L, W = 7 mg and V_B = 200 μL,

$$C = (LLD \times (W + V_B))/W \text{ mg/kg}$$

$$= (0.05 \times (200 + 7))/7 \text{ mg/kg}$$

$$= 1.4 \text{ mg/kg}$$

Table I shows hypothetical concentrations of a drug in small tissue samples. As all of the measurable amounts of antimicrobial shown in this example are very close to the LLD of the assay, it would be prudent to redesign this study to ensure that the tissue weight was in excess of 7 mg or reduce the volume of buffer added to the tissue sample to the minimum practically possible.

It is generally agreed that because antimicrobials bind to different degrees depending on tissue type (Nix et al., 1991a), assay calibrators should be prepared in a homogenate of a drug-free sample of the same tissue as that under investigation (Cars & Ögren, 1985). If possible, normal antimicrobial-free tissue should be obtained at operation or, if this is not possible, then post-mortem tissue obtained within 24 h of death may be used (Lorian, 1980). Appropriate amounts of antimicrobial solution should be added to antimicrobial-free tissue before

Table I. Hypothetical concentrations to be assayed when processing small tissue samples. Lower limit of detection of the assay taken as 0.05 mg/L

Time after dose (h)	Concentration (mg/L)		
	serum	in tissue assuming 200% penetration	in homogenate[a]
2	0.79	1.58	0.05
4	0.73	1.46	0.05
6	1.1	2.2	0.07
8	0.9	1.8	0.06
12	0.78	1.56	0.05
24	0.43	0.86	(0.03)[b]

[a]In 7 mg of tissue homogenized in 200 μL of buffer
[b]Below lower limit of detection of the assay.

homogenizing and then these calibrators should be treated in exactly the same way as the test tissues. This procedure for the preparation of calibrators is recommended since addition of antimicrobial solution after homogenization may underestimate the total binding to whole-tissue components.

In the case of small biopsy samples it is often impracticable to obtain sufficient supplies of antimicrobial-free tissue and therefore samples are sometimes homogenized in buffer and assayed using calibrators prepared in an identical buffer (Honeybourne et al., 1988). If it is not possible routinely to prepare calibrators in antimicrobial-free tissue, it is essential to demonstrate at assay validation that recovery from the tissue sample does not vary with concentration (binding may be different with varying volumes of buffer added to a fixed amount of tissue). If recovery proves to vary with concentration it may be necessary to find a tissue 'substitute', with a similar composition to the test tissue. In addition to determining recovery it is also essential to demonstrate that antimicrobial-free tissue extracts have no inhibitory effect in the assay.

Correction for blood contamination by haemoglobin estimation

As already mentioned, the marker usually used to determine the blood content of tissue samples is haemoglobin. The method which is much favoured for its simplicity is that of Lowry & Hastings (1942). The formula for determining a factor (x) for adjusting for blood is:

$$x = (C_S \times Hb_T)/Hb_{WB}$$

Where C_S is the serum concentration of antimicrobial at the time the tissue was sampled (mg/L), and Hb_T and Hb_{WB} are the haemoglobin contents of the tissue and

whole blood respectively. The true tissue concentration adjusted for the presence of blood (C_{at}, in mg/kg) is then derived from the equation $C_{at} = C_t - x$, where C_{at} is the measured tissue concentration (mg/L) and x is the factor calculated above.

Interpretation of results

Tissue concentrations of drug are usually expressed as mg/kg of tissue (see above), assuming that the specific gravity of the tissue is 1.0. The true tissue concentration (C_{at}) is then used to determine the tissue to serum ratio (TSR) from the equation TSR $= C_{at}/C_s$.

These procedures express the concentration of drug within the tissue homogenate and do not give any indication of where the drug may have been located within that tissue. It has been suggested by Schentag (1989) that tissue concentration data should take into consideration the distribution of antimicrobials within the blood and the extracellular and intracellular fluids. Nix et al. (1991a) have suggested that data from tissue homogenates should be analysed looking at the distribution of the drug throughout the tissue in the interstitial blood and extracellular and intracellular fluid.

Assays in small (<100 mg) tissue samples

Homogenization procedure

1. Keep the excised tissue samples in a humidifier before weighing to prevent loss of moisture by evaporation (loss of moisture results in an overestimation of the true antimicrobial concentration).

2. Weigh each tissue in a pre-weighed glass tube (plastic tubes may not be suitable for use with ultrasonicators), discarding any heavily bloodstained tissue.

3. Predict the smallest amount of antimicrobial which can be measured for varying weights of tissue and volumes of diluent (see above).

4. Add chilled diluent (a measured volume determined by the weight of the tissue obtained at biopsy).

5. Homogenize the tissue by ultrasonication on ice (for example using an Ultrasonics Heat Systems W225 (Science Laboratories Ltd, Luton, UK), 50% duty cycle pulsed) for 1 min.

6. Preferably assay immediately (if a microbiological assay is to be performed) or process further as necessary if a non-biological assay method is to be used.

Assay validation

When sufficient quantities of antimicrobial-free tissue are not likely to be available for preparing repeated sets of calibrators, preliminary recovery studies should be performed, using what drug-free tissue is available, to identify an alternative suitable matrix.

1. Prepare calibrators by weighing appropriate material (antimicrobial reference powder or solution) into both drug-free tissue homogenate and phosphate buffer.

2. 'Spike' some drug-free tissues to give a range of concentrations within and just above the range of the calibrators.

3. Process both sets of calibrators and the spiked samples using the normal procedure.

4. Determine recovery (results for spiked tissues read against buffer calibrators). The percentage recovery is the assayed concentration divided by the assigned concentration and multiplied by 100. If the recovery exceeds 90% and is reproducible, the procedure is acceptable. Check that recovery is not concentration-dependent and is linear (i.e. concentrations above the top calibrator are accurately assayed when the sample is diluted).

5. If recovery is poor the stability of the drug during the procedure should be checked by comparing results obtained when samples in phosphate buffer are assayed before and after exposure to the procedure. Gross loss of antimicrobial may be due to overheating during ultrasonication and for this reason ultrasonication is always performed on ice.

6. Compare the results obtained for spiked samples using calibrators prepared in tissue extract and tissue-free diluent. The results should agree within ±10%. If the results are outside this limit then calibrators prepared in tissue-free diluent cannot be used and an alternative matrix (e.g. some other easily obtained tissue homogenate) will need to be evaluated.

7. If using a microbiological assay ensure that antimicrobial-free tissue does not inhibit the indicator organism.

8. Check whether the antimicrobial is degraded by enzymes present in tissue extracts; for example, enzymes present in red blood cells are known to breakdown cefotaxime and streptogramins.

9. Check that the drug is not susceptible to light degradation.

Calculation of concentration in tissue

Because with small tissue samples it is often not possible to measure the amount of blood in the tissue extract, no correction for blood can be made. The concentration of antimicrobial (C_t) in tissue in mg/kg is calculated from the equation:

$$C_t = (C_A \times (V_B + W))/W$$

Where C_A is the assayed concentration (in mg/L), V_B is the volume of buffer added to tissue (µL) and W is the weight of biopsy (mg).

Assay of antimicrobials in tissues other than small samples

Sample processing

1. Ensure that the sample is transported in a humidifier to prevent loss of moisture from the sample.

2. Examine the tissue for the presence of superficial blood contamination. If blood is present, either (i) rinse the surface very quickly with buffer and then dry with gauze or (ii) remove as much blood as possible by trimming away any blood-stained tissue. Note, as stated previously, that the samples should only represent one type of tissue.

3. Mix equal amounts of tissue and chilled diluent (w/v assuming 1 g of tissue is equivalent to 1 mL of diluent) and homogenize on ice (Ultraturex TP/18, Janke & Kunkel, GMbH & Co. Kg, Staufen, Sweden) for approximately 2 min (this time depends on the resilience of the tissue to be homogenized).

4. Centrifuge at 2000**g** at 4°C for 10 min.

5. Dilute the supernatant further, if necessary, in diluent.

6. Assay as quickly as possible. Store at +4°C in the dark until assayed or process further as necessary if a non-microbiological assay is to be performed.

Preparation of calibrators

1. Obtain antimicrobial-free normal tissue. Remove superficial blood as described above.

2. Trim tissue into seven 1 g pieces. Place in seven separate glass containers and label 1–7.

3. Spike each sample with 1 mL of appropriate antimicrobial solutions to prepare appropriate calibrators and controls (e.g. five calibrators at doubling concentrations of 0.125, 0.25, 0.5, 1 and 2 mg/L and controls of 1.5 and 0.2 mg/L) (see Table II).

4. Extract as described above. Assay the clear supernatant, the extracted calibrators and internal controls.

The following points should be noted:

• If binding occurs to the antimicrobial-free normal tissue used for the preparation of calibrators, then the lower limit of detection of the assay may be reduced.

• If long-term storage of tissue before assay is envisaged, ensure that the tissue does not become dehydrated thus concentrating the antimicrobial.

• If prostate tissue is to be assayed, the procedure for obtaining the tissue sample should be designed to prevent contamination with urine.

The following studies should be performed to confirm assay validity:

1. Check that the antimicrobial-free supernatant (harvested after centrifugation from antimicrobial-free control

Table II. Preparation of calibrators for assay of tissue samples by mixing drug-free tissue with 1 mL of antimicrobial solution

Code no.	Weight of tissue (g)	Antimicrobial solution concentration (mg/L)	Final concentration (mg/L) 1:1 (w/v) dilution
1	1	2	1
2	1	1	0.5
3	1	0.5	0.25
4	1	0.25	0.13
5	1	0.13	0.06
6	1	1.5 (control)	0.75
7	1	0.2 (control)	0.1

tissue) has no inhibitory effect on the indicator organism.

2. Check the stability of the antimicrobial during the homogenizing and centrifugation procedures as described below:

   (i) Take some tissue known to contain an unknown concentration (y mg/kg) of antimicrobial and divide it into six roughly equal pieces.

   (ii) To three of these add additional antimicrobial to a known concentration (x mg/kg). Add nothing to the other three.

   (iii) Process and assay all six samples.

The amount of antimicrobial lost during extraction is calculated as follows. The antimicrobial concentration measured in the three spiked tissues, z_1, is equal to $y + x$ mg/kg. The concentration measured in the unspiked samples, z_2, is equal to y. If there is 100% recovery then $z_2 + x = z_1$. If $z_2 + x > z_2$ then the difference is the amount of antimicrobial lost in the procedure.

Calculation of results

When calculating the true tissue concentrations the correction for blood contamination as described earlier should be applied.

Assay of intracellular concentrations of antimicrobial

Bronchoalveolar lavage (BAL), a standard procedure used for research and diagnostic purposes, involves instilling successive volumes of sterile 0.9% saline down the working channel of a bronchoscope followed by immediate gentle aspiration. The first aliquot is discarded to avoid contamination of the sample with fluid and cells from the larger airways; the remaining aspirations are pooled for analysis. To minimize the number of macrophages lost by adherence to collection containers, Teflon tubes and silanized glassware should be used throughout. After removal

of a small amount of the sample for cell counting and differential microscopy, the remainder of the sample is centrifuged immediately to separate the cell pellet, containing macrophages, from the supernatant-containing epithelial lining fluid. After ultrasonication of the pellet, the concentration of antimicrobial associated with the macrophages can be assayed (see below). The result obtained is a measure of the total macrophage-associated drug and does not identify its intracellular location (the drug may be bound to the membrane, cytoplasmic or nuclear components or lysosomes). Although high concentrations of cell-associated drug may be found, it does not necessarily imply that the drug is located at the site where the organism is found. A method identifying the precise location of the intracellular site of a drug may be a better predictor of efficacy, because the environment may have a marked effect on the efficacy of the drug; for example, the efficacy of quinolones, macrolides and aminoglycosides which are located in the acid environment of the lysosome may be reduced (Desnottes & Diallo, 1992; Garcia et al., 1992).

Preliminary studies

1. Determine the recovery of antimicrobial from antimicrobial-free macrophage preparations spiked with known concentrations of drug.

2. If recovery is <90%, determine if the reduced recovery is related to instability of the antimicrobial during processing or binding to cell components.

3. Determine that the recovery is linear over an antimicrobial concentration range.

4. Determine the smallest amount of antimicrobial that can be measured from the number of macrophages likely to be harvested.

Method for the assay of antimicrobials in BAL macrophages

1. Place the BAL in pre-weighed Teflon containers.

2. Determine a total and differential cell count.

3. Centrifuge at 400g for 5 min.

4. Remove the supernatant into a pre-weighed container.

5. Re-weigh the original Teflon container and record its weight. Add buffer to the sample (volume is dependent on the sensitivity of the assay).

6. Ultrasonicate on ice to disrupt the cells. (Ultrasonics Heat Systems W225, 50% duty cycle pulsed) for 2 min.

7. Assay sample or extract it further if necessary.

Calculate the results as follows. Total cell volume (TCV) = $A \times B \times V_M$, where A is the volume of lavage (mL), B is the number of macrophages/mL and V_M is the mean macrophage volume (mL). Macrophage volume may be determined by velocity-gradient centrifugation (Johnson et al., 1980) but for human BAL macrophages a working volume of 2.48 μL/10^6 cells may be used. Antimicrobial concentration in a macrophage (C_M, in mg/L) is determined from: $C_M = C_A \times D$/TCV where C_A is the assayed concentration in the macrophage homogenate (mg/L), D is the volume of macrophages and buffer (μL) and TCV is the total cell volume (μL).

Assay of antimicrobial concentrations in bone

The value of measuring concentrations of antimicrobials in bone has been questioned for many years because of the many methodological problems involved. As with other tissue samples, it is important to assay only one type of bone; for example, hard sclerotic bone and soft trabecular bone should be kept separate for assay.

Calculation of bone penetration may also be misleading when comparing by weight (mg/kg) the concentration of antimicrobial in spongy bone with that of cortical bone because of the high specific gravity of cortical bone, which results from its high mineral content. To avoid any confusion, some authors have expressed their results as mg/L bone, correcting for the density of bone being assayed (Meissner et al., 1990).

The presence of blood within the bone sample which originates from the intravascular bone compartment and also any superficial blood acquired at operation and its effect on the assay result are similar to those described previously for tissue samples. Superficial blood may be removed by rinsing in buffer (Leigh et al., 1985) or by trimming (Grimer et al., 1986), but there is concern that washing in buffer may remove antimicrobial from the bone sample. Plaue et al. (1980) calculated the error resulting from intravascular blood to be 5–10%, whereas Wittmann et al. (1980) found only 2–3% error. Nevertheless, measurement of haemoglobin is usually undertaken and it is usual to make a correction for the amount of antimicrobial present attributed to blood contamination. Another methodological problem is the uncertainty that the concentration of drug measured in the eluate is representative of the true concentration in the bone or whether binding to bone, or perhaps inactivation (Rosdahl et al., 1979) of the drug during processing results in an underestimation of bone concentration. Studies have been performed to determine the amount of drug recovered from bone samples. These may involve repeating the procedure several times and assaying the supernatants to see if any drug is detected in the later supernatants (Mertes et al., 1992), or samples may be spiked with an additional amount of the antimicrobial under investigation and the amount recovered from the spiked sample after processing is then compared with the amount recovered from a similar sample not spiked with drug (Cunha et al., 1984). For the complete recovery of quinolones from bone it has been found necessary to use multiple prolonged elutions with acidic solutions (Meissner et al., 1990).

Processing method for bone

1. If there is superficial blood on the outside of the bone, remove it by rinsing or, preferably, trimming.

2. Disrupt bone by one of the following methods. (i) Using a hammer and chisel or diamond bone-saw, trim the bone into 1 cm squares. Crush these bone squares in a pestle and mortar (Leigh et al., 1985). Alternatively, (ii) place the bone pieces into liquid nitrogen and then reduce them to a fine powder with a Spex 6700 crusher (Spex Industries, Edison, NJ, USA) (Massias et al., 1992).

3. Place the bone sample into a suitable diluent; buffer of pH 7 may be appropriate for β-lactams, while acidic solutions may be more appropriate for quinolones which are known to bind to bone (A. Dalhoff, Bayer AG, personal communication).

Preliminary studies

1. Determine the degree of binding of the drug to bone by performing recovery experiments, if possible, with normal antimicrobial-free bone.

2. Check the stability of the drug during processing and storage if prolonged storage of bone before assay is envisaged. It is important to ensure that the bone does not become desiccated during storage. To avoid this the bone may be processed as soon as possible after collection and the extracts stored for assay.

Assays in body fluids other than blood

Measurements of the concentrations of antimicrobials in extravascular sites can be made (i) from natural fluids, such as CSF, sputum, urine and peritoneal fluid or (ii) in 'models of extravascular sites' such as blisters induced chemically (by cantharides) or mechanically (by suction or dermabrasion), or surgically induced fluids (cotton threads) or implanted tissue cages (Nix et al., 1991b).

Assays in sputum

In normal individuals, airway secretions are transported from the lower respiratory tract via the cilia and then swallowed, a process which usually goes unnoticed by the individual. The product of an abnormal respiratory tree is sputum, which contains a mixture of bronchial secretions, cells, cellular debris, organisms and saliva. Bronchial secretions (from normal and abnormal airways) may also be diluted by transudated fluid and surfactant from the alveoli. Studies of antimicrobial penetration into sputum and bronchial secretions demonstrate great variation and are often difficult to interpret because of the variation in study design and assay methodology. Not withstanding the difficulties, because of the ease of collection and the relatively large amount available for analysis, many studies have measured the concentration of antimicrobials in sputum and bronchial secretions.

Sample preparation

A known volume of sputum and a known volume of diluent (buffer or mucolytic agent such as Sputolysin (Oxoid, Basingstoke, UK)) are mixed and disrupted either by ultrasonicating on ice in a glass tube (Ultrasonics Heat System w225, 50% duty cycles pulsed) or by mixing on a vortex mixer or by using a Stomacher (Lab Blender 400, Seward Medical, Blackfriars, UK). After centrifugation, the supernatant is assayed.

Preliminary studies

1. Assay calibrators should be prepared in antimicrobial-free sputum. If this is not routinely possible then check for non-specific binding to sputum by spiking a variety of antimicrobial-free sputum samples (i.e. purulent, mucopurulent, salivary) and measuring recovery after extraction against buffer calibrators.

2. If no significant binding occurs then assay calibrators may be prepared in buffer.

3. Ensure that drug-free sputum homogenate does not inhibit the assay.

4. Check the stability of the antimicrobial during processing.

The following points should be noted.

- If the antimicrobial is known to concentrate in phagocytic cells, then pus cells and macrophages may be disrupted during the homogenization, releasing antimicrobial into the homogenate. This must be taken into account when interpreting results.

- If the antimicrobial to be assayed is known to accumulate in erythrocytes, then the presence of blood in the sample may falsely elevate the apparent penetration.

- If a β-lactam is being studied, it is necessary to establish that there are no enzymes present in the homogenate which are capable of destroying the drug. For example, bacterial β-lactamase or (in the case of cefotaxime) desacetylating enzyme from blood cells.

Assays in urine

The measurement of antimicrobials in urine usually presents few problems as most drugs are concentrated in urine and samples are usually large. The major problems associated with assaying urine samples are as follows:

- Dilution of the urine sample before assay: accurate dilutions must be made using volumetric glassware (often urine needs to be diluted 1 in 50 to 1 in 500 to bring the drug concentration into the assay range).

- If a microbiological assay is being used, it must be confirmed that the drug to be assayed is not excreted in the urine as (or partially as) a microbiologically active metabolite.

- The stability of the antimicrobial in urine over long collection time intervals (usually 4 h collections for pharmacokinetic studies) may present problems. Quinolones are unstable when exposed to light and β-lactams are thermally unstable. As a precaution it is advisable to place urine collection containers in a 'cooler box' containing cooling packs or in a refrigerator to protect them from heat and light. For long-term storage of samples, freezing below –40°C is preferred.

- Antibacterial activity can also be lost as a result of the drug binding to plastic collection containers. Teicoplanin, for example, binds strongly to some plastics. It is advisable, therefore, to collect antimicrobial-free urine, and spike it with concentrations of antimicrobial likely to be present in the test samples. Store these samples under simulated test conditions and determine recovery.

Assays in epithelial lining fluid

Epithelial lining fluid (ELF) is the fluid that bathes the terminal bronchi and alveoli. Using a conventional BAL technique (Reynolds *et al.*, 1977) the median volume of lavage recovered is 62.8 mL (range 27.1–108 mL). The volume of ELF can be calculated using the urea dilution method of Rennard *et al.* (1986). This method assumes that the blood urea and ELF urea concentrations are the same and back-calculates the ELF volume from the urea concentration in the BAL fluid. Using this method the median volume of ELF is 1.52% (range 0.53–4.06%) (Baldwin *et al.*, 1991) of the BAL fluid volume. This means that there is a large dilution of the ELF in BAL. In order to measure the concentration of antimicrobial it is usually necessary to freeze-dry the lavage sample and then to reconstitute to one-tenth of its original volume before assay.

Sample preparation

1. After centrifugation of the lavage sample, remove the supernatant and place in a container suitable for freeze-drying.
2. Weigh the BAL collected to determine its volume; assume that 1 g = 1 mL.
3. Remove 1 mL of lavage for urea estimation.
4. Freeze the remainder of the sample at $-70°C$.
5. Freeze-dry the sample. Ensure that sample is protected from light during this process.
6. Reconstitute to one-tenth of the original volume with distilled water or other appropriate diluent.
7. Prepare assay calibrators in 9% sodium chloride (lavage fluid comprises 0.9% sodium chloride; after freeze drying and reconstitution with distilled water to one-tenth the original volume, the lavage sample will be equivalent to 9% sodium chloride).

Calculation of results

The concentration of antimicrobial in the ELF (C_{ELF}, in mg/L) is determined from: $C_{ELF} = C_A \times U_B/U_L$ where C_A is one-tenth the assayed concentration (in mg/L), and U_B and U_L are the concentrations (in mmol/L) of urea in the blood and the lavage, respectively. Note that controls of 0.9% sodium chloride spiked with appropriate concentrations of antimicrobial and freeze-dried with the test samples should always be included.

Assays in small volumes of body fluids

This covers assay of volumes of <100 μL uncontaminated by blood and includes such specimens as peritoneal fluid, wound exudate, joint fluid, burns fluid, etc.

In studies where very small volumes of fluid are obtained or at sites where fluid is not easily collected, 6 mm blotting-paper discs can be used to absorb fluid (Wise *et al.*, 1983) or microlitre amounts can be pipetted on to discs which are then placed on the surface of microbiological assay plates in order to measure the concentration of drug in the fluid. The volume of fluid absorbed by the disc can be calculated either by re-weighing pre-weighed discs as soon as possible after removal from the sample site or by noting the volume applied to the disc. The principle of the assay is to determine the concentration of drug on a disc containing a variable volume of sample by comparison with a calibration curve prepared from calibrators where a fixed volume is applied to the disc.

The true concentration in the fluid (C_F mg/L) is calculated from the assayed concentration C_A using the equation $C_F = C_A \times V_S/V_F$ where V_S is the volume of calibrator pipetted on to the disc (μL) and V_F is the volume of fluid absorbed or pipetted on to the disc (μL).

Procedure using a disc to absorb fluid

1. Determine the protein content of the fluid.
2. Prepare assay calibrators in a diluent containing an equivalent amount of protein (e.g. the equivalent amount of protein found in peritoneal fluid obtained from non-inflamed tissue is 20% that of serum). The concentration range of the calibrators depends on the sensitivity of assay and the expected concentrations.
3. Determine blood contamination as follows:
   (i) Prepare 1%, 2%, 3%, 4%, 5% and 10% solutions of blood in phosphate buffer pH 7 (use blood that is negative for HIV and hepatitis).
   (ii) Pipette 25 μL of each concentration on to 6 mm blotting paper discs.
   (iii) Visually compare the appearance of test discs with the discs containing the various blood concentrations. Discard any test disc visibly appearing to contain more than 10% blood.
4. Follow one of the alternative procedures below either (a) adding an unknown volume to the disc or (b) adding a measured volume to the disc.
5. Carry out either step (a) or step (b).
   (a) Determine the amount of fluid absorbed by disc as follows: (i) Record the dry weight of each individual disc immediately on receipt into the laboratory re-weigh each disc and determine the volume of fluid absorbed by the disc by difference. Assume that 1 mL = 1 g. (ii) Prepare at least three replicates for each sample if possible. (iii) Ensure that disc uptake is at least 15 μL (15 mg).

   or

   (b) Divide sample into three and pipette on to three 6 mm discs. Record each volume in μL.
6. Calculate results as follows. If the assayed concentration for each replicate = C_A mg/L, the volume of fluid on the calibrator discs = V_C μL and the volume of fluid on the test fluid disc = V_F μL then the concentration in the fluid (C_F mg/L) is derived from the equation $C_F = C_A \times V_C/V_F$.

Assays in larger volumes of fluid

These include uncontaminated fluids where a reasonable volume is available, e.g. ascitic fluid, pleural fluid, bile, pus from abscess cavities, CAPD fluid, inflammatory fluid, CSF.

Assay calibrators should be prepared in a medium containing an equivalent amount of protein (albumin) to that found in the test sample. Mean inflammatory fluid protein content is 38.12 g/L (*c.*70% that of plasma); CSF protein content is 0.1–4 g/L. In some instances it may be necessary to measure the levels of acute-phase proteins such as α_1 acid glycoprotein to which some antimicrobials (for example trimethoprim and some macrolides) bind selectively.

In studies looking at the penetration of drugs into mild inflammatory exudate, calibrators have been prepared in human serum diluted to 70% of full-strength with buffer (Kavi *et al.*, 1988); in animal studies, muscle tissue fluid has been assayed using 50% rabbit serum (Ryan and Cars, 1983). In the case of CSF, even an extreme albumin concentration of 4 g/L is still only 7% of the total albumin found in human plasma and therefore this may have little effect on the assay of most antimicrobials.

Procedure

1. Measure the albumin content of the fluid (if this is not possible consult Documenta Geigy Scientific Tables (Lentner, 1981) to find a typical value).

2. Prepare assay calibrators in a diluent containing the same protein concentration.

3. Dilute the samples, if necessary, in the same medium as that used for the preparation of the calibrators.

References

Baldwin, D. R., Wise, R., Andrews, J. M. & Honeybourne, D. (1991). Microlavage: a technique for determining the volume of epithelial lining fluid. *Thorax* **46**, 658–62.

Baldwin, D. R., Wise, R., Andrews, J. M. & Honeybourne, D. (1992). Quantitative morphology and water distribution of bronchial biopsy samples. *Thorax* **47**, 504–7.

Barza, M. (1990). Approaches to the measurement of antibiotic concentrations in tissues. *Research and Clinical Forums* **12**, 15–22.

Benoni, G., Cuzzolin, L., Bertrand, C., Puchetti, V. & Velo, G. (1987). Imipenem kinetics in serum, lung tissue and pericardial fluid in patients undergoing thoracotomy. *Journal of Antimicrobial Chemotherapy* **20**, 725–8.

Bouvet, O., Bressole, F., Courtieu, C. & Galtier, M. (1992). Penetration of pefloxacin into gynaecological tissues. *Journal of Antimicrobial Chemotherapy* **29**, 579–87.

Brouwer, W. K. W., Kuiper, K. M. & Langbroek, W. (1984). The efficacy of prophylaxis with cefuroxime and metronidazole in gynaecological surgery. *Journal of International Biomedical Information and Data* **5**, 37–43.

Cars, O. (1981). Tissue distribution of ampicillin: assays in muscle tissue and subcutaneous tissue cage fluid from normal and nephrectomized rabbits. *Scandinavian Journal of Infectious Diseases* **13**, 283–9.

Cars, O. & Ögren, S. (1981). A microtechnique for determination of antibiotics in muscle. *Journal of Antimicrobial Chemotherapy* **8**, 39–48.

Cars, O. & Ögren, S. (1985). Antibiotic tissue concentrations: methodological aspects and interpretation of results. *Scandinavian Journal of Infectious Diseases, Suppl.* 44, 7–15.

Cunha, B. A., Gossling, H. R., Pasternak, H. S., Nightingale, C. H. &

Quintiliani, R. (1984). Penetration of cephalosporins into bone. *Infection* **12**, 80–4.

Davey, P. G., Precious, E. & Winter, J. (1991). Bronchial penetration of ofloxacin after single and multiple oral dosage. *Journal of Antimicrobial Chemotherapy* **27**, 335–41.

Desnottes, J. F. & Diallo, N. (1992). Cellular uptake and intracellular bactericidal activity of RP 59500 in murine macrophages. *Journal of Antimicrobial Chemotherapy* **30**, Suppl. A, 107–15.

Fox, K. E. (1989). Total extraction of aminoglycosides from guinea pig and bullfrog tissues with sodium hydroxide or trichloroacetic acid. *Antimicrobial Agents and Chemotherapy* **33**, 448–51.

Garcia, I., Pascual, A., Guzman, M. C. & Perea, E. J. (1992). Uptake and intracellular activity of sparfloxacin in human polymorpho-nuclear leukocytes and tissue culture cells. *Antimicrobial Agents and Chemotherapy* **36**, 1053–6.

Grimer, R. J., Karpinski, M. R. K., Andrews, J. M. & Wise, R. (1986). Penetration of amoxycillin and clavulanic acid into the bone. *Chemotherapy* **32**, 185–91.

Hearse, D. J. (1984). Microbiopsy metabolite and paired flow analysis: a new rapid procedure for homogenisation, extraction and analysis of high energy phosphates and other intermediates without any errors from tissue loss. *Cardiovascular Research* **18**, 384–90.

Honeybourne, D., Andrews, J. M., Ashby, J. P., Lodwick, R. & Wise, R. (1988). Evaluation of the penetration of ciprofloxacin and amoxycillin into the bronchial mucosa. *Thorax* **43**, 715–9.

Hoogkamp-Korstanje, J. A. A., van Oort, H. J., Schipper, J. J. & van der Wal, T. (1984). Intraprostatic concentration of ciprofloxacin and its activity against urinary pathogens. *Journal of Antimicrobial Chemotherapy* **14**, 641–5.

Johnson, J. D., Hand, L. W., Francis, J. B., King-Thompson, N. & Corwin, R. W. (1980). Antibiotic uptake by alveolar macrophages. *Journal of Laboratory and Clinical Medicine* **95**, 429–39.

Kavi, J., Andrews, J. M., Ashby, J. P., Hillman, G. & Wise, R. (1988). Pharmacokinetics and tissue penetration of cefpirome, a new cephalosporin. *Journal of Antimicrobial Chemotherapy* **22**, 911–6.

Kinzig, M., Sorgel, F., Brismar, B. & Nord, C. E. (1992). Pharmacokinetics and tissue penetration of tazobactam and piperacillin in patients undergoing colorectal surgery. *Antimicrobial Agents and Chemotherapy* **36**, 1997–2004.

Leigh, D. A., Griggs, J., Tighe, C. M., Powell, H. D. W., Church, J. C. T., Wise, R. *et al.* (1985). Pharmacokinetic study of ceftaxidime in bone and serum of patients undergoing hip and knee arthroplasty. *Journal of Antimicrobial Chemotherapy* **16**, 637–42.

Lentner, C. (Ed.) (1981). *Geigy Scientific Tables Volume 1, Units of Measurement, Body fluids, Composition of the Body, Nutrition.* Ciba-Geigy, Basle, Switzerland.

Lorian, V. (1980). Determination of antibiotic concentration from tissue and fluids other than blood. In *Antibiotics in Laboratory Medicine*, (Lorian, V. Ed.), pp. 252–5. Williams & Wilkins, Baltimore, MD.

Lowry, O. H. & Hastings, A. R. (1942). Histochemical changes associated with ageing. 1. Methods and calculations. *Journal of Biological Chemistry* **145**, 257–69.

Massias, L., Dubois, C., de Lentdecker, P., Brodaty, O., Fischler, M. & Farinotti, R. (1992). Penetration of vancomycin in uninfected sternal bone. *Antimicrobial Agents and Chemotherapy* **36**, 2539–41.

Meissner, A., Borner, K. & Koeppe, P. (1990). Concentrations of ofloxacin in human bone and in cartilage. *Journal of Antimicrobial Chemotherapy* **26**, Suppl. D, 69–74.

Mertes, P. M., Jehl, F., Burtin, P., Dopff, C., Pinelli, G. Villemot, J. P. *et al.* (1992). Penetration of ofloxacin into heart valves, myocardium, mediastinal fat and sternal bone marrow in humans. *Antimicrobial Agents and Chemotherapy* **36**, 2493–6.

Muller-Serieys, C., Bancal, C., Dombret, M. C., Soler, P., Murciano, G., Aubier, M. *et al.* (1992). Penetration of cefpodoxime proxetil in lung parenchyma and epithelial lining fluid of noninfected patients. *Antimicrobial Agents and Chemotherapy* **36**, 2099–103.

Nix, D. E., Goodwin, S. D., Peloquin, C. A., Rotella, D. L. & Schentag, J. J. (1991a). Antibiotic tissue penetration and its relevance: models of tissue penetration and their meaning. *Antimicrobial Agents and Chemotherapy* **35**, 1947–52.

Nix, D. E., Goodwin, S. D., Peloquin, C. A., Rotella, D. L. & Schentag, J. J. (1991b). Antibiotic tissue penetration and its relevance: impact of tissue penetration on infection response. *Antimicrobial Agents and Chemotherapy* **35**, 1953–9.

Philpott, A. D., Crawford, E. D. & Miller, G. J. (1991). A new method for assaying antimicrobials in the prostate. *Infection* **19**, *Suppl. 3*, 150–3.

Plaue, R., Bethke, R. O., Fabricius, K. & Müller, O. (1980). Kritische Untersuchungen zur Methodik von Antibiotika-spiefeibestimmungen in menschlichen Geweben. *Arzneimillelforschung Drug Research* **30**, 1–5.

Rennard, S. I., Basset, G., Lecossier, D., O'Donnel, K. M., Pinkston, P., Martin, P. *et al.* (1986). Estimation of volume of epithelial lining fluid recovered by lavage using urea as a marker of dilution. *Journal of Applied Physiology* **60**, 532–8.

Reynolds, H. Y., Fulmer, J. D., Kazmierowski, J. A., Roberts, W. A., Frank, M. M. & Crystal, R. G. (1977). Analysis of cellular protein content of broncho-alveolar lavage fluid from patients with idiopathic pulmonary fibrosis and chronic hypersensitivity pneumonitis. *Journal of Clinical Investigation* **59**, 165–75.

Rosdahl, V. T., Sorensen, T. S. & Colding, H. (1979). Determination of antibiotic concentrations in bone. *Journal of Antimicrobial Chemotherapy* **5**, 275–80.

Ryan, D. M. & Cars, O. (1983). A problem in the interpretation of β-lactam antibiotic levels in tissues. *Journal of Antimicrobial Chemotherapy* **12**, 281–4.

Schentag, J. J. (1989). Clinical significance of antibiotic tissue penetration. *Clinical Pharmacokinetics* **16**, *Suppl. 1*, 25–31.

Shyu, W. C., Reilly, J., Campbell, D. A., Wilber, R. B. & Barbhaiya, R. H. (1993). Penetration of cefprozil into tonsillar and adenoidal tissues. *Antimicrobial Agents and Chemotherapy* **37**, 1180–3.

Thurmann-Nielsen, E., Jetlund, O. & Walstad, R. A. (1990). Different approaches to the analysis of tonsillar antibiotic levels. *Research and Clinical Forums* **12**, 75–80.

Waldren, R., Arkell, D. G., Wise, R. & Andrews, J. M. (1986) The intraprostatic penetration of ciprofloxacin. *Journal of Antimicrobial Chemotherapy* **17**, 544–5.

Wise, R., Donovan, I. A., Drumm, J., Andrews, J. M. & Stephenson, P. (1983). The penetration of amoxycillin/clavulanic acid into peritoneal fluid. *Journal of Antimicrobial Chemotherapy* **11**, 57–60.

Wittmann, D. H., Schassan, H. H. & Freitag, V. (1980). Pharmacokinetic studies and results of a clinical trial with cefotaxime (HR-756). In *Current Chemotherapy and Infectious Disease* (Nelson, J. D. & Grassi, C., Eds), pp. 114–6. American Society of Microbiology, Washington, DC.

Chapter 7

Quality assurance

Les O. White[a] and Jenny M. Andrews[b]

[a]Department of Medical Microbiology, Southmead Hospital, Bristol BS10 5NB; [b]Department of Medical Microbiology, City Hospital NHS Trust, Birmingham B18 7QH, UK

Introduction

Assays provide information in the form of numbers. It is the aim of quality assurance procedures to convince the analyst, the recipient and now commonly a third party, in the form of an accreditation agency, that the numbers produced are timely, meaningful and of value because they are a sufficiently accurate reflection of the truth. Just how accurate the assay results need to be depends on the context in which they are to be used. For therapeutic monitoring a serum gentamicin result may be perfectly acceptable if it is reliably within ±25% of the true concentration in the patient, but it must be produced in a matter of hours rather than days. If the result is required for a pharmacokinetic study then no more than ±5% variation from the true concentration might be required; it might, however, be perfectly acceptable if the sample takes several weeks to process.

The quality assurance of assays comprises five aspects: assay validation, assay protocols, internal quality control (IQC), external quality assessment (EQA), and accreditation; these will all be discussed in this chapter. Firstly, however, some of the commonly used terms need to be defined.

Definitions

Accuracy

Accuracy is a measure of how close an assay result is to the true concentration. If, when a sample is assayed repeatedly, the results are always less than the true concentration, the assay is said to have *negative bias*. If the results are always too high then the assay shows *positive bias*. Such errors are referred to as *systematic errors*. Systematic bias may be constant (the same fixed amount at every concentration), proportional (the fixed amount increases or decreases with the concentration) or variable (for example a fixed negative value at one point on the curve and a fixed positive value on another part of the curve). The latter type of bias may be a result of poor curve fitting, (see below). Degree of inaccuracy is frequently expressed as percentage error (%Error) using the formula:

$$\%\text{Error} = (x - t)/t \times 100$$

where x is the assay result and t is the target concentration. Accuracy depends on several factors, including: use of appropriate calibration material, matrix effects, use of appropriate curve fitting procedures, sample stability, assay specificity and random error.

Precision

Precision (sometimes referred to as imprecision, reproducibility or repeatability) is a measure of the closeness of agreement between results when the same sample is assayed several (at least four but preferably six or more) times. Errors due to poor precision are referred to as *random errors*. Precision is often expressed in terms of sample standard deviation (S.D.) using the formula:

$$\text{S.D.} = [\Sigma(x - mean)^2/n - 1]$$

where x represents each individual observation, *mean* is the mean result and n is the number of observations. Relative precision at different concentrations is best compared using the *percentage coefficient of variation* (%CV) (sometimes called the relative S.D.) calculated from:

$$\%\text{CV} = (\text{S.D.}/mean) \times 100$$

Where assay results are normally distributed about the mean, the S.D. or %CV can provide useful information regarding the precision of assay results. Sixty-eight per cent of results will lie within ± 1 S.D. of the mean and 95% of results will lie within ± 2 S.D. of the mean. Therefore if a particular assay technique produces %CVs of 5%, then 95% (19 of every 20 results) will lie within 10% (2 S.D.) of the true result.

When assays are validated then two measures of precision are usually determined. *Within-assay (or intra-assay) reproducibility* is determined by making replicate assays in a single analytical run, and the more stringent *between-assay (or inter-assay) reproducibility* is determined by making the replicate assays in different analytical runs (usually on different days).

Specificity

This is a measure of an assay's ability to measure the correct analyte and not measure (in addition to or instead of) some other compound in the sample. Any other compound which erroneously produces a result is said to interfere with the assay. Assay specificity is usually determined during validation by processing samples containing known concentrations of any other compounds which are likely to be present in real samples. When checking for interfering compounds it is also recommended that a number of samples from patients known not to be taking the drug to be assayed are processed. In this way unexpected interference is sometimes encountered. Poor specificity may give rise to a positive bias error.

Sensitivity

This term is often used without clear definition to indicate the minimum quantity of a substance which can be assayed. If a sample contains concentrations of drug below the level of assay sensitivity, a false negative result will be obtained. Assay results of zero concentration should therefore be recorded as 'not detected, 'no detectable level' or 'below level of assay sensitivity'. Various terms to define sensitivity have been used including, minimum detectable quantity (MDQ), lower limit of quantification (LLQ) and detection limit (DL). For microbiological assays it is safest to define the LLQ as the concentration of the lowest calibrator although the DL will be the lowest concentration of antimicrobial which will produce a zone of inhibition. For a microbiological assay the DL will always be greater than but close to the MIC of the assay indicator microorganism and this information may help in the choice of an appropriate strain. For HPLC assays the MDQ may be defined as the concentration of substance giving a response three to five times greater than baseline noise. However, such low concentrations may only be assayed with very poor accuracy and reproducibility. The better limit of sensitivity for pharmacokinetic studies would be the LLQ defined as the smallest concentration that can be accurately measured with a %CV better than (say) 10%.

Calibrators

Calibrators are the materials used to calibrate an assay procedure and to calculate the result of an analysis. Most antimicrobial assays will use more than one (and up to six) calibrators and each will comprise a known amount of antimicrobial substance dissolved in an appropriate matrix. The matrix will have been defined during validation and will usually be similar to the matrix of the samples to be assayed. For good assay accuracy, appropriate antimicrobial reference powder should be used. Some antimicrobials (e.g. most aminoglycosides) have international reference powders which can be purchased for the purposes of calibration. In the preparation of calibrator solutions it is essential to use an appropriate sensitive and accurately calibrated analytical balance and certified volumetric

glassware. Variable-volume calibrated pipettes (e.g. those made by Gilson) may be used in many procedures but, since these are mechanical devices which must be regularly serviced, to ensure assay accuracy and precision they should be used correctly and their performance checked regularly (see Appendix).

Calibration

Calibration is the set of operations which establish, under specified conditions, the relationship between the values indicated by a measuring instrument or system (e.g., zone sizes, peak heights, absorbance measurements, polarization units) and the corresponding values of a quantity realized by reference to a calibrator or calibrators of known concentration.

Internal quality control

Internal quality control (IQC) is a procedure which utilizes the results obtained by only one laboratory for quality control purposes. IQC materials are normally processed along with the other samples. They are usually called internal controls or just controls.

External quality assessment

External quality assessment (EQA) is a procedure which utilizes, for quality control purposes, the results of several (preferably many) laboratories which assay the same quality control materials.

Assay validation

Assay validation is only part of a total procedural validation which is essential if assay results are to be reliable and useful. It is pointless for the analyst to validate painstakingly an assay if the overall procedure from administration of the drug to the patient or volunteer, taking the samples, storing the samples, transporting the samples, and so on, is not properly controlled and validated. For example, some drugs (including many β-lactams) are thermally unstable and so immediate separation of the serum, and its transport and storage at low temperatures, are essential. Other drugs may be stable in the dark yet rapidly degraded when exposed to sunlight (e.g., chloramphenicol, fluoroquinolones). A third possibility is chemical or enzymic degradation *in vitro*. For example, red blood cells contain an esterase which rapidly desacetylates cefotaxime. If red cell lysis occurs in the sample then rapid in-vitro desacetylation of cefotaxime will follow. Thus procedures for taking samples for cefotaxime assay (especially tissue samples) must be rigorously validated.

Sample collection, storage and transport are often not the responsibility of the analytical laboratory and may take place on a different site or even in a different country. It is the responsibility of the analytical laboratory to ensure that samples are collected in such a way to assure the validity of the analytical results. The analyst should be fully involved in the design of study protocols and procedures to assure the quality of the information they are being asked to provide.

Appropriateness

Appropriateness is very often given insufficient attention. If it is necessary to assay a drug and metabolites then HPLC or GLC might be the most appropriate method. On the other hand, if a non-metabolized antimicrobial is to be assayed in a large number of samples from healthy volunteers given a single agent, then microbiological assay could provide a rapid, inexpensive means of assay. Immunoassays may be the most appropriate assays for therapeutic monitoring because they provide rapid and reliable results. Nevertheless they may be inappropriate for other studies. For example, the Abbott TDX assay for vancomycin cross-reacts with vancomycin degradation products and would therefore be unacceptable for use in a study of the stability of vancomycin in, say, CAPD peritoneal fluid. In choosing an assay methodology the following aspects should be considered:

- degree of accuracy and reproducibility demanded
- total number of samples to be assayed
- time scale, including acceptable time between taking the samples and providing the results
- likelihood of other drugs being present
- source of samples (e.g., volunteers or patients; tissues, urine or blood)
- lowest concentration to be quantified (amount of anti-microbial likely to be present in the sample and the amount it is theoretically possible to assay; this may not be the lowest concentration in the sample since very small samples may require considerable dilution)
- stability of sample (size of sample, drug stability, feasibility of storage for batch assay)
- local resources (equipment, human)
- legal or other requirements
- budgetary constraints

Accuracy

Calibrators. Calibrators should be prepared from the purest material available for laboratory usage. For new compounds this should be available from the manufacturers, with more established drugs reference materials are available from the British Pharmacopoeia Commission, the National Institute for Biological Standards and Controls, the European Pharmacopoeia Commission, the US Pharmacopoeia, relevant pharmaceutical companies or commercial suppliers of chemicals (e.g., Aldrich or Sigma). Reference substances should always be stored and handled according to the information provided by the supplier. If none is given the analyst should ask for the relevant information in writing.

Many reference substances come with a stated potency; for example, ampicillin trihydrate is equivalent to 881 μg/mg ampicillin. International reference substances may have their potency stated in International Units (IU) and this can be a source of confusion. Gentamicin sulphate reference substance may contain 641 IU/mg gentamicin, but since gentamicin assays are always quoted in units of mg/L then it is necessary to know the factor to convert from IU to mg. In some cases (e.g., amikacin and netil-micin international standards) there has yet to be international agreement on what the conversion factor should be and it will be necessary to use an interim conversion factor which could, by international consensus, be changed at some future date. If an interim conversion factor is used it is essential to document the source of the information. Failure to allow for potency, properly or at all, will produce a constant bias error. Potency may be >100% or >1000 μg/mg. This is the case for vancomycin where the original potency calculations were made on an impure fermentation product.

The reference material must be accurately weighed unless it is supplied as an accurately weighed amount (usually freeze-dried in a vial). Most laboratory balances require at least 50 mg to be weighed to give an accuracy of ±0.2%. If only tiny quantities of substance are available then a microbalance will be required. The analyst weighing the powder should be made fully aware of the rarity and value of the substance. There are numerous anecdotal stories of people weighing out and dissolving 50 mg of a new drug only to be told at a later date that there was only 52 mg available worldwide! It might be advisable to weigh tiny quantities directly into appropriate volumetric glassware. Powders should always be brought to room temperature before they are removed from their containers since cold powder may attract condensation, leading to inaccurate weighing. Some powders may rapidly absorb moisture and the manufacturer or supplier should provide information for the appropriate handling and storage of such substances.

Some powders will be insoluble or poorly soluble in water or buffer. Initial solution in a solvent such as dimethylsulphoxide (DMSO) or methanol may be required. Some substances may be initially dissolved in a weak acid such as lactic or phosphoric acid (e.g., trimethoprim) or a weak alkali (e.g., some quinolones, sulphonamides) followed by rapid dilution with water or buffer. It is advisable to seek guidance first from the manufacturer or supplier when solubility problems are encountered. Once the substance is in solution dilutions should be made in Class A volumetric glassware and should be thoroughly mixed, with frothing kept to a minimum. Protection from light, especially bright sunlight, may be necessary for many substances. Further dilutions into the appropriate matrix for calibrators or control samples should also be made with accurate volumetric pipettes. Series of calibrators which may be spaced at two-fold concentration increases should *never* be prepared by doubling dilution since this technique will amplify any small error in the dilution with every subsequent dilution and the calibrators will become progressively more inaccurate. This is an example of a proportional bias error.

Matrix effects and recovery. In microbiological assays the matrix in which the calibrators are prepared can have a significant influence on assay accuracy. Calibrators prepared in human serum may give smaller zones of inhibition than similar calibrators prepared in phosphate buffer, so calibrators must be prepared in serum if serum samples are being assayed. In microbiological assays the samples and calibrators should be of the same pH since the activity of some antimicrobials is very pH-dependent. Stored human serum often has a higher pH than fresh material and this may need to be corrected for, for example by exposing the serum to CO_2. Animal serum may differ from human serum considerably with regard to such things as protein content and protein binding and should only be used as a matrix if it has been compared with human serum and shown to give identical results. In contrast, serum calibrators will be inappropriate for the assay of other fluids such as urine or CSF. Calibrators are usually prepared in buffer if urine or CSF samples are being assayed. Occasionally calibrators may be prepared in 'pooled human urine' but urine is a highly variable matrix, difficult to standardize. Since antimicrobial concentrations in urine are often very high, urine samples will usually require considerable dilution before assay, making buffer calibrators acceptable. If, however, urine components—for example divalent cations—are going to exert large effects in a microbiological assay, then another technique such as HPLC may be indicated.

With chromatographic assays there may be loss of substance during sample preparation procedures. (Note: the loss may also apply to any internal standard being used, see *Chapter 5*.) It is therefore possible that calibrators will need to be prepared in a similar matrix to, and subject to an identical sample preparation procedure to, the samples to ensure assay accuracy.

The measurement of the proportion of a compound which remains after a particular sample preparation procedure in often referred to as *recovery*. In HPLC, recovery is calculated by comparing the peak height (or area) of various drug concentrations in water, buffer, or other solvent with those obtained after identical concentrations in an appropriate matrix have been passed through a specific sample preparation procedure. Percentage recovery ($\%R$) is then determined from:

$$\%R = (H_{aq} - H_{proc})/H_{aq} \times 100$$

where H_{aq} is the peak height (or area) of the aqueous solution and H_{proc} is the peak height (or area) of the processed sample with the same initial drug concentration. For good assay accuracy it is essential that recovery is reproducible and preferably relatively high. Note also that this applies to the internal standard as well as the analyte. If recovery is virtually 100% then aqueous calibrators may be used. Note, however, that most HPLC sample preparation procedures other than simple protein precipitation will give <100% recovery.

In microbiological assay the reduction of zone sizes due to matrix effects produces a similar effect to the loss of drug during HPLC sample preparation, although for different reasons. As a measure of the degree of matrix effect, the equation above could be used for validation of plate assays by using zone sizes in the place of peak heights.

Another potential problem is loss of substance by adsorption to glass or plastic surfaces. To eliminate or minimize losses of this sort containers should always be filled to at least three-quarters capacity. Loss by adsorption is easily tested by measuring the concentration of drug in a solution before and after exposure to the surface or container being tested. The large plastic containers used to collect urine samples should be tested for adsorption effects. Urine passed soon after dosing may often be in large volumes and contain relatively high concentrations of drug; adsorption may not be a problem since a small fixed amount rather than a proportion will usually become adsorbed. In contrast urines collected 12–24 h after the dose may contain only low drug concentrations and loss from adsorption may be significant. If significant loss from adsorption is likely the problem must be eliminated. Teicoplanin (*Chapter 10*) is of interest since it readily adsorbs to plastic and glass and the minor components adsorb on more readily than the major component.

Loss due to adsorption to plastics such as pipette tips can be eliminated by filling the pipette and discarding the first filling, a procedure which saturates the binding sites. A similar procedure could be used with storage containers.

It is good practice never to collect small volumes of sample in large containers. If such a practice is unavoidable then possible losses due to binding must be carefully investigated during assay validation.

Sample stability. Although this information is necessary to validate assay results, clearly it can only be obtained when a suitable assay has already been developed. Sometimes unexpected sample instability is only detected when a suitable assay has been developed. Close attention must be paid to the stability of the drug in the samples being assayed, both when stored and when being processed, and likewise in the calibrators. If samples cannot be assayed straight away, then appropriate information on storage is essential. Many drugs will remain stable in biological fluids

Table I. Suggested protocol for a sample stability study.

Storage time	Storage temperature for biological specimens				
	RT (in dark)	+4°C	−20°C	−40°C	−70°C
0 h (pre-storage)	yes	yes	yes	yes	yes
1 h	yes	yes			
2 h	yes	yes			
8 h	yes	yes			
12 h	yes	yes			
24 h	yes	yes			
1 week		yes	yes	yes	yes
1 month			yes	yes	yes
2 months			yes		
6 months			yes	yes	yes
12 months			yes	yes	yes

RT, room temperature.
'yes' indicates the time/temperature combinations to be tested.

for many months at −20°C but others will not and may require storage at −40°C or −70°C. Freezers allowing storage at −140°C are now available if necessary. The matrix may well have an effect on stability; for example, extremes of pH should be avoided. Some drugs may require the samples to be stabilized in some way by the addition of buffers or anti-oxidants. Appropriate storage conditions can only be determined experimentally by assaying samples in biological fluids which have been stored under various conditions for various lengths of time. Table I gives a suggested protocol which could be followed. All assays evaluating sample stability must be done with a calibration curve derived with freshly prepared calibrators. All samples should be replicated at least five or six times.

Sample stability after any sample preparation procedure must also be studied but over a limited time-scale. If samples are to be assayed immediately after sample preparation, then relative instability may not be a problem. If, however, prepared samples are going to sit for up to 18 h in an autosampler before they are processed, then sample stability for at least 24 h should be confirmed experimentally. Spiked control samples may be included with test samples during the study period and during storage and/or transport; assay of these will reveal evidence of any sample degradation.

Curve fitting. A variable which can have a marked effect on accuracy is the mathematical curve fitting procedure used to construct the calibration curve (see *Chapters 4* and *5*). Curve fitting of HPLC assays is discussed in greater detail by Philips *et al.* (1990) and fitting of microbiological assays by Hewitt & Vincent (1989). Immunoassay curve fitting is usually determined by the kit manufacturers and/or programmed into the analytical instrument. However, it should be remembered that curve-fitting procedures on

some instruments may fit some data relatively poorly (White & Reeves, 1998). Poor curve fitting may cause a variable bias error and poor linearity (see below).

Specificity. Poor specificity will affect accuracy by producing a positive bias error if interfering drugs are present.

Random error. Random error will produce unpredictable negative or positive bias on individual results but the effect can be minimized if appropriate numbers of replicates are assayed.

Precision

Within-assay precision should be determined by assaying samples containing a high, medium and low concentration of drug each replicated at least five or six times. S.D. and %CV should be determined.

Between-assay precision should be determined by similarly assaying three samples on at least six and up to 20 different occasions. Microbiological assays and chromatographic procedures would be recalibrated on each successive occasion. Commercial immunoassays, however, might only require calibration at the beginning of the study since their calibration curves can be very stable. In this case between-assay precision studies may also become studies of calibration curve stability (see 'Internal quality control').

Sensitivity

With microbiological assay the LLQ will be the concentration of the lowest calibrator, assuming that replication of this concentration gives acceptable reproducibility. With microbiological assay samples giving zones smaller than the lowest calibrator or greater than the highest calibrator cannot be assayed with any confidence (high concentrations may of course be diluted and reassayed).

With chromatographic assays the LLQ will be the lowest concentration which gives a peak discernible from baseline noise with acceptable accuracy and reproducibility. The MDQ will be less than this but since concentrations around the MDQ will probably show rather poor reproducibility such concentrations should be reported as being 'less than LLQ' in pharmacokinetic studies. Clearly the LLQ will depend on the sensitivity and signal-to-noise ratio of the detector. Therefore the equipment used must be clearly stated when LLQ or DL values are quoted. For example, fluorescence detectors may vary enormously in their sensitivity and signal-to-noise ratio. Consequently a published method with a claimed detection limit of (say) 0.001 mg/L may give nowhere near that value when performed with a less sensitive or very 'noisy' detector.

Specificity

This is tested by analysing samples containing any other drugs or substances liable to be present in the test samples and which might interfere with the assay. With microbiological assay other antimicrobials are usually tested and specificity is usually achieved by either using an indicator organism resistant to other antimicrobials or removing, inactivating or diluting out the other antimicrobials. However, many non-antimicrobials, including anticancer drugs, barbiturates, beta-blockers, diuretics and non-steroidal antiinflammatory agents, have antimicrobial activity (Wright & Matsen, 1980; Cederlund & Mårdh, 1993). If samples are from patients then it may be necessary to evaluate such agents for interference.

With chromatographic assays any UV-absorbing or fluorescent substance is a potential source of interference. With fluorescence detection it might be prudent also to test non-fluorescent compounds for interference in the presence of a known low concentration of the drug being assayed since a non-fluorescent compound co-eluting with a fluorescent compound may quench the fluorescence and cause an underestimate of the true concentration. With chromatography it is also necessary to ensure there is no interference with the internal standard, if used, as well as the drug being assayed.

The drugs tested for causing interference should be prepared in a similar matrix to the samples being tested, initially at a concentration equivalent to a high therapeutic concentration. It is also important to ensure that the correct compound is tested since many drugs may be chemically changed after administration. For example chloramphenicol succinate (the parenteral form of chloramphenicol) will not interfere with a microbiological assay since it has no antimicrobial activity. If drugs are known to be extensively metabolized then the metabolites should also be tested for interference. For example, cefotaxime is partially metabolized to desacetylcefotaxime, and itraconazole is extensively metabolized to microbiologically active hydroxyitraconazole; both metabolites may interfere with microbiological assays for the parent drug (White et al., 1981; Hostetler et al., 1993).

In validating an assay for a clinical trial it is also recommended that appropriate samples, such as blood, sputum, urine, CSF, from similar patients not taking the drug under investigation (possibly pre-treatment blood taken for routine biochemical investigations) are assayed to test assay specificity. It is not uncommon for interfering substances to be encountered this way, especially in urine and sputum.

In trials a blank pre-treatment sample should be obtained from each patient. This is particularly important since sputum samples are often inhibitory and serum samples are occasionally inhibitory to assay indicator organisms. In this way assay specificity can be validated for each individual subject. However, it should always be borne in mind that a pre-treatment blood sample may very occasionally be forgotten and actually be taken just after the first dose of drug has been given. Clear, concise study protocols and check-lists should minimize this likelihood.

Linearity

The term linearity is sometimes used to describe the statistical measure of how close the actual points on a calibration curve fit to the curve generated by the curve-fitting mathematics. A method should only pass validation if a good curve-fit is reproducibly obtained. A simple way to assess curve fit graphically is visually to compare the line of best fit with the actual curve. In addition, or alternatively, back-calculate the concentration of the calibrators from the fitted curve. If they lie outside some pre-determined limits, for example, 5% or 10% of their true concentration, linearity should be considered unacceptable and the source of error traced or an alternative curve-fitting procedure evaluated.

The term linearity may also be used in a different way to describe the accuracy of an assay when a sample of known concentration is diluted 1:1 or 1:3, assuming of course that the diluted concentration falls within the calibrator range. Clearly a sample diluted 1:1 should give half the concentration of the undiluted sample and a 1:3 dilution should give a quarter of value of the undiluted sample. If when it is tested this is not the case, the source of the problem should be determined and remedied. If the dilutions are not made in an appropriate matrix then changes in pH or protein binding may produce non-linearity. Loss of sample by adsorption to glass or plastic surfaces will also produce non-linearity. It is important to validate this property of an assay if it is known that samples containing concentrations higher than or close to the top calibrator are going to be encountered and dilution is going to be frequently used.

Assay protocols and standard operating procedures

Once an assay has been validated, the procedure should be written as a protocol which may or may not be part of a standard operating procedure (SOP). The protocol should describe the materials and procedures used clearly and accurately, and also include expected performance in terms of reproducibility, specificity, sensitivity, linearity and, if a comparison has been made, correlation with another technique. Vague terms such as 'phosphate buffer pH 7.2' or 'nutrient agar' are not acceptable in assay protocols. The phrase '0.5 M potassium phosphate buffer pH 7.2' may be acceptable but a better, more easily followed, protocol would list the weights of the different constituents used and state how the pH value of 7.2 was achieved. HPLC protocols should state the type of column and its dimensions. The phrase 'a C_{18} column' is acceptable if any C_{18} column will genuinely suffice but specific information such as 'Hypersil 5 μm ODS (4 $\times$ 150 mm)' is more useful. If retention times are laid down in a protocol or SOP then it is essential to state the solvent flow rate. It is more useful to state the solvent flow rate than a retention time since the former can influence the performance of several components of an HPLC system all of which will contribute to the value of the retention time. Some accreditation standards demand that protocols and SOPs are very detailed and insist on revalidation of the whole assay procedure if any part is changed.

The elements of assays given in the other chapters of this book do not, on the whole, give sufficient detail to be considered as protocols or SOPs. Rather, this book aims to give sufficient background information about assays in general and sufficient information about specific assays to enable a competent laboratory worker to set up and validate an assay for a particular purpose and then write a protocol tailored to their own laboratory.

Internal quality control

IQC is the long-term continual monitoring of the accuracy and precision of an assay. The NCCLS (1991) states that the purpose of IQC is to 'monitor analytical performance relative to medical goals and alert analysts to unsatisfactory analytical performance'. In the relatively short-term validation of an assay, accuracy and precision data are determined, but continual IQC is essential to ensure that the assay continues to perform acceptably in the long term and during routine use. For assays that are calibrated each time, IQC can produce data similar to those for inter-assay precision. With assays that are calibrated only infrequently, IQC is the tool that alerts the analyst to the need to recalibrate as well as to problems of loss of precision. IQC is practised by assaying one or more stable internal control samples in each batch of assays and statistically comparing the results obtained with these samples with some predetermined values.

IQC samples (controls)

Preparation. Although the use of pooled samples from patients may be appropriate in some instances, for most drug assays internal controls will be prepared by adding a known amount of drug to an appropriate drug-free matrix. The matrix should be of a composition similar or identical to the samples being analysed. For serum samples, controls will usually be in serum unless extensive validation procedures have shown that an alternative matrix is acceptable. Ideally the drug powder (or solution) used to prepare the controls should be from a separate source to that used to prepare the calibrators and should be stored under suitable conditions in a different place from the powder used to prepare calibrators. The definition of 'separate source' may vary and may be laid down by an accreditation agency. At the very least it means 'from a different batch or lot number' and it may mean 'from raw materials from different vendors'. Separate-source chemicals for use as calibrators and controls are available for some analytes from some suppliers but at the time of writing these did not include antimicrobials. For some assays (e.g., aminoglycosides, vancomycin) internal controls may be purchased commercially. They should ideally be obtained from a supplier other than the supplier of the assay. Abbott supply internal controls for use with their TDX and FLX assays and, although these provide a simple method of IQC, ideally samples from an alternative (third-party; separate-source) supplier should be used. The use of third-party internal controls is mandatory in some countries for some assays. Internal controls for most routine antimicrobial assays can be obtained from various suppliers. In the USA Baxter Scientific Products supply so called 'survey-validated controls'. These are prepared from the same pool of control material that is distributed as part of the College of American Pathologists' proficiency testing programme (see 'External quality assessment' section and Table II). The target concentrations of these controls are based on the mean peer-group returns of over 7000 survey subscribers. One supplier (Bio-Rad) runs a free interlaboratory scheme for laboratories that use their 'Lyphochek' controls. Internal control results are returned to Bio-Rad every month. The pooling of these worldwide data from many laboratories all using the same internal controls provides a simple way for individual laboratories to compare their performance with the whole pool of users.

Internal controls must be stable. Therefore commercial controls should be stored exactly according to manufacturers' instructions and the manufacturers' expiry dates should be adhered to. If in-house controls are being prepared then stability studies will have been performed during assay validation. Some antimicrobials may be assayed only infrequently and may be relatively unstable in solution (e.g. some β-lactams). In this situation calibrators and controls must be freshly prepared each time the assay is run. If this is the case the controls should be prepared by someone other than the analyst who prepares the calibrators, and their target values should not be revealed until after the assay has been performed.

Concentration range. Internal controls should cover one to three concentrations within the normal concentration-range of the assay. Immunoassay calibration curves should be validated with controls at the low and high ends and the middle of the calibration curve. Microbiological assays too should be validated with at least two (low and high) and preferably three different controls. One control may be sufficient to validate HPLC assays and some chemical assays but if large batches are being processed controls of two or three different but appropriate concentrations are recommended; if the assays are for therapeutic monitoring the concentration of the controls should be near those concentrations where critical medical decisions are made.

Internal controls are available commercially but usually contain multiple analytes. Some of these will contain antimicrobials which cross react in immunoassays. Gentamicin and netilmicin are an example. In such a situation the target concentration for any particular analyte will be method-dependent relying as it does on the degree of cross-reactivity between antibodies. If a particular internal control does not state a target value for the method in use then this should be determined in-house by assaying at least 20 replicates in runs where existing controls give acceptable results. Commercial multi-analyte internal controls may cause problems with HPLC assays since they contain many analytes some of which may not have been tested for interference during assay validation, making the possibility of interference with the drug or internal standard peak high. Commercial multi-analyte controls are, in general, unlikely to be of use with microbiological assays since they may contain seven or more different antimicrobials and may also contain a preservative such as sodium azide.

Frequency of use of internal controls

In general, every batch of assays should contain at least one internal control. Every microbiological assay plate should contain at least one and preferably two internal controls (one near the top and one near the bottom of the calibrator range) as well as the calibrators and unknowns. A large batch of samples for HPLC assay should have several internal controls interspersed among the tests.

A possible exception to the above rule is the TDX or FLX analyser for which some analysts consider it is only necessary to run internal controls once or twice a day. Since this assay is inherently very reliable, this practice may be acceptable and minimize costs. However, more cautious practitioners may prefer to include at least one control in every batch of TDX or FLX assays. If controls are run only in the first and last batches of the day there is the risk of unacceptable control results necessitating the rejection of the entire day's work.

With most immunoassays and HPLC assays the last sample assayed should always be a control sample. Using this practice, if reagents become depleted before the run is complete this will be revealed as an inaccurate result for the last (control) sample. With some assays (e.g. TDX) the analyser will not process a batch of samples unless it has sufficient reagents. Having an internal control as the last sample is still sensible since a correct result indicates that the analyser was working correctly throughout the run. Controls may also be placed randomly within every batch (especially if the batches are large). Random placement of controls is preferable to fixed placement. It is possible that certain positions on a TDX/FLX carousel could become dirty or contaminated in some way (for example by liquid spillages), causing the results for samples at the positions to be inaccurate. If control samples were always placed in the same positions such dirty positions might never be recognized.

Setting acceptable and unacceptable control limits

Control limits of acceptability are set to suit the needs to which the results are to be put. Often the S.D. determined at assay validation will be taken and results will be considered acceptable if they are within 2 S.D. and unacceptable if they are outside this value. This is good practice but may be much too stringent a test for clinical assays. Many immunoassays have %CVs of 3% or better.

For a gentamicin assay result to be clinically useful it is generally considered that ±25% variation from the true concentration is the maximum acceptable. For this reason many laboratories will define the acceptability of their results using an arbitrary value rather than a value related to the inherent reproducibility of the technique they are using. Common practice is for clinical laboratories to take ±10% as acceptable for medium and high internal controls and ±15% as acceptable for the low control. These ranges are, in many instances, somewhat greater than the inherent reproducibility of many assays in use today and could represent three or even four S.D.s for some techniques.

Representation of internal quality control results

Graphical plots are a simple, straightforward way to keep day-to-day IQC data which will help alert the analyst to unacceptable performance or (more importantly perhaps) impending problems.

Cusum plot. A plot of the cumulative value of the individual control results or the cumulative value of the difference between the target value and the result obtained are simple graphical plots which shows clearly when results become progressively inaccurate (often meaning an instrument needs recalibrating). Figure 7.1(a) shows a plot of some results for a 2 mg/L internal control over a 1 month period. The point where the assay begins to lose accuracy is marked with the arrow. After this point the line deviates progressively from the expected line. This loss of accuracy is shown even more clearly when the cumulative difference between the target value and the result are plotted (Figure 7.1(b)).

Shewhart plot. This graphical representation plots the results obtained on the *y* axis against the day on the *x* axis. The target concentration is at the centre of the *y* axis and lines representing degrees of deviation (e.g. one or two S.D.s, 10% error, 15% error, etc.) are drawn either side of the target (Figure 7.2). Results falling outside the limits become immediately apparent.

Rules and rule violation

Where the assay is calibrated at the time as the samples are assayed (microbiological assay; HPLC) then the calibration should be validated before the results of any internal controls are considered. For example, it is unacceptable to approve the results of a microbiological plate assay because the internal control is acceptable if the statistics of the calibration curve-fit are poor. In such a case the assay may be accurate at the level of the internal control by chance. In the case of immunoassays where the instrument is calibrated infrequently, this consideration does not apply; most instruments will either accept or reject a calibration themselves by reference to certain values held in memory or will prompt the operator to accept or reject

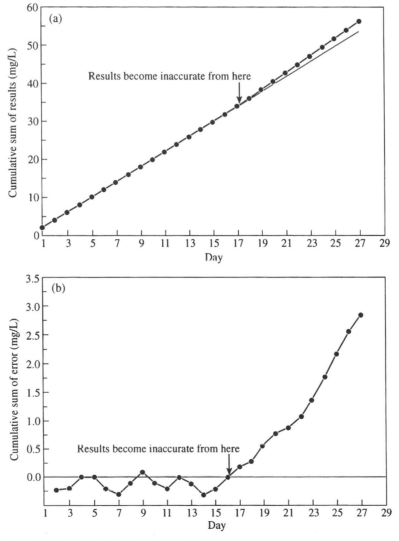

Figure 7.1 Cusum plots for a 2 mg/L internal control.

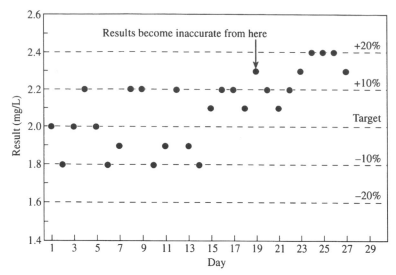

Figure 7.2 Shewhart plot for a 2 mg/L internal control.

the calibration curve themselves by reference to certain limits of acceptability.

The laboratory should have written guidelines for the procedure to be followed when internal controls fall outside the predetermined limits of acceptability. Individual internal control results may themselves be taken in isolation to accept or reject a set of assay values; alternatively, sets of results may be taken together to make decisions about the acceptability of the assay results. The rules most often used to make decisions about acceptability of internal control data are those of Westgard (Westgard *et al.*, 1977, 1981; Koch *et al.*, 1990). These rules have been given names based on a letter or number (e.g. 'R' for range error, '1' for a rule based on one control result, '10' for a rule based on ten control results) with a subscript suffix (e.g. '2S' or '4S', indicating two and four s.d.s, respectively). Some of the most useful rules are outlined below. Rule violation may be determined visually from inspection of Shewhart plots or laboratory computer systems into which internal control data are input can be programmed to flag rule violations.

Westgard rule 1_{2S} One internal control is outside the ±2 s.d. limit. This type of result may warn of an impending problem. Other controls in the run and the controls in the previous run should be examined before accepting or rejecting the results.

Westgard rule 1_{3S} One internal control is outside the ±3 s.d. limits. Violation of this rule may result from random or systematic error. The results should not be accepted if one control violates this rule.

Westgard rule 2_{2S} Two consecutive internal controls are both outside either the +2 s.d. or the –2 s.d. limits. Violation detects systematic error.

Westgard rule R_{4S} The results of two consecutive internal controls are >4 s.d. apart. (For example, one is just over +2 s.d. and the next is –2 s.d.). This rule detects random error.

Westgard rule 4_{1S} This rule is violated when four consecutive controls are either +1 s.d. or –1 s.d. This rule detects systematic error. Rejection of the results is not mandatory. Violation of this rule may indicate that recalibration or instrument maintenance is indicated.

Westgard rule 10_x This rule is violated when ten consecutive internal controls are all on the same (+ or –) side of the expected value. This rule detects systematic error. Rejection of the results is not mandatory. Violation of this rule may indicate that recalibration or instrument maintenance is indicated.

Figure 7.3 shows some simulated control data and indicates where particular rules have been violated.

External quality assessment (EQA)

All laboratories should perform the IQC and, in addition, should participate in external quality assessment (EQA) where it is available. In EQA laboratories assay specimens of undivulged concentrations submitted to them by an outside agency. Such specimens are circulated through EQA programmes organized by voluntary, commercial or

Table II. Some schemes distributing EQA samples for antimicrobial assay

College of American Pathologists (USA)

The Therapeutic Drug Monitoring Surveys ZM and ZZM include gentamicin only and Surveys Z and ZZ include amikacin, gentamicin, tobramycin and vancomycin. Five 5 mL freeze-dried specimens are shipped four times per year. Performance is judged against target concentrations based on peer-group means. Gentamicin assay only is regulated by Federal requirements and acceptable performance is based on target concentration ±25%. For the other analytes performance is based on target concentration ±2 s.d. or ±10% (whichever is the greater).

KKGT (The Netherlands)

Five times per year KKGT distributes freeze-dried calf serum samples containing netilmicin, amikacin and flucytosine (I) and gentamicin, tobramycin and vancomycin (II) (two samples of each representing a pre- and post-dose sample). Participation is voluntary and performance is judged against weighed-in target concentrations.

UK NEQAS (UK)

The UK NEQAS distributes one liquid human serum sample per month (12 per year) for each of the following: amikacin, chloramphenicol, flucytosine, gentamicin, netilmicin, tobramycin and vancomycin assay. Participation is voluntary and performance is judged against weighed-in target concentrations based on the mean % error and s.d. about this mean for the six most recent distributions. Acceptable performance is based on a modulus mean % error + 2 s.d. of <30.

INSTAND (Germany)

Four times per year INSTAND distributes freeze-dried serum samples containing netilmicin alone (I) and gentamicin, tobramycin, amikacin and vancomycin together (II) (two samples of each). Target concentrations are set by expert laboratories. Participation is not mandatory and performance is assessed according to rules laid down by the Bundesärztekammer.

EQAS for Antibiotic Assay (Denmark)

This scheme distributes four shipments of three specimens each year.

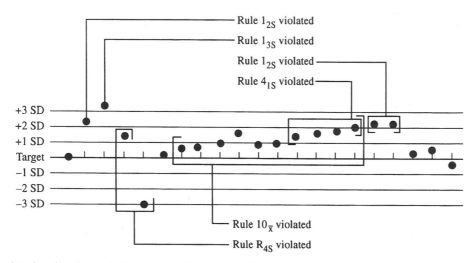

Figure 7.3 Shewhart plot showing the violation of some Westgard rules.

governmental agencies. EQA samples should be processed in the same manner as patient samples. In some countries where EQA distributions are used for the proficiency testing of laboratories, a signature may be required on the result form confirming that the EQA samples were treated in exactly the same way as patient samples.

EQA schemes

Many countries run their own national schemes and in some countries participation is mandatory for laboratory proficiency testing. Some schemes are funded centrally but most require the payment of an annual subscription.

Examples of some schemes which include antimicrobials are given in Table II. It should be noted, however, that very few of the antimicrobials referred to in this book are covered by EQA schemes. At the time of writing the only antimicrobials covered routinely were amikacin, gentamicin, netilmicin, tobramycin, chloramphenicol, flucytosine and vancomycin.

Providers of EQA usually circulate samples of pooled normal human serum spiked with known concentrations of the appropriate antimicrobials. Most, but not all, schemes circulate freeze-dried samples because these are more convenient to store and distribute than liquid samples. Frequency of distributions will vary from one per month to two or three per year. Some schemes ship batches of freeze-dried samples twice a year and the laboratory is required to reconstitute and assay the appropriate samples one at a time on a specified date. Distributions would normally include samples covering the entire range of therapeutically relevant concentrations. Some schemes may distribute samples with clinical histories to aid in the interpretation of the results. Some schemes will have each individual analyte in a separate vial, while others may combine them (as with some IQC samples), making the use of a microbiological assay impossible.

Methods of assessment for individual samples

Laboratories assay the sample received and submit the result obtained, usually by a certain deadline. The returns are processed and individual laboratory performance and mean performance are determined. Individual laboratory performance is determined based on a target concentration derived in one of three ways: (i) consensus or peer-group mean; (ii) weighed-in concentration; or (iii) results obtained by 'expert' laboratories.

Laboratory performance assessment based on consensus or peer-group means. Mean returns are determined and then mathematically 'trimmed' using one of several available methods. In trimming, any results lying a long way from the mean (outliers) are excluded and the mean recalculated. In this way a trimmed mean is calculated which gives a consensus value for the sample, free from bias due to a few very inaccurate results. Peer-group trimmed means are similarly calculated but here the results are first split into different peer-groups based on the method of assay. An individual laboratory is then assessed in comparison with these trimmed means.

The advantage of this method is that like is compared with like. The major disadvantages are that (i) if all members of the same peer-group get the same wrong result, the error goes undetected; and (ii) a relatively large number of participants is required to provide sound statistical data. The College of American Pathologists requires at least ten laboratories in a peer-group.

Laboratory performance assessment based on a weighed-in target concentration. The result of an individual laboratory is scored against the weighed-in target concentration. The advantages are that (i) the standard of performance is not controlled by any particular manufacturer or method; and (ii) useful data con be obtained even with only two participants. The disadvantages are that (i) this method of scoring is of no use if sample preparation, storage or shipping procedures cause the drug to be degraded; (ii) multiple analytes cannot be distributed in the same sample if there is any possibility of cross-reactivity with any of the assay techniques likely to be used.

Laboratory performance assessment based on results obtained by expert laboratories. Certain laboratories are chosen by the EQA provider to be designated expert laboratories. These laboratories assay the EQA samples by a reference method or acceptable routine method before they are distributed. When target concentrations obtained with the EQA samples may be method-dependent then a routine method of known reliability rather than reference methods will be used by the expert laboratory. The results obtained by the expert laboratories are used to set the target concentrations. This method of assessment is useful where sample preparation and manufacture procedures make it impossible to lay down reliable weighed-in target concentrations and/or where peer-groups may be too small (<10 laboratories) or too variable to be used for setting targets. The advantage is that target concentrations are known to be reliable. The disadvantage is that laboratories using the method used by the expert laboratories might be expected to perform better than those using other methods.

Overall performance assessment

Whatever method is used, determining laboratory performance on the basis of a single result may be useful if the result is highly inaccurate but is of little value if the result is not so, since an accurate result may be obtained purely by chance. Consequently overall performance is often assessed statistically on the results for three or more samples. Such an assessment allows accuracy (e.g. percentage error (sometimes called bias) of individual results) and precision (e.g. S.D. of the mean percentage error) to be determined.

The UK NEQAS for Antibiotic Assays—an example of how laboratory performance is assessed

UK NEQAS (UK National External Quality Assessment Schemes) is a consortium of UK providers of EQA schemes. The UK NEQAS for Antibiotic Assays distributes every month seperate samples for amikacin, chloramphenicol, flucytosine, gentamicin, netilmicin, tobramycin or vancomycin assay. They comprise liquid samples (*c*.1 mL) of pooled human serum spiked with a predetermined concentration of a single analyte. The laboratory receives the samples by post and is given 14 days to perform the assays and return the results. The

Table III. Simulated EQA returns from six consecutive monthly samples

	Percentage error of the returned result			
	laboratory A	laboratory B	laboratory C	laboratory D
Month 1	−23	−5	−17	+6.8
Month 2	−22	+3.5	−18	+7.1
Month 3	−31	−4.5	−21	+6.2
Month 4	−23	+3.1	+21	+7.3
Month 5	+2	+0.5	+18	+5.1
Month 6	+3.1	−5.5	+17	+9.9
Mean % error	−15.65	−1.32	0	+7.07
S.D.	14.47	4.18	20.53	1.6
Modulus of mean + 2 S.D (at month 6)	44.59	9.67	41.07	10.26

Laboratory A: Large negative bias corrected in month 5.
Laboratory B: Small negative bias, acceptable reproducibility.
Laboratory C: Large negative bias changing to large positive bias.
Laboratory D: Positive bias but very good reproducibility.

return for each analyte is scored against the chosen target concentration and the percentage error of each return is calculated. Laboratory performance is assessed statistically, not on the result for a single assay, but on the mean performance with the last six monthly samples. The returns for the six most recent monthly distributions are examined and if at least three (but preferably all six) returns have been received the mean percentage error (and the S.D. about this mean) is calculated. The modulus of this mean percentage error + 2 S.D. (MEAN +2 S.D.) is determined. A MEAN+2 S.D. of ≤30 is considered acceptable performance for a clinical laboratory. Table III shows some simulated EQA returns for four hypothetical laboratories. Laboratory A has a large negative mean percentage error and a large S.D. Looking at the individual months it appears that a problem producing a negative error of around 20% had been correct between months 4 and 5 and the returns for months 5 and 6 were much more accurate. If this trend to improved accuracy continues over months 7, 8 and 9, the mean error and the S.D. will continue to improve. The performance of Laboratory A is currently unacceptable but looks set to improve in the future. Laboratory B has a small negative mean error and a relatively small S.D. These returns would be considered very good for clinical antimicrobial assays. Laboratory C has a mean error of zero but a very large S.D. Looking at the individual returns it seems that attempts to cure a large negative bias of around 20% converted it to a large positive bias. Clearly this laboratory has problems which it needs to address. Performance at present does not look set to improve. Laboratory D uses a method which produces a reproducible positive bias of around 7%. Because the reproducibility is very good (small S.D.), the performance of this laboratory is acceptable.

Oversight mechanisms

In the UK, accredited EQA schemes report poor performance to relevant National External Quality Assessment Advisory Panels (NQAAPs) These EQA Panels comprise professionals representing various relevant professional bodies and with expertise in the area of the particular EQA scheme. Laboratories identified as poor performers are offered help and advice but, unless the problem is unresolved, their identity remains undisclosed to the NQAAP. UK NEQASs consider their role to be educational rather than punitive. Some other countries take a different approach to EQA; participation may be mandatory and poor performance may mean a laboratory is legally obliged to stop providing the assay service in question. In the USA the Medicare Act (Public Law 89-97 as amended), the Clinical Laboratory Improvement Act of 1967 (CLIA '67, Public Law 90-174) and the Clinical Laboratory Improvement Amendments of 1988 (CLIA '88, Public Law 100-578) require that laboratories successfully participate in an approved proficiency testing programme. Successful participation is defined as an overall score of at least 80% for each analyte tested. Unsuccessful participation in one distribution requires the laboratory to undertake appropriate training programmes to correct the problem responsible for the failure. Failure in two successive or two of three consecutive distributions may result in suspension of licence and/or termination of Medicare approval.

Laboratories concerned about the quality of the results they produce may find it beneficial to participate in more than one EQA scheme. The cost of EQA participation is small in relation to the costs of possible lost contracts, litigation or the resultant hidden costs associated with the release of inaccurate/unreliable results. Where no EQA schemes exist for a particular analyte interested laboratories would be well advised to collaborate in organizing smaller (perhaps local but possibly covering a wide geographical area) exchanges of samples for the purposes of EQA. Laboratories are also encouraged to participate in experimental or pilot EQA distributions especially as they are usually free.

Accreditation

With the increasing need, whether voluntarily or as a result of legislation, to provide objective assessment of the quality of any service offered, many laboratories performing assays will be working to achieve some form of quality standard laid down by a third party. In the UK an important quality standard for the operation of large and small businesses is that laid down by the British Standards Institute (BSI) in the form of BS 5750 or ISO 9000; many laboratories may comprise part of a company working to this standard. There are, however, quality standards aimed specifically at laboratories. Clinical laboratories in the UK will most probably be accredited by Clinical Pathology Accreditation (UK) Ltd (CPA). The College of American Pathologists has its own Laboratory Accreditation Program for laboratories in the USA. Alternatively the assay laboratory may be required to work to the standards of Good Laboratory Practice (GLP) for some, if not all, investigations. GLP is one of the most stringent quality standards to which a laboratory can work and is discussed below.

Good Laboratory Practice

Good Laboratory Practice (GLP) is concerned with the organizational processes and the conditions under which laboratory studies are planned, performed, monitored, recorded and reported. The concept of GLP began in the 1970s and the idea is now almost universally accepted. A good review of GLP is given by Broad & Dent (1990). Adherence to GLP principles ensures that studies are properly planned and that there are adequate facilities to carry them out and, in addition, facilitates the proper conduct of studies and promotes their full and accurate reporting together with a means whereby the integrity of studies can be verified. The application of GLP should ensure the quality and integrity of the data generated. In the UK it was first applied in 1982 to laboratories generating toxicology and other health and environmental safety data for submission to regulatory authorities. Many laboratories are required to work to GLP standards in the health and environmental safety testing of agrochemicals, cosmetics, food additives and pharmaceuticals. GLP is monitored by the UK GLP Compliance Monitoring Unit. Laboratories wishing to be inspected for GLP compliance will be inspected at their request or at the request of a

regulatory authority. Clinical laboratories in the UK are not required to work to GLP standards. However, a laboratory may wish or be asked to carry out a study to GLP standards. This is possible as long as the principles of GLP are adhered to. Some of the principles specifically relevant to assays are listed below:

Management. Scientists responsible for particular aspects of a study should be suitably qualified and experienced. Supporting staff should be similarly competent. The study director should ensure that the study is performed according to plan and any modifications are recorded. All data generated should be fully documented and the study plan, the final report, all raw data and specimens suitably archived at the end of the study.

Quality assurance. Each laboratory should have a formal arrangement assuring that facilities, equipment, staff, methods, procedures and documentation conform to GLP requirements. To achieve this the management should appoint a person to be responsible for the regular monitoring of the laboratory's operations and procedures and the periodic monitoring of each study for GLP compliance. The person concerned should be separate and independent of the study personnel and should report directly to the study director and management any problems likely to affect the integrity of the study.

Facilities. Storage areas for test, control and reference substances should be separated from each other and be adequate to preserve their identity, concentration, purity and stability. To avoid confusion similar laboratory operations from different studies should be separated in space or time. Adequate suitable space should be provided for secure storage of all data and specimens.

Equipment. All equipment should be properly calibrated and maintained. Records of repair and maintenance should be kept. Calibrations where possible should be traceable to national or international standards of measurement.

Standard operating procedures (SOPs). Written SOPs should be approved by the management and be adequate to ensure the quality and accuracy of the data generated. SOPs should cover: receipt, identification, characterization, handling, formulation and storage (including expiry date) of test, control and reference substances; preparation of samples; analyses; collection and identification of specimens; use, maintenance, cleaning and calibration of equipment (includes balances); identification of and instructions for use of relevant computer hardware and software; data collection, handling, storage and retrieval; report preparation; quality assurance procedures. All reagents and solutions used in the study should be labelled to indicate identity, concentration, storage requirements and shelf life. Adequate information on the source and the chemical identity of the test substance, its physical form, physical properties, degree of purity (and identity of significant impurities) and stability should be made available before the study commences.

Study plan and conduct. The written study plan should include: title; statement of purpose; name of sponsor; starting and completion dates; justification for the assay system used; characterization of the test system; description of experimental design including details of randomization; statistical methods used; records to be maintained; dated signature of Study Director.

Data storage. Specimens from the study should be retained for as long as it could reasonably be expected that the quality of the specimen would permit re-evaluation. All raw data must be kept in an indexed archive to facilitate rapid retrieval.

References

Broad, R. D. & Dent, N. J. (1990). An introduction to good laboratory practice (GLP). In *Good Laboratory and Clinical Practices* (Carson, P. A. & Dent, N. J., Eds), pp. 3–15. Heinemann Newnes, London.

Cederlund, H. & Mårdh, P.-A. (1993) Antibacterial activities of non-antibiotic drugs. *Journal of Antimicrobial Chemotherapy* **32**, 355–65.

Hewitt, W. & Vincent, S. (1989). *Theory and Application of Microbiological Assay.* Academic Press, London.

Hostetler, J. S., Heykants, J., Clemons, K. V., Woestenborghs, R., Hanson, L. H. & Stevens, D. A. (1993). Discrepancies in bioassay and chromatography determinations explained by metabolism of itraconazole to hydroxyitraconazole: studies of interpatient variations in concentrations. *Antimicrobial Agents and Chemotherapy* **37**, 2224–7.

Koch, D. D., Oryall, J. J., Quam, E. F., Feldbruegge, D. H., Dowd, D. E. & Barry, P. L. (1990). Selection of medically useful quality control procedures for individual tests done in a multitest analytical system. *Clinical Chemistry* **36**, 230–3.

National Committee for Clinical Laboratory Standards. (1991). *Internal Quality Control Testing: Principles and Definitions.* Approved Guideline C24-A. NCCLS, Villanova, PA.

Phillips, L. J., Alexander, J. & Hill, H. M. (1990) Quantitative characterization of analytical methods. In *Analysis for Drugs and Metabolites, Including Anti-infective Agents* (Reid, E. & Wilson, I. D., Eds), pp. 23–36. Royal Society of Chemistry, Cambridge.

Westgard, J. O., Groth, T., Aronsson, T & de Verdier, C.-H. (1977). Combined Shewhart–cusum control chart for improved quality control in clinical chemistry. *Clinical Chemistry* **23**, 1881–7.

Westgard, J. O., Barry, P. L., Hunt, M. R. & Groth, T. (1981). A multi-rule Shewhart chart for quality control in clinical chemistry. *Clinical Chemistry* **27**, 493–501.

White, L. O., Reeves, D. S., Holt, H. A. & Bywater, M. J. (1981). Microbiological assay of cefotaxime. *Journal of Antimicrobial Chemotherapy* **7**, 308–9.

White, L. O. & Reeves, D. S. (1998) Antibiotic assays. In *Quality Control, Principles and Practice in the Microbiological Laboratory*, 2nd edn (Snell, J. J. S., Farrell, I. D. & Roberts, C., Eds), Public Health Laboratory Service, London. In press.

Wright, D. N. & Matsen, J. M. (1980). Bioassay of antibiotics in fluids from patients receiving cancer chemotherapeutic agents. *Antimicrobial Agents and Chemotherapy* **17**, 417–22.

Appendix: The use of Gilson variable-volume pipettes

This information is included because correct usage of variable volume pipettes is essential for obtaining reliable assay results.

Normal operation

1. Use a new tip each time
2. Select the volume required. When increasing the volume it is necessary to turn the wheel to just over the required setting and then turn it back slowly to the correct position.
3. Use the appropriate colour tip as follows:

Volume to be measured (μL)	Tip colour	Test volumes (μL)
2–20	yellow	2, 10, 20
20–100	yellow	20, 50, 100
50–200	yellow	50, 100, 200
200–1000	blue	200, 500, 1000
1000–5000	white	1000, 2000, 5000

4. Place the tip firmly on to the shaft with a slight twisting motion to ensure an air-tight seal.
5. Depress the plunger to the first stop and place 3–4 mm into the liquid to be sampled. Release the plunger slowly and allow the liquid to rise into the pipette. Remove the pipette and wipe the tip on the side of the container.
6. Expel the liquid back into the container by pressing to the first stop and then pressing down hard to the second stop. This procedure pre-rinses the tip.
7. Repeat step 5.
8. Place the tip into the receiving vessel and touch the side, expel the liquid slowly by pressing to the first stop and then press down hard to the second stop to ensure full delivery.
9. Remove tip from vessel, release the plunger and discard the tip.

Reverse-pipetting using the second stop

This method should be used when it is important not to introduce air bubbles into the sample (e.g. loading a TDX or Solaris analyser).

1. Proceed as in steps 1–4 above.
2. Depress the plunger all the way to the second stop, place the tip in the liquid and draw up liquid slowly. Wipe the tip against the side of the container.

3. Expel the liquid into the receiving vessel by depressing the plunger to the first stop, touching the side of the vessel with the tip. Do **not** depress to second stop.
4. As step 9 above.

Quality control of variable-volume pipettes

Pipettes should be checked at least once every 6 months, or at the start of a new study, and the results kept in a log. The accuracy and precision of each pipette are assessed using three different volumes of distilled water (see *Normal operation*, step 3, above)

1. Note the pipette's serial number.
2. Place a weighing boat on an accurately calibrated analytical balance.

3. Pipette the correct test volume of distilled water into the weighing boat and immediately record the weight.
4. Replicate each volume 10 times and calculate the mean weight, S.D. and CV%. Assume that 1 μL of distilled water weighs 1 mg.

The mean volume should be within 2% of the set volume and the CV% should be <1 (except for the Gilson P20 at 2 μL, which should give a mean within 5% of 2 and a CV% of <1.5. Pipettes failing to achieve this standard of performance should be professionally serviced and recalibrated.

Chapter 8

β-Lactams

Les O. White[a] and Jenny M. Andrews[b]

[a]*Department of Medical Microbiology, Southmead Hospital, Bristol BS10 5NB;* [b]*Department of Medical Microbiology, City Hospital NHS Trust, Birmingham B18 7QH, UK*

Introduction

The era of true antibiotics (naturally occurring anti-microbial agents which act in low concentration) began in 1929 when Alexander Fleming observed the anti-staphylococcal activity of *Penicillium notatum*. This activity was due to penicillin (benzyl penicillin), the first β-lactam antimicrobial to be discovered. Sixteen years later, another β-lactam antibiotic, cephalosporin C, was isolated from *Cephalosporium acremonium* and the foundations for the development of these agents were complete. These agents all share the four-membered, nitrogen-containing, β-lactam ring which is essential for their antimicrobial activity. Numerous naturally occurring and semi-synthetic β-lactams have been described and many of these are used clinically. Most are antimicrobials but some are inhibitors of β-lactamases, those bacterial enzymes which inactivate β-lactams by hydrolysing the β-lactam ring. The newer β-lactams have improved antimicrobial spectrum and/or improved pharmacokinetic properties and/or improved β-lactamase stability compared with the early agents.

The β-lactams act by inhibiting bacterial cell wall synthesis (to which they owe their selective toxicity). The site of action is a transpeptidation reaction which cross-links cell wall peptide chains with the elimination of D-alanine from a peptide side-chain with a terminal D-alanyl-D-alanine residue. The β-lactams inhibit transpeptidases, enzymes which remove the D-alanine (D,D-carboxy-peptidases) and which form the cross-link. It is proposed that they do this by being structural analogues of acyl-D-alanyl-D-alanine (see Bryan & Godfrey, 1991).

Chemistry and nomenclature

A β-lactam can be a single ring structure (azetidin-2-one) but most have two central rings and there are some multi-cyclic compounds under development; they are weak acids and some are amphoteric. Penicillins (penams, Figure 8.1a) have a sulphur-containing, five-membered, thiazolidine ring adjacent to the β-lactam ring. If this ring contains a double bond, the compound is a penem (Figure 8.1b). The only clinically important penems are the carbapenems which have carbon in place of the sulphur. If the sulphur is

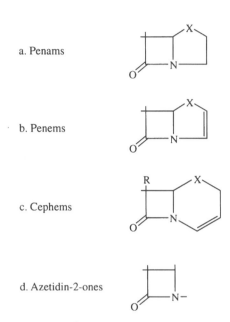

a. Penams

b. Penems

c. Cephems

d. Azetidin-2-ones

Figure 8.1 Types of β-lactam. X is usually sulphur but can be oxygen (oxa-) or carbon (carba-). R is hydrogen in cephalosporins, and OCH₃ in cephamycins.

replaced by an oxygen the compound is a clavam. Cephems (Figure 8.1c) have a sulphur-containing, six-membered, dihydrothiazine ring and these compounds include cephalosporins and cephamycins which differ in that the latter have a methoxy substitution (R in Figure 8.1c). Latamoxef, which is no longer in clinical usage, is an oxacephem. The monocyclic azetidin-2-ones include the monobactams.

Cephalosporins

Cephalosporins, of which there are many in clinical usage, are usually further classified and several schemes have been suggested (O'Callaghan, 1979; Bryskier *et al.*, 1990). Those in clinical usage are usually divided into rather ill-defined 'generations' on the basis of their structure, antimicrobial spectrum and resistance to degradation by β-lactamases:

- 'First-generation' cephalosporins: the original cephalosporins, such as cephalothin and cephalexin.

- 'Second-generation' cephalosporins: later compounds with β-lactamase resistance and better activity against Gram-negative bacteria; an example is cefuroxime.
- 'Third-generation' cephalosporins: compounds with increased activity, extended antimicrobial spectrum and β-lactamase resistance (e.g. cefotaxime or ceftazidime). Bryskier *et al.* (1990) defined a third-generation cephalosporin as any compound having two or more of the following properties: (i) a 2-aminothiazolyl ring; a broad antimicrobial spectrum; (ii) a MIC_{90} for Enterobacteriaceae (excluding certain cephalosporinase-producing organisms) of <1 mg/L; (iii) antipseudomonal activity; (iv) high resistance to plasmid-mediated β-lactamases excluding extended-spectrum β-lactamases.
- 'Fourth-generation' cephalosporins: the more recent compounds with improved anti-Gram-positive activity and activity against strains resistant to third-generation agents (e.g. cefpirome or cefepime).

One agent which does not fit into this classification is the narrow-spectrum anti-pseudomonal cephalosporin, cefsulodin.

The spelling of the names of the various cephems depends on when the agent was developed: those described before 1975 have names beginning with 'ceph', while later ones begin with 'cef'.

Penicillins

Penicillins are usually classified as follows:

- narrow-spectrum agents (e.g. benzyl penicillin);
- wide-spectrum agents (e.g. ampicillin);
- agents resistant to staphylococcal β-lactamase (e.g. flucloxacillin);
- agents with anti-pseudomonal activity (e.g. piperacillin).

Dosage forms

The β-lactams are available for either oral or parenteral administration; a few (amoxycillin, ampicillin, flucloxacillin, cefuroxime and cephradine) have both oral and parenteral preparations. The preparations available in the UK at the time of writing are listed below. Drugs no longer available for clinical use in the UK include carbenicillin, methicillin and mezlocillin.

Agents with only oral formulations

Those agents for oral-only use usually come as tablets, capsule, syrups and/or suspensions containing additional inactive ingredients. Individual preparations include:

- cefaclor (Figure 8.3a): available in all forms and as a special slow-release tablet to help maintain sustained blood levels;

- cefadroxil (Figure 8.3b): available as the monohydrate in capsules and suspension;
- cefixime (Figure 8.3f): available as tablet and powder;
- cefpodoxime (Figure 8.3q): available as the 1-isopropyl-oxycarbonyloxy-ethyl ester (proxetil) in tablet form 130 mg equivalent to 100 mg cefpodoxime;
- ceftibuten (Figure 8.3t): available as the dihydrate in capsule or powder;
- cephalexin (Figure 8.3y); available as the monohydrate in various forms.

Agents with only parenteral formulations

Parenteral agents come for iv or im usage and some may contain additional ingredients. Individual preparations include:

- azlocillin (Figure 8.2d): available in 1 or 2 g amounts as the sodium salt;
- aztreonam (Figure 8.4a): available in 500 mg to 2 g vials in combination with 780 mg/g L-arginine;
- benzyl penicillin (Figure 8.2e): available in 600 mg (1 megaunit) vials as the sodium salt and in 60/300 mg vials as a mixture of sodium and procaine formulations;
- cefotaxime (Figure 8.3l): available in 500 mg to 2 g vials as the sodium salt;
- cefoxitin (Figure 8.3o): available in 1–2 g vials as the sodium salt;
- cefpirome (Figure 8.3p): available as the sulphate together with 242 mg/g sodium carbonate in 250 mg to 2 g vials;
- ceftazidime (Figure 8.3s): available as the pentahydrate together with 118 mg/g sodium carbonate in 250 mg to 3 g vials;
- ceftizoxime (Figure 8.3u): available in 1 g vials as the sodium salt;
- ceftriaxone (Figure 8.3v): available as the disodium monohydrate in 250 mg to 2 g vials;
- cephamandole (Figure 8.3c1): available as the nafate salt equivalent to 1 g cephamandole with 63 mg/g sodium carbonate;
- cephazolin (Figure 8.3d1): available as the sodium salt in 500 mg to 1 g vials;
- imipenem (Figure 8.4d): available as the monohydrate 1:1 with cilastatin sodium in vials containing 250–500 mg imipenem;
- meropenem (Figure 8.4e): available as the trihydrate with 208 mg/g sodium carbonate in 250 mg to 1 g vials;
- piperacillin (Figure 8.2m): available as the sodium salt with or without tazobactam sodium (one part tazobactam (Figure 8.4g) to eight parts piperacillin) in 1–4 g vials;
- temocillin (Figure 8.2n): available as the sodium salt in 1 g vials;

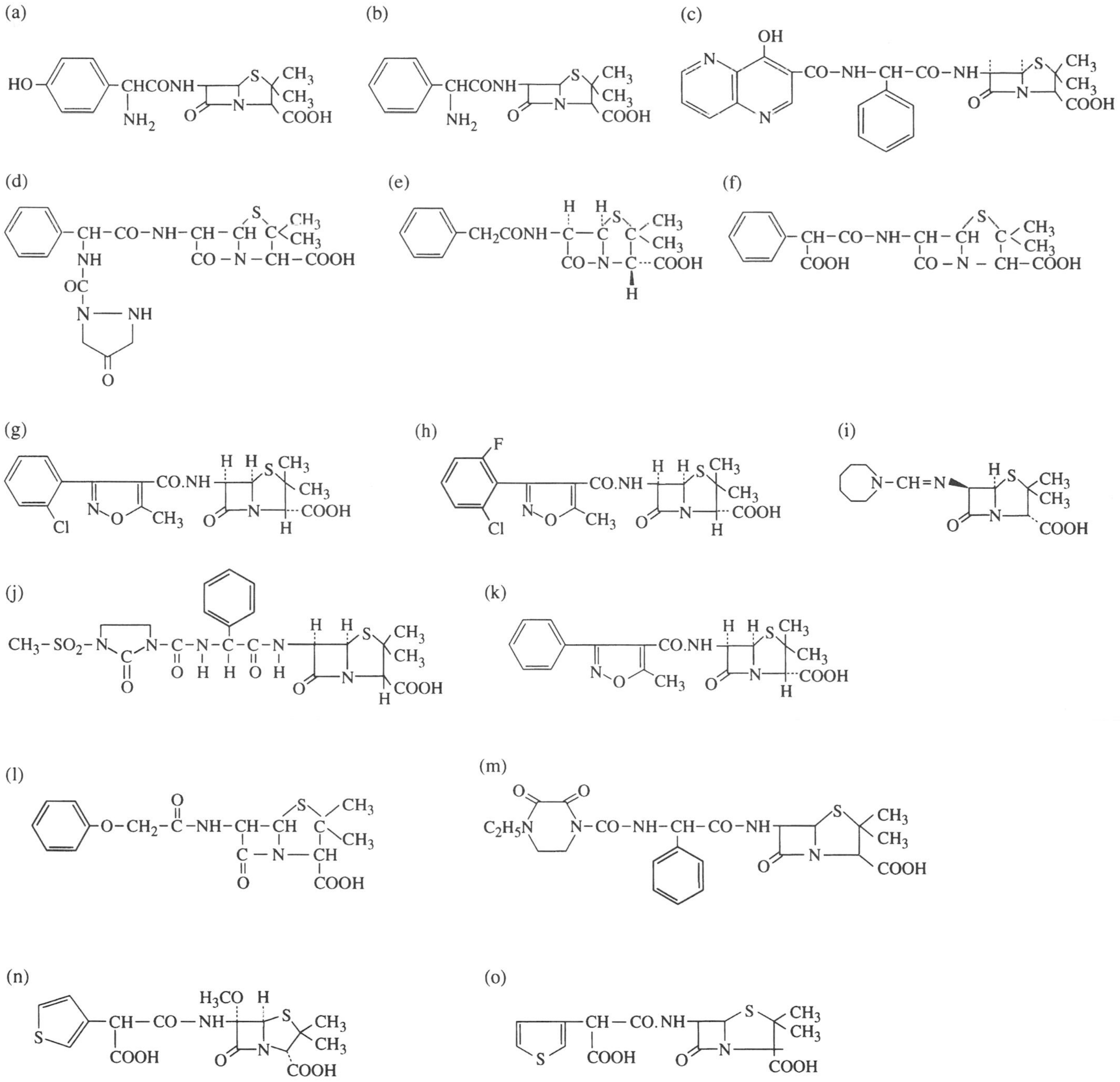

Figure 8.2 Structure of some penam antimicrobials (penicillins). (a) Amoxycillin, (b) ampicillin, (c) apalcillin, (d) azlocillin, (e) benzyl penicillin (penicillin G), (f) carbenicillin, (g) cloxacillin, (h) flucloxacillin, (i) mecillinam (amdinocillin), (j) mezlocillin, (k) oxacillin, (l) phenoxymethyl penicillin (penicillin V), (m) piperacillin, (n) temocillin, (o) ticarcillin.

• ticarcillin (Figure 8.2o): available as the sodium salt with potassium clavulanate (15:1) in vials containing 1.5 or 3 g ticarcillin.

Agents with both oral and parenteral formulations

These formulations include:

• amoxycillin (Figure 8.2a): available as the trihydrate for oral formulations and the sodium salt for parenteral

use. It is available alone or in combination with potassium clavulanate, as five parts amoxycillin to one part clavulanic acid (Figure 8.4c) for parenteral use or in ratios of 250:125 (tablets) and 250:62 (suspension);

• ampicillin (Figure 8.2b): available as the trihydrate or pivaloyloxymethyl ester (pivampicillin) for oral use and the sodium salt for parenteral use; it also comes in combination with cloxacillin or flucloxacillin (see below);

• cefuroxime (Figure 8.3w): available as the sodium salt

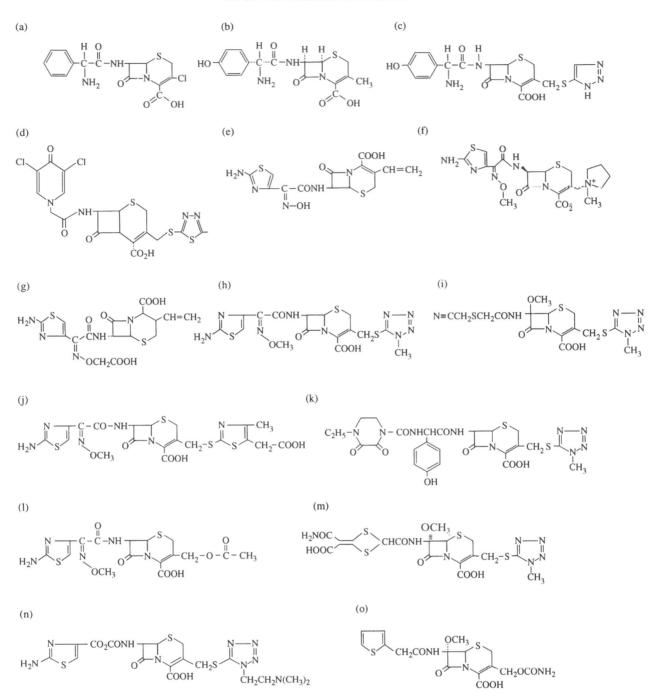

Figure 8.3 Structure of some cephem antimicrobials (cephalosporins, cephamycins, oxacephems). (a) Cefaclor, (b) cefadroxil, (c) cefatriazine, (d) cefazedone, (e) cefdinir, (f) cefepime, (g) cefixime, (h) cefmenoxime, (i) cefmetazole, (j) cefodizime, (k) cefoperazone, (l) cefotaxime, (m) cefotetan, (n) cefotiam, (o) cefoxitin, (p) cefpirome, (q) cefpodoxime, (r) cefsulodin, (s) ceftazidime, (t) ceftibuten, (u) ceftizoxime, (v) ceftriaxone, (w) cefuroxime, (x) cephacetrile, (y) cephalexin, (z) cephaloglycin, (a1) cephaloridine, (b1) cephalothin, (c1) cephamandole, (d1) cephazolin, (e1) cephradine, (f1) latamoxef.

in 250 mg to 1.5 g vials for parenteral use or as the axetil ester in oral formulations;

- cephradine (Figure 8.3e1): available as oral formulations or in 500 mg to 1 g vials in combination with L-arginine for parenteral use;
- cloxacillin (Figure 8.2g): available as the sodium salt only in combination with ampicillin trihydrate for oral

use (1:1 or 2:1 ampicillin:cloxacillin) or ampicillin sodium (1:1 or 2:1 ampicillin:cloxacillin in 75–500 mg vials) for parenteral use;

- flucloxacillin (Figure 8.2h); the sodium salt is used in both oral and parenteral (250 mg to 1 g vials) formulations and flucloxacillin sodium is also available in combination with ampicillin trihydrate for oral use or

(p)

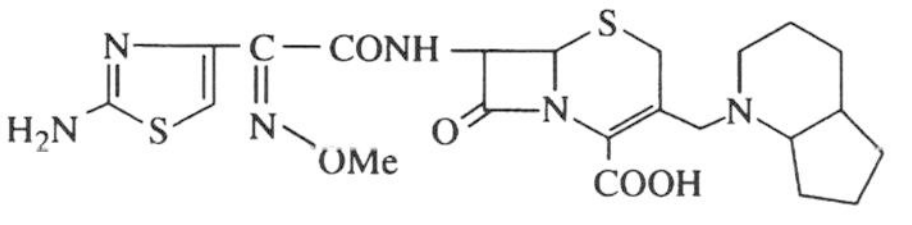

(q)

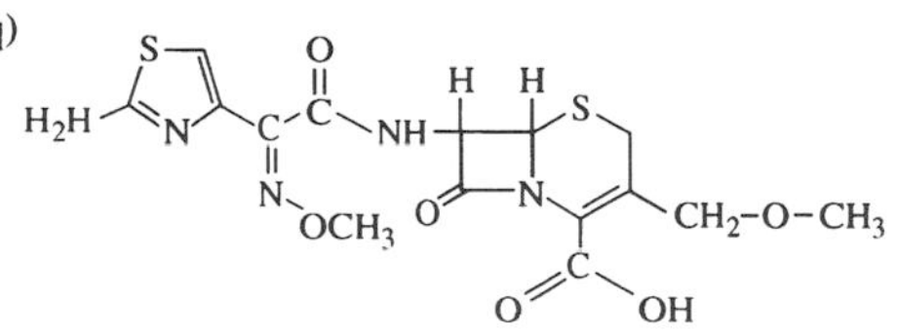

(r)

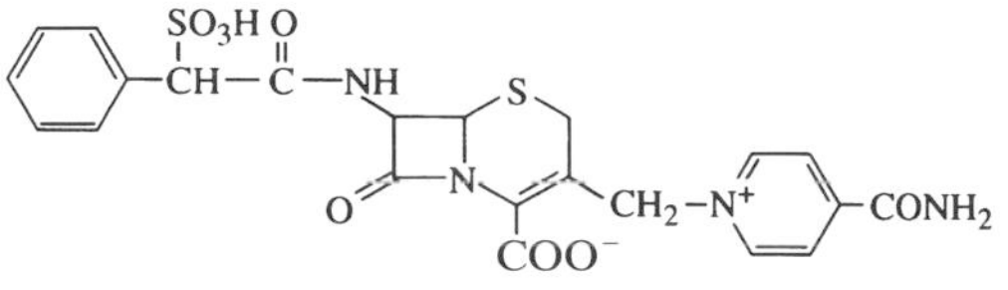

(s)

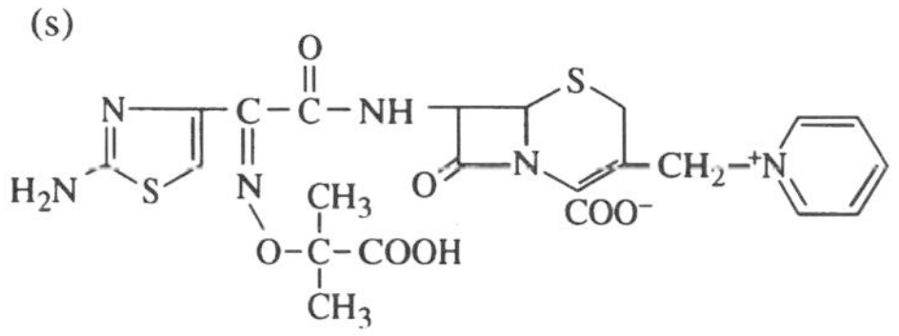

(t)

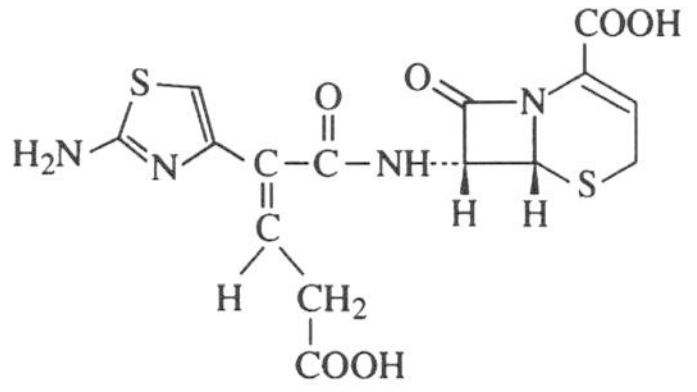

(u)

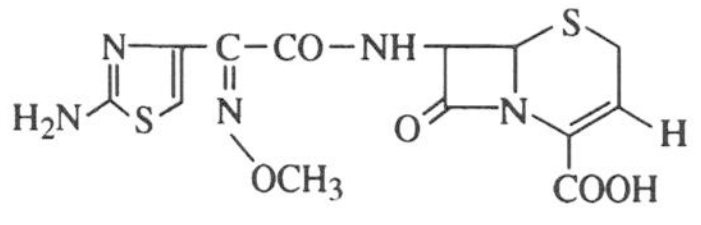

(v)

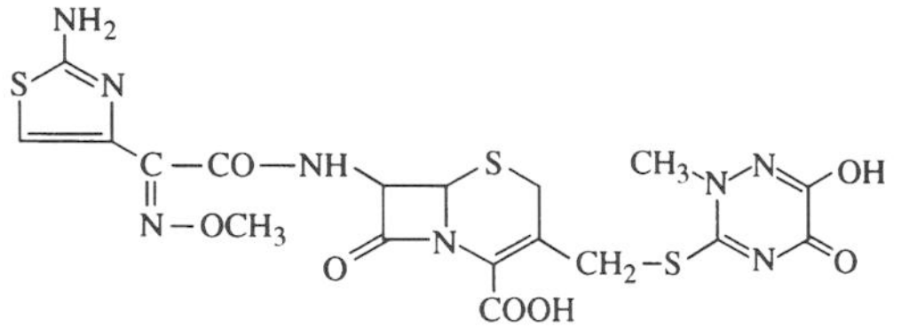

(w)

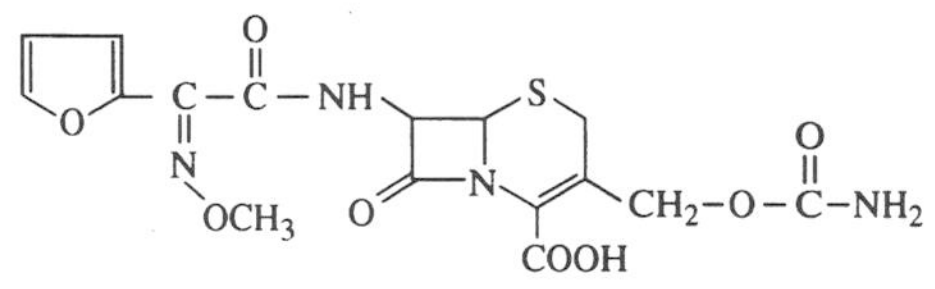

(x)

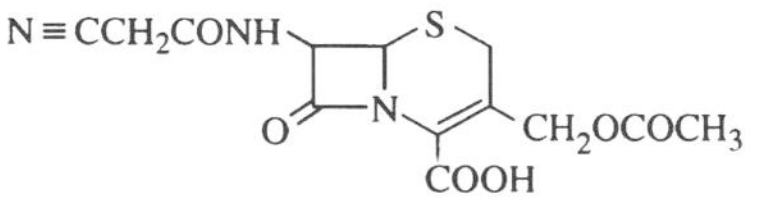

(y)

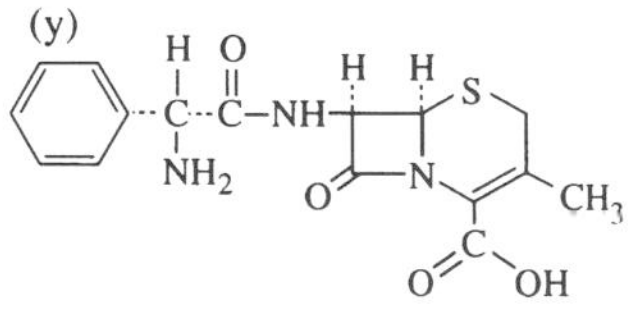

(z)

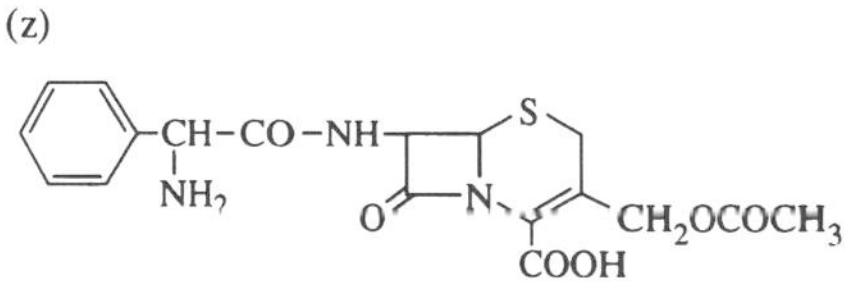

(a1)

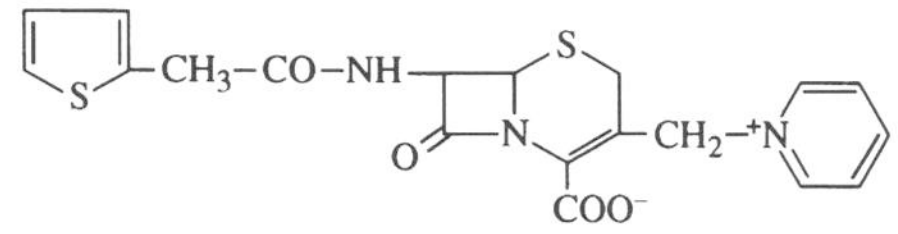

(b1)

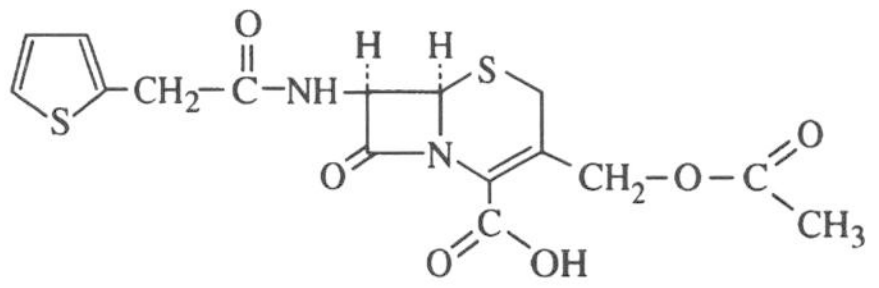

(c1)

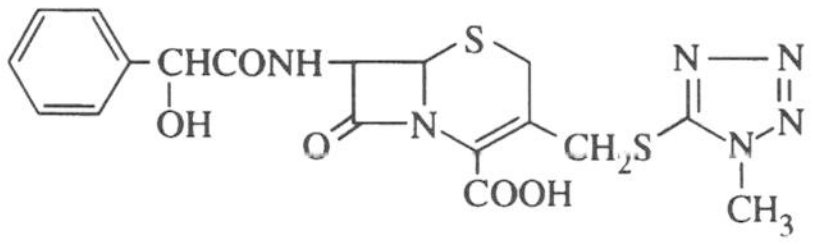

(d1)

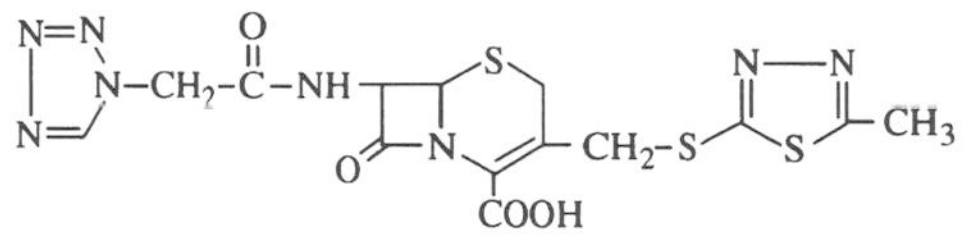

(e1)

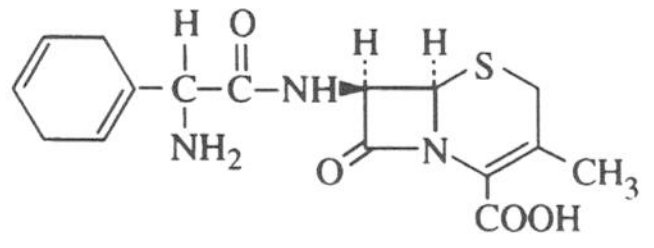

(f1)

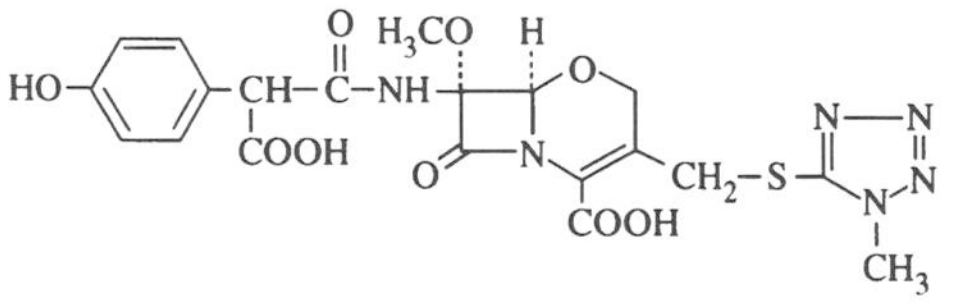

ampicillin sodium for parenteral use (1:1 ampicillin: flucloxacillin).

Pharmacokinetics

Absorption

Some β-lactams can be administered orally (see above, as parent compound or as a prodrug), whereas others (e.g. benzyl penicillin, ceftazidime) cannot by reason of acid instability or poor lipophilicity or both. Food enhances absorption of some compounds and hinders the absorption of others. Ampicillin absorption is poor and variable compared with amoxycillin or ampicillin esters. Bioavailability can vary from 40% (ampicillin) to 90% (cephalexin). Assays may occasionally be useful when moving from parenteral to oral therapy to confirm acceptable bioavailability.

Distribution

These drugs distribute into most extracellular fluids well but do not significantly enter cells. Penetration into CSF is

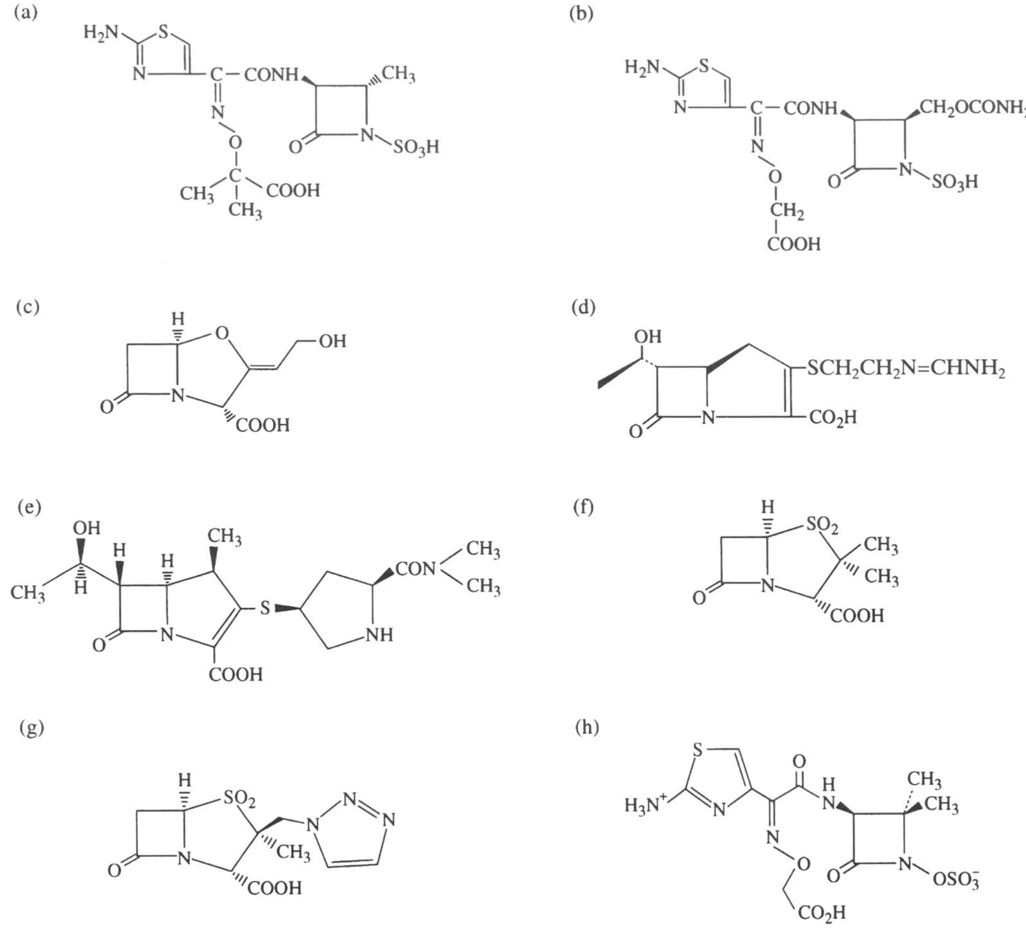

Figure 8.4 Structures of some other β-lactam antimicrobials and β-lactamase inhibitors. (a) Aztreonam, (b) carumonam, (c) clavulanic acid, (d) imipenem, (e) meropenem, (f) sulbactam, (g) tazobactam, (h) tigemonam.

usually poor, although high-dosage cefotaxime can be used to treat meningitis. Protein binding is low to moderate for most compounds (Table I), but high (95%) for ceftriaxone and flucloxacillin.

Elimination

All β-lactams are excreted to a large extent in the urine by glomerular filtration and, for penicillins and some cephalosporins, tubular excretion. Co-administration of probenecid with drugs excreted by tubular secretion can prolong blood levels by increasing the elimination half-life ($t_{1/2}$). The $t_{1/2}$ values for most β-lactams are short, thus necessitating frequent administration (ceftriaxone is an exception with a $t_{1/2}$ of >6 h) and prolonged by renal impairment (see *Chapter 3*) where accumulation will occur if appropriate dosage modification is not made.

Many β-lactams are also metabolized to a greater or lesser extent. Penicillins are converted to their microbiologically inactive '-oic' acid derivative by cleavage of the β-lactam ring by liver enzymes and/or thermal degradation. These breakdown products, which are excreted in the urine, have no antimicrobial activity but

may play a role in mediating some adverse reactions. Cleavage of the β-lactam ring in aztreonam also forms a stable inactive metabolite which can be assayed by high-performance liquid chromatography (HPLC) (Lindner & Pilkiewicz, 1986) but cephalosporin molecules tend to break up once the β-lactam ring is cleaved. One exception is cefotaxime: the open-ring metabolites of desacetylcefotaxime lactone are stable and accumulate in the serum of renally impaired patients (Reeves *et al.*, 1980). Carbapenems, too, are metabolized by β-ring cleavage by a dipeptidase enzyme found in the kidneys (dehydropeptidase I). This metabolism is so great in the case of imipenem that it is always co-administered with cilastatin, an inhibitor of dehydropeptidase I.

A few compounds are metabolized to a significant degree to microbiologically active derivatives. Cephalosporins with an acetoxy group, such as cephalothin or cefotaxime, are the important examples as they undergo significant desacetylation. Desacetylcefotaxime is a highly active drug, but not as active as cefotaxime, and with a narrower spectrum. Serum cefotaxime can only be assayed accurately by bioassay if an indicator organism resistant to desacetylcefotaxime is used. The enzyme(s) responsible

Table I. Physicochemical and pharmacokinetic data for β-lactams

Class/drug	Molecular weight	pKa	Dosage forms	Elimination half-life (h)	Protein binding (%)	Metabolites	Urinary excretion
Carbapenems							
imipenem	299.4	3.2, 9.9	monohydrate	1	20–25	β-lactam ring cleaved by renal DHP-I	70% when metabolism inhibited by cilastatin. Unstable in solution
meropenem	383.46	2.9, 7.4	trihydrate	1	5	open-ring	70% unchanged; 20% metabolite
Cephalosporins							
cefaclor	367.8	1.5, 7.2	monohydrate	0.5–1	25	unstable in serum	50–60% unchanged; 30% inactive
cefadroxil	363.4	3.6, 7.5, 10	monohydrate	1.3–1.6	20	none	90% unchanged
cephamandole	462.5	2.6–2.9	sodium nafate	0.8	c.75	nafate hydrolyses to cephamandole	80% unchanged
cephazolin	454.5	2.1	sodium salt	1.6–2.2	80	none	90% unchanged
cefepime	517.5		hydrochloride monohydrate	2–2.5	16	7-epimer, n-methyl-pyrrolidine (NMP) NMP oxide (NMPO)	88% unchanged NMPO, 7%; 7-epimer, 2.5%; NMP, <1%
cefixime	453.4	2–2.5	trihydrate	2.5–3.8	65–70	none	12–34%, oral dose
cefmenoxime	511.55	2.8, 3.2	hydrochloride	1	43–75	none	85% unchanged
cefodizime	584.65	2.9, 3.4, 4.2	disodium	2–8	73–89	none	80% unchanged
cefoperazone	645.65	2.8, 10	sodium salt	1.7–2	c.90	A–F	<30%
cefotaxime	455.5	3.8	sodium salt	0.9–1.3	40	UP1 and UP2 (open-ring metabolites) and active desacetylcefotaxime	up to 60% unchanged 20–25% desacetyl metabolite c. 6% as UP1 or UP2
cefpodoxime	427.45		proxetil	2.2–2.7	40	none	80% unchanged
cefpirome sulphate	610.6	2.51, 2.81		2–2.2	c.10	none	80% unchanged
cefsulodin	532.5	3.3–4.8	sodium salt	1.5	25–35	none	90% unchanged
ceftazidime	546.6	1.8, 2.7, 4.1	pentahydrate	1.8–2.2	10–17	none	90% of dose, all unchanged
ceftibuten	410.4	2.17, 3.67, 4.07	dihydrate	2.4–2.5	60–62	conversion to trans-ceftibuten	59% unchanged 10% trans
ceftizoxime	383.4	2.9	sodium salt	1.3–1.6	c.30	none	90% of dose, all unchanged
ceftriaxone	554.6		disodium salt	6–9	95	inactive metabolites in faeces	56% of dose, all unchanged
cefuroxime	424.4	2.5	sodium salt axetil (oral)	1.2	30–35	axetil rapidly hydrolysed	≥90% of dose, >95% unchanged
cephalexin	347.4	3.6, 5.3, 7.3	monohydrate (oral)	0.9–1	6–15	none	90% unchanged
cephaloridine	415.5	3.4		1.4	20	none	80% unchanged
cephalothin	396.4	2.2–2.4	sodium salt	0.3–0.9	65–72	desacetyl metabolite (active)	65% unchanged, remainder metabolite
cephradine	349.4	2.6, 7.3		0.3–3[a]	<10	none	90% unchanged
Cephamycins							
cefmetazole	471.52	2.8		1	85	none	75% unchanged
cefotetan	575.6	2.1, 3.3	disodium salt	2.8–4	c.88	none but epimers and a tautomer interchange	64–84% unchanged
cefoxitin	427.4	3.5	sodium salt	0.9–1	65–80	decarbamyl-cefoxitin	85% up to 5% as metabolite

Table I. *Continued*

Class/drug	Molecular weight	pK$_a$	Dosage forms	Elimination half-life (h)	Protein binding (%)	Metabolites	Urinary excretion
Monobactams							
aztreonam	435.4	−0.5, 2.6, 3.7	combined with L-arginine	1.7	50–56	open-ring and minor unidentified metabolites	66% unchanged, 7% open ring metabolite, 3% unidentified
Oxacephems							
latamoxef	520.5	2.4, 3.5, 9.9	disodium salt	1.3–2.6	40–50	none (?)	75% unchanged
Penicillins							
amoxycillin	365.4	2.4, 7.4, 9.6	trihydrate and sodium salt	1–1.2	c.20	penicilloic acid	60% (oral) to 75% (iv) with approx. 10% as metabolite
ampicillin	349.4	2.7, 7.2	trihydrate, sodium salt and various esters	1–2	c.20	penicilloic acid	30% (oral) to 75% (iv) with approx. 10% as metabolite
azlocillin	461.5		sodium salt	1–1.5	28–40	penicilloic and penilloic acids	50–70% of dose with up to 20% as metabolites
benzylpenicillin (penicillin G)	334.4	2.8	sodium or potassium salts	0.3–1	50–65	penicilloic acid	95%, 19% as metabolite
carbenicillin	378.4	2.6, 3.3	disodium salt	1–1.5	45–50	little or none reported	>90% unchanged
cloxacillin	435.9	2.7	sodium salt monohydrate	0.5–1	94	see flucloxacillin	56% unchanged, 10.6% as (1) 12.3% as (2), 2.2% as (3)
flucloxacillin	453.9	2.7	sodium salt monohydrate	0.6–1.3	95	(1) penicilloic acid, (2) active 5-hydroxy-methyl metabolite and (3) its penicilloic acid	65% unchanged, 3.8% as (1) 10.5% as (2), 1% as (3)
mecillinam	325.4	3.4, 8.9	acid pivmecillinam hydrochloride (oral) free base	1	5–10	spontaneously hydrolyses to penicilloic acid and active *N*-formyl-6-amino penicillanic acid	60% but unstable
mezlocillin	539.6	2.7	sodium salt monohydrate	0.9–1.2^b	27–55	penicilloic acid and decarboxylated penicilloic acid	70% (10–15% as metabolites)
phenoxymethyl-penicillin (penicillin V)	350.4	2.7	potassium salt	0.5	80	penicilloic acid	70% (approx. half as metabolite)
piperacillin	517.6		sodium salt	0.6–1.3^b	16–22	none	80% unchanged
ticarcillin	384.4	2.4, 3.6	disodium salt	1–1.3	45–65	penicilloic acid	95% (14% as metabolite)
β-Lactamase inhibitors							
clavulanic acid	199.2		potassium salt	1	22–30%	not known	up to 40%
tazobactam	300.3	2.1	sodium salt with piperacillin	<1	20–25	open-ring metabolite	50–60%, 25% as metabolite

^{a} Depending on route administered.
^{b} Dose-dependent.

for this desacetylation are found in the liver, red blood cells and, in the case of neonates, serum. In blood samples taken from neonates, marked desacetylation of cefotaxime can therefore occur after the sample has been taken (Crooks *et al.*, 1984); the same is true of a badly haemolysed sample from an adult.

Some pharmacokinetic data for *β*-lactams are summarized in Table I and in *Chapter 3*.

Toxicity

The *β*-lactams are considered to be very well tolerated and are often given in large doses. Side-effects are various and usually mild and reversible. Some are due to allergy and cannot be related to high blood levels of the drug. Other toxicities may be dose-related but there are few data quoting specific blood or fluid levels. Often 'high' blood levels are thought to be responsible for observed adverse reactions and this can only be interpreted as 'levels higher than those seen after normal dosages'.

The major serious side-effect of *β*-lactam therapy is anaphylaxis, which is most commonly seen with benzyl penicillin. Other less severe allergic reactions of immediate or delayed appearance may also be seen with most *β*-lactams. Penicillin allergy should be assumed to cross-react with carbapenems but <10% cross-reactivity with cephalosporins is seen (Sutherland, 1997; Wise 1997).

In addition, ampicillin and amoxycillin produce a rash, which may be severe, in a small proportion of patients (or 95% of patients with infectious mononucleosis). It may be dose-related and is thought to be due to the formation of polymers.

Large doses of penicillin, especially in patients with renal impairment, can cause neurotoxicity and convulsions; CSF levels of benzyl penicillin >10 mg/L may be associated with the latter (see Sutherland, 1997). Imipenem and, to a lesser extent, meropenem may show neurotoxicity, especially in renally impaired patients.

Prolonged bleeding times may be seen in patients treated with carbenicillin, ticarcillin or latamoxef, especially those with renal impairment. This may be due to a concentration-dependent action on platelets.

There are some specific problems with drugs bearing a methylthiotetrazole (MTT) constituent (e.g. cephamandole, cefotetan, cefmenoxime or latamoxef), associated with the release of this side-chain. A disulphram-like reaction due to circulating MTT dimers has been seen, and severe hypoprothrombinaemia, reversible by vitamin K, may occur especially in malnourished or renally impaired patients.

Cephaloridine and, to a considerably lesser extent, cephazolin show renal toxicity.

Assay methods

Whatever assay is used, *β*-lactams should be considered as labile compounds. Samples should be rapidly collected and stored at low temperature. Some compounds, notably some early drugs such as cefaclor, cefatrizine and mecillinam, are unstable in serum and may continue to degrade even at low temperatures. Stability studies are essential if reliable data are not available. Some compounds require special treatment. Imipenem is unstable and plasma, serum or urine samples must be stabilized before storage. Plasma or serum should be mixed with an equal volume of 1 M 4-morpholine-ethanesulphonic acid buffer. Urines should be mixed with an equal volume of 4-morpholine-propanesulphonic acid buffer (Gravallese *et al.*, 1984). Samples may then be stored at −70°C to −80°C.

Microbiological plate assay or HPLC are the methods most likely to be used to assay a *β*-lactam in a clinical sample. Immunoassays have been reported but have not made a serious impact in the clinical field. An immunoassay for amoxycillin is described below for the reader's information. Any antiserum for use in an immunoassay must be specific for the parent compound. Raising an antiserum to a penicillin is likely to be difficult since degraded penicillin (penicilloic acid) is likely to be present in the preparation used to raise the antiserum. The described method for amoxycillin suffered from cross-reactivity with amoxicilloic acid.

Miscellaneous assay methods

Here details of an immunoassay and an enzymatic assay are described. Both suffer from problems which could be circumvented by the use of microbiological assay or HPLC methods.

ELISA assay of amoxycillin

This method (Hill *et al.*, 1992) uses an anti-amoxycillin antibody produced in-house by inoculation of rabbits with an amoxycillin–bovine serum albumin (BSA) conjugate and in-house sheep anti-rabbit IgG. It is likely to require modification if other antisera are used.

Peroxidase substrate. Dissolve 10 mg of *ortho*-phenylenediamine hydrochloride (Sigma, Poole, UK) in 20 mL of citrate–phosphate buffer pH 5.0. and add 20 μL hydrogen peroxide (BDH, Poole, UK).

Procedure
1. Bind amoxycillin to the wells of a microtitre plate by filling each well with 200 μL of amoxycillin 100 mg/L in 0.05 M bicarbonate buffer pH 9.6. Leave at room temperature for 2 h and then wash three times with phosphate-buffered saline containing 1% BSA and 1% Tween 20 (PBSBT).
2. Mix 100 μL of calibrators (25–250 μg/L) and samples with 100 μL of antibody diluted 1/400 in PBSBT and place into the microtitre wells. Also prepare control wells comprising: (i) antigen/antibody free (blank); (ii)

antibody only (maximum binding); (iii) non-immune control. Incubate for 1.5 h at 4°C.

3. After incubation wash three times with PBSBT and add 200 μL of diluted sheep anti-rabbit IgG. Incubate for 2 h at 4°C.

4. Wash three times again and add 200 μL of peroxidase substrate. After an appropriate time stop the reaction with 50 μL of 0.5 M citric acid.

5. Plot percent binding (taking the maximum binding as 100%) for each calibrator against log concentration. Interpolate unknowns from this curve. With sputum samples a procedure to compensate for overestimation is required.

Performance expectations. Within-assay CVs of 10% were reported. Substantial specificity problems were encountered. The assay cross-reacted with temocillin (50%) and slightly with cephamandole, but >100% with amoxycilloic acid. The assay overestimated sputum concentrations of amoxycillin even after the drug was converted to amoxycilloic acid (with β-lactamase) before assay. The reasons for this were not determined and a modified procedure in which 50 μg/L amoxycillin was added to one of duplicate samples was utilized.

Enzymatic assay of sulbactam

This assay (Sotto *et al.*, 1995) uses nitrocefin as an enzyme substrate and determines its rate of hydrolysis by measuring change in UV absorbance at 482 nm. TEM-I β-lactamase at an activity of 0.9 IU/mL is used.

Procedure. Prepare serum calibrators containing 0.5, 1, 2, 4, 8, 16, 32 and 50 mg/L sulbactam and a serum blank. Mix 5 μL of each calibrator, blank, and each sample with 10 μL β-lactamase and 170 μL 0.1 M phosphate buffer pH 7. Incubate for 5 min at 37°C. After this time add 25 μL of 1 mM nitrocefin and continue incubation for a further 30 min. Add 10 μL of 0.35 M cloxacillin to stop the reaction. Read absorbance at 482 nm. Unknown concentrations are interpolated from the calibration curve.

Performance expectations. The serum blank should remain stable throughout the assay. Amoxycillin did not interfere. Between-assay CVs of 12% (at 2 mg/L) to 5% (at 16 mg/L) were obtained and the lowest concentration which could be accurately assayed to ±10% was 1 mg/L. The assay correlated well with an HPLC method ($y = 1.3x - 0.56$; $r = 0.99$). Unfortunately it may now be impossible to obtain nitrocefin commercially.

Microbiological assays

The large plate method (*Chapter 4*) is recommended and consideration should be given to the indicator organism, the medium and the pH of the samples.

The assay organism should be one which produces sharp

zone edges and has sufficient sensitivity. However, where there are microbiologically active metabolites then the need for an organism which is insensitive to the metabolites may compromise these other properties. Although the organism produces less-than-optimal zone edges, White *et al.* (1981) used a strain of *Morganella morganii* resistant to desacetylcefotaxime to assay cefotaxime specifically in body fluids. Burnett & Sutherland (1970) found that Gram-positive organisms were unsuitable for the assay of carbenicillin because they were more sensitive to the small amount of benzyl penicillin contaminating commercial carbenicillin than they were to carbenicillin itself. This unexpected type of problem should always be borne in mind when dealing with new compounds.

The medium can influence zone size and here pH is important since the activity of β-lactams is pH-dependent. Rich complex media of high osmolality can interfere with drug activity; some compounds may produce larger zones when a medium of pH of 6 is used (O'Callaghan & Kirby, 1978). Finally the pH of the samples may be critical. Loss of carbon dioxide from serum samples will raise the pH above normal physiological levels and may affect zone sizes.

Some suggested microbiological assay conditions are detailed in Table II and some specific procedures are outlined below. The first method is designed to assay the enzyme inhibitor clavulanic acid based on its ability to render an indicator organism sensitive to a β-lactam incorporated in the medium. The second is an assay for benzyl penicillin in serum which could be used for therapeutic monitoring.

Bioassay for clavulanic acid in the presence of amoxycillin or ticarcillin (method of SmithKline Beecham Laboratories)

Clavulanic acid has very little intrinsic antimicrobial activity and cannot be assayed directly. This assay uses an indicator organism growing in a sub-inhibitory concentration of benzyl penicillin. Clavulanate diffusing into the agar inhibits the β-lactamase of the indicator strain, rendering it susceptible to inhibition by the penicillin and producing a zone of inhibition.

Indicator organism. Klebsiella pneumoniae NCTC 11228 grown overnight at 37°C in Tryptone Soya Broth (Lab-M, Bury, UK).

Medium. Nutrient agar (Lab-M).

Calibrators. Prepare a 50 mg/L solution of clavulanate in Sörensen's citrate buffer pH 6.5. From this prepare seven serum calibrators of 5, 2.5, 1.25, 0.625, 0.313, 0.156 and 0.078 mg/L.

Procedure
1. Add 12 mL of indicator organism culture to 400 mL of molten agar at 50–55°C and also add benzyl penicillin to give a final concentration of 60 mg/L.

Table II. Conditions for the microbiological assay of β-lactams

Drug	Medium[a]	Indicator organism	Calibrator concentration		Incubation temp. (°C)	Applicator (size in mm)	LLD[b]	Internal controls		Typical CV%
			plasma	tissue/BAL				plasma	tissue/BAL	
Amoxycillin	Ab 1	M. lutea ATCC 9341	0.03–1	0.03–1	37	well (5)	0.03	0.8/0.08	0.8/0.08	11.4
Amoxycillin	Ab 2	B. subtilis	0.5–8		37	well (7)	0.1	6/0.75		
Ampicillin	Ab 1	M. lutea ATCC 9341	0.03–1	0.03–1	37	well (5)	0.03	0.8/0.08	0.8/0.08	11.4
Ampicillin	Ab 2	B. subtilis	0.5–8		37	well (7)	0.1	6/0.75		
Azlocillin	Ab 1	P. aeruginosa NCTC 10701	4–64		37	well (8)	4	50/6		
Aztreonam	Ab 1	E. coli NIHJ SM7338	5–80		37	disc (6)	1.25	60/8		
Benzyl penicillin	Ab 1	B. subtilis (spores)	0.25–40		37	disc (6)	0.1	3/0.4		
Carbenicillin	Ab 1	P. aeruginosa NCTC 10701	4–64		37	well (8)	4	50/6		
Cefaclor	Ab 1	B. subtilis	2.5–40		37	disc (6)	1	30/4		
Cephamandole	Ab 1	B. subtilis	2.5–40		37	disc (6)	1	30/4		
Cefdinir	Ab 1	Proteus retgeri DRH-j294	0.03–0.5	0.03–0.5	30	well (5)	0.03	0.4/0.04	0.4/0.04	
Cefepime	Ab 1	E. coli SCH 12655	1–16		37	disc (6)	0.25	12/1.5		
Cefixime	Ab 1	Providencia stuartii DRH-K166	0.05–0.8	0.015–0.25	30	disc (6)	0.015	0.6/0.08	0.2/0.02	7.4
Cefotaxime	Ab 1	M. morganii Southmead	2–40			disc (6)	2	30/3		
Cefotetan	Ab 1	E. coli NIHJ SM7338	10–80		37	well (10)	5	60/15		
Cefoxitin	Ab 1	B. subtilis	0.06–1		37	disc (6)	0.03	0.8/0.1		
Cefpirome	Isosensitest	E. coli ATCC 25922	2–32	0.24–4	37	disc (6)	0.25	25/3	3/0.4	7.9
Cefpodoxime	Ab 1	P. retgeri UG12186 323UC2		0.02–0.64	37	well (5)	0.02		0.5/0.05	9.6
Cefpodoxime	Ab 2	M. morganii IF03848 313UC3	0.5–4		37	well (5)	0.1	3/0.4		7.3

Table II. *Continued*

Drug	Medium[a]	Indicator organism	Calibrator concentration plasma	Calibrator concentration tissue/BAL	Incubation temp. (°C)	Applicator (size in mm)	LLD[b]	Internal controls plasma	Internal controls tissue/BAL	Typical CV%
Cefprozil	Ab 1	*M. lutea* ATCC 9341	0.12–4		37	well (5)	0.1	3/0.2		8.3
Ceftazidime	Ab 1	*E. coli* NIHJ SM7338	4–64		37	disc (6)	2	20/5		
Ceftibuten	Ab 1	*P. stuartii* DRH-K166	0.05–0.8	0.05–0.8	37	disc (6)	0.05	0.6/0.08	0.6/0.08	5.7
Ceftizoxime	Ab 1	*E. coli* SCH 12655	0.25–4		37	disc (6)	0.12	3/0.4		
Ceftriaxone	Ab 1	*E. coli* Roche-1346	0.5–8		37	disc (6)	0.5	6/0.75		
Cefuroxime	Isosensitest + 20 mg/L NAD, 5% horse blood	*S. pneumoniae* ATCC 6303	0.25–4	0.05–0.8	37 4–6% CO_2	well (5)	0.04	3/0.4	0.6/0.08	10.2
Cefuroxime	Pen 1	*B. subtilis* MB325DR	2–32		37	disc (6)	1	20/3		10.5
Cefuroxime	Pen 2	*B. subtilis* MB325DR	0.5–8		37	well (5)	0.25	6/0.8		
Cephalexin	Ab 1	*B. subtilis* (spores)	2–32		37	well (7)	1	20/3		
Cephaloridine	Ab 1	*B. subtilis* (spores)	2–32		37	well (7)	1	20/3		
Cephalothin	Ab 1	*B. subtilis*	2.5–40		37	disc (6)	1	30/4		
Cephradine	Ab 1	*B. subtilis* (spores)	2–32		37	well (7)	1	20/3		
Clavulanate	Ab 2 + 80 mg/L piperacillin	*K. aerogenes* BRL1003 ATCC 29665	0.08–1.28	0.02–0.32 0.16–2.56	37	disc (6) well (5)	0.02	1/0.1		11.6
Cloxacillin	Ab 1	*M. lutea* ATCC 9341	2.5–40		37	disc (6)	1	30/4		
FK037	Ab 1	*E. coli* NIHJ SM7338	0.5–8		30	disc (6)	0.5	6/0.8		6.3
Flucloxacillin	Ab 1	*B. subtilis* NCTC 8236	6–64		37	disc (6)	2	40/6		
Imipenem	Ab 1	*E. coli* NIHJ SM7338	5–80		37	disc (6)	4	60/8		

β-Lactams

Drug	Medium[a]	Test organism	Range	Assay	Temp			
Imipenem	Ab 1	B. sublilis spores	0.25–4		37	0.2	3/0.75	
Latamoxef	Ab 1	E. coli NIHJ SM7338	2–80	well (7)	37	1.25	60/3	
Loracarbef	Ab 1	M. lutea ATCC 9341	0.25–4	disc (6)	37	0.25	3/0.3	7.2
Mecillinam	Ab 1	E. coli NIHJ SM7338	2–32	disc (6)	37	1	20/3	
Meropenem	Isosensitest	E. coli NIHJ SM7338	2–32 and 0.25–4	disc (6)	30	0.1	20/3 and 3/0.4	6.0
Methicillin	Ab 1	M. lutea ATCC 9341	2.5–40	disc (6)	37	1	30/4	
Mezlocillin	Ab 1	P. aeruginosa NCTC 10701	4–64	well (8)	37	4	50/6	
Piperacillin	Ab 1	P. aeruginosa NCTC 10701	4–64	well (5)	37	4	50/6	4.8
Tazobactam	Ab 2	K. aerogenes BRL1003, ATCC 29665	0.5–8	well (5)	37	0.25	6/0.8	7.5
Ticarcillin	At 1	P. aeruginosa NCTC 10701	2–32	well (8)	37	1	20/3	

All concentrations are in mg/L.
[a] Ab 1, Oxoid CM327; Ab 2, Oxoid CM335.
[b] LLD, Lower limit of detection.

105

2. Pour assay plates and cool.

3. Punch 7 mm wells in the agar.

4. Dilute any samples which might have a concentration of >5 mg/L in an appropriate matrix to approximately 0.6 mg/L.

5. Fill the wells with sample, calibrator or control, allow 30 min for diffusion and then incubate overnight at 37°C.

Performance expectations. The assay can measure clavulanate concentrations in the presence of amoxycillin up to a ratio of 1:100 and ticarcillin up to a ratio of 1:50. For samples with no penicillin the inoculum of indicator strain is reduced to 1 mL and the concentration of benzyl penicillin in the agar is only 5 mg/L. For urine samples calibrators should be prepared in buffer.

Assay of benzyl penicillin in serum
Indicator organism. *Bacillus subtilis* NCTC 10400 spore suspension.

Medium. Difco (Detroit, MI, USA) Penassay agar (100 mL) molten and cooled to 55–60°C.

Calibrators. Benzyl penicillin in human serum at 0.5, 1, 2, 4 and 8 mg/L.

Procedure
1. Inoculate the molten agar with 0.5 mL spore suspension and pour into a large plate.

2. Allow to set and then place at 37°C until the surface is dry.

3. Soak each sample, calibrator, and control into a paper disc in triplicate and stand each disc on a filter paper to soak up excess sample.

4. Apply the discs to the plate using a predetermined random pattern (see *Chapter 4*).

5. Place the plate under a UV sterilizing light for 10 min.

6. Incubate overnight at 37°C. Measure the zone diameters the next day.

Performance expectations. Assay reproducibility should be ±15% or better. Co-administered antimicrobials should be removed or inactivated as described in *Chapter 4*. Since benzyl penicillin is relatively unstable, calibrators and controls should be freshly prepared each day.

Non-microbiological assays (HPLC)

HPLC is ideally suited to the assay of β-lactams. Most are weak acids of varying lipophilicities and can be maintained in the non-ionized form in solutions of low pH which are ideal for reversed-phase mobile phase formulations. Some β-lactams are amphoteric but these too are easily assayed by HPLC. Huang *et al.* (1991) investigated the HPLC behaviour of five amphoteric β-lactams: cephradine, cephalexin, cefaclor, ampicillin and amoxycillin. In mobile phases coupled to C_{18} or phenyl reversed-phase columns, retention times were shortest at pH 4–6. Below pH 3, ampicillin and cephradine eluted with identical retention times from a C_{18} column whereas they remained separated on a phenyl column. Ion pair reagents such as tetra-ethylammonium or tetrabutylammonium salts only formed ion pairs above pH 5.5–6. Heptanesulphonic acid was only useful below pH 5.

The β-lactams also absorb UV radiation sufficiently to make it ideal for their detection. In general, cephems (which have a double bond in the nucleus) absorb strongly around 254 nm whereas penicillins require wavelengths of 210–214 nm for adequate sensitivity. Some compounds, e.g. meropenem, show strong absorption at higher wavelengths (c. 300 nm) due to specific side-chains. An exception is clavulanic acid which has poor UV absorption, is present in serum in relatively low concentrations and has

Table III. UV absorbance ratios of some β-lactams

Drug	UV absorbance ratio $A_{214\,nm}/A_{254\,nm} \times 100$	Drug	UV absorbance ratio $A_{214\,nm}/A_{254\,nm} \times 100$
Ampicillin	3800	Cefuroxime	76
Azlocillin	9900	Cephalexin	177
Aztreonam	92	Cephalothin	97
Benzyl penicillin	3675	Cephamandole	200
Carbenicillin	4000	Cephazolin	106
Cefadroxil	153	Cephradine	113
Cefotaxime	67	Flucloxacillin	300
Cefotetan	194	Imipenem	350
Cefoxitin	100	Methicillin	780
Cefsulodin	153	Piperacillin	300
Ceftazidime	68	Ticarcillin	2297

a low affinity for reversed-phase packings. Derivatization with imidazole can overcome these problems and an example assay method is outlined later in the chapter.

The ratios of absorption at 214 nm and 254 nm can be used to confirm specificity and aid peak identification. Table III lists typical ratios for a series of *β*-lactams.

White (1986) reviewed cephalosporin assay methodologies and found that reversed-phase chromatography, most commonly on a C_{18} stationary phase with a mobile phase consisting of water and an organic solvent (methanol or acetonitrile) buffered at pH 2–4, was suitable for almost any compound. Swaisland (1986) found a similar picture for penicillins with the addition of ion-pair methods. Three HPLC methods (for benzyl penicillin, cefotaxime and clavulanic acid) suitable for clinical samples are outlined below. A summary of methods for cephems is to be found in Table IV, for penicillins in Table V and for other *β*-lactams and mixtures in Table VI.

Assay of benzyl penicillin in serum

This is an unpublished method used for therapeutic monitoring in the senior author's (LOW) laboratory. It is very similar to the method of Rumble & Roberts (1985) and could be modified to measure penicilloic acid levels also.

Stationary phase. Hypersil 5 ODS 10×4 cm (HPLC Technology, Macclesfield, UK).

Mobile phase. Methanol:water:phosphoric acid (35:64:1 by volume) pumped at 1 mL/min.

Detection. UV absorbance at 214 nm.

Procedure. Mix serum sample with an equal volume of acetonitrile, shake and sediment the precipitate by centrifugation. Inject 20 μL of the clear supernatant.

Calibration. The assay does not use an internal standard. Single-point calibration is by use of an aqueous solution of benzyl penicillin 50 mg/L subjected to the same sample preparation procedure.

Performance expectations. Assay CVs are 5% or better. The limit of detection depends on the quality of the detector but should be at least 1–2 mg/L. Calibrators and controls should be freshly prepared each day.

Assay of cefotaxime and metabolites in serum

This is an ion-pair method. Cefotaxime can be separated from serum components without the need for ion pairing but it is essential to ensure separation of the metabolites (desacetyl cefotaxime, UP1 and UP2), all of which have been found in serum (Reeves *et al.*, 1980).

Stationary phase. Microbondapak C_{18} (Waters).

Mobile phase. Water:methanol (97:7 v/v) containing 0.005 M heptane sulphonic acid pumped at 2 mL/min.

Detection. UV absorbance at 254 nm.

Procedure. Mix serum sample with an equal volume of acetonitrile, shake and sediment the precipitate by centrifugation. Inject 5–20 μL of the clear supernatant.

Calibration. The assay does not use an internal standard. Calibration is by spiked serum samples subjected to the same procedure.

Performance expectations. CVs of better than 5% should be obtained; the limit of sensitivity is around 0.5 mg/L. The elution order is: desacetyl cefotaxime, UP1, UP2, cefotaxime. Haemolysis in the serum will rapidly desacetylate the cefotaxime and produce spurious results.

Assay of clavulanic acid in serum

This is the method of Foulstone & Reading (1982); it uses pre-column derivatization with imidazole reagent to produce a product with strong absorbance at 311 nm. Note that the same mobile phase can be used to assay the amoxycillin present with the clavulanic acid.

Stationary phase. Microbondapak C_{18} (Waters).

Mobile phase. 0.1 M phosphate buffer pH 3.2 plus 6% methanol pumped at 2.5 mL/min.

Detection. UV absorbance at 311 nm.

Procedure. Mix serum sample with an equal volume of 0.1 M phosphate buffer pH 7 and ultrafilter through a suitable membrane for removing serum proteins (e.g. Amicon YMT membrane). Mix 400 μL of ultrafiltrate with 100 μL imidazole reagent (8.25 g imidazole in 24 mL water plus 2 mL of 5 M hydrochloric acid; adjust to pH 6.8 then make up to 40 mL with water) and incubate at ambient temperature for 10 min. Inject 50 μL.

Calibration. The assay does not use an internal standard. Calibration is by spiked serum samples subjected to the same procedure.

Performance expectations. Buffering the serum is essential; if it is not done, clavulanate is unstable in the ultrafiltrate. The limit of detection was 0.1 mg/L.

Table IV. HPLC methods for cephamycins, cephalosporins and oxacephems

Drug	Stationary phase	Mobile phase	Detection (nm)	Matrix	Comments	Reference
Cefaclor	Lichrosorb C$_8$	MeOH:0.01 M sodium acetate buffer pH 5.2 (30:70)	UV 266	serum	protein precipitation	Rotschafer *et al.* (1982)
Cefaclor	Merck C$_{18}$	0.05 M citrate buffer pH 4: MeOH:dioxane (80:13:7)	UV 254	serum	protein precipitation with perchloric acid: ACN:dioxane	Ullmann (1980)
Cefadroxil	μBondapak C$_{18}$	MeOH:0.01 M acetate buffer pH 4.8 (5:95)	UV 240	plasma	protein precipitation with ACN	Brisson & Fourtillan (1982b)
Cefadroxil	ODS	ACN:water:HAC (5:85:1)	UV 235	serum, saliva	protein precipitation with TCA	Windorfer & Bauer (1982)
Cefatriazine	R Sil C$_{18}$	0.025 M phosphate buffer pH 7 + 7.5–10% ACN	PCD and fluorescence	serum, urine	protein precipitation with TCA	Crombez *et al.* (1979)
Cefazedone	Lichrosorb RP8	ACN:buffer gradient	UV 278	biological fluids	protein precipitation with EtOH + HCl	Sailer *et al.* (1979)
Cefdinir	Nova-Pak C$_{18}$	phosphate buffer pH 3.1 or 3.3; ACN (89:11 or 88:12)	UV 287	plasma, urine or blister	protein precipitation with ACN	Richer *et al.* (1995)
Cefepime	Partisil 5 ODS-3RAC (urine)	0.01 M SDS (pH 3):MeOH: 5% TCA:0.85 M phosphoric acid:tetrahydrofuran (1500:1846:144:28:197) (urine)	UV 280	urine, plasma	urines diluted in 200mM Na acetate buffer pH 4.25	Barbhaiya *et al.* (1987)
	Nova-Pak C$_{18}$ (plasma)	ACN 0.005 M OSA (12:88) (plasma)			extraction procedure for plasma	Barbhaiya *et al.* (1987)
Cefepime	Nucleosil 5 C$_{18}$	5 mM octane sulphonate:ACN 90:10 (plasma) MeOH:tetrahydrofuran: 10 mM SDS (90:15:150) (urine)	UV 280	plasma, urine	extraction procedure (plasma); urines diluted in acetate buffer pH 4.25	Cronqvist *et al.* (1992)
Cefixime	Techsphere C$_{18}$ 3 μm	170 mL ACN, 1.36 g monobasic sodium phosphate, 2 mL phosphoric acid, 828 mL water	UV 313	serum and CSF	protein precipitation with TCA	White *et al.* (1993)
Cefmenoxime	μBondapak C$_{18}$	ACN:0.2 M acetate buffer pH 5.3 (13:87)	UV 254	plasma	MeOH, SDS and ultrafiltration	Granneman *et al.* (1982)
Cefmenoxime	Nucleosil C$_{18}$	water:ACN:HAC (50:10:1)	UV 254	serum, urine	protein precipitation with MeOH	Itakura *et al.* (1982)
Cefmenoxime	μBondapak Phenyl	14% ACN, 0.2% phosphoric acid in water	UV 254	plasma, urine	protein precipitation with ACN	Noonan *et al.* (1983)
Cefmenoxime	Lichrosorb RP18	0.05 M phosphate buffer pH 6.6: tetrahydrofuran (20:1)	UV 254	biological fluids	protein precipitation with ACN	Yamaoka *et al.* (1983)
Cefmetazole	μBondapak C$_{18}$	0.005 M citrate buffer pH 5.4 + 10–15% ACN	UV 254	serum	protein precipitation with TCA and MeOH	Sekine *et al.* (1982)
Cefminox	Nova-Pak C$_{18}$	2% HAC:MeOH:ACN (92:5:3) pH 4	UV 270	plasma	protein precipitation with 6% TCA	Aguilar *et al.* (1994)

Compound	Column	Mobile phase	Detection	Sample	Procedure	Reference
Cefodizime	μBondapak C_{18}	MeOH:ammonium acetate	UV 280	plasma, urine	extraction procedure	Bryskier et al. (1990)
Cefoperazone	μBondapak C_{18}	MeOH:0.01 M acetate buffer pH 4.8 (15:85)	UV 240	biological fluids	extraction procedure	Brisson & Fourtillan (1981a)
Cefoperazone	Lichrosorb RP18	0.05 M phosphate buffer pH 6.6: MeOH (2:1)	UV 254	biological fluids	protein precipitation with ACN	Yamaoka et al. (1983)
Cefoperazone and metabolite A	μBondapak C_{18}	10 mM acetate buffer pH 4:ACN (90:10 or 92:8 (bile))	UV 254	serum, urine, bile	protein precipitation with ACN	Kemmerich et al. (1983a)
Cefoperazone and metabolites	μBondapak C_{18}	triethylamine + HAH:ACN gradient	UV	serum		Bosso et al. (1983)
Cefoperazone and metabolites	μBondapak C_{18}	triethylamine + HAC gradient	UV 254	serum, urine	protein precipitation with MeOH	Dokladolova et al. (1983)
Ceforanide	μBondapak C_{18}	MeOH:0.05 M ammonium acetate pH 4 (10:90)	UV 254	serum, urine	protein precipitation with TCA and ACN	Dajani et al. (1982)
Cefotaxime	μBondapak C_{18}	MeOH:water:HAC (14:85:1)	UV 254	plasma, tissue	protein precipitation	Mullany et al. (1982)
Cefotaxime	Lichrosorb RP8	MeOH:2 mM phosphoric acid (28:72)	UV 310	serum	protein precipitation with TCA	Bergan & Solberg (1981)
Cefotaxime	μBondapak C_{18}	MeOH:0.01 M acetate buffer pH 4.8 (15:85)	UV 234	biological fluids	extraction procedure	Brisson & Fourtillan (1981a)
Cefotaxime	ODS	ACN:0.15% HAC (13:87) pH 2.8	UV 270	biological fluids	anion exchange procedure	Fasching & Peterson (1982)
Cefotaxime	μBondapak C_{18}	ACN:10 mM acetate buffer pH 4.5 (5:95)	UV	serum	protein precipitation with ACN	Kemmerich et al. (1983b)
Cefotaxime and metabolites	μBondapak C_{18}	MeOH:water:HAC:HEPSA (250:750:5:1.1 g)	UV 254	uraemic serum, urine, PD fluid	protein precipitation with ACN	Reeves et al. (1980)
Cefotetan and its tautomer	Nucleosil C_{18}	ACN:0.1 M phosphate buffer pH 3 (8:92)	UV 280	uraemic serum	protein precipitation with ACN	Ohkawa et al. (1983)
Cefotetan and its tautomer	Hypersil 5 ODS	0.1 M NaH_2 phosphate:ACN: phosphoric acid (910:90:1)	UV 280	plasma, urine	protein precipitation with TCA	Yates et al. (1983)
Cefotetan epimers and tautomer	Hypersil 5 ODS	MeOH:water:phosphoric acid (20:79:1)	UV 254	uraemic serum, PD fluid	protein precipitation with ACN	L. O. W. (unpublished)
Cefotiam	Nucleosil C_{18}	0.1 M acetate buffer:ACN (95:5)	UV 254	serum, urine	protein precipitation with MeOH	Itakura et al. (1982)
Cefotiam	Lichrosorb RP18	0.05 M phosphate buffer pH 3: tetrahydrofuran (10:1)	UV 254	biological fluids	protein precipitation with ACN	Yamaoka et al. (1983)
Cefotiam and Δ3-cefotiam	C_{18}	phosphate buffer pH 7.7 + ACN (88:12)	UV 254	plasma	column switching enrichment	Yamashita et al. (1992)
Cefoxitin	μBondapak C_{18}	MeOH:0.01 M acetate buffer pH 4.8 (15:85)	UV 254	biological fluids	extraction procedure	Brisson & Fourtillan (1981a)
Cefoxitin	μBondapak C_{18}	0.05 M phosphate buffer pH 3: ACN (80:20)	UV 254	plasma, urine	protein precipitation with TCA	Charles & Ravenscroft (1984)
Cefoxitin	ODS	ACN:15% HAC (13:87) pH 2.8	UV 270	biological fluids	anion-exchange procedure	Fasching & Peterson (1982)

Table IV. *Continued*

Drug	Stationary phase	Mobile phase	Detection (nm)	Matrix	Comments	Reference
Cefoxitin	μBondapak C$_{18}$	MeOH:water:HAC (42:57:1)	UV 254	serum, urine	protein precipitation with ACN	Reeves *et al.* (1981)
Cefoxitin decarbamyl metabolite and lactone	Zipax SAX	0.25 M sodium acetate pH 5	UV 245	urine		Buhs *et al.* (1974)
Cefpirome	μBondapak C$_{18}$	tetrahydrofuran:ammonium phosphate buffer pH 5.2 (2.5:97.5)	UV 275	serum, CSF	protein precipitation with ACN	Nix *et al.* (1992)
Cefpirome	μBondapak C$_{18}$	acetate buffer:MeOH (86:14) pH 4.2	UV 270	plasma, urine	protein precipitation with ACN	Paradis *et al.* (1992)
Cefpodoxime	C$_{18}$	acetate buffer pH 3.8:MeOH:ACN	UV 235	plasma, sinus mucosa	solid-phase extraction procedure	Camus *et al.* (1994)
Cefpodoxime	Nucleosil 5 C$_{18}$	acetate buffer pH 4:ACN (91:9)	UV 260	serum, urine	protein precipitation with ACN	Saathoff *et al.* (1992)
Cefpodoxime	μBondapak C$_{18}$	0.05 M acetate buffer pH 4.6:ACN (93:7) (plasma) or citrate buffer pH 3.5:ACN (92:8) (urine)	UV 254	plasma, urine	solid-phase extraction procedure	Tremblay *et al.* (1990)
Cefroxadine	Lichrosorb RP8	1 mM phosphoric acid:MeOH (18:7)	UV	serum	protein precipitation	Bergan & Solberg (1982)
Cefsulodin	Nucleosil C$_{18}$	0.1 M acetate buffer:ACN (92:8)	UV 254	serum, urine	protein precipitation with MeOH	Itakura *et al.* (1982)
Cefsulodin	Zorbax BP C$_8$	0.035 M ammonium acetate:ACN (95.5:4.5) pH 5.2	UV 265	serum, urine	protein precipitation with MeOH	Reed *et al.* (1984)
Ceftazidime	Hypersil 5 ODS	0.05 M ammonium dihydrogen phosphate + 6–7% ACN and 0.01% formic acid	UV 257	serum, urine	protein precipitation with perchloric acid	Ayrton (1981)
Ceftazidime	Zorbax Sil	0.1 M acetate buffer:MeOH:ACN (87:9:4)	UV 254	uraemic serum, urine	protein precipitation with perchloric acid	Welage *et al.* (1984)
Ceftezole	Nucleosil C$_{18}$	0.1 M acetate buffer:ACN (95:5)	UV 254	serum, urine	protein precipitation with MeOH	Itakura *et al.* (1982)
Ceftibuten	μBondapak C$_{18}$	0.05 M ammonium acetate:ACN (98:2)	UV 254 (urine) UV 262 (serum)	serum, urine	extraction procedure using Nonidet P-40	Kearns *et al.* (1991)
Ceftibuten and *trans*-ceftibuten	μBondapak C$_{18}$	ammonium acetate:ACN (98:2)	UV 254	plasma, urine, dialysate	serum proteins not precipitated	Kelloway *et al.* (1991)
Ceftibuten and *trans*-ceftibuten	μBondapak C$_{18}$	ACN:0.05 M ammonium acetate (2:98)	UV 254	plasma, urine	sample mixed with 0.2 M phosphate buffer pH 7 and injected	Lin *et al.* (1995)
Ceftizoxime	μBondapak C$_{18}$	HAC (1.5 mL/L):ACN (87:13) pH 2.8	UV 270	serum	extraction procedure	Fasching *et al.* (1982)

Compound	Column	Mobile phase	Detection	Sample	Sample preparation	Reference
Ceftizoxime	μBondapak C18	ACN:water:HAC (130:842:28)	UV 310	uraemic serum	protein precipitation with ACN	McCormick et al. (1984)
Ceftriaxone	Lichrosorb NH2	ACN:water:10% ammonium carbonate (70:26:4)	UV 274	plasma, urine, saliva	protein precipitation with ACN	Ascalone & Dal Bo (1983)
Ceftriaxone	Supelcosil C8 at 40°C	water:ACN (72:28) + 4 g/L TBAB	UV 254	serum, tissue	extraction procedure	Bryan et al. (1984)
Ceftriaxone	Chromegabond C18	ACN:10 g/L HTAB:phosphate buffer pH 7:water (60:30:1:9)	UV 280	serum	protein precipitation with ACN	Patel et al. (1981)
Ceftriaxone	Chromegabond C18	ACN:10 g/L TOAB:phosphate buffer pH 7:water (44:35:1.2:19.8)	UV 280	urine	protein precipitation with ACN	Patel et al. (1984)
Ceftriaxone	ODS	TBAHS in phosphate buffer: MeOH (80:20)	electrochemical detection	uraemic serum	protein precipitation with MeOH	Ti et al. (1984)
Ceftriaxone	Lichrosorb RP18	ACN plus various basic ion-pair reagents	UV 274 or 300	biological fluids	protein precipitation with EtOH	Trautmann & Haefelfinger (1981)
Cefuroxime	μBondapak C18	MeOH:0.01 M acetate buffer pH 4.8 (15:85)	UV 254	biological fluids	extraction procedure	Brisson & Fourtillan (1981a)
Cefuroxime	μBondapak C18 at 40°C	0.1 M acetoacetate buffer pH 4.8: ACN (89:11)	UV 254	serum, urine	protein precipitation with TCA	Brisson & Fourtillan (1981b)
Cefuroxime	Lichrosorb RP8	phosphate buffer pH 4.6:EtOH (425:75)	UV 278	serum, urine	protein precipitation with perchloric acid	van Dalen et al. (1979)
Cefuroxime (a); its axetil (b)	Hypersil 5 ODS	Water:ACN (50:50) for (a) 0.05 M ammonium dihydrogen phosphate:ACN (85:15) for (b)	UV 273	serum, urine	protein precipitation with ACN	Harding et al. (1984)
Cephacetrile	Nucleosil C18	0.1 M acetate buffer:ACN (92:8)	UV 254	serum, urine	protein precipitation with MeOH	Itakura et al. (1982)
Cephalexin	ODS	various proportions of MeOH; ammonium carbonate	UV 254	serum, urine	direct injection	Carroll et al. (1977)
Cephalexin	μBondapak C18	MeOH:water:HAC (20:80:5)	UV 254	plasma, urine, saliva	extraction procedure	Nahata (1981)
Cephalexin	μBondapak C18	MeOH:water with 0.5% HAC (1:8 for urine or 1:5 for plasma)	UV 254	plasma, urine	protein precipitation with MeOH	Nakagawa et al. (1978)
Cephalexin	ODS	5 mM SPS in 70% ACN pH 3	UV 254	serum	protein precipitation with MeOH	Sutsumi et al. (1982)
Cephaloglycine (a), desacetyl metabolite/lactone (b), benzoyl formic acid (c), phenyl glycine (d)	Lichrosorb RP18	as below* (a & b); MeOH:water (1:3) containing 0.01 M TBAB (c); MeOH:water (1:12) containing 0.005 M HEPSA (d)	UV 254	urine		Haginaka & Nakagawa (1980)
Cephaloglycine and desacetyl metabolite/lactone	μBondapak C18	*MeOH:water containing 0.004 M HEPSA	UV 254	urine	acidification (desacetyl metabolite becomes lactone)	Haginaka & Nakagawa (1979)
Cephaloridine	Phenyl Corasil	18–20% MeOH in 0.2 M ammonium acetate	UV 254	serum, tissue	protein precipitation with TCA	Wold & Turnipseed (1977b)
Cephalothin	AS-Pellionex SAX at 50°C	0.01 M phosphate buffer with 0.01 M sodium nitrate pH 4.8	UV 254	serum, urine	ion-pair extraction procedure	Rolewicz et al. (1977)

Table IV. *Continued*

Drug	Stationary phase	Mobile phase	Detection (nm)	Matrix	Comments	Reference
Cephalothin and desacetyl metabolite	μBondapak C$_{18}$	MeOH:0.01 M acetate buffer pH 4.8 (15:85)	UV 240	biological fluids	extraction procedure	Brisson & Fourtillan (1981a)
Cephalothin and desacetyl metabolite	Zipax SAX	0.25 M sodium acetate pH 5.0	UV 254	urine		Buhs *et al.* (1974)
Cephalothin and desacetyl metabolite	μBondapak C$_{18}$	MeOH:1% HAC (40:60)	UV 254	uraemic serum	protein precipitation with DMF	Nilsson-Ehle & Nilsson-Ehle (1979)
Cephalothin and desacetyl metabolite lactone	Phenyl Corasil	10–13% MeOH in 0.2 M ammonium acetate	UV 254	serum	protein precipitation with TCA	Wold & Turnipseed (1977a)
Cephamandole	μBondapak C$_{18}$	MeOH:water:HAC (33:66:1)	UV 254	plasma, tissue		Mullany *et al.* (1982)
Cephamandole	μBondapak C$_{18}$	MeOH:0.01 M acetate buffer pH 4.8 (15:85)	UV 270	biological fluids	extraction procedure	Yamaoka *et al.* (1983)
Cephapirin	ODS	ACN:0.15% HAC pH 2.8 (13:87)	UV 270	biological fluids	anion-exchange extraction procedure	Fasching & Peterson (1982)
Cephazolin	μBondapak C$_{18}$	MeOH:0.01 M acetate buffer pH 4.8 (15:85)	UV 275	biological fluids	extraction procedure	Brisson & Fourtillan (1981a)
Cephazolin and cephalothin	Phenyl Corasil	10–13% MeOH in 0.2 M ammonium acetate	UV 254	serum	protein precipitation with TCA	Wold & Turnipseed (1977a)
Cephazolin and latamoxef	μBondapak Phenyl	22% ACN + 0.5 N TBAP in 0.1 M phosphate buffer + 77% water	UV 270	serum	protein precipitation with isopropanol]	Polk *et al.* (1981)
Cephradine	μBondapak C$_{18}$	0.01 M phosphate buffer pH 6.8: MeOH (40:10)	UV	biological fluids	protein precipitation	Hayashi (1982)
Cephradine	Merck C$_8$	0.05 N citrate buffer pH 4:MeOH: dioxane (80:13:7)	UV 254	serum	protein precipitation	Ullmann (1980)
Latamoxef	Chromegabond C$_{18}$ (serum) Zorbax + TMS (urine)	0.1 M ammonium acetate:ACN (95:5) (serum) / 5 mM heptylamine in MeOH: water (11:89) pH 6 (urine)	UV 270 (serum) / UV 280 (urine)	serum, urine	protein precipitation with MeOH	Miner *et al.* (1981)
Latamoxef	Rad-Pak A	ACN:0.05 M TBAP (40:60)	UV 280	plasma, CSF	extraction procedure	Modai *et al.* (1982)
Latamoxef	μBondapak Phenyl	ACN:0.5 M TBAP:water (22:1:77)	UV 270	serum	protein precipitation with isopropanol]	Polk *et al.* (1981)
Latamoxef	Rad-Pak C$_{18}$	Waters PIC-A:ACN (32:68)	UV 270	serum	extraction procedure	Romagnoli *et al.* (1982)
Latamoxef epimers	Hypersil 5 ODS	MeOH:1% nitric acid (14:86)	UV 254	serum	protein precipitation with saturated ammonium sulphate	Wise *et al.* (1981)

ACN, acetonitrile; DMF, dimethylformamide; EtOH, ethanol; HAC, glacial acetic acid; HEPSA, heptane sulphonic acid; HTAB, hexadecyl trimethyl ammonium bromide; MeOH, methanol; OSA, octanesulphonic acid; SDS, sodium dodecyl-sulphate; SPS, sodium propane-2-sulphonate; TBAB, tetrabutyl ammonium bromide; TBAHS, tetrabutyl ammonium hydrogen sulphate; TBAP, tetrabutyl ammonium phosphate; TCA, trichloracetic acid; TOAB, tetraoctyl ammonium bromide.

Table V. HPLC assay methods for penicillins

Drug	Stationary phase	Mobile phase	Detection (nm)	Matrix	Comments	Reference
Amoxycillin	Lichrosorb RP8	phosphate buffer pH 8:MeOH (92:8)	UV 310	plasma, urine	post-column derivatization (imidazole)	Carlqvist & Westerlund (1979)
Amoxycillin	CP Microsphere C$_{18}$	MeOH + paired ion reagents	fluorescence ex. 372, em. 470	biological fluids	column switching and post-column derivatization with fluorescamine	Carlqvist & Westerlund (1985)
Amoxycillin	Spherisorb ODS-II	50 mM phosphate buffer, 10 mM thiosulphate pH 4.6, 15% ACN	UV 328	preterm infant serum	imidazole derivatization procedure	Huisman et al. (1995)
Amoxycillin and ampicillin	Lichrosorb RP8	MeOH:buffer	UV 225	plasma, urine, saliva	protein precipitation with perchloric acid	Vree et al. (1978)
Amoxycillin in co-amoxiclav	µBondapak C$_{18}$	0.1 M phosphate buffer + 4–6% MeOH	UV 227	serum, urine	protein removed by ultrafiltration	Foulstone & Reading (1982)
Ampicillin and metampicillin	Lichrosorb RP-8, Ultracarb 5 ODS-30	ACN:phosphate buffer gradient	UV	biological fluids	column switching	Lee et al. (1995)
Ampicillin and mecillinam	Lichrosorb RP8	phosphate buffer pH 8:MeOH (70:30)	UV 310	plasma, lymph, urine	post-column derivatization (imidazole)	Westerlund et al. (1979)
Ampicillin, amoxycillin, and metabolites	Lichrosorb RP8	phosphate buffer pH 4.6 +/– MeOH (425:75)	UV 225	plasma, saliva, urine	protein precipitated with perchloric acid	Vree et al. (1978)
Apalcillin	µBondapak C$_{18}$	Water:ACN:MeOH:triethylamine pH 3.4 (600:340:150:4)	UV 258	plasma, urine	extraction procedure	Demotes-Mainard et al. (1985)
Apalcillin	Lichrosorb RP18	acetate buffer pH 3.15:ACN (80:20)	UV 254	body fluids	protein precipitation with ACN	Lode et al. (1984)
Apalcillin and metabolites	Lichrosorb RP18	ACN:buffer gradient elution	UV 254/315	serum, urine	protein precipitation with ACN	Borner et al. (1982)
Azlocillin	µBondapak C$_{18}$	phosphate buffer pH 6.4: ACN (87:13)	UV 220	serum, urine	protein precipitation with ACN	Lode et al. (1983)
Azlocillin	µBondapak C$_{18}$	phosphate buffer pH 3.5 + 45–55% MeOH	UV 254/220	serum	protein precipitation with ACN	Weber et al. (1983)
Benzyl penicillin (penicillin G)	µBondapak C$_{18}$	phosphate buffer pH 7:MeOH (72:30)	UV 214	plasma, urine	protein precipitation with ACN	Rumble & Roberts (1985)
Benzyl penicillin (penicillin G)	Hypersil C$_{18}$	MeOH:ACN:TBAHB buffer	UV 231	body fluids	protein precipitation with ACN	Van Gulpen et al. (1986)
Carbenicillin (1994)	Microsorb C$_{18}$ 3 µm	ACN:TBAP pH 6.6	UV 208	serum	extraction procedure	Naidong et al.
Carbenicillin and sulbenicillin	µBondapak C$_{18}$	MeOH:0.1 M TBAB	UV 254	urine	filter	Yamaoka et al. (1979)
Carbenicillin and ticarcillin	Develosil ODS-5 at 40°C	TBAB buffer:ACN (1.8:1)	UV 327/331	serum, urine	pre-column derivatization with triazole	Haginaka & Wakai (1985)

Table V. *Continued*

Drug	Stationary phase	Mobile phase	Detection (nm)	Matrix	Comments	Reference
Flucloxacillin	Spherisorb 5 ODS	phosphate buffer pH 7:MeOH (55:45)	UV 220	serum	dichloromethane extraction	Bergan *et al.* (1986)
Mecillinam (amdinocillin)	not stated	not stated	not stated	serum, urine	not stated	Barrière *et al.* (1982)
Mecillinam and ampicillin	Lichrosorb RP8	phosphate buffer pH 8:MeOH (70:30)	UV 310	plasma, lymph, urine	post-column derivatization (imidazole)	Westerlund *et al.* (1979)
Mezlocillin	Spherisorb 5 ODS	phosphate buffer pH 7.5:ACN (81:19)	UV 215	serum	solvent extraction	Colaizzi *et al.* (1986)
Mezlocillin	μBondapak C_{18}	phosphate buffer pH 7:ACN (73:27)	UV 220	serum, urine	solid-phase extraction	Fiore *et al.* (1984)
Mezlocillin and piperacillin	μBondapak C_{18}	phosphate buffer pH 6:ACN (75:25)	UV 229	serum	protein precipitation with ACN	Martens *et al.* (1987)
Oxacillin and metabolites	Lichrosorb RP18	acetate buffer pH 6:MeOH (2:1)	UV 254	urine	dilution and filtration	Murai *et al.* (1981)
Phenoxymethyl penicillin (penicillin V)	μBondapak C_{18}	acetate buffer pH 6.5:ACN (80:20)	UV 215	serum	ether extraction	Lindberg *et al.* (1984)
Piperacillin	μBondapak C_{18}	acetate buffer pH 4.8:MeOH (60:40)	UV 254	plasma	chloroform extraction	Brisson & Fourtillan (1982a)
Piperacillin	μBondapak C_{18}	phosphate buffer pH 3.5:ACN (750:230)	UV 210/254	plasma	protein precipitation with ACN	Wilson *et al.* (1982)
Piperacillin and mezlocillin	μBondapak C_{18}	phosphate buffer pH 6:ACN (75:25)	UV 229	serum	protein precipitation with ACN	Martens *et al.* (1987)
Temocillin	Spherisorb 5 ODS II	phosphate buffer pH 7 + 2–8% MeOH	UV 233	serum, urine	extraction procedure	Guest *et al.* (1985)
Ticarcillin	μBondapak C_{18}	acetate buffer pH 4:MeOH (85:15)	UV 242	serum	extraction procedure	Shull & Dick (1985)
Ticarcillin	μBondapak C_{18}	phosphate buffer pH 2.1:MeOH (70:30)	UV 229	body fluids		Syrogiannopoulos *et al.* (1987)
Ticarcillin and carbenicillin	Develosil ODS-5 at 40°C	TBAB buffer:ACN (1.8:1)	UV 327/331	serum, urine	pre-column derivatization with triazole	Haginaka & Wakai (1985)

ACN, acetonitrile; EtOH, ethanol; HAC, glacial acetic acid; HEPSA, heptane sulphonic acid; MeOH, methanol; SPS, sodium propane-2-sulphonate; TBAB, tetrabutyl ammonium bromide; TBAHS, tetrabutyl ammonium hydrogen sulphate; TBAP, tetrabutyl ammonium phosphate; TCA, trichloracetic acid.

114

Table VI. HPLC assay methods for miscellaneous β-lactams and β-lactam mixtures

Drug	Stationary phase	Mobile phase	Detection (nm)	Matrix	Comments	Reference
β-Lactams (various)	Ultrasphere ODS	various depending on drug	UV	serum	protein precipitation with ACN	Jehl et al. (1987)
Azlocillin, mezlocillin, ceftizoxime	Waters C₁₈	acetate buffer pH 4.8:MeOH (87:13 to 60:40)	UV 220 or 270	body fluids	extraction procedure	Bamberger et al. (1986)
Aztreonam	Bio-Sil ODS-5S	0.005 M TBAHS:ACN (83:17)	UV 293	faeces	samples dissolved and filtered	Ehret et al. (1987)
Aztreonam	μBondapak C₁₈	0.005 M TBAHS pH 3:ACN (85:15)	UV 293	patient's serum	protein precipitation with ACN	Jones et al. (1984)
Aztreonam	μBondapak C₁₈	0.05M TBAP:ACN (83:17)	UV 280	CSF	extraction procedure	Modai et al. (1986)
Aztreonam	μBondapak C₁₈	ACN:TBAHS buffer (proportions optimized for species)	UV 293	serum and urine from various species	protein precipitation with ACN	Pilkiewicz et al. (1983)
Aztreonam	μBondapak C₁₈	0.005 M TBAHS pH 3	UV 293	body fluids	protein precipitation with ACN	Swabb et al. (1981)
Aztreonam and metabolite	μBondapak C₁₈	0.005 M TBAHS:ACN (80:20)	UV	β-lactamase hydrolysis mixtures		Bush et al. (1982)
Aztreonam and metabolite	μBondapak C₁₈	0.005M TBAHS:ACN (15:85)	UV 293	serum and urine	protein precipitation with ACN: methylene chloride	Mihindu et al. (1983)
Aztreonam and metabolite	μBondapak C₁₈	0.005 M TBAHS pH 3:ACN (75:15)	UV 293	urine	solid-phase extraction	Swabb et al. (1983)
Carumonam and metabolite	Nucleosil 5 C₁₈	0.005 M TBAHS pH 3:ACN (88:12)	UV 313	urine	dilution	Kita et al. (1986)
Clavulanic acid	Develosil ODS-5	water:MeOH (1:2)	fluorescence ex. 386; em. 456	serum and urine	benzaldehyde derivatization	Haginaka et al. (1986)
Clavulanic acid in co-amoxiclav	μBondapak C₁₈	phosphate buffer pH 3.2 + 4–6% ACN	UV 311	body fluids	imidazole derivatization	Foulstone & Reading (1982)
Clavulanic acid in timentin	μBondapak C₁₈	phosphate buffer pH 2.1: MeOH (5:95)	UV 313	body fluids	imidazole derivatization	Syrogiannopoulos et al. (1987)
Clavulanic acid in timentin		THAHS:MeOH	UV 322	body fluids	imidazole derivatization	Walstad et al. (1986)
Imipenem	Hypersil 5 ODS	MeOH:borate buffer pH 7.2 (2:98)	UV 299	serum	protein precipitation with ACN	Freij et al. (1985)
Imipenem	μBondapak C₁₈	acetate buffer:MeOH (98:2) pH 6.0	UV 298	plasma, urine (stabilized in MES or MOPS)	samples ultrafiltered	Paradis et al. (1992)
Imipenem	Micro Pak MCH 10	MeOH:water (7:93)	UV 300	serum	sera stored in MES pH 6. Deproteination by ultrafiltration	Myers & Blumer (1984)

Table VI. *Continued*

Drug	Stationary phase	Mobile phase	Detection (nm)	Matrix	Comments	Reference
Imipenem	μBondapak C$_{18}$	phosphate buffer pH 6.21	UV 300	saline and serum	protein precipitation with ACN	Swanson *et al.* (1986)
Meropenem	Hypersil 3 ODS	[a](a) MeOH:0.005 M TBAP (12:88); (b) phosphate buffer pH 7:MeOH (85:15)	UV 296	plasma	solid-phase extraction procedure	Kelly *et al.* (1995)
Meropenem	Hypersil 3 ODS	[a](a) ACN:phosphate buffer pH 7.4 (6:100); (b) ACN: phosphate buffer pH 7 (7:100)	UV 296	urine	solid-phase extraction procedure	Kelly *et al.* (1995)
Meropenem Sulbactam	Hypersil 5 ODS Beckman ODS-5	MeOH:H$_3$PO$_4$:water (25:1:74) phosphate buffer pH 6.1:ACN (89:11)	UV 296 UV 313	serum plasma, urine, tissues	protein precipitation with ACN imidazole derivatization	Lovering *et al.* (1995) Bawdon & Madsen (1986)
Sulbactam Sulbactam	μBondapak C$_{18}$ Magnusphere C$_{18}$	water:MeOH + 5% HAC ammonium phosphate:MeOH: phosphoric acid (99.9:0:0.1) (urine) or 95:55:0.1	UV 225 UV 225	body fluids serum, saliva, urine	ether extraction procedure extraction procedure	Fredj *et al.* (1986) Rogers *et al.* (1983)
Tigemonam	C$_{18}$ 5 μm IBM	ACN:0.005 M TBAHS pH 3 (25:75)	UV 293	body fluids	protein precipitation with ACN	Clark *et al.* (1987)

ACN, acetonitrile; EtOH, ethanol; HAC, glacial acetic acid; HEPSA, heptane sulphonic acid; MeOH, methanol; SPS, sodium propane-2-sulphonate; TBAB, tetrabutyl ammonium bromide; TBAHS, tetrabutyl ammonium hydrogen sulphate; TBAP, tetrabutyl ammonium phosphate; TCA, trichloracetic acid; THAHS, tetraheptyl ammonium hydrogen sulphate

[a]Option (a) was used in one part of the study and option (b) in the other.

References

Aguilar, L., Esteban, C., Frias, J., Perez-Balcabao, I., Carcas, A. J. & Dal-Re, R. (1994). Cefminox: correlation between in-vitro susceptibility and pharmacokinetics and serum bactericidal activity in healthy volunteers. *Journal of Antimicrobial Chemotherapy* **33**, 91–101.

Ascalone, V. & Dal Bo, L. (1983). Determination of ceftriaxone, a novel cephalosporin, in plasma, urine and saliva by high-performance liquid chromatography on an NH_2 bonded-phase column. *Journal of Chromatography* **273**, 357–66.

Ayrton, J. (1981). Assay of ceftazidime in biological fluids using high-pressure liquid chromatography. *Journal of Antimicrobial Chemotherapy* **8**, *Suppl. B*, 227–31.

Bamberger, D. M., Peterson, L. R., Gerding, D. N., Moody, J. A. & Fasching, C. E. (1986). Ciprofloxacin, azlocillin, ceftizoxime and amikacin alone and in combination against Gram-negative bacilli in an infected chamber model. *Journal of Antimicrobial Chemotherapy* **18**, 51–63.

Barbhaiya, R. H., Forgue, S. T., Shyu, W. C., Papp, E. A. & Pittman, K. A. (1987). High-pressure liquid chromatographic analysis of BMY-28142 in plasma and urine. *Antimicrobial Agents and Chemotherapy* **31**, 55–9.

Barré, J. (1990). Pharmacokinetics of cefodizime: a review of the data on file. *Journal of Antimicrobial Chemotherapy* **26**, *Suppl. C*, 95–101.

Barrière, S. L., Gambertoglio, J. G., Lin, E. T. & Conte, J. E. (1982). Multiple-dose pharmacokinetics of amdinocillin in healthy volunteers. *Antimicrobial Agents and Chemotherapy* **21**, 54–7.

Bawdon, R. E. & Madsen, P. O (1986). High-pressure liquid chromatographic assay of sulbactam in plasma, urine, and tissue. *Antimicrobial Agents and Chemotherapy* **30**, 231–3.

Bergan, T. & Solberg, R. (1981). Assay of cefotaxime by high-pressure-liquid chromatography. *Chemotherapy* **27**, 155–65.

Bergan, T. & Solberg, R. (1982). Comparison of microbiological and high-pressure-liquid chromatographic assays of the new cephalosporin cefroxadine. *Analytical Abstracts* **43**, 42.

Bergan, T., Engeset, A., Olszewski, W., Ostby, N. & Solberg, R. (1986). Extravascular penetration of highly protein-bound flucloxacillin. *Antimicrobial Agents and Chemotherapy* **30**, 729–32.

Borner, K., Lode, H. & Elvers, A. (1982). Determination of apalcillin and its metabolites in human body fluids by high-pressure liquid chromatography. *Antimicrobial Agents and Chemotherapy* **22**, 949–53.

Bosso, J. A., Chan, G. M. & Matsen, J. M. (1983). Cefoperazone pharmacokinetics in pre-term infants. *Antimicrobial Agents and Chemotherapy* **23**, 413–5.

Brisson, A. M. & Fourtillan, J. B. (1981a). Determination of cephalosporins in biological material by reversed-phase liquid column chromatography. *Journal of Chromatography* **223**, 393–9.

Brisson, A. M. & Fourtillan, J. B. (1981b). Pharmacokinetique du cefuroxime chez l'homme. *Therapie* **36**, 143–9.

Brisson, A. M & Fourtillan, J. B. (1982a). High-performance liquid chromatographic determination of piperacillin in plasma. *Antimicrobial Agents and Chemotherapy* **21**, 664–5.

Brisson, A. M. & Fourtillan, J. B. (1982b). Pharmacokinetic study of cefadroxil following single and repeated doses. *Journal of Antimicrobial Chemotherapy* **10**, *Suppl. B*, 11–5.

Bryan, C. S., Morgan, S. L., Jordan, A. B., Smith, C. W., Sutton, J. P. & Gangemi, J. D. (1984). Ceftriaxone levels in blood and tissue during cardiopulmonary bypass surgery. *Antimicrobial Agents and Chemotherapy* **25**, 37–9.

Bryan, L. E. & Godfrey, A. J. (1991). *β*-Lactam antibiotics: mode of action and bacterial resistance. In *Antibiotics in Laboratory Medicine*, 3rd edn (Lorian, V., Ed.), pp. 599–664. Williams and Wilkins, Baltimore, MD.

Bryskier, A., Procyk, T. & Labro, M. T. (1990). Cefodizime, a new 2-aminothiazolyl cephalosporin: physicochemical properties, toxicology and structure–activity relationships. *Journal of Antimicrobial Chemotherapy* **26**, *Suppl. C*, 1–8.

Buhs, R. P., Maxim, T. E., Allen, N., Jacob, T. A. & Wolf, F. J. (1974). Analysis of cefoxitin, cephalothin and their deacetylated metabolites in human urine by high-performance liquid chromatography. *Journal of Chromatography* **99**, 609–18.

Burnett, J. & Sutherland, R. (1970). Procedures for the assay of carbenicillin in body fluids. *Applied Microbiology* **19**, 264–7.

Bush, K., Freudenberger, J. S. & Sykes, R. B. (1982). Interaction of azthreonam and related monobactams with beta-lactamases from Gram-negative bacteria. *Antimicrobial Agents and Chemotherapy* **22**, 414–20.

Camus, F., Deslandes, A., Harcouet, L. & Farinotti, R. (1994). High-performance liquid chromatographic method for the determination of cefpodoxime levels in plasma and sinus mucosa. *Journal of Chromatography B: Biomedical Applications* **656**, 383–8.

Carlqvist, J. & Westerlund, D. (1979). Determination of amoxicillin in body fluids by reversed-phase liquid chromatography coupled with a post-column derivatization procedure. *Journal of Chromatography* **164**, 373–81.

Carlqvist, J. & Westerlund, D. (1985). Automated determination of amoxycillin in biological fluids by column switching in ion-pair reversed-phase liquid chromatographic systems with post-column derivatization. *Journal of Chromatography* **344**, 285–96.

Carrol, M. A., White, E. R., Jancsik, Z. & Zarembo, J. E. (1977). The determination of cephradine and cephalexin by reverse phase high performance liquid chromatography. *Journal of Antibiotics* **30**, 397–403.

Charles, B. G. & Ravenscroft, P. J. (1984). Rapid HPLC analysis of cefoxitin in plasma and urine. *Journal of Antimicrobial Chemotherapy* **13**, 291–4.

Clark, J. M., Olsen, S. J., Weinberg, D. S., Dalvi, M., Whitney, R. R., Bonner, D. P. *et al.* (1987). *In vivo* evaluation of tigemonam, a novel oral monobactam. *Antimicrobial Agents and Chemotherapy* **31**, 226–9.

Colaizzi, P. A., Coniglio, A. A., Poynor, W. J., Vishniavsky, N., Karnes, H. T. & Polk, R. E. (1986). Comparative pharmacokinetics of two multiple-dose mezlocillin regimens in normal volunteers. *Antimicrobial Agents and Chemotherapy* **30**, 675–8.

Crombez, E., Van der Weken, G., Van den Bosche, W. & De Moerloose, P. (1979). Quantitative liquid chromatographic determination of cefatrizine in serum and urine by fluorescence detection after post-column derivatization. *Journal of Chromatography* **177**, 323–32.

Cronqvist, J., Nilsson-Ehle, I., Oqvist, B. & Norrby, S. R. (1992). Pharmacokinetics of cefepime dihydrochloride arginine in subjects with renal impairment. *Antimicrobial Agents and Chemotherapy* **36**, 2676–80.

Crooks, J., White, L. O., Burville, L. J., Speidel, B. D. & Reeves, D. S. (1984). Pharmacokinetics of cefotaxime and desacetyl-cefotaxime in neonates. *Journal of Antimicrobial Chemotherapy* **14**, *Suppl. B*, 97–101.

Dajani, A. S., Thirumoorthi, M. C., Bawdon, R. E., Buckley, J. A., Pfeffer, M., van Harken, D. R. *et al.* (1982). Pharmacokinetics of intramuscular ceforanide in infants, children and adolescents. *Antimicrobial Agents and Chemotherapy* **21**, 282–7.

Demotes-Mainard, F. M., Vincon, G. A., Jarry, C. H., Bourgeois, G. L. & Albin, H. C. (1985). High-performance liquid chromatographic determination of apalcillin in plasma and urine. *Journal of Chromatography* **342**, 234–40.

Dokladalova, J., Quercia, G. T. & Stankewich, J. P. (1983). High-performance liquid chromatographic determination of cefoperazone in human serum and urine. *Journal of Chromatography* **276**, 129–37.

Ehret, W., Probst, H. & Ruckdeschel, G. (1987). Determination of aztreonam in faeces of human volunteers: a comparison of reversed-phase high pressure liquid chromatography and bioassay. *Journal of Antimicrobial Chemotherapy* **19**, 541–9.

Falkowski, A. J., Look, Z. M., Noguchi, H. & Siber, B. M. (1987). Determination of cefixime in biological samples by reversed-phase high-performance liquid chromatography. *Journal of Chromatography* **422**, 145–52.

Fasching, C. E. & Peterson, L. R. (1982). Anion-exchange extraction of cephapirin, cefotaxime, and cefoxitin from serum for liquid chromatography. *Antimicrobial Agents and Chemotherapy* **21**, 628–33.

Fasching, C. E., Peterson, L. R., Bettin, K. M. & Gerding, D. N. (1982). High-pressure liquid chromatographic assay of ceftizoxime with an anion-exchange extraction technique. *Antimicrobial Agents and Chemotherapy* **22**, 336–7.

Fiore, D., Auger, F. A., Drusano, G. L., Dandu, V. R. & Lesko, L. J. (1984). Improved micromethod for mezlocillin quantitation in serum and urine by high-pressure liquid chromatography. *Antimicrobial Agents and Chemotherapy* **26**, 775–7.

Foulstone, M. & Reading, C. (1982). Assay of amoxicillin and clavulanic acid, the components of Augmentin, in biological fluids with high-performance liquid chromatography. *Antimicrobial Agents and Chemotherapy* **22**, 753–62.

Fredj, G., Paillet, M., Aussel, F., Brouard, A., Barreteau, H., Divine, C. *et al.* (1986). Determination of sulbactam in biological fluids by high-performance liquid chromatography. *Journal of Chromatography* **383**, 218–22.

Freij, B. J., McCracken, G. H., Olsen, K. D. & Threlkeld, N. (1985). Pharmacokinetics of imipenem–cilastatin in neonates. *Antimicrobial Agents and Chemotherapy* **27**, 431–5.

Granneman, G. R., Sennello, L. T., Steinberg, F. J. & Sonders, R. C. (1982). Intramuscular and intravenous pharmacokinetics of cefmenoxime, a new broad-spectrum cephalosporin, in healthy subjects. *Antimicrobial Agents and Chemotherapy* **21**, 141–5.

Gravallese, D. A., Musson, D. G., Pauliukonis, L. T. & Bayne, W. F. (1984). Determination of imipenem (*N*-formimidoyl thienamycin) in human plasma and urine by high-performance liquid chromatography, comparison with microbiological methodology and stability. *Journal of Chromatography* **310**, 71–84.

Guest, E. A., Horton, R., Mellows, G., Slocombe, B., Swaisland, A. J. & Tasker, T. C. G. (1985). Human pharmacokinetics of temocillin (BRL 17421) side chain epimers. *Journal of Antimicrobial Chemotherapy* **15**, 327–36.

Haginaka, J., Nakagawa, T. & Uno, T. (1979). Acidic degradation of cephaloglycin and high performance liquid chromatographic determination of deacetylcephaloglycin in human urine. *Journal of Antibiotics* **32**, 462–7.

Haginaka, J., Nakagawa, T. & Uno, T. (1980). Chromatographic analysis and pharmacokinetic investigation of cephaloglycin and its metabolites in man. *Journal of Antibiotics* **33**, 236–43.

Haginaka, J. & Wakai, J. (1985). High-performance liquid chromatographic assay of carbenicillin, ticarcillin and sulbenicillin in serum and urine using pre-column reaction with 1,2,4-triazole and mercury(II) chloride. *Analyst* **110**, 1185–8.

Haginaka, J., Yasuda, H., Uno, T. & Nakagawa, T. (1986). High-performance liquid chromatographic assay of clavulanate in human plasma and urine by fluorimetric detection. *Journal of Chromatography* **377**, 269–77.

Harding, S. M., Williams, P. E. O. & Ayrton, J. (1984). Pharmacology of cefuroxime as the 1-acetoxyethyl ester in volunteers. *Antimicrobial Agents and Chemotherapy* **25**, 78–82.

Hayashi, Y. (1982). High-performance liquid chromatographic microassay of cephradine in biological fluids. *Analytical Abstracts* **43**, 42.

Hill, S. L., Burnett, D., Lovering, A. L. & Stockley, R. A. (1992). Use of an enzyme-linked immunosorbent assay to assess penetration of amoxicillin into lung secretions. *Antimicrobial Agents and Chemotherapy* **36**, 1545–52.

Huang, H. S., Wu, J. R. & Chen, M. L. (1991). Reversed-phase high-performance liquid chromatography of amphoteric beta-lactam antibiotics: effects of columns, ion-pairing reagents and mobile phase pH on their retention times. *Journal of Chromatography* **564**, 195–203.

Huisman-de Boer, J. J., van den Anker, J. N., Vogel, M., Goessens, W. H. F., Schoemaker, R. C. & de Groot, R. (1995). Amoxicillin pharmacokinetics in preterm infants with gestational ages of less than 32 weeks. *Antimicrobial Agents and Chemotherapy* **39**, 431–4.

Itakura, K., Mitani, M., Aoki, I. & Usui, Y. (1982). High performance liquid chromatographic assay of cefsulodin, cefotiam and cefmenoxime in serum and urine. *Chemical and Pharmaceutical Bulletin* **30**, 622–7.

Jehl, F., Birckel, P. & Monteil, H. (1987). Hospital routine analysis of penicillins, third-generation cephalosporins and aztreonam by conventional and high-speed high-performance liquid chromatography. *Journal of Chromatography* **413**, 109–19.

Jones, P. G., Bodey, G. P., Swabb, E. A., Ho, D. H., Fainstein, V. & Pasternak, J. (1984). Clinical pharmacokinetics of aztreonam in cancer patients. *Antimicrobial Agents and Chemotherapy* **26**, 455–61.

Kearns, G. L., Reed, M. D., Jacobs, R. F., Ardite, M., Yogev, R. D. & Blumer, J. L. (1991). Single-dose pharmacokinetics of ceftibuten (SCH 39720) in infants and children. *Antimicrobial Agents and Chemotherapy* **35**, 2078–84.

Kelloway, J. S., Awni, W. M., Lin, C. C., Lim, J., Affrime, M. B., Keand, W. F. *et al.* (1991). Pharmacokinetics of ceftibuten-*cis* and its *trans* metabolite in healthy volunteers and in patients with chronic renal insufficiency. *Antimicrobial Agents and Chemotherapy* **35**, 2267–74.

Kelly, H. C., Hutchison, M. & Haworth, S. J. (1995). A comparison of the pharmacokinetics of meropenem after administration by intravenous injection over 5 min and intravenous infusion over 30 min. *Journal of Antimicrobial Chemotherapy* **36**, *Suppl. A*, 35–41.

Kemmerich, B., Lode, H., Borner, K., Belmega, D., Jendroschek, T., Koeppe, P. & Goertz, G. (1983a). Biliary excretion and pharmacokinetics of cefoperazone in humans. *Journal of Antimicrobial Chemotherapy* **12**, 27–37.

Kemmerich, B., Lode, H., Belmega, D., Jendroschek, T., Borner, K. & Koeppe, P. (1983b). Comparative pharmacokinetics of cefoperazone, cefotaxime and moxalactam. *Antimicrobial Agents and Chemotherapy* **23**, 429–34.

Kita, Y., Fugono, T. & Imada, A. (1986). Comparative pharmacokinetics of carumonam and aztreonam in mice, rats, rabbits, dogs, and cynomolgus monkeys. *Antimicrobial Agents and Chemotherapy* **29**, 127–34.

Lee, H., Lee, J. S. & Lee, H. S. (1995). Simultaneous determination of ampicillin and metampicillin in biological fluids using high-performance liquid chromatography with column switching. *Journal of Chromatography B: (Biomedical Applications)* **664**, 335–40.

Lin, C., Radwanski, E., Affrime, M. & Cayen, M. N. (1995). Multiple-dose pharmacokinetics of ceftibuten in healthy volunteers. *Antimicrobial Agents and Chemotherapy* **39**, 356–8.

Lindberg, R. L., Huupponen, R. K. & Huovinen, P. (1984). Rapid high-pressure liquid chromatographic method for analysis of phenoxymethylpenicillin in human serum. *Antimicrobial Agents and Chemotherapy* **26**, 300–2.

Lindner, K. R. & Pilkiewicz, F. G. (1986). High pressure liquid chromatography assay of aztreonam. In *High Performance Liquid Chromatography in Medical Microbiology* (Reeves, D. S. & Ullmann, U., Eds), pp. 75–82. Fischer Verlag, Stuttgart.

Lode, H., Elvers, A., Koeppe, P. & Borner, K. (1984). Comparative pharmacokinetics of apalcillin and piperacillin. *Antimicrobial Agents and Chemotherapy* **25**, 105–8.

Lode, H., Madey, V., Dzwillo, G., Borner, K. & Koeppe, P. (1983). Serum bactericidal activity and kinetics of azlocillin and moxalactam after single and combined administration. *Journal of Antimicrobial Chemotherapy* **11**, Suppl. B, 121–6.

Lovering, A. M., Vickery, C. J., Watkin, D. S., Leaper, D., McMullin, C. M., White, L. O. et al. (1995). The pharmacokinetics of meropenem in surgical patients with moderate or severe infections. *Journal of Antimicrobial Chemotherapy* **36**, 165–72.

Luthy, R., Blaser, J., Bonetti, A., Simmen, H., Wise, R. & Siegenthaler, W. (1981). Comparative multiple-dose pharmaco-kinetics of cefotaxime, moxalactam and ceftazidime. *Antimicrobial Agents and Chemotherapy* **20**, 567–75.

Martens, M. G., Faro, S., Feldman, S., Cotton, D. B., Dorman, K. & Riddle, G. D. (1987). Pharmacokinetics of the acylureido-penicillins piperacillin and mezlocillin in the postpartum patient. *Antimicrobial Agents and Chemotherapy* **31**, 2015–7.

McMormick, E. M., Echols, R. M. & Rosano, T. G. (1984). Liquid chromatographic assay of ceftizoxime in sera of normal and uraemic patients. *Antimicrobial Agents and Chemotherapy* **25**, 336–8.

Mihindu, J. C., Scheld, W. M., Bolton, N. D., Spyker, D. A., Swabb, E. A. & Bolton, W. K. (1983). Pharmacokinetics of aztreonam in patients with various degrees of renal dysfunction. *Antimicrobial Agents and Chemotherapy* **24**, 252–61.

Miner, D. J., Coleman, D. L., Shepherd, A. M. M. & Hardin, T. C. (1981). Determination of moxalactam in human body fluids by liquid chromatographic and microbiological methods. *Antimicrobial Agents and Chemotherapy* **20**, 252–7.

Modai, J., Wolff, M., Lebas, J., Meulemans, A. & Manuel, C. (1982). Moxalactam penetration into cerebrospinal fluid in patients with bacterial meningitis. *Antimicrobial Agents and Chemotherapy* **21**, 551–3.

Modai, J., Vittecoq, D., Decazes, J. M., Wolff, M. & Meulemans, A. (1986). Penetration of aztreonam into cerebrospinal fluid of patients with bacterial meningitis. *Antimicrobial Agents and Chemotherapy* **29**, 281–3.

Mullany, L. D., French, M. A., Nightingale, C. H., Low, H. B., Ellison, L. H. & Quintiliani, R. (1982). Penetration of ceforanide and cephamandole into the right atrial appendage, pericardial fluid, sternum and intercostal muscle of patients undergoing open heart surgery. *Antimicrobial Agents and Chemotherapy* **21**, 416–20.

Murai, Y., Nakagawa, T., Yamaoka, K. & Uno, T. (1981). High performance liquid chromatographic analysis and pharmaco-kinetic investigation of oxacillin and its metabolites in man. *Chemical and Pharmaceutical Bulletin* **29**, 3290–7.

Myers, C. M. & Blumer, J. L. (1984). Determination of imipenem and cilastatin in serum by high-pressure liquid chromatography. *Antimicrobial Agents and Chemotherapy* **26**, 78–81.

Nahata, M. C. (1981). High-performance liquid chromatographic determination of cephalexin in human plasma, urine and saliva. *Journal of Chromatography* **225**, 532–8.

Naidong, W., Dzerk, A. M. & Lee, J. W. (1994). Development and validation of an LC method for the quantitation of carbenicillin in human serum. *Journal of Pharmaceutical and Biomedical Analysis* **12**, 845–50.

Nakagawa, T., Haginaka, J., Yamaoka, K. & Uno, T. (1978). High speed liquid chromatographic determination of cephalexin in human plasma and urine. *Journal of Antibiotics* **31**, 769–75.

Nilsson-Ehle, I. & Nilsson-Ehle, P. (1979). Pharmacokinetics of cephalothin: accumulation of its deacetylated metabolite in uraemic patients. *Journal of Infectious Diseases* **139**, 712–6.

Nix, D. E., Wilton, J. H., Velasquez, N., Budny, J. L., Lassman, H. B., Mitchell, P. et al. (1992). Cerebrospinal fluid penetration of cefpirome in patients with non-inflamed meninges. *Journal of Antimicrobial Chemotherapy* **29**, Suppl. A, 51–7.

Noonan, I. A., Gambertoglio, J. G., Barrière, S. L., Conte, J. E. & Lin, E. T. (1983). High-performance liquid chromatographic determination of cefmenoxime (AB-50912) in human plasma and urine. *Journal of Chromatography* **273**, 458–63.

O'Callaghan, C. H. (1979). Description and classification of the newer cephalosporins with their relationships with established compounds. *Journal of Antimicrobial Chemotherapy* **5**, 635–71.

O'Callaghan, C. H. & Kirby, S. M. (1978). Cephalosporins. In *Laboratory Methods in Antimicrobial Chemotherapy* (Reeves, D. S., Phillips, I., Williams, J. D. & Wise, R, Eds), pp. 181–91. Churchill Livingstone, Edinburgh.

Ohkawa, M., Hirano, S., Tokunaga, S., Motoi, I., Shoda, R., Ikeda, A. et al. (1983). Pharmacokinetics of cefotetan in normal subjects and patients with impaired renal function. *Antimicrobial Agents and Chemotherapy* **23**, 31–5.

Paradis, D., Vallee, F., Allard, S., Bisson, C., Daviau, N., Drapeau, C. et al. (1992). Comparative study of pharmacokinetics and serum bactericidal activities of cefpirome, ceftazidime, ceftriaxone, imipenem and ciprofloxacin. *Antimicrobial Agents and Chemotherapy* **36**, 2085–92.

Patel, I. H., Chen, S., Parsonnet, M., Hackman, M. R., Brooks, M. A., Konikoff, J. et al. (1981). Pharmacokinetics of ceftriaxone in humans. *Antimicrobial Agents and Chemotherapy* **20**, 634–41.

Patel, I. H., Sugihara, J. G., Weinfeld, R. E., Wong, E. G. C., Siemsen, A. W. & Berman, S. J. (1984). Ceftriaxone pharmaco-kinetics in patients with various degrees of renal impairment. *Antimicrobial Agents and Chemotherapy* **25**, 438–42.

Pilkiewicz, F. G., Remsburg, B. J., Fisher, S. M. & Sykes, R. B. (1983). High-pressure liquid chromatographic analysis of aztreonam in sera and urine. *Antimicrobial Agents and Chemotherapy* **23**, 852–6.

Polk, R. E., Kline, B. J. & Markowitz, S. M. (1981). Cefazolin and moxalactam pharmacokinetics after simultaneous intravenous infusion. *Antimicrobial Agents and Chemotherapy* **20**, 576–9.

Reed, M. D., Stern, R. C., Yamashita, T. S., Ackers, I., Myers, C. M. & Blumer, J. L. (1984). Single-dose pharmacokinetics of cefsulodin in patients with cystic fibrosis. *Antimicrobial Agents and Chemotherapy* **25**, 579–81.

Reeves, D. S., Bullock, D. W., Bywater, M. J., Holt, H. A., White, L. O. & Thornhill, D. P. (1981). The effect of probenecid on the pharmacokinetics and distribution of cefoxitin in healthy volunteers. *British Journal of Clinical Pharmacology* **11**, 353–9.

Reeves, D. S., White, L. O., Holt, H. A., Bahari, D., Bywater, M. J. & Bax, R. P. (1980). The human metabolism of cefotaxime. *Journal of Antimicrobial Chemotherapy* **6**, Suppl. A, 93–101.

Richer, M., Allard, S., Manseau, L., Vallee, F., Pak, R. & LeBel, M. (1995). Suction-induced blister fluid penetration of cefdinir in healthy volunteers following ascending oral doses. *Antimicrobial Agents and Chemotherapy* **39**, 1082–6.

Rogers, H. J., Bradbrook, I. D., Morrison, P. J., Spector, R. G., Cox, D. A. & Lees, L. J. (1983). Pharmacokinetics and bioavailability of sultamicillin estimated by high performance liquid chromatography. *Journal of Antimicrobial Chemotherapy* **11**, 435–45.

Roholt, K. (1977). Pharmacokinetic studies with mecillinam and pivmecillinam. *Journal of Antimicrobial Chemotherapy* **3**, Suppl. B, 71–81.

Rolewicz, T. F., Mirkin, B. L., Cooper, M. J. & Anders, M. W. (1977). Metabolic disposition of cephalothin and deacetylcephalothin in children and adults: comparison of high-performance liquid chromatographic and microbial assay procedures. *Clinical Pharmacology and Therapeutics* **22**, 928–35.

Romagnoli, M. F., Flynn, K., Siber, G. R. & Goldmann, D. A. (1982). Moxalactam pharmacokinetics in children. *Antimicrobial Agents and Chemotherapy* **22**, 47–50.

Rotschafer, J. C., Crossley, K. B., Lesar, T. S., Zaske, D. & Miller, K. (1982). Cefaclor pharmacokinetic parameters: serum concentrations determined by a new high-performance liquid chromatographic technique. *Antimicrobial Agents and Chemotherapy* **21**, 170–2.

Rumble, R. H. & Roberts, M. S. (1985). Determination of benzyl penicillin in plasma and urine by high-performance liquid chromatography. *Journal of Chromatography* **342**, 436–41.

Saathoff, N., Lode, H., Neider, K., Depperman, K. M., Borner, K. & Koeppe, P. (1992). Pharmacokinetics of cefpodoxime proxetil and interactions with an antacid and an H₂ receptor antagonist. *Antimicrobial Agents and Chemotherapy* **36**, 796–800.

Sailer, H., Diekmann, H. W., Faro, H. P. & Garbe, A. (1979). Pharmacokinetics and metabolism of cefazedone in Wistar rat, beagle dog and rhesus monkey. *Arzneimittel-forschung* **29**, 404–11.

Sekine, M., Sasahara, K., Kojima, T. & Morioka, T. (1982). High-performance liquid chromatographic method for determination of cefmetazole in human serum. *Antimicrobial Agents and Chemotherapy* **21**, 740–3.

Shull, V. H. & Dick, J. D. (1985). Determination of ticarcillin levels in serum by high-pressure liquid chromatography. *Antimicrobial Agents and Chemotherapy* **28**, 597–600.

Sotto, A., Peray, P., Geny, F., Brunschwig, C., Carrière, C., Galtier, M. *et al.* (1995). An enzymatic method for assaying sulbactam in human serum: comparison with high performance liquid chromatography. *Journal of Antimicrobial Chemotherapy* **35**, 429–33.

Sutherland, R. (1997). β-Lactams: penicillins. In *Antibiotic and Chemotherapy*, 7th edn (O'Grady, F., Lambert, H. P., Finch, R. G. & Greenwood, D., Eds), pp. 256–306. Churchill Livingstone, Edinburgh.

Sutsumi, K., Kubo, H. & Kinoshita, T. (1982). Determination of serum cephalexin by high-performance liquid chromatography. *Analytical Abstracts* **43**, 157.

Swabb, E. A., Leitz, M. A., Pilkiewicz, F. G. & Sugerman, A. A. (1981). Pharmacokinetics of the monobactam SQ 26,776 after single intravenous doses in healthy subjects. *Journal of Antimicrobial Chemotherapy* **8**, Suppl. E, 131–40.

Swabb, E. A., Sugerman, A. A. & McKinstry, D. N. (1983). Multiple-dose pharmacokinetics of the monobactam azthreonam (SQ 26,776) in healthy subjects. *Antimicrobial Agents and Chemotherapy* **23**, 125–32.

Swaisland, A. J. (1986). The HPLC assay of penicillins in body fluids. In *High Performance Liquid Chromatography in Medical Microbiology* (Reeves, D. S & Ullmann, U., Eds), pp. 45–60. Fischer Verlag, Stuttgart.

Swanson, D. J., DeAngelis, C., Smith, I. L. & Schentag, J. J. (1986). Degradation kinetics of imipenem in normal saline and in human serum. *Antimicrobial Agents and Chemotherapy* **29**, 936–7.

Syrogiannopoulos, G. A., Al-Sabbagh, A., Olsen, K. D. & McCracken, G. H. (1987). Pharmacokinetics and bacteriological efficacy of ticarcillin–clavulanic acid (timentin) in experimental *Escherichia coli* K-1 and *Haemophilus influenzae* type b meningitis. *Antimicrobial Agents and Chemotherapy* **31**, 1296–300.

Ti, T. Y., Fortin, L., Kreeft, E. J. H., East, D. S. Ogilvie, R. I. & Somerville, P. J. (1984). Kinetic disposition of intravenous ceftriaxone in normal subjects and patients with renal failure on haemodialysis or peritoneal dialysis. *Antimicrobial Agents and Chemotherapy* **25**, 83–7.

Trautmann, K. H. & Haefelfinger, P. (1981). Determination of the cephalosporin Ro13-9904 in plasma, urine and bile by means of ion-paired reversed phase chromatography. *Journal of High Resolution Chromatography and Chromatography Communications* **4**, 54–9.

Tremblay, D., Dupront, A., Ho, C., Coussediere, D. & Lenfant, B. (1990). Pharmacokinetics of cefpodoxime in young and elderly volunteers after single doses. *Journal of Antimicrobial Chemotherapy* **26**, Suppl. E, 21–8.

Ullmann, U. (1980). High-pressure liquid chromatography and microbiological assay in the determination of serum levels using cefradine and cefaclor. *Zentralblatt für Bakteriologie-1-Abt-Originale-A: Medizinische Mikrobiologie, Infektionskrankheiten und Parasitologie* **248**, 414–21.

van Dalen, R., Vree, T. B., Hafkencheid, J. C. M. & Gimbrere, J. S. F. (1979). Determine of plasma and renal clearance of cefuroxime and its pharmacokinetics in renal insufficiency. *Journal of Antimicrobial Chemotherapy* **5**, 281–92.

Van der Auwera, P. & Santella, P. J. (1993). Pharmacokinetics of cefepime: a review. *Journal of Antimicrobial Chemotherapy* **32**, Suppl. B, 103–15.

van Gulpen, C., Brokerhof, A. W., van der Kaay, M., Tjaden, U. R. & Mattie, H. (1986). Determination of benzylpenicillin and probenecid in human body fluids by high-performance liquid chromatography. *Journal of Chromatography* **381**, 365–72.

Vree, T. B., Hekster, Y. A., Baars, A. M. & Van der Kleijn, E. (1978). Rapid determination of amoxycillin (clamoxyl) and ampicillin (penbritin) in body fluids of many by means of high-performance liquid chromatography. *Journal of Chromatography* **145**, 496–501.

Walstad, R. A., Hellum, K. B., Thurmann-Nielsen, E. & Dale, L. G. (1986). Pharmacokinetics and tissue penetration of timentin: a simultaneous study of serum, urine, lymph, suction blister and subcutaneous thread fluid. *Journal of Antimicrobial Chemotherapy* **17**, Suppl. C, 71–80.

Weber, A., Opheim, K. E., Wong, K. & Smith, A. L. (1983). High-pressure liquid chromatographic quantitation of azlocillin. *Antimicrobial Agents and Chemotherapy* **24**, 750–3.

Welage, L. S., Schultz, R. W. & Schentag, J. J. (1984). Pharmacokinetics of ceftazidime in patients with renal insufficiency. *Antimicrobial Agents and Chemotherapy* **25**, 201–4.

Westerlund, D., Carlqvist, J. & Theodorsen, A. (1979). Analysis of penicillins in biological material by reversed phase liquid chromatography and post-column derivatization. *Acta Pharmaceutica Suedica* **16**, 187–214.

White, L. O. (1986). The assay of cephems by HPLC. In *High Performance Liquid Chromatography in Medical Microbiology* (Reeves, D. S & Ullmann, U., Eds), pp. 91–121. Fischer Verlag, Stuttgart.

White, L. O., Reeves, D. S., Holt, H. A. & Bywater, M. L. (1981). Microbiological assay of cefotaxime. *Journal of Antimicrobial Chemotherapy* **7**, 308–9.

White, L. O., Reeves, D. S., Lovering, A. M. & MacGowan, A. P. (1993). HPLC assay of cefixime in serum and CSF. *Journal of Antimicrobial Chemotherapy* **31**, 450–1.

Williams, J. D., Andrews, J., Mitchard, M. & Kendall, M. J. (1976). Bacteriology and pharmacokinetics of the new amino penicillin-mecillinam. *Journal of Antimicrobial Chemotherapy* **2**, 61–9.

Wilson, C. B., Koup, J. R., Opheim, K. E., Adelman, L. A., Levy, J., Stull, T. L. *et al.* (1982). Piperacillin pharmacokinetics in pediatric patients. *Antimicrobial Agents and Chemotherapy* **22**, 442–7.

Windorfer, A. & Bauer, P. (1982). Pharmacokinetics and clinical studies with cefadroxil in paediatrics. *Journal of Antimicrobial Chemotherapy* **10**, *Suppl. B*, 85–91.

Wise, R. (1990). The pharmacokinetics of the oral cephalosporins—a review. *Journal of Antimicrobial Chemotherapy* **26**, *Suppl. E*, 13–20.

Wise, R. (1997). *β*-Lactams: cephalosporins. In *Antibiotic and Chemotherapy*, 7th edn (O'Grady, F., Lambert, H. P., Finch, R. G. & Greenwood, D., Eds), pp, 205–55. Churchill Livingstone, Edinburgh.

Wise, R., Wills, P. J. & Bedford, K. A. (1981). Epimerase of moxalactam in vitro comparison of activity and stability. *Antimicrobial Agents and Chemotherapy* **20**, 30–2.

Wold, J. S. & Turnipseed, S. A. (1977a). The simultaneous quantitative determination of cephalothin and cephazolin in serum by high-pressure liquid chromatography. *Clinica Chimica Acta* **78**, 203–7.

Wold, J. S. & Turnipseed, S. A. (1977b). Determination of cephaloridine in serum and tissue by high-performance liquid chromatography. *Journal of Chromatography* **136**, 170–3.

Yamaoka, K., Nakagawa, T. & Uno, T. (1983). Moment analysis for disposition kinetics of several cephalosporin antibiotics in rats. *Journal of Pharmacy and Pharmacology* **35**, 19–22.

Yamaoka, K., Narita, S., Nakagawa, T. & Uno, T. (1979). High-performance liquid chromatographic analyses of sulbenicillin and carbenicillin in human urine. *Journal of Chromatography* **168**, 187–93.

Yamashita, K., Motohashi, M. & Yashiki, T. (1992). Automated high-performance liquid chromatographic method for the simultaneous determination of cefotiam and delta 3-cefotiam in human plasma using column switching. *Journal of Chromatography* **577**, 174–9.

Yates, R. A., Adam, H. K., Donnelly, R. J., Houghton, H. L., Charlesworth, E. A. & Laws, E. A. (1983). Pharmacokinetics and tolerance of single intravenous doses of cefotetan disodium in male caucasian volunteers. *Journal of Antimicrobial Chemotherapy* **11**, *Suppl.*, 185–91.

Chapter 9

Aminoglycosides

Andrew M. Lovering

Department of Medical Microbiology, Southmead Hospital, Bristol BS10 5NB, UK

Introduction

Aminoglycosides are characterized by the possession of an amino sugar and an aminocyclitol joined by a glycosidic link. They are subdivided into three groups depending on whether the aminocyclitol moiety is streptidine, 2-deoxystreptamine or fortamine (Figure 9.1). Streptomycin is the only major aminoglycoside in the first group (Figure 9.2). The other important compounds belong to the second group. The 2-deoxystreptamine may be 4,5-disubstituted (Figures 9.3 and 9.4) or 4,6-disubstituted (Figures 9.5 and 9.6) and most clinically used compounds fall into the latter classification. A veterinary drug, apramycin, is a 4-substituted 2-deoxystreptamine compound (Figure 9.7). The third group includes fortimicins A and B, sporaricins A and B, istamycins A and B and dactimycin. The numbering convention for the carbon atoms is illustrated in *Chapter 5* and will not be further commented on here.

The aminoglycosides exhibit such similarities in chemical properties, mode of action, pharmacokinetics, spec-

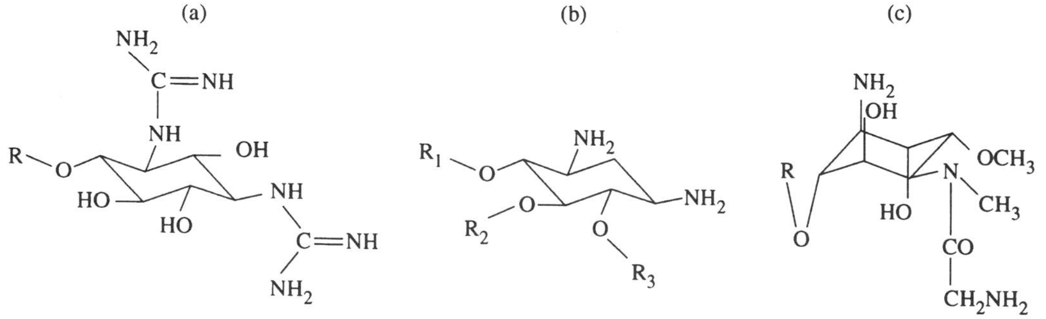

Figure 9.1. The structures of (a) streptidine, (b) 2-deoxystreptamine and (c) fortamine rings.

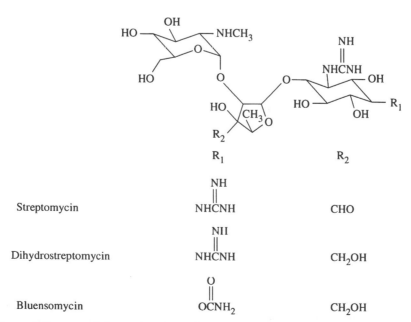

Figure 9.2. The structure of streptidine-containing aminoglycosides.

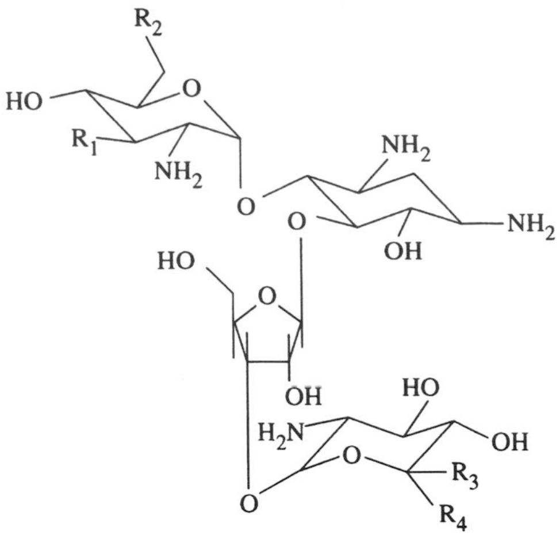

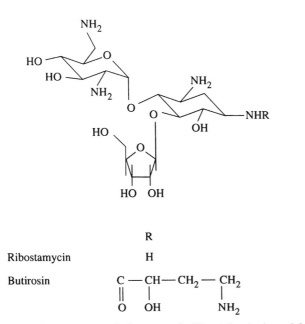

Figure 9.3. The structure of neomycin-like 4,5-substituted 2-deoxystreptamine-containing aminoglycosides.

	R_1	R_2	R_3	R_4
Neomycin B	OH	NH_2	H	CH_2NH_2
Neomycin C	OH	NH_2	CH_2NH_2	H
Paromomycin I	OH	OH	H	CH_2NH_2
Paromomycin II	OH	OH	CH_2NH_2	H
Lividomycin B	H	OH	H	CH_2NH_2

Figure 9.5. The structure of kanamycin-like 4,6-substituted 2-deoxystreptamine-containing aminoglycosides.

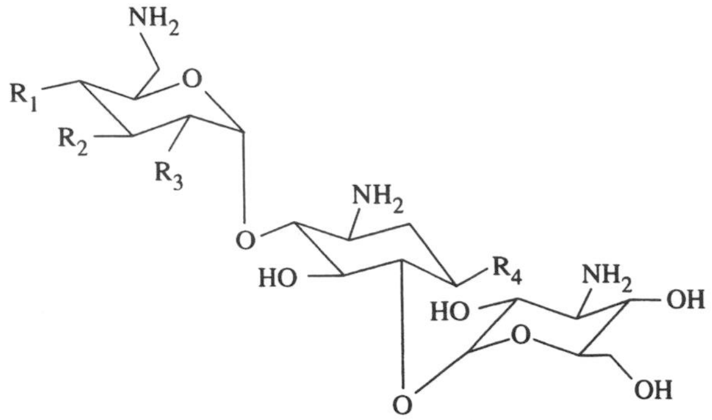

	R_1	R_2	R_3	R_4
Kanamycin A	OH	OH	OH	NH_2
Kanamycin B	OH	OH	NH_2	NH_2
Tobramycin	OH	H	NH_2	NH_2
Dibekacin	H	H	NH_2	NH_2
Habakacin	H	H	NH_2	$NHCHOHCH_2CH_2NH_2$
Amikacin	OH	OH	OH	$NHCHOHCH_2CH_2NH_2$

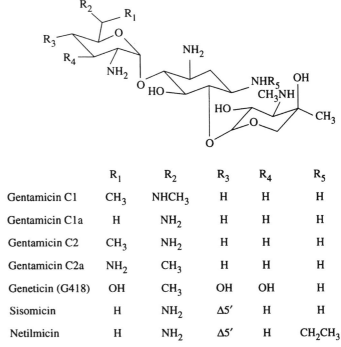

	R
Ribostamycin	H
Butirosin	C—CH—CH$_2$—CH$_2$ with ‖O, OH, NH$_2$

Figure 9.4. The structure of ribostamycin-like 4,5-substituted 2-deoxystreptamine-containing aminoglycosides.

Figure 9.6. The structure of gentamicin-like 4,6-substituted 2-deoxystreptamine-containing aminoglycosides.

	R_1	R_2	R_3	R_4	R_5
Gentamicin C1	CH_3	$NHCH_3$	H	H	H
Gentamicin C1a	H	NH_2	H	H	H
Gentamicin C2	CH_3	NH_2	H	H	H
Gentamicin C2a	NH_2	CH_3	H	H	H
Geneticin (G418)	OH	CH_3	OH	OH	H
Sisomicin	H	NH_2	$\Delta 5'$	H	H
Netilmicin	H	NH_2	$\Delta 5'$	H	CH_2CH_3

trum of activity and nature of toxicity, that in this chapter they will be discussed collectively, and only individually where properties differ from the group as a whole. Currently, within the UK, the aminoglycoside antibiotics used in cases of severe and life-threatening infection and, consequently, those most frequently requested for assay are gentamicin, tobramycin, netilmicin and amikacin (Phillips, 1982). In addition, streptomycin, which is now being increasingly used in the treatment of tuberculosis, and, less frequently, neomycin (or framycetin), which is often prescribed in either oral or topical preparations, may require assay. A number of workers have suggested that

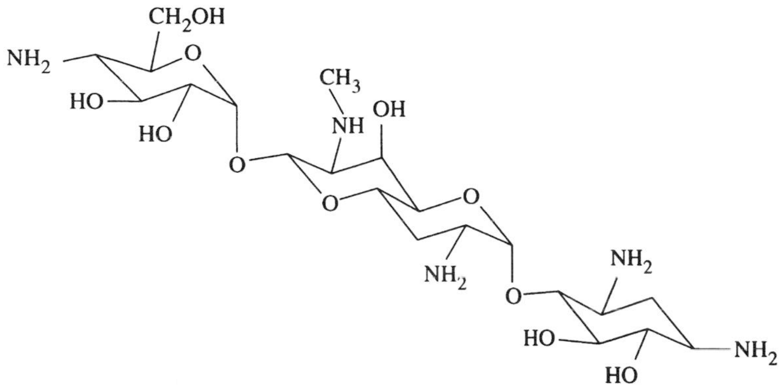

Figure 9.7. The structure of apramycin, a 4-substituted 2-deoxystreptamine-containing aminoglycoside.

the aminocyclitol spectinomycin should also be considered along with the aminoglycosides. However, since instances where there is a need to assay spectinomycin are rare, this antibiotic will not be discussed further in this chapter.

A number of other aminoglycosides are licensed for use outside of the UK, but globally the pattern of aminoglycoside usage is little different from that seen within the UK, with only sisomicin, dibekacin, habekacin, astromycin, isepemycin and, perhaps, kanamycin used to any extent (Price, 1986).

In recent years a number of new aminoglycoside derivatives have been described. Most of these are less susceptible to degradation by aminoglycoside-modifying enzymes than the aminoglycosides in current usage, yet retain a similar spectrum of activity and degree of toxicity to them. During the late 1970s and early 1980s there was considerable concern that an increasing prevalence of bacterial resistance would negate the value of aminoglycosides. However, within the UK, and much of Europe, aminoglycoside resistance is not common, typically <4% (Lovering et al., 1988; Spencer et al., 1990; MacGowan et al., 1993), so these newer derivatives appear to offer little advantage over the currently available aminoglycosides. Furthermore, since the mid 1980s the usage of aminoglycosides has tailed off, and generic products have become available, resulting in a loss of revenue for pharmaceutical companies marketing aminoglycosides (Price, 1986). It is probable that in the foreseeable future the only aminoglycosides that laboratories will be asked to assay are those currently available; this chapter will, therefore, deal mainly with methods available for the assay of these compounds.

Physicochemical properties

Aminoglycosides are white or off-white hygroscopic powders which should be stored desiccated. In the event that aminoglycosides have not been stored desiccated, drying by heating at 110°C in a hot air oven for 2 h will be required. Normally, the stated potency values for aminoglycosides will be the potency after drying, but some aminoglycoside powders will have so-called 'as is' potency

values and in this case drying should not be performed. Repeated heatings will result in decomposition leading to a loss of potency and we would not recommend that samples are dried more than once. Aminoglycosides are usually supplied as the sulphate, since this causes least discomfort on injection or, as in the case of amikacin, improves solubility, but laboratory material may be the free base. The appropriate correction for potency must be made when preparing calibrators or controls, otherwise a large error will ensue; potencies are usually in the range 600–750 μg/mg. Although some aminoglycosides are pure substances, for which molecular weights can be quoted (Table I), several are mixtures of a number of components, the exact proportions of which vary from batch to batch. For example, gentamicin (Figure 9.6) is composed of four major components, C1, C1a, C2 and C2a (White et al., 1983b), kanamycin (Figure 9.5) two components, A and B (Adams et al., 1997), and neomycin (Figure 9.3) at least three components, A, B and C (Roets et al., 1995) (neomycin A is a degradation product of neomycins B and C, and has little antimicrobial activity).

Because they contain four to six NH_2 groups, aminoglycosides are strong bases and exhibit a number of pK_a values, dependent upon the number of NH_2 groups, in the pH range 5–8. All aminoglycosides are freely soluble in aqueous solution, but only poorly so in organic or alcoholic solution (<10 mg/L). Solutions of aminoglycosides exhibit little or no absorbance in the UV or visible spectrum.

Aqueous aminoglycoside solutions are extremely stable (>3 months at 4°C (McBride et al., 1991)) and may be autoclaved or stored frozen for several years without loss of activity; as a result, calibrator solutions may be prepared in batches and stored frozen. However, solutions of streptomycin are marginally less stable than those of other aminoglycosides and we recommended that these are stored at –70°C and not kept for >6 months.

Because of their large positive charge, aminoglycosides bind to some plastics or glasses, so when preparing calibrators it is important to ensure that the containers used for storing solutions do not absorb aminoglycosides. In addition, aminoglycosides may react and be inactivated

Table I. The physicochemical and pharmacokinetic properties of aminoglycosides

	Composition	Molecular weight	$t_{1/2}\beta$ (h)	% protein bound	$V_d{}^a$ (L/kg)	Cl^b (mL/min/kg)	pK_a
Amikacin	pure	585.6	2–3	4	0.2	1.0	8.1
Kanamycin	97% A, 3% B	NA	2–3	3	0.3	1.4	7.2
Gentamicin	variable C1, C1a, C2 and C2a content	NA	1–3	13	0.2	1.1	NA
Netilmicin	pure	475.6	2.7	<3	0.2	1.0	8.1
Neomycin	>70% B, <30% C	614.6	2	NA	0.25	NA	NA
Streptomycin	pure	581.6	2.5	35	–	–	4.7–7
Tobramycin	pure	467.5	2.5	<3	0.3	1.5	6.7, 8.3, 9.9

NA, not applicable. $^a V_d$, volume of distribution. $^b Cl$, clearance.

by other antibiotics, or other drugs, and therefore care should be taken in the preparation of solutions containing more than one antimicrobial agent (McLaughlin & Reeves, 1971; Daly *et al.*, 1997).

Dosage forms

Amikacin is available as vials for injection containing amikacin sulphate equivalent to 500 mg or, for paediatric use, 100 mg of amikacin in 2 mL of solution without preservative. At the time of writing a liposomal amikacin preparation (Mikosome, NeXstar) was undergoing clinical trials.

Gentamicin is available in ointments and eyedrops, which will contain additional ingredients, and as solutions for injection which contain gentamicin sulphate equivalent to 160, 80, 20 (paediatric use) or 5 (intrathecal use) mg gentamicin in 1 or 2 mL of solution, which may contain methylhydroxybenzoate and propylhydroxybenzoate as preservatives.

Neomycin is available in many oral and topical formulations with or without other antibacterials such as polymyxin B and bacitracin. A tablet formulation contains 350,000 international units (approximately 500 mg) of neomycin sulphate per tablet.

Netilmicin for injection comes as vials of netilmicin sulphate aqueous solution (2, 1.5 and 1 mL vials) equivalent to 100, 50 or 10 mg netilmicin/mL.

Streptomycin is available as sterile streptomycin sulphate powder in vials equivalent to 1 g streptomycin.

Tobramycin is available as an eyedrop solution and also for injection as solutions of tobramycin sulphate equivalent to 40 mg (in 1 mL), 80 mg (in 2 mL) or 20 mg (in 2 mL) tobramycin. The preparations also contain phenol, sodium bisulphite and disodium edetate as preservatives.

Pharmacokinetics

Following oral administration of an aminoglycoside, little or no absorption occurs, with <1% of the dose recoverable in the urine (Finegold, 1959). As a result, oral administration is sometimes used for gut decontamination before surgery or in the treatment of enteritis (Garrod *et al.*, 1981). However, in patients with colitis who are also suffering from renal insufficiency, appreciable and, occasionally, toxic blood levels may be obtained following oral therapy. Similarly, although the transdermal absorption of aminoglycoside is low, following topical application where the drug is administered over large areas, such as in burns patients or neonates, appreciable blood levels may result.

The usual route of administration of aminoglycosides is by either im or iv injection, although absorption from the peritoneal cavity may also give acceptable blood levels (Bodey *et al.*, 1975). Peak blood concentrations are observed 30–90 min following im injection, and are usually measured at 60 min. Following iv administration, the distribution or α-phase lasts about 15–30 min and again concentrations are usually measured 60 min after a dose. The serum concentration of gentamicin 60 min after dose is the same irrespective of whether the drug has been given by the im or iv route (Edwards *et al.*, 1992). In recent years, the use of once-daily aminoglycoside therapy has increased in popularity, but protocols for monitoring patients on such therapy are only poorly developed and controversy still exists on the value and timing of the post-dose samples (Nicolau *et al.*, 1992; Parker & Davey, 1993; Reeves *et al.*, 1995).

Aminoglycosides distribute well into most body fluids with a volume of distribution approximating to the volume of the extracellular fluid (Table I). Although good concentrations are rapidly achieved in synovial, peritoneal, ascitic and pleural fluids, only slow distribution is observed into bile, faeces, prostatic and amniotic fluid while concentrations in the CNS and eye are frequently therapeutically inadequate (Chisholm *et al.*, 1968). However, recent studies have cast doubt on a number of these early tissue penetration studies, since the simple buffer protocols used in early studies to extract aminoglycosides from tissue have been found to give only partial extraction of drug and significantly higher levels are obtained when sodium hydroxide or trichloroacetic acid extraction is used (Fox, 1989; Brown *et al.*, 1990). Low protein binding has been reported for aminoglycosides, with <10% recorded for most of the members of the group (Table I). Excretion is almost exclusively via glomerular filtration, with 85–95% of the administered dose recovered in the urine (Gyselynck *et al.*, 1971). No in-vivo metabolism has been reported to date for the aminoglycosides, although in-vitro metabolism has been reported (Crann *et al.*, 1992), and it is probable that other routes of elimination, such as biliary, account for the rest of the dose. Although the plasma β-phase elimination half-lives ($t_{1/2}\beta$) of aminoglycosides are frequently quoted as 2–4 h (Table I), considerable variation has been reported, even in patients with normal renal function (Zaske, 1980). The most frequent reason for an increased $t_{1/2}\beta$ is decreased renal function: in complete renal failure half-lives may exceed 40 h. Recent pharmacokinetic studies conducted in patients receiving once-daily aminoglycoside therapy have additionally shown the existence of a γ-phase, which becomes significant when the β-phase concentrations become very low (Prins *et al.*, 1993; MacGowan & Reeves, 1994). Estimations of the half-life determined for the γ-phase range from 5–15 h and 24 h post-dose concentrations have been found to be higher than expected from simple extraction of the β-phase of the blood concentration–time curve.

Toxicity

The two most frequently reported toxic effects of aminoglycosides are ototoxicity and nephrotoxicity, although other side effects including neuromuscular blockade, hypersensitivity, gastrointestinal upset and CNS effects have been reported (Wilson & Ramsden, 1977; Garrod *et al.*, 1981). The nephrotoxic effects of aminoglycosides are related to their specific action on the cells lining the proximal tubule, and although the damage caused to individual cells is permanent, regeneration of new cells from the basement membrane means that the damage is reversible. In contrast, ototoxic effects resulting from toxicity to the auditory and vestibular branches of the VIII cranial nerve, although not fully understood, are irreversible. At present there is substantial confusion and controversy regarding the incidence of aminoglycoside toxicity, partly resulting from the different patient populations and criteria applied (Lerner *et al.*, 1983). However, conservative reports suggest that toxic effects, mainly subclinical, occur in 2–10% of all patients receiving aminoglycosides (Smith *et al.*, 1980; Kahlmeter & Dahlager, 1984; Shrimpton *et al.*, 1993). Although the exact relationship between serum concentrations and toxicity has not been firmly established, a higher risk of ototoxic and nephrotoxic effects has been associated with elevated trough serum concentrations.

Recent experimental evidence suggests that it is no more toxic, and probably less so, to give an equivalent dose of an aminoglycoside once a day rather than divided into smaller doses and given more often (De Broe *et al.*, 1991). Results from clinical trials appear to support this as increases in serum creatinine, an indicator of renal toxicity, are similar in frequency (Nordstrom *et al.*, 1990; Rozalzinski *et al.*, 1993) or are less frequent (Prins *et al.*, 1993) in patients receiving once-daily therapy than in those with twice-daily dosing. Although there are only relatively few data, no evidence has been found from audiological examination to suggest that patients receiving once-daily dosing with aminoglycosides are at a higher risk of ototoxicity than patients receiving the conventional twice-daily dosing.

A number of factors influence the serum aminoglycoside concentrations attained in a particular patient, of which the major factors are the dose given, the frequency of dosing, the volume of distribution and the rate of excretion. As already indicated, the volume of distribution of aminoglycosides approximates the volume of extracellular fluid, which may vary from 20–25% of body weight in healthy volunteers up to 50–70% of body weight in neonates. Furthermore, the volume of distribution can vary both between patients (as a result of factors such as heart failure, peritonitis or obesity) and within the same patient, depending upon the state of hydration of that patient (Mann *et al.*, 1987). Variability in the half-life of aminoglycosides has been reported in healthy volunteers and in patients with renal impairment, often caused as a side effect of aminoglycoside therapy, where considerably extended half-lives are observed. Consequently, it is difficult to predict aminoglycoside serum concentrations with any degree of accuracy following a particular dose. Therapeutic monitoring of aminoglycosides is discussed fully in *Chapter 1*.

Liposomal preparations

There have been a number of recent publications describing liposomal preparations of streptomycin, dihydrostreptomycin, kanamycin, tobramycin, netilmicin, gentamicin and amikacin (Ladigina & Vladimirsky, 1986; Swenson *et al.*, 1990; Tomioka *et al.*, 1991; Nightingale *et al.*, 1993; Khalil *et al.*, 1996; Omri & Ravaoarinoro, 1996; Sanderson & Jones, 1996). The possible advantages of liposomal drug delivery systems include enhanced penetration into sites of infection, reduced toxicity and improved pharmacokinetics allowing less frequent dosing.

In general, liposomal preparations of aminoglycosides exibit much lower volumes of distribution and longer elimination half-lives than conventional preparations. Serum levels are 20–30 times higher and half-lives are typically >24 h, allowing dosing intervals of 1–5 days. Despite the high serum levels, liposomal preparations appear to be free from nephrotoxicity and ototoxicity. Liposomal preparations show two- to eight-fold less antimicrobial activity than the parent compound *in vitro*, yet studies *in vivo* suggest greater antimicrobial activity. Liposomal preparations are clearly an interesting development and will raise many questions about the pharmacokinetics and clinical usage of these compounds.

The assay of aminoglycoside antibiotics

It was recognized shortly after the introduction of aminoglycoside antibiotics into clinical use that a need existed for their assay. Since then probably more methods have been reported for the assay of these agents than any other class of antibiotic. Published evaluations of these methods, usually included in the report describing the method, indicate that—when performed 'correctly'—all of the methods return perfectly acceptable results. However, these evaluations have been performed under ideal conditions and under routine laboratory usage performance frequently falls short of that originally reported. Although there are well-established assay procedures for conventionally dosed aminoglycosides, there is little information on the application of these procedures when liposomal preparations are being used.

Although many assay methods have been reported, relatively few have entered widespread usage. A number of factors may be identified as influencing the acceptance of a particular method. These may be: (i) intrinsic to the method; such as time taken to return a result (speed), the performance of the method under 'real' conditions (robustness), the flexibility of the method to assay other antibiotics or difficulties created by undisclosed antimicrobials (specificity); (ii) technical factors; such as the level of skill and degree of training required of the operator, the ease of operation of the method or the amount of operator time required per assay; (iii) financial factors; such as the cost of reagents, equipment (including maintenance) and operator time; or (iv) idiosyncratic factors; such as personal preferences and/or the skills and experience available in the laboratory.

Microbiological assay

Microbiological techniques were the first methods to be described for the assay of aminoglycoside antibiotics and are still significant methods for their assay. They are still the statutory methods in use by industry to determine the potency of drug powders and solutions, although the use of microbiological methods in routine clinical laboratories is now minimal. In terms of the assay of aminoglycoside antibiotics, the plate diffusion assay has a number of advantages and disadvantages over other currently available techniques. Probably the most significant advantage of this method is that little or no capital equipment is required and the reagents used are both readily available and cheap. However, because a degree of expertise is required to ensure results of sufficient and consistent accuracy, staff need a significant level of training (Reeves & Bywater, 1976). A further advantage is that the method is suitable for the assay of all aminoglycosides and, with the minimum of alteration, most other antimicrobials. The method has two major disadvantages in the clinical setting. The first is the length of time taken to return a result because, although methods have been described whereby results can be obtained in under 5 h (Shanson & Hince, 1977b; Yamamoto & Pinto, 1996), in practice they are rarely available earlier than 18 h after receipt of sample. The second disadvantage is poor specificity in the presence of concurrently administered antimicrobials which may, unless measures are taken to negate their activity, lead to an overestimation of the concentration of aminoglycoside present. This problem is becoming increasingly acute as poly-antimicrobial therapy is common practice and with the continuing introduction of new antimicrobials (e.g. fluoroquinolones) the traditional ways of dealing with the mixtures may prove inadequate.

The popularity of the method has decreased dramatically, as shown by UK NEQAS returns. From 93% usage in 1973, popularity declined to just 7% in 1988 and <1% in 1998; a decrease mirrored by the increased use of immunoassays.

A number of different variations on the plate diffusion assay have been described for the assay of aminoglycoside antibiotics. For example, different media are used, including Diagnostic Sensitivity Test, Isosensitest and Mueller–Hinton agar. Although there is little to suggest that any one medium is better than another (Shanson & Hince, 1977a), media with high calcium or magnesium contents will produce smaller zones of inhibition. Whichever medium is used, it should be buffered to the same pH as the samples (pH 7.3). Likewise, the use of various indicator organisms has been reported, including *Bacillus subtilis, Bacillus pumilis, Staphylococcus aureus,* various Gram-negative species and even *Vibreo natriegans*. These vary in suitability depending on the sensitivity, speed and precision required of the method, and also on possible interference from other antimicrobials present in the sample. Currently the enterobacteria afford a good compromise between these factors with a multi-resistant organism, such as *Klebsiella edwardsii* NCTC 10896, being a suitable choice. With small sample volumes, such as those from neonates, insufficient sensitivity may be encountered with *K. edwardsii*, and *Bacillus pumilis* NCTC 90315 may be used as an alternative but with a loss in specificity since the isolate is not multiply resistant. Some methods of inactivating or removing potentially interfering antibiotics are outlined in *Chapter 4*. These two indicator organisms and the described inactivation or removal procedures described were routinely used in our laboratory for many years. Procedures for microbiological plate assays are described below. These assays should be performed following the general guidelines in *Chapter 3*.

Plate assay with a Gram-negative indicator organism
This method is suitable for sample volumes of ≥0.3 mL.

Medium. The agar composition (in g/L) is: protease peptone, 10.0; veal infusion solids, 10.0; glucose, 2.0; sodium chloride, 3.0; disodium phosphate, 2.0; sodium acetate, 1.0; adenine sulphate, 0.01; guanine hydrochloride, 0.01; uracil, 0.01; xanthine, 0.01; aneurine, 0.00002; agar, 12.0; the volume should be made up with distilled or deionized water and the pH adjusted to 7.3. This can be purchased as Diagnostic Sensitivity Test (DST) agar (Oxoid).

Indicator strain. Klebsiella edwardsii var. *atlantae* NCTC 10896.

Calibrators. It is recommended that calibrators are prepared in pooled antibiotic-free human serum at concentrations of 0.5, 1, 2, 4 and 8 mg/L or 1, 2, 4, 8 and 16 mg/L for the assay of gentamicin, tobramycin, netilmicin or neomycin, and at 2, 4, 8, 16 and 32 mg/L for amikacin and streptomycin. Samples containing concentrations that fall above the standard curve should be diluted with antibiotic-free serum and re-assayed. It is recommended that internal controls are prepared and used as outlined in *Chapter 7*.

Procedure. The assay should be performed in large (25 cm × 25 cm) assay plates which have been sterilized by UV irradiation and then levelled on a suitable apparatus, such as an adjustable tripod.

1. Pour 100 mL of molten agar into the plate to give about a 2 mm depth of agar.
2. After drying at 37°C for 10 min, inoculate the surface with 20 mL of a 1 in 100 dilution (in sterile water) of an overnight broth culture of the indicator strain.
3. Ensure the entire surface of the plate is covered and drain off the excess inoculum.
4. Further dry the plate at 37°C for 15 min.
5. Using a 30- or 45-well template, punch wells using a sterile no. 5 cork borer (9 mm). Remove the agar plugs with a scalpel tip and discard into disinfectant.
6. Fill the wells to the top (so that a flat/slightly convex meniscus forms) with the appropriate samples, calibrators and controls in triplicate according to a chosen random code (see *Chapter 4*).
7. Incubate overnight at 37°C.
8. Measure the diameter of the zones of inhibition with a magnifying zone reader. **Note**: although it is possible to determine zone sizes after 5 h incubation, greater precision is obtained if the plates are incubated for at least 8 h.
9. Calculate the aminoglycoside concentrations in the samples by either (a) interpolation from a plot of the (mean zone diameter)^2 obtained for each calibrator against the log concentration of the calibrators with the best-fit straight line drawn through the points or (b) comparison of the mean zone diameter versus log calibrator concentration using a suitable curve fitting procedure such as described in *Chapter 4*. Both of these procedures may be performed either graphically or using a suitable computer program.

Plate assay with a Gram-positive indicator organism
This assay is suitable for sample volumes of <0.3 mL, such as capillary blood samples from babies or children.

Medium. DST agar (see above for composition).

Indicator strain. Bacillus pumilis NCTC 90315 spore suspension. The spore suspension is prepared by inoculating an overnight nutrient broth culture of *B. pumilis* on to a 300 mL layer of nutrient agar in a 2 L Roux bottle. After incubation at 37°C for 1 week the spores are harvested by washing the slope with 100 mL of saline. Vegetative cells are killed by heating at 30 min at 65°C. The spore suspension is prepared by resuspending the cell pellet, following centrifugation, in 20 mL of saline.

Calibrators. These are prepared as described above for the Gram-negative assay protocol.

Procedure. The assay should be performed in large (25 cm × 25 cm) assay plates which have been sterilized by UV irradiation and then levelled on a suitable apparatus such as an adjustable tripod.

1. Cool 100 mL of molten DST agar to 50°C and then inoculate with 100 μL of aqueous spore suspension before pouring the plate. This is sufficient to produce semi-confluent growth after 18 h at 37°C. Pour the agar into the plate and spread it over the entire plate by tilting before the agar sets.
2. After the plate has set, dry at 37°C for 15 min and then surface sterilize by placing the plate under a UV light source for 10 min.
3. Apply each sample, calibrator and control to thick filter paper discs of Whatman no. 1 chromatography paper (300 g/m^2), approximately 1 mm thick and 6.5 mm in diameter by dipping and then touching against a dry tissue for 2 s to remove excess fluid. Such discs can be purchased from a number of manufacturers and are commonly sold as antibiotic assay (AA) discs.
4. Apply each sample, calibrator and control in triplicate to the surface of the plate according to a 30 or 45 random code.
5. Incubate at 37°C overnight.
6. Measure the zone diameters and perform the calculations as described for the Gram-negative procedure.

Immunoassay

The original principles of immunoassays and special considerations in their usage are discussed in *Chapter 5;*

for specific operating procedures readers are advised to consult the current recommendations of individual manufacturers.

As already indicated, the dramatic decrease in usage of microbiological assay techniques within the UK (as revealed by UK National External Quality Assessment Scheme (UK NEQAS) returns) has been mirrored by an increase in usage of immunoassays. A number of factors may be identified as contributing to this change:

(a) Ease of operation: since with the majority of immuno-assay techniques at least partial automation is employed and sample manipulations are limited.

(b) Time taken to return a result: with most immunoassays it is possible to return a result within 20 min of sample receipt. This not only affords a significant clinical advantage over microbiological methods but also renders the technique much more acceptable for an out-of-hours service or the assay of samples arriving late in the working day.

(c) The high specificity of antibody used: antisera (often monoclonal) are highly specific, which avoids problems caused by other concurrently administered antimicrobials. Where cross-reactivity is found this is between agents which would not normally be co-administered, for example gentamicin and netilmicin, or tobramycin and amikacin.

(d) Sample size required: most immunoassay techniques require only 50 μL or less of sample, making them ideally suited for the assay of samples from neonates. Recent work has shown that assays could be developed which require only dried blood spots (Fujimoto et al., 1989).

(e) Quality of results obtained: returns to the UK NEQAS for Antibiotic Assays over the past 5 years show that laboratories reporting results by immunoassay methods generally and consistently perform well, with blunders being the main reason for poor performance.

Although a number of laboratories have reported the development of in-house aminoglycoside immunoassays (Wills & Wise, 1979; White et al., 1980), the preparation of such kits is beyond the scope of most clinical laboratories. Consequently only commercially available methods will be discussed.

Currently within the UK, and probably also on a worldwide basis, the principal immunoassay in use for aminoglycoside assays is the fluorescence polarization immunoassay (FPIA) designed to be run on the Abbott TDX, FLX or Axsym analysers (including third-party FPIA kits designed to run on TDX or FLX analysers). Other assay systems, including EMIT (Behring), the Opus solid-phase assay (Behring), the Bayer latex agglutination assay, the Beckman nephelometric rate inhibition immunoassay (NIIA) assay and Roche FPIA kits designed to run on Roche analysers, are also being used to varying

Table II. Changes in the methods used for returns to the UK NEQAS for Gentamicin Assays 1980–1995, showing the percentage of participants returning results by a particular method

Method	1980	1985	1990	1995
EMIT	10.3	45.7	43.7	12.7
FIA	1.7	15.0	4.3	–
FPIA	–	17.0	54.4	82.1
Gram-positive plate	6.6	0.3	–	–
Gram-negative plate	77.2	13.3	3.2	1.9
NIIA	–	0.7	–	1.9
OPUS	–	–	–	0.7
RIA	2.3	0.3	–	–
Transferase	1.7	–	–	–
Urease	1.0	0.3	–	–
Others[a]	1.0	7.3	–	–
No. of participants	302	300	279	268

Abbreviations: EMIT, enzyme immunoassay; FIA, fluoroimmunoassay (marketed by Ames); FPIA, polarization fluoroimmunoassay; LAI, latex agglutination immunoassay; NIIA, nephelometric rate inhibition assay; RIA, radioimmunoassay. OPUS is a thin-film FIA.
[a]Other methods include tube dilution, GLC, HPLC and quenching immunoassay but mainly the Macrovue latex agglutination which was withdrawn in 1985.

but lesser extents (Table II). For all of these methods, cross-evaluation studies have been undertaken to show that both accuracy and precision are acceptable for clinical monitoring (Ratcliff et al., 1981; White et al., 1983a; Witebsky et al., 1983; Andrews & Wise, 1984; Araj et al., 1986; Boyce et al., 1989; Joos et al., 1989; Van der Bijl et al., 1990; Miglioli et al., 1993; Holt et al., 1994). However, as with all immunoassay systems, there is a tendency for decreased accuracy at the extremes of the standard curve and care is needed in the interpretation and internal quality control at such levels. For example, decreased accuracy has been reported at the top end of the standard curve for EMIT and Beckman NIIA (Ratcliff et al., 1981; White et al., 1983a) and at the bottom end of the standard curve for Abbott FPIA (Andrews and Wise, 1984; Araj et al., 1986). We have recently found that the dilution protocol of the Abbott TDX or FLX FPIA analyser may be used in reverse, and an increased rather than reduced sampling volume is programmed into the analyser. Both accuracy and precision are improved at gentamicin concentrations of <0.5 mg/L and streptomycin concentrations of <5 mg/L (Ryder & White, 1994; White et al., 1994). Other workers have reported that if the assay parameters on the Abbott TDX are changed to permit the use of calibrators containing 0, 25, 50, 75, 100 and 150 μg/L, sample and calibrator volumes increased to 20.0 and the span decreased to 50, reproducible assay results down to 10 μg/L are possible for gentamicin (Govantes et al., 1995). With the increasing use of once-daily dosing for aminoglycosides, where pre-dose levels are low, and the need to monitor aminoglycoside concentrations in tissue fluids, such as bronchio-alveolar lavage, the ability to monitor low concentrations using such procedures is of increasing importance.

Lovering *et al.* (1997) recently evaluated the suitability of three FPIA kits (Abbott, Innofluor and Sigma) for the assay of serum amikacin concentrations in patients treated with a liposomal preparation (MiKasome, NeXstar). Although all three kits gave a linear response over the tested concentration range, the Abbott and Innofluor kits showed very low recovery (Abbott, $18.0 \pm 1.3\%$; Innofluor, $20.2 \pm 0.6\%$) compared with the Sigma kit ($94.5 \pm 2.2\%$). Heating the serum samples at $56°C$ for 30 min improved recovery with the Abbott (to 80%) and Innofluor (to 76%) kits but reduced recovery with the Sigma kit (down to 85%). Heating had no effect on the recovery of non-liposomal amikacin or liposomal amikacin in the absence of serum. Accordingly we advise caution in the assay of samples from patients treated with liposomal amikacin.

Although all of the major immunoassay systems produce perfectly acceptable results for clinical monitoring purposes, there is a growing volume of literature to suggest that there is a slight method-specific bias between the different systems for the assay of gentamicin (Boyce *et al.*, 1989; Miglioli *et al.*, 1993). It must be stressed that, in the opinion of the author, any slight bias inherent to a particular method is unlikely to be of relevance in a clinical setting, apart from where a laboratory is operating two assay systems and alternating their use, such as using one system during the day and another out of hours. Although it has been suggested that the slight difference between assay methodologies may influence the parameters determined in pharmacokinetic studies, we doubt whether any such differences are likely to be significant. These findings do, however, support the case for the use of a single assay method when conducting a pharmacokinetic study. It is difficult to see why there should be a slight systematic bias associated with a particular assay system, although it has been postulated that patient factors may be responsible (Miglioli *et al.*, 1993). In the case of gentamicin it is probable that the explanation relates to differing cross-reactivity of the antibody used in the kit and the different gentamicin C components. If the antibody used does not have 100% cross-reactivity to all of the major C-components of gentamicin, a slight bias will result depending on the component ratio present in the sample as opposed to the calibrator.

In recent years laboratories have come under increasing financial pressure to reduce costs, including those of antibiotic assays. Although a substantial number of UK laboratories poorly understand the fundamentals of costing antibiotic assays, various approaches have been taken to the reduction of assay costs; in a recent study prices were found to range over a 1200% difference (Vacani *et al.*, 1993). These approaches have included: (i)

attempts to reduce the number of inappropriate assay requests (Li *et al.*, 1992); (ii) increasing the number of assays per kit by reducing reagent volumes (this can be done, for example, by running EMIT assays on some auto-analysers); (c) extending the interval between calibrations (Joos *et al.*, 1989; Zaninotto *et al.*, 1992); (iv) centralization or rationalization of services; (e) the use of third-party kits (this is evident in the UK where aminoglycoside FPIA kits manufactured by Sigma and Oxis are available for the Abbott TDX or FLX analysers).

However, a number of these approaches involve deviation from the manufacturers' accepted working practices and place an increased reliance on adequate, and effective, quality assurance within the laboratory to maintain the quality of results. Some modifications may be impossible to instigate in laboratories working to quality standards for the purposes of accreditation. The use of third-party kits may offer a saving over the use of proprietary kits, given equivalence in quality (which should be demonstrated in objective evaluations published in peer-reviewed journals). Any laboratory planning a change to an alternative supplier purely to cut costs will need fully to assess potential savings and legal implications, such as where equipment is on loan, before changing. However, it is becoming increasingly difficult under the current fiscal conditions to argue against such changes and it is likely that the major suppliers of immunoassays will come under increasing pressure to reduce prices in order to maintain market share.

Radioenzymatic assay (transferase assay)

The principles of this technique and considerations in its usage are discussed in *Chapter 5*. The transferase assay has never been popular in the UK and in recent years its usage has been superseded by non-isotopic immunoassay techniques. The reasons for this lack of popularity can be identified as: (i) the need to designate a specific part of the laboratory for radio-isotope work (including the disposal of waste); (ii) prejudices of staff against working with radio-isotopes; (iii) the labour-intensive nature of the method combined with a lack of automation; and (iv) the requirement for specialized equipment (γ- or β-counters).

The transferase assay is a highly sensitive and specific technique. Although it has no advantages over the commercially available immunoassay techniques for the routine assay of aminoglycosides, there are isolated cases where the use of the transferase assay is indicated. For instance, it might be used for assaying an aminoglycoside for which an immunoassay kit is either unavailable or too expensive for a one-off assay and which has, perhaps, been co-administered with other antibiotics rendering microbiological assay impossible. For example, we were able recently to assay a toxic serum neomycin concentration in a patient with an acutely inflamed bowel and renal insufficiency. This patient was receiving oral neomycin, colistin, nystatin and vancomycin, and parenteral ciprofloxacin, teicoplanin, sulphamethoxazole and trimethoprim.

The transferase method used is based upon aminoglycoside acetyltransferase enzymes (AAC) and we used the enzyme AAC(3)IV (produced by *Escherichia coli* JR 225) for the assay of all aminoglycosides except amikacin, for which we used AAC(6′)I (produced by *E. coli* W677/R5) (see *Chapter 5*). The method is as described below.

Acetyltransferase assay

Enzyme production. A starter culture of the enzyme-producing strain (*E. coli* JR 225 or *E. coli* W677/R5) grown overnight in 3 mL of brain–heart infusion broth is inoculated into 100 mL of nutrient broth. This is then incubated at 37°C, with aeration if available, until the culture reaches late-logarithmic phase (about 6 h), when the cells are harvested by centrifugation at 10,000**g** for 10 min. The cell pellet is then washed by resuspending it in 50 mL of saline and recentrifuged at 10,000**g** for 15 min. To the resultant cell pellet 5 mL of MgCl$_2$/dithiothreitol solution (MgCl$_2$ 5 mM, dithiothreitol 1 mM) is added and enzymes are liberated by ultrasonic disruption for 30 s at 5 A. The enzyme extract is then centrifuged at 30,000**g** for 15 min, divided into 500 μL volumes and stored at –20°C until use (up to 6 months).

Procedure. Acetylation reactions are carried out in microtitre trays. To duplicate wells:

1. Add 50 μL of acetyl coenzyme A/buffer solution (100 mM Tris hydrochloride pH 7.5 containing 30 mM MgCl$_2$ and 12.5 μM [1-^{14}C]acetyl coenzyme A (specific activity 10 Ci/mol)).

2. Add 50 μL of enzyme extract.

3. Add 50 μL of appropriate serum calibrator, patient sample or internal control.

4. Cover and incubate at 37°C for 30 min.

5. Transfer 100 μL volumes of the reaction mixtures to 2 cm × 2 cm squares of P81 phosphocellulose paper (Whatman).

6. Wash these twice in phosphate buffered saline (500 mL) then once in tap water (1 L).

7. Dry in a hot air oven (60°C).

8. Count radiolabelled aminoglycoside bound to the P81 papers in a liquid scintillation counter after placing each piece of paper into a suitable scintillant.

9. The concentration of aminoglycoside present in the patient samples is determined from a calibration curve obtained from plotting the count obtained with each calibrator against the concentration of that calibrator.

10. The recommended calibrator concentrations are 0, 0.5, 1, 2, 4 and 8 mg/L for gentamicin, tobramycin and netilmicin and 0, 1, 2,4, 8 and 16 mg/L for neomycin, kanamycin and amikacin.

High-performance liquid chromatography (HPLC)

At physiological concentration aminoglycosides neither absorb UV light nor fluoresce sufficiently to permit detection. Consequently for the detection of these antibiotics it is necessary to derivatize them to a product which can be detected, by either UV absorption or fluorimetry. The derivatization process can be carried out either before the sample is injected into the chromatograph (pre-column derivatization) or after the sample has been separated on the column but before reaching the detector (post-column derivatization); both methods have their merits. Of the various groups present in an aminoglycoside molecule, only the primary amino groups lend themselves to simple derivatization. However, serum contains many amino-containing compounds (for example, albumin) and these must be removed before derivatization to minimize possible interference. The simplest method of removing proteinaceous material is by precipitation using acetonitrile. This is best performed under buffered-alkaline conditions to increase the recovery of any protein-bound fractions. After separation from the precipitate, a back-extraction with methylene chloride will remove most of the added acetonitrile and prevent undue sample dilution.

Dansyl chloride, fluorescamine, benzene sulphonyl-chloride, 1-fluoro-2,4-dinitrobenzene, *ortho*-phthalaldehyde (OPA), 9-fluoroenylmethyl chloroformate and *ortho*-phthalicdicarboxaldehyde have all been used to derivatize aminoglycoside antibiotics (Peng *et al.*, 1977; Anhalt & Brown, 1978; Barends *et al.*, 1981; Larson *et al.*, 1980; Walker & Coates, 1981; D'Souza & Ogilvie, 1982; Stead & Richards, 1996; Bethune *et al.*, 1997; Forest *et al.*, 1997). The methods described vary in their separation of different aminoglycosides and it is only with methods using pre-column derivatization with OPA that it is possible to resolve the four major components of gentamicin. With some of these methods serum deproteinization has been performed on small columns of Sephadex, salicic acid or Amberlite resin, with derivatization being effected on the column before extraction and assay. The close chemical similarity of the clinically important aminoglycosides allows these assays to be adapted for other members of the group. However, amikacin (Figure 9.5) contains an amino-hydroxybutyramide side-chain, and multiple derivatization to a variety of products is frequently encountered, causing chromatographic difficulties. While many HPLC methods have been reported, in our opinion, they are not suited for use in a clinical assay service because they are so complex; only three laboratories have ever returned results to UK NEQAS using HPLC and there have been no HPLC returns for aminoglycosides for over 12 years. There are, however, instances where the assay of aminoglycosides by HPLC may be indicated, such as in the determination of the ratio of the various components in pharmaceutical preparations of gentamicin or in pharma-cokinetic and pharmacological studies of the individual gentamicin components. The methods that we would recommend in such instances are, for pharmaceutical preparations, the method reported by Freeman *et al.* (Freeman *et al.*, 1979; White *et al.*, 1983b) and, for biological samples, the method reported by Marples & Oates (1982) or that by Maloney & Awai (1990), depending on what instrumentation is available.

HPLC method of Freeman *et al.* (1979) for gentamicin

Stationary phase. Hypersil 5 ODS.

Mobile phase. Acetic acid:water:methanol 50:250:700 (by volume) containing 5 g/L heptane sulphonic acid.

Detection. UV at 330 nm.

Procedure. Samples are derivatized with *ortho*-phthalaldehyde–thioglycollic acid (50 μL of sample, 50 μL of propan-2-ol, 50 μL of reagent comprising 80 mM OPA in 1 M boric acid and 250 mM thioglycollic acid pH 10.4) for 10 min at 60°C. After centrifugation 20 μL of clear supernatant is injected. Nonylamine is used as an internal standard but the internal standard peak is relatively unstable compared with the aminoglycoside peaks. Better reproducibility may therefore be obtained if the internal standard is not used. Gentamicins C1, C2, C1a and C2a are separated. The method may also be used for tobramycin if the mobile phase composition is 50:285:665 (White *et al.*, 1983b).

HPLC method of Marples & Oates (1982) for gentamicin, netilmicin and tobramycin in serum

Stationary phase. Spherisorb 5 ODS.

Mobile phase. Methanol:water:EDTA (80 g/L), 80:15:5 (by volume), pH 7.2

Detection. Fluorescence: excitation, 340 nm; emission, 455 nm.

Procedure. Serum samples (500 μL) are derivatized with OPA on an Amberlite G-50 resin column made in a Pasteur pipette. The derivatized sample is eluted and 40 μL is injected. For gentamicin or netilmicin assays, tobramycin is used as internal standard; for tobramycin assays netilmicin is the internal standard. The method gives good resolution of the gentamicin C components.

References

Adams, E., Dalle, J., De Bie, E., De Smedt, I., Roets, E. & Hoogmartens, J. (1997). Analysis of kanamycin sulfate by liquid chromatography with pulsed electrochemical detection. *Journal of Chromatography A* **766**, 133–9.

Andrews, J. M. & Wise, R. (1984). A comparison of the homogenous enzyme immunoassay and polarization fluoroimmunoassay of gentamicin. *Journal of Antimicrobial Chemotherapy* **14**, 509–20.

Anhalt, J. P & Brown, S. D. (1978). High-performance liquid-chromatographic assay of aminoglycoside antibiotics in serum. *Clinical Chemistry* **24**, 1940–7.

Araj, G. F., Khattar, M. A., Thulesius, O. & Pazhoor, A. (1986). Measurements of serum gentamicin concentrations by a biological method, fluorescence polarization immunoassay and enzyme multiplied immunoassay. *International Journal of Clinical Pharmacology, Therapy and Toxicology* **24**, 542–5.

Barends, D. M., Zwaan, C. L. & Hulshoff, A. (1981). Improved microdetermination of gentamicin and sisomicin in serum by high-performance liquid chromatography with ultraviolet detection. *Journal of Chromatography* **222**, 316–23.

Bethune, C., Bui, T., Liu, M. L., Kay, M. A. & Ho, R. J. Y. (1997). Development of a high-performance liquid chromatographic assay for G418 sulfate (geneticin). *Antimicrobial Agents and Chemotherapy* **41**, 661–4.

Bodey, G. P., Chang, H.-Y., Rodriguez, V. & Stewart, D. (1975). Feasibility of administering aminoglycoside antibiotics by continuous intravenous infusion. *Antimicrobial Agents and Chemotherapy* **8**, 328–33.

Boyce, E. G., Lawson, L. A., Gibson, G. A. & Nachamkin, J. (1989). Comparison of gentamicin immunoassays using univariate and multivariate analyses. *Therapeutic Drug Monitoring* **11**, 97–104.

Brown, S. A., Newkirk, D. R., Hunter, R. P., Smith, G. G. & Sugimoto, K. (1990). Extraction methods for quantification of gentamicin residues from tissues using fluorescence polarization immunoassay. *Journal of the Association of Official Analytical Chemists* **73**, 479–83.

Chisholm, G. D., Calnan, J. S., Waterworth, P. M. & Reis, N. D. (1968). Distribution of gentamicin in body fluids. *British Medical Journal* **2**, 22–4.

Crann, S. A., Huang, M. Y., McLaren, J. D. & Schacht, J. (1992). Formation of a toxic metabolite from gentamicin by a hepatic cytosolic fraction. *Biochemical Pharmacology* **43**, 1835–9.

Daly, J. S., Dodge, R. A., Glew, R. H., Keroack, M. A., Bednarek, F. J. & Whalen, M. (1997). Effect of time and temperature on inactivation of aminoglycosides by ampicillin at neonatal dosages. *Journal of Perinatology* **17**, 42–5.

De Broe, M. E., Verbist, L. & Verpooten, G. A. (1991). Influence of dosage schedule on renal cortical accumulation of amikacin and tobramycin in man. *Journal of Antimicrobial Chemotherapy* **27**, Suppl. C, 41–7.

D'Souza, J. & Ogilvie, R. I. (1982). Determination of gentamicin components C1a, C2 and C1 in plasma and urine by high-performance liquid chromatography. *Journal of Chromatography* **232**, 212–8.

Edwards, C., Bint, A. J., Venables, C. W. & Scott, D. K. (1992). Sampling time for serum gentamicin serum levels. *Journal of Antimicrobial Chemotherapy* **29**, 575–80.

Finegold, S. M. (1959). Kanamycin. *Archives of Internal Medicine*, **104**, 15–280.

Forest, J. M., Mailhot, C., Cartier, P. & Sirois, G. (1997). Comparison of a fluorescence polarization immunoassay of netilmicin in plasma, peritoneal dialysate, and urine with a high-performance liquid chromatographic method. *Therapeutic Drug Monitoring* **19**, 74–8.

Fox, K. E. (1989). Total extraction of aminoglycosides from guinea pig and bullfrog tissues with sodium hydroxide or trichloroacetic acid. *Antimicrobial Agents and Chemotherapy* **33**, 448–51.

Freeman, M., Hawkins, P. A., Lorian, J. S. & Stead, J. A. (1979). Analysis of gentamicin sulphate in pharmaceutical specialities by high performance liquid chromatography. *Journal of Liquid Chromatography* **2**, 1305–17.

Fujimoto, T., Tsuda, Y., Tawa, R. & Hirose, S. (1989). Fluorescence polarization immunoassay of gentamicin or netilmicin in blood spotted on filter paper. *Clinical Chemistry* **35**, 867–9.

Garrod, L. P., Lambert, H. P. & O'Grady, F. (1981). *Antibiotics and Chemotherapy*, 5th edn, Chapter 5. pp. 115–54. Churchill Livingstone, London.

Govantes, C., Carcas, A. J., Garcia-Stue, J. L., Zapater, P. & Frias, J. (1995). Measurement of tobramycin in bronchoalveolar fluid by a modified fluorenscence polarization immunoassay. *Journal of Antimicrobial Chemotherapy* **36**, 1111–3.

Gyselynck, A. M., Forrey, A. & Cutler, R. (1971). Pharmacokinetics of gentamicin: distribution and plasma and renal clearance. *Journal of Infectious Diseases* **124**, *Suppl.*, S70–6.

Hoey, L. L., Tschida, S. J., Rotschafer, J. C., Guay, D. R. & Vance-Bryan, K. (1997). Wide variation in single, daily-dose aminoglycoside pharmacokinetics in patients with burn injuries. *Journal of Burn Care and Rehabilitation* **18**, 116–24.

Holt, H. A., White, L. O., Bedford, K. A., Bowker, K. E., Reeves, D. S. & MacGowan, A. P. (1994). An evaluation of three new immunoassays for determination of serum gentamicin concentrations. *Journal of Antimicrobial Chemotherapy* **34**, 747–54.

Joos, B., Lüthy, R. & Blaser, J. (1989). Long-term accuracy of fluorescence polarization immunoassays for gentamicin, tobramycin, netilmicin and vancomycin. *Journal of Antimicrobial Chemotherapy* **24**, 797–803.

Kahlmeter, G. & Dahlager, J. I. (1984). Aminoglycoside toxicity—a review of clinical studies published between 1975 and 1982. *Journal of Antimicrobial Chemotherapy* **13**, *Suppl. A*, 9–22.

Khalil, R. M., Murad, F. E., Yehia, S. A., El-Ridy, M. S. & Salama, H. A. (1996). Free versus liposome-entrapped streptomycin sulfate in treatment of infections caused by *Salmonella enteritidis*. *Pharmazie* **51**, 182–4.

Ladigina, G. A. & Vladimirsky, M. A. (1986). The comparative pharmacokinetics of ³H-dihydrostreptomycin in solution and liposomal form in normal and *Mycobacterium tuberculosis* infected mice. *Biomedicine and Pharmacotherapy* **40**, 416–20.

Larson, N. E., Marinelli, K. & Heilesen, A. M. (1980). Determination of gentamicin in serum using liquid column chromatography. *Journal of Chromatography* **221**, 182–7.

Lerner, A. M., Reyes, M. P., Cone, L. A., Blair, D. C., Jansen, W., Wright, G. E. *et al.* (1983). Randomised, controlled trial of the comparative efficacy, auditory toxicity, and nephrotoxicity of tobramycin and netilmicin. *Lancet i*, 1123–6.

Li, S. C., Ioannides-Demos, L. L., Spicer, W. J., Spelman, D. W., Tong, N. & McLean, A. J. (1992). Prospective audit of an aminoglycoside consultative service in a general hospital. *Medical Journal of Australia* **157**, 308–11.

Lovering, A. M., Bywater, M. J., Holt, H. A., Champion, H. & Reeves, D. S. (1988). Resistance of bacterial pathogens to four aminoglycosides and six other antibacterials and prevalence of aminoglycoside modifying enzymes, in 20 UK centres. *Journal of Antimicrobial Chemotherapy* **22**, 823–39.

Lovering, A. M., White, L. O., MacGowan, A. P. & Reeves, D. S. (1997). Difficulties in the assay of liposomal amikacin (MiKasome) in serum. In *Proceedings of the Thirty-Seventh Interscience Conference on Antimicrobial Agents and Chemotherapy, Toronto*, Abstract F38, p.152. American Society for Microbiology, Washington, DC.

MacGowan, A. P. & Reeves, D. S. (1994). Serum monitoring and practicalities of once-daily aminoglycoside dosing. *Journal of Antimicrobial Chemotherapy* **33**, 349–50.

MacGowan, A. P., Brown, N. M., Holt, H. A., Lovering, A. M., McCulloch, S. Y. & Reeves, D. S. (1993). An eight-year survey of the antimicrobial susceptibility patterns of 85,971 bacteria isolated from patients in a district general hospital and the local community. *Journal of Antimicrobial Chemotherapy* **31**, 543–57.

Maloney, J. A. & Awni, W. M. (1990). High-performance liquid chromatographic determination of isepamicin in plasma, urine and dialysate. *Journal of Chromatography* **526**, 487–96.

Mann, H. J., Fuhs, D. W., Awang, R., Ndemo, F. A. & Cerra, F. B. (1987). Altered aminoglycoside pharmacokinetics in critically ill patients with sepsis. *Clinical Pharmacy* **6**, 148–53.

Marples, J. & Oates, M. D. G. (1982). Serum gentamicin, netilmicin and tobramycin assays by high performance liquid chromatography. *Journal of Antimicrobial Chemotherapy* **10**, 311–8.

McBride, H. A., Martinez, D. R., Trang, J. M., Lander, R. D. & Helms, H. A. (1991). Stability of gentamicin sulfate and tobramycin sulfate in extemporaneously prepared ophthalmic solutions at 8°C. *American Journal of Hospital Pharmacy* **48**, 507–9.

McLaughlin, J. E. & Reeves, D. S. (1971). Clinical and laboratory evidence for inactivation of gentamicin by carbenicillin. *Lancet i*, 261–4.

Miglioli, P. A., Pea, F., Mazzo, M., Berti, T. & Lanzafame, P. (1993). Possible influence of assay methods in studies of the pharmacokinetics of antibiotics. *Journal of Chemotherapy* **5**, 27–31.

Nicolau, D., Quintiliani, R. & Nightingale, C. H. (1992). Once-daily aminoglycosides. *Connecticut Medicine* **56**, 561–3.

Nightingale, S. D., Saletan, S. L., Swenson, C. E., Lawrence, A. J., Watson, D. A., Pilkiewicz, F. G., Silverman, E. G. & Cal, S. X. (1993). Liposome-encapsulated gentamicin treatment of *Mycobacterium avium–Mycobacterium intracellulare* complex bacteremia in AIDS patients. *Antimicrobial Agents and Chemotherapy* **37**, 1869–72.

Nordstrom, L., Ringberg, H., Cronberg, S., Tjernstrom, O. & Walder, M. (1990). Does administration of an aminoglycoside in a single daily dose affect its efficacy and toxicity? *Journal of Antimicrobial Chemotherapy* **25**, 159–73.

Omri, A. & Ravaoarinoro, M. (1996). Comparison of the bactericidal action of amikacin, netilmicin and tobramycin in free and liposomal formulation against *Pseudomonas aeruginosa*. *Chemotherapy* **42**, 170–6.

Parker, S. E. & Davey, P. G. (1993). Practicalities of once-daily aminoglycoside dosing. *Journal of Antimicrobial Chemotherapy* **31**, 4–8.

Peng, G. W., Gadalla, M. A. F., Peng, A., Smith, V. & Chiou, W. L. (1977). High-pressure liquid-chromatographic method for determination of gentamicin in plasma. *Clinical Chemistry* **23**, 1838–44.

Phillips, I. (1982). Aminoglycosides. In *Good Antimicrobial Prescribing*, pp. 51–62. Passmore, London.

Price, K. E. (1986). Aminoglycoside research 1975–1985: prospects for the development of improved agents. *Antimicrobial Agents and Chemotherapy* **29**, 543–8.

Prins, J. M., Buller, H. R., Kuijper, E. J., Tange, R. A. & Speelman, P. (1993). Once versus thrice daily gentamicin in patients with serious infections. *Lancet* **341**, 335–9.

Ratcliff, R. M., Mirelli, C., Moran, E., O'Leary, D. & White, R. (1981). Comparison of five methods for the assay of serum gentamicin. *Antimicrobial Agents and Chemotherapy* **19**, 508–12.

Reeves, D. S. & Bywater, M. J. (1976). Assay of antimicrobial agents. In *Selected Topics in Clinical Bacteriology* (de Louvoir, J., Ed.), pp. 21–78. Bailliere Tindal, London.

Reeves, D. S., MacGowan, A. P., Holt, H. A., Lovering, A. M., Warnock, D. W. & White, L. O. (1995). Therapeutic monitoring of antimicrobials: a summary of the information presented at the UK NEQAS for Antibiotic Assays Meeting for participants, October 1993. *Journal of Antimicrobial Chemotherapy* **35**, 213–26.

Roets, E., Adams, E., Muriithi, I. G. & Hoogmartens, J. (1995). Determination of the relative amounts of the B and C components of neomycin by thin-layer chromatography using fluorescence detection. *Journal of Chromatography A* **696**, 131–8.

Rozdzinski, E., Kern, W. V., Reichle, A., Moritz, T., Schmeiser, T., Gaus, W. *et al.* (1993). Once-daily versus thrice-daily dosing of netilmicin in combination with β-lactam antibiotics as empirical therapy for febrile neutropenic patients. *Journal of Antimicrobial Chemotherapy* **31**, 585–98.

Ryder, D. & White, L. O. (1994). Assay of streptomycin trough concentrations by fluorescence polarization immunoassay. *Journal of Antimicrobial Chemotherapy* **33**, 1067–8.

Sanderson, N. M. & Jones, M. N. (1996). Encapsulation of vancomycin and gentamicin within cationic liposomes for inhibition of growth of *Staphylococcus epidermidis*. *Journal of Drug Targeting* **4**, 181–9.

Shanson, D. C. & Hince, C. J. (1977a). Factors affecting plate assay of gentamicin. II. Media. *Journal of Antimicrobial Chemotherapy* **3**, 17–23.

Shanson, D. C. & Hince, C. (1977b). Serum gentamicin assays of 100 clinical serum samples by a rapid 40°C klebsiella method compared with overnight plate diffusion and acetyltransferase assays. *Journal of Clinical Pathology* **30**, 521–5.

Shrimpton, S. B., Milmoe, M., Wilson, A. P., Felmingham, D., Drayan, S., Barrass, C. *et al.* (1993). Audit of prescription and assay of aminoglycosides in a UK teaching hospital. *Journal of Antimicrobial Chemotherapy* **31**, 599–606.

Smith, C. R., Lipsky, J. J., Laskin, O. L., Hellmann, D. B., Mellits, E. D., Longstreth, J. *et al.* (1980). Double-blind comparison of the nephrotoxicity and auditory toxicity of gentamicin and tobramycin. *New England Journal of Medicine* **302**, 1106–9.

Spencer, R. C., Wheat, P. F., Magee, J. T. & Brown, E. H. (1990). A three year survey of clinical isolates in the United Kingdom and their antimicrobial susceptibility. *Journal of Antimicrobial Chemotherapy* **26**, 435–46.

Stead, D. A. & Richards, R. M. (1996). Sensitive fluorimetric determination of gentamicin sulfate in biological matrices using solid-phase extraction, pre-column derivatization with 9-fluorenyl-methyl chloroformate and reversed-phase high-performance liquid chromatography. *Journal of Chromatography B: Biomedical Applications* **675**, 295–302.

Swenson, C. E., Stewart, K. A., Hammett, J. L., Fitzsimmons, W. E. & Ginsberg, R. S. (1990) Pharmacokinetics and in vivo activity of liposome-encapsulated gentamicin. *Antimicrobial Agents and Chemotherapy* **34**, 235–40.

Tomioka, H., Saito, H., Sato, K. & Yoneyama, T. (1991) Therapeutic efficacy of liposome-encapsulated kanamycin against *Mycobacterium intracellulare* infection induced in mice. *American Review of Respiratory Disease* **144**, 575–9.

Vacani, P. F., Malek, M. M. & Davey, P. G. (1993). Cost of gentamicin assays carried out by microbiology laboratories. *Journal of Clinical Pathology* **46**, 890–5.

Van der Bijl, P., Lawrance, J. F., Uebel, R. A., Brits, D. A. & Kotze, T. J. (1990). Assay of amikacin by two methods and four instruments. *Journal of Antimicrobial Chemotherapy* **26**, 860–2.

Walker, S. E. & Coates, P. E. (1981). High-performance liquid chromatographic methods for determination of gentamicin in biological fluids. *Journal of Chromatography* **223**, 131–8.

White, L. O., Scammell, L. M. & Reeves, D. S. (1980). An evaluation of a gentamicin fluoroimmunoassay kit; correlation with radio immunoassay, acetyl transferase and microbiological assay. *Journal of Antimicrobial Chemotherapy* **6**, 267–73.

White, L. O., Bywater, M. J. & Reeves, D. S. (1983a). Assay of netilmicin in serum by substrate labelled fluoroimmunoassay. *Journal of Antimicrobial Chemotherapy* **12**, 403–6.

White, L. O., Lovering, A. & Reeves, D. S. (1983b). Variations in gentamicin C1, C1a, C2 and C2a content of some preparations of gentamicin sulphate used clinically as determined by high-performance liquid chromatography. *Therapeutic Drug Monitoring* **5**, 123–6.

White, L. O., MacGowan, A. P., Lovering, A. M., Holt, H. A., Reeves, D. S. & Ryder, D. (1994). Assay of low trough serum gentamicin concentrations by fluorescence polarization immunoassay. *Journal of Antimicrobial Chemotherapy* **33**, 1068–70.

Wills, P. J. & Wise, R. (1979). Rapid, simple enzyme immunoassay for gentamicin. *Antimicrobial Agents and Chemotherapy* **16**, 40–2.

Wilson, P. & Ramsden, R. T. (1977). Immediate effects of tobramycin on human cochlea and correlation with serum tobramycin levels. *British Medical Journal* **i**, 259–61.

Witebsky, F. G., Sliva, C. A., Selepak, S. T., Ruddel, M. E., MacLowry, J. D., Johnson, E. E. *et al.* (1983). Evaluation of four gentamicin and tobramycin assay procedures for clinical laboratories. *Journal of Clinical Microbiology* **18**, 890–4.

Yamamoto, C. H. & Pinto, T. J. A. (1996). Rapid determination of neomycin by a microbiological agar diffusion assay using triphenyltetrazolium chloride. *Journal of AOAC International* **79**, 434–40.

Zaninotto, M., Secchiero, S., Paleari, C. D. & Burlina A. (1992). Performance of a fluorescence polarization immunoassay system evaluated by therapeutic monitoring of four drugs. *Therapeutic Drug Monitoring* **14**, 301–5.

Zaske, D. E. (1980). Aminoglycosides. In *Applied Pharmacokinetics* (Evans, W., Schentag, J. & Jusko, J., Eds), pp. 331–381. Applied Therapeutics, San Francisco, CA.

Chapter 10

Glycopeptides: vancomycin and teicoplanin

David Felmingham

GR Micro Ltd, 7–9 William Road, London NW1 3ER, UK

Introduction

The glycopeptides comprise a group of naturally occurring antibacterial compounds, principally produced by species of the Actinomycetes. Chemically, individual members of the group are characterized as cyclic structures formed from linked amino acid residues with attached sugars. Currently, two representatives of the glycopeptides are available for parenteral administration to humans, namely, vancomycin and teicoplanin, and these are the subject of this chapter.

Vancomycin and teicoplanin share a similar spectrum of antibacterial activity which is restricted to Gram-positive species. Bacterial isolates requiring an MIC of 4 mg/L vancomycin, or less, are considered clinically susceptible. In general, teicoplanin is two to four times more active than vancomycin against susceptible strains. However, teicoplanin is less active than vancomycin against some isolates of coagulase-negative *Staphylococcus* spp., especially *Staphylococcus haemolyticus*. For these strains, MICs of teicoplanin, but not vancomycin, are affected substantially by the medium and inoculum used in their determination (Felmingham *et al.*, 1987). Debate continues regarding the appropriate susceptibility breakpoint concentration. Thus in the USA the National Committee for Clinical Laboratory Standards (NCCLS, 1998) has set the value at 8 mg/L whilst in Europe 4 mg/L is favoured (Comité de l'Antibiogramme, 1996; Working Party, 1996; Swedish Reference Group, 1997). Both compounds are slowly bactericidal for most susceptible bacteria. However, against isolates of *Enterococcus* spp., some α-haemolytic streptococci (e.g. *Streptococcus sanguis*) and *S. haemolyticus* their bactericidal activity is diminished, with 1–10% of the original inoculum remaining viable after 24 h exposure to concentrations of four or eight times the MIC (O'Hare *et al.*, 1989; Felmingham, 1993).

Both vancomycin and teicoplanin exert antibacterial activity by interfering with the second stage of cell wall synthesis when UDP-*N*-acetylmuramyl pentapeptide and *N*-acetylglucosamine are sequentially polymerized to form peptidoglycan. This occurs when the glycopeptides bind tightly to the D-alanyl-D-alanine terminus at the free carboxyl end of the pentapeptide side-chain on *N*-acetylmuramic acid, resulting in steric hindrance of the binding of enzymes responsible for elongation of the cell wall polymer (Reynolds, 1985).

Inherent resistance to the antibacterial activity of the glycopeptides is observed consistently with isolates of *Lactobacillus*, *Leuconostoc* and *Pediococcus* spp. Various patterns of acquired resistance to both vancomycin and teicoplanin (VanA resistance) or to vancomycin only (VanB and VanC resistance) are now well-documented amongst isolates of *Enterococcus* spp. VanA resistance is associated with a substantial alteration in the binding affinity of the glycopeptides for the free carboxyl ends of the pentapeptide side-chains which are modified from D-alanyl-D-alanine to D-alanyl-D-lactate. The modification is inducible and the genes are present on a transposon, Tn*1546*, or related elements which may be present on transferable plasmids (Arthur & Courvalin, 1993; Quintiliani *et al.*, 1993; Arthur *et al.*, 1996). VanB-type resistance is mediated by a gene cluster similar to that controlling VanA-type resistance (Evers & Courvalin, 1996). VanC-type constitutive resistance is characterized by the production of D-alanine–D-serine in place of D-alanine–D-alanine with a resulting decrease in the affinity for vancomycin. This mechanism results in only low-level resistance (Navarro & Courvalin, 1994). More recently a new type of constitutive resistance, designated VanD, has been reported in *E. faecium* characterized by an MIC of 64 mg/L vancomycin and of 4 mg/L teicoplanin (Perichon *et al.*, 1997). In *Staphylococcus* spp. acquired low-level glycopeptide resistance is mediated by a different mechanism from that seen in enterococci. The exact mechanism is still unclear; however, in a strain of methicillin-resistant *Staphylococcus aureus* it is associated with thickened cell walls resulting from the increased production of PBP2 and PBP2' (Hiramatsu *et al.*, 1997). This observation supports the hypotheses of others (Billot-Klein *et al.*, 1998).

Physico-chemical properties

Vancomycin (Figure 1) was isolated originally from the fermentation broth of a strain of *Amycolatopsis orientalis* (McCormick *et al.*, 1956). Structurally, it is a tricyclic glycopeptide in which two chlorinated β-hydroxytyrosine

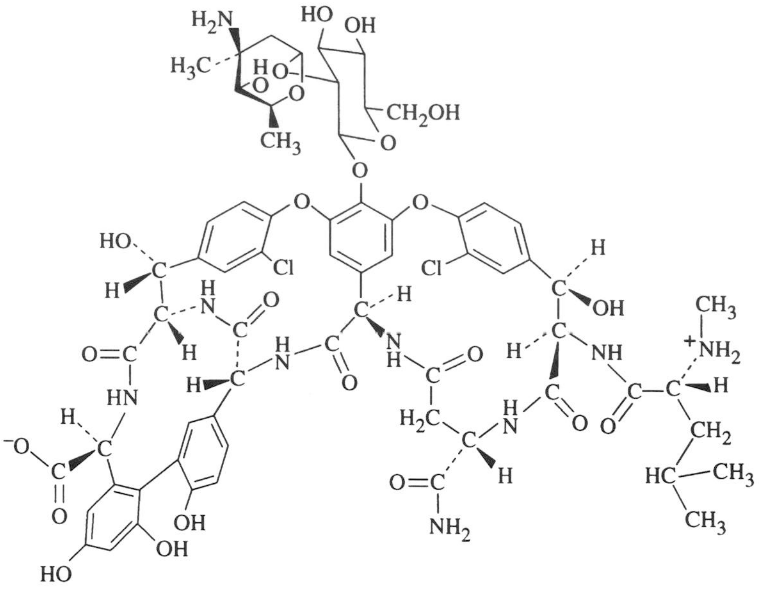

Figure 10.1. Vancomycin.

units, three substituted phenylglycine systems, *N*-methyl-leucine and aspartic acid amide, are linked to form a seven-membered peptide chain. A disaccharide, composed of glucose and vancosamine (a unique amino-sugar), is attached to one of the phenylglycine residues. Vancomycin base has a molecular weight of 1449.3 and an empirical formula of $C_{66}H_{75}Cl_2N_9O_{24}$ (Pfeiffer, 1981). The molecular weight of the hydrochloride is 1484.5. It is moderately soluble in methanol but relatively insoluble in other organic solvents. Aqueous solubility is a function of pH and is highest in acid conditions, decreasing at neutrality (pH 7) to a minimum solubility of 15 g/L. It is soluble but unstable at alkaline pH.

Teicoplanin was isolated from the fermentation products of another actinomycete, *Actinoplanes teichomyceticus* (Parenti *et al.*, 1978). It is not a pure substance but a complex of antibacterial molecules with five major components (designated A2-1 to A2-5) each having the same linear heptapeptide containing aromatic amino acids with attached α-D-mannose, *N*-acetyl-β-D-glucosamine and one of five different *N*-acyl-β-D-glucosamine fatty acids (Figure 2). In addition, there is a hydrolysis product (A3-1) found in the purified drug (but not the fermentation broth) which lacks the *N*-acyl-glucosamine group. A2-2 is the major component. The molecular weights are: 1564.3 (A3-1), 1877.7 (A2-1), 1879.7 (A2-2 and A2-3) and 1893.7 (A2-4 and A2-5). The manufactured form of teicoplanin comes as the sodium salt (molecular weight 1993.0). The pK_a values of teicoplanin are 5.0, 7.1 and 9–12.5. The teicoplanin A2 components each have similar antibacterial potency (Borghi *et al.*, 1984). Teicoplanin is freely soluble in dimethyl formamide, dimethylsulphoxide and water at pH 7.4, and is moderately soluble in methanol. Like vancomycin it is relatively insoluble (<0.01%) in other organic solvents but it is more lipophilic than vancomycin with lipophilicity increasing from A2-1 to A2-5.

Pharmacokinetics

Although not absorbed from the gastro-intestinal tract, both vancomycin and teicoplanin distribute widely, reaching therapeutic concentrations in most body compartments following parenteral administration (Campoli-Richards *et al.*, 1990). However, neither glycopeptide penetrates appreciably into the cerebrospinal fluid of subjects with either normal or inflamed meninges (Schaad *et al.*, 1980; Stahl *et al.*, 1987).

The plasma half-life of vancomycin varies widely between patients with normal renal function with values of 3–13 h reported (Rotschafer *et al.*, 1982), whereas that of teicoplanin is considerably longer, ranging from 33 h to 190 h depending upon the pharmacokinetic model used for analysis and the last sampling time (Falcoz *et al.*, 1987). In practice, this results in two, three or, occasionally, four times daily dosing of vancomycin compared with once daily (following a loading dose such as two doses at 12 h intervals during the first 24 h) dosing of teicoplanin.

Approximately 55% of vancomycin and 90% of teicoplanin is bound to human serum albumin (Krogstad *et al.*, 1980; Assandri & Bernareggi, 1987). Steady-state volume of distribution estimates for vancomycin range from 0.5 to 0.9 L/kg and those for teicoplanin from 0.3 to 1.1 L/kg. Teicoplanin, being more lipophilic, penetrates tissues to a greater degree.

Vancomycin and teicoplanin are removed from the body predominantly by glomerular filtration in the kidneys with no evidence to suggest significant tubular secretion or

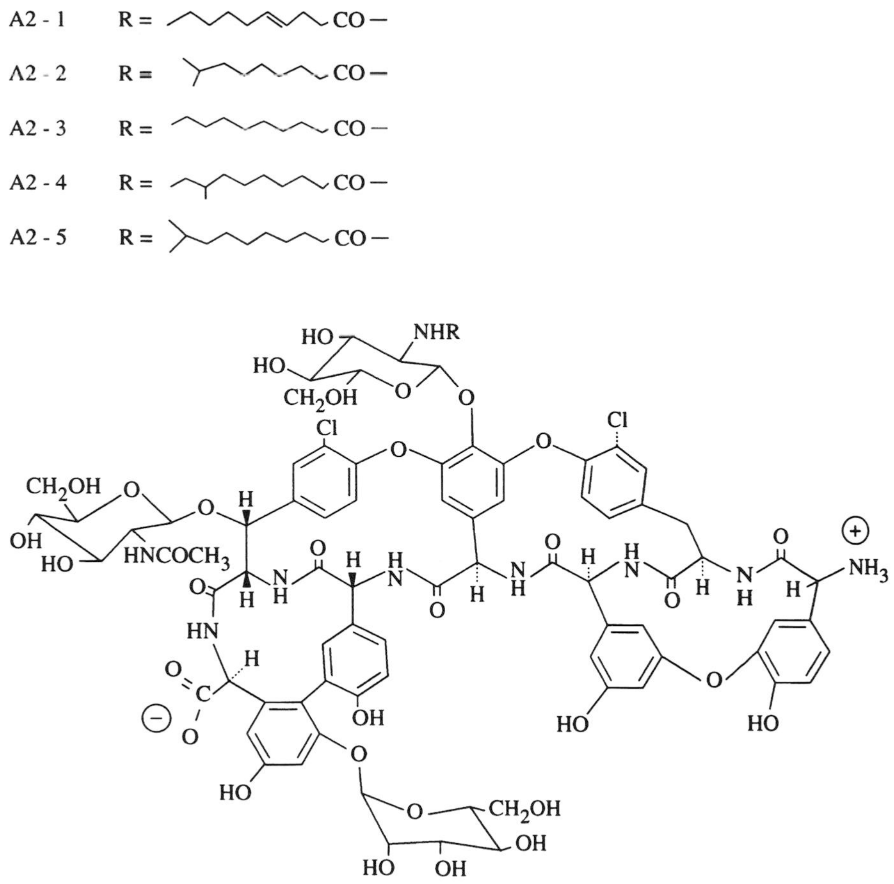

Figure 10.2. Teicoplanin.

reabsorption. The pharmacokinetics of the individual components of teicoplanin have been studied and, although there are only very slight differences in elimination half-life values, there are decreases in the volume of distribution at steady state and clearance with increasing lipophilicity (Rowland, 1990). However, clinically these differences between the teicoplanin components are unimportant. The elimination half-lives of both teicoplanin and vancomycin are increased substantially in patients with impaired renal function, necessitating modification to dosage, particularly in the case of vancomycin, which may be predictable to some extent by creatinine clearance values but which can be optimized, adequately, only by monitoring plasma concentrations. In this regard, the relative behaviour of the teicoplanin components is affected only slightly in renal failure and, consequently, dosage adjustment can be based on assay of total teicoplanin concentration (Falcoz et al., 1987). Neither of the glycopeptides is removed efficiently from the body during either haemodialysis or haemofiltration (Campoli-Richards et al., 1990). Therefore, patients treated by these procedures who require antibacterial therapy with a glycopeptide should be given an appropriate loading dose and plasma levels of the drug monitored to determine the frequency of further doses. Vancomycin and teicoplanin are removed to some extent by peritoneal dialysis. However, the relative inefficiency of this method of removal implies that dosage adjustment in patients undergoing this form of management of renal dysfunction also requires monitoring of plasma concentrations.

Renal clearance of teicoplanin may be more rapid in iv drug abusers and plasma concentrations should be monitored in such patients being treated for endocarditis or other life-threatening sepsis (Rybak et al., 1991). Renal clearance of vancomycin in drug abusers with a variety of infections has been reported also to be higher than in control subjects (Rybak et al., 1990a).

Although mainly removed by the renal route, a prolonged vancomycin elimination half-life has been observed in some patients with hepatic failure (Brown et al., 1983). Monitoring of plasma concentrations is indicated in these patients.

Toxicity

Vancomycin is considered to be potentially both nephrotoxic and ototoxic. Increased risk of nephrotoxicity has been associated with length of treatment, concurrent therapy with an aminoglycoside, and trough plasma concentrations in excess of 10 mg/L (Rybak et al., 1990b). Teicoplanin appears to possess considerably less nephrotoxic potential which has not been related to dose, plasma concentration or concomitant therapy with an amino-glycoside (Kureishi *et al.*, 1991).

During the early years of its use, vancomycin was associated with ototoxicity, particularly in patients with plasma levels in excess of 80 mg/L (Woodley & Hall, 1961). However, a more recent retrospective study failed to find evidence of ototoxicity in the records of 98 patients treated with more purified preparations of vancomycin between 1974 and 1981 (Farber & Moellering, 1983). No dose-associated ototoxicity has been reported for teicoplanin.

Recommended plasma concentrations

Both vancomycin and teicoplanin have a linear dose–serum concentration relationship following iv administration, over the usual therapeutic range. The following trough and peak serum concentrations have been associated with minimum risk of toxicity and optimum therapeutic efficacy (Geraci, 1977; Wilson *et al.*, 1994), although the value of therapeutic monitoring of glycopeptides remains an area for debate (see *Chapter 1*).

Vancomycin

Trough (pre-dose): 5–10 mg/L; peak (30–60 min after completion of an iv infusion of at least 1 h duration): 20–40 mg/L or (2 h after the completion of the infusion) 18–26 mg/L. If peak and trough levels are being used to calculate the elimination half-life it is important to take the peak at least 2 h post-dose since the distribution phase may not be complete in some patients if the sample is taken earlier.

Teicoplanin

Trough (pre-dose): 15–20 mg/L or, in monotherapy of staphylococcal endocarditis, >20 mg/L; peak (1 h after iv injection over 5 min): 25–40 mg/L. When using teicoplanin as monotherapy for the treatment of *S. aureus* endocarditis in iv drug abusers, it is recommended that trough serum concentrations should be maintained above 20 mg/L (Wilson *et al.*, 1994). Teicoplanin may be administered by the im route. When this route is used, peak serum concentrations are observed 3 h after administration and may be <50% of the concentration found with the same dose 1 h after iv administration.

Assay methods

Types of assay available

Microbiological, immunological and HPLC methods are available for the assay of both vancomycin and teicoplanin in blood and other body fluids.

The microbiological method (agar plate diffusion assay) is least applicable to vancomycin because the dosing inter-val (usually 8 or 12 h) is considerably less than the incubation period of the assay (usually 18–24 h) resulting in at least one, and often two, doses being administered before an assay result is available. In the case of teicoplanin, which in the great majority of patients is administered once-daily, there is generally ample time to provide assay results to the clinician within the dose interval, using a microbiological method.

In addition to the problem of the length of time required to produce an assay result, the microbiological method lacks specificity. This is a serious shortcoming of this procedure since both vancomycin and teicoplanin are frequently administered together with other unrelated antimicrobials which may have activity against the assay indicator strain. It is therefore of paramount importance that the co-administered compounds are declared by the requesting clinican and their effect removed either by prior inactivation (e.g. β-lactamase treatment of samples) or the use of assay indicator microorganisms exhibiting appropriate antimicrobial resistance.

Problems associated with turnaround time, specificity and small sample volume may be overcome by the use of either immunoassays, which are available for both glycopeptides, or possibly HPLC. This is particularly advantageous for vancomycin bearing in mind the shorter dosing intervals and greater need for assay when compared with teicoplanin, even though the apparatus and reagent costs are considerably greater than those of the microbiological method. For teicoplanin, the single commercial immunoassay, Innoflour FPIA (Oxis, Portland, OR, USA) for use with Abbott TDX or FLX analysers, available at present, is expensive and more difficult to justify, although undoubtedly very convenient. Both vancomycin and teicoplanin can be assayed by HPLC.

Teicoplanin binds to glass and plastic and care should be taken during sample preparation and sample transfer procedures to minimize this loss. Transporting or storing small volumes of sample in relatively large containers should be avoided. Bottles or vials should, if possible, be filled more than half full. The loss of the minor components by binding is significantly greater than the loss of the major component. White & Reeves (1991) reported the following percentage recoveries when a solution of teicoplanin 10.5 mg/L was serially transferred three times into plastic microcentrifuge tubes: A2-2, 47.9%; A2-4, 22.4%; A2-5, 21.4%. Binding can be minimized by discarding the first filling of pipettes as no binding occurs in the second filling. Pre-exposure of plastic surfaces to human serum or peritoneal dialysate reduces teicoplanin binding by 60% (Wilcox *et al.*, 1994).

Presentation and availability of pure substances

Pharmaceutical preparations of vancomycin are available as lyophilized, chromatographically purified vancomycin hydrochloride with no added preservative equivalent to 250, 500 or 1000 mg (250,000, 50,000 or 1,000,000 international units) of vancomycin. When reconstituted in water, a clear solution at pH 2.8–4.5, depending upon mass and volume, results. Vancomycin is also available as an oral formulation as Matrigel capsules containing the same chromatographically purified vancomycin hydrochloride equivalent to 125 or 250 mg vancomycin.

Teicoplainin comes as 200 and 400 mg vials of lyophilized sodium salt, without preservative, for injection. However, the vials contain a measured overage to ensure the patient receives the full dose. It is important to avoid foaming when dissolving teicoplanin since the drug concentrates in the foam. Therefore, during reconstitution only gentle rolling of the vial should be employed. If frothing does occur, the solution should be allowed to stand for 15–30 min before use. The composition of teicoplanin varies only slightly between batches. Jehl *et al.* (1988) reported the mean (S.E.M.) composition of six batches to be: A3, 9.2% (1%); A2-1, 5.1% (0.9%); A2-2, 53.9% (2%); A2-3, 10.5% (1.8%); A2-4, 10.6% (1.1%); and A2-5, 11.8% (1.2%).

Vancomycin analytical reference standard can be obtained from Eli Lilly, and WHO and USP primary reference standards are also available. The material available from Eli Lilly is a secondary reference standard and its potency may depend on whether your laboratory adheres to WHO or USP standards. For example, lot RS0124 (expiry date September 1994) comes as vials containing 87.1 mg of powder equivalent to 82.4 mg vancomycin hydrochloride. The package insert states that potency should be taken as 97.0 mg of vancomycin base/vial if using USP standards or 105.3 mg vancomycin base/vial if using WHO standards. This situation is rather confusing and the potency value of >100% relates to the fact that early preparations of vancomycin were rather impure cocktails compared with the chromatographically purified material currently available.

Teicoplanin analytical reference standard can be obtained from Hoechst Marion Roussel (Denham, UK). It comes in a glass vial containing a known amount (approximately 50 mg) of teicoplanin expressed as A2-2 equivalent. The entire vial should be reconstituted to 50 mL in 1% methanol in water.

Microbiological plate diffusion assay for vancomycin or teicoplanin

The principles and practice of the microbiological plate assay are discussed in *Chapter 3* and should be applied rigorously to the assay of glycopeptides by this method, as with other antimicrobials. The following method is essentially that described by Patton *et al.* (1987).

Medium. Antibiotic Medium No. 1 (Oxoid, Basingstoke, UK) with supplementary sodium chloride (30 g/L) and adjusted to pH 5.6–5.7 by the addition of 1.25 mL of 1 M HCl to 1 L of medium after sterilization and cooling to 45°C, can be used for the microbiological assay of both vancomycin and teicoplanin over the normal therapeutic range.

Assay indicator strains. For the assay of either glycopeptide in the absence of other antimicrobials with activity against Gram-positive bacteria, use *Bacillus subtilis* NCTC 10400 (ATCC 6633) obtainable as a standardized spore suspension from Difco Laboratories, West Molesey, UK. For use, add the contents of a 1 mL vial of spore suspension to 4 mL sterile distilled water and seed 100 mL molten, sterile agar with 1 mL of this diluted suspension before pouring.

As an alternative, *S. aureus* may be used as indicator organism. The advantage of using this species is that both the fully susceptible 'Oxford' strain (NCTC 6571) and clinical isolates expressing resistance to antimicrobials commonly prescribed in combination with glycopeptides (e.g. aminoglycosides, rifampicin, erythromycin, co-trimoxazole) can be maintained in the laboratory. For use, prepare a suspension of the appropriate strain in sterile distilled water equivalent in opacity to a 0.5 McFarland tube. Use this inoculum to seed the surface of an agar plate by flooding and subsequent removal of excess fluid.

Calibrators. Prepare glycopeptide calibrators in pooled human serum, previously screened for antibacterial activity against the chosen indicator strains, at concentrations of 4, 8, 16, 32 and 64 mg/L of either vancomycin or teicoplanin. These may be prepared in advance and stored at −70°C, for at least 3 months, without loss of activity.

Preparation of patient samples. When β-lactams are given in combination with either glycopeptide, they can be inactivated by hydrolysis with a broad-spectrum β-lactamase mixture (Genzyme Biochemicals Ltd, Maidstone, UK). To use, reconstitute one vial of lyophilized β-lactamase mixture in 5 mL of sterile distilled water. Add 0.2 mL of this solution to 0.8 mL of serum sample, incubate at room temperature for 5 min and then apply the sample to the assay plate. Untreated patient sample and both a treated and an untreated solution of the co-administered β-lactam, at an appropriate concentration, should be assayed at the same time as controls of the inactivation procedure.

Note: results obtained with the inactivated sample after this treatment must be adjusted to 120% of the observed concentration to allow for dilution with the β-lactamase.

Removal of aminoglycosides by adsorption on to an ion-exchange material such as phosphocellulose is inappropriate for use with samples containing either glycopeptide since both, but particularly teicoplanin, may be removed also. Use of a suitable aminoglycoside-resistant strain of *S. aureus* as indicator organism is recommended instead. Whatever procedure is used, appropriate aminoglycoside controls must be used.

Screening of serum samples for undeclared antibacterials co-administered with the glycopeptide can be simultaneously performed with *Escherichia coli* NCTC 10418 (insusceptible to glycopeptides). Remember, however, that this will only show the presence of agents with activity against *E. coli*.

Incubation conditions. After application of samples, calibrators and internal controls, incubate the plates in air for 18–24 h at 35–37°C. After incubation measure zone sizes and determine concentrations using the general analytical and statistical procedures described in *Chapter 3*.

Immunoassay of vancomycin and teicoplanin

Vancomycin. Immunoassay reagents for determining concentrations of vancomycin in serum and other body fluids are available commercially from Behring (EMIT assay for use with the Solaris and other analysers), Abbott (fluorescence polarization immunoassay (FPIA) for use with the Abbott TDX or FLX analysers), Roche (FPIA for use on the Cobas Phara analyser), Behring (thin-film assay for use on the Opus analyser) and Biostat (Innofluor FPIA reagents produced by Oxis (Portland, OR, USA) for use on the Abbott TDX analyser). In the UK most laboratories use either EMIT or FPIA for vancomycin assays. Recent examples of how these techniques perform in the UK NEQAS for Antibiotic Assays are shown in Table I.

A word of caution is necessary with regard to the use of FPIA assay for estimating vancomycin concentrations in patients with severe renal impairment. White *et al.* (1988) have shown that the antibody used in Abbott kits cross-reacts with a microbiologically inactive breakdown product of vancomycin which does not possess antibacterial activity. These workers suggest that if breakdown product accumulated in the blood of anuric patients, the Abbott FPIA method would overestimate the concentration of

Table I. UK NEQAS returns for vancomycin assays

Sample no.	Target (mg/L)	Method	Method mean (mg/L)	S.D.	CV%	No.
778	41.5	EMIT	38.0	2.6	6.9	13
		bioassay[a]	41.0			1
		FPIA (Roche)	36.5			1
		FPIA (TDX)	40.3	2.2	5.6	133
769	16.7	EMIT	17.2	0.8	4.7	11
		bioassay	16.0			1
		HPLC	17.6			1
		FPIA (Roche)	14.7			1
		FPIA (TDX)	17.7	1.1	6.2	133
765	79.0	EMIT	76.0	6.8	9.0	12
		bioassay	82.0	5.7	6.9	2
		FPIA (Roche)	71.6			1
		FPIA (TDX)	77.4	5.2	6.8	133
760	24.6	EMIT	23.2	2.2	9.6	13
		bioassay	25.0	0.0	0.0	2
		HPLC	27.4			1
		FPIA (Roche)	23.5			1
		FPIA (TDX)	25.1	1.4	5.5	134
753	35.6	EMIT	32.5	3.4	10.6	12
		bioassay	34.2	0.2	0.6	2
		HPLC	35.5			1
		FPIA (Roche)	32.6			1
		FPIA (TDX)	34.6	1.5	4.4	134
748	4.4	EMIT	3.8	1.2	30.6	12
		bioassay	4.8	1.3	28.3	2
		HPLC	5.3			1
		FPIA (Roche)	4.9			1
		FPIA (TDX)	4.6	0.6	12.7	137

[a] 'Bioassay' = microbiological plate assay with Gram-positive indicator strain.

biologically active vancomycin, as has been reported in patients undergoing peritoneal dialysis (Morse *et al.*, 1987a & b). The antibody used in the EMIT system does not cross-react with the vancomycin breakdown product. Innofluor FPIA reagents give similar results to Abbott FPIA reagents and show similar cross-reactivity with degraded vancomycin (White *et al.*, 1997). In 1993 the UK NEQAS for Antibiotic Assays distributed two serum samples containing vancomycin which were identical except that one had been incubated at 37°C for 7 days to simulate conditions of an anuric patient receiving weekly vancomycin. The returns (Table II) clearly demonstrated the difference between EMIT and FPIA results when breakdown products are present. At the time of writing commercial vancomycin assay kits that did not cross-react with breakdown products were under development.

Teicoplanin. An FPIA has been developed by Oxis for the determination of teicoplanin concentrations in serum (Mastin *et al.*, 1993). These reagents are formulated for use on the Abbott TDX or FLX analyser and the assay covers the concentration range 5–100 mg/L. McMullin *et al.* (1994) compared FPIA performance with that of the HPLC method of White & Shi (1990) and a microbiological assay. It performed well. Its major drawback is that the kits are very expensive. In an experimental European EQA circulation FPIA was the most popular and best-performing method (White *et al.*, 1996).

HPLC of vancomycin and teicoplanin

Several HPLC methods have been described for both glycopeptides (Tables III and IV).

HPLC assay for vancomycin

HPLC is seldom routinely applied to the assay of vancomycin in clinical samples in view of its technical demands, the apparatus needed, the ready availability of comparatively inexpensive immunoassay reagents and the need for a rapid, same-day results service. Nevertheless, the author is aware of one hospital in the UK which submitted results to the UK NEQAS for Antibiotic Assays using an adaptation of the HPLC method described by Rosenthal *et al.* (1986) which was shown to correlate closely with results obtained by FPIA.

The basic features of this adapted method are as follows:

Sample preparation. Protein precipitation with methanol containing 3,4,5-trimethoxyphenylacetonitrile as internal standard.

Chromatographic separation. Isocratic elution using a mobile phase of phosphate buffer pH 2.6 containing 25% acetonitrile and 5% methanol with a flow rate of 1 mL/min, on a Supelco LC-18 column (Supelco Inc., Bellefonte, PA, US; 5 μm particle size, 4.6 mm × 150 mm). In the method described this resulted in retention times of approximately 3.7 min for vancomycin and 7.1 min for the internal standard.

Detection. Vancomycin and the internal standard were detected by UV absorbance at 230 nm.

Dose–response. The method showed a linear relationship between peak height ratios and concentration over the range 5–80 mg/L vancomycin.

HPLC assay for teicoplanin

HPLC methods for the determination of teicoplanin in serum and other body fluids have been published by several groups (Table IV) (Riva *et al.*, 1987; Georgopoulos *et al.*, 1989). These procedures were, in the main, developed to support early pharmacokinetic studies, to determine the concentration of the six major components of teicoplanin separately, especially in pharmacokinetic studies in patients with impaired renal function and, as clinical use increased, to provide a specific assay method for use in patients receiving polyantimicrobial therapy. However, as the perceived need to separate the six individual components has diminished and the FPIA has become available, it seems likely that HPLC methods, especially gradient elution methods, will be used less frequently. Two methods are described here in some

Table II. Differences beweeen assay methods when vancomycin degradation products are present in the sample

Method	Sample 791			Sample 791A (degraded)		
	mean (mg/L)	*n*	S.D.	mean (mg/L)	*n*	S.D.
All methods	23.4	149	1.3	21.2	148	1.4
EMIT	23.1	12	2.4	17.2	11	1.2
Bioassay	21.0	1	–	16.8	1	–
HPLC	20.2	1	–	15.1	1	–
Roche FPIA	22.7	1	–	21.4	1	–
Abbott TDX	23.5	136	1.2	21.4	138	1.2

Table III. Some published HPLC methods for the assay of vancomycin

Mobile phase	Stationary phase	Detection	Matrix	Comments	Reference
0.05 M phosphate buffer: acetonitrile (91:9 v/v)	Microbondapak C_{18}	UV 210 nm	serum	solid-phase extraction with ristocetin internal standard.	Uhl & Anhalt (1979)
Acetonitrile: 2 M ammonium acetate:water (9:10:81 by volume)	Ultrasphere ODS	UV 214 nm	serum	protein precipitation with isopropanol: acetonitrile; no internal standard.	Jehl et al. (1985)
0.05 M potassium phosphate: acetonitrile (90:10 v/v)	Zorbax C_{18}	UV 210 nm	serum of CAPD patients	similar to method of Uhl & Anhalt (1979).	Whitby et al. (1987)
Acetonitrile:water: 2 M ammonium acetate (9:81:10 by volume)	Zorbax C_{18}	UV 214 nm	infusion solution	no internal standard.	Greenberg et al. (1986)
0.1 M heptane sulphonic acid:acetonitrile (12:88 v/v)	Micropak C_{18}	UV 210 nm	serum of CAPD patients	protein precipitation; no internal standard; septrin interfered.	Morse et al. (1987a)
0.05 M potassium dihydrogen phosphate:acetonitrile (90:10 v/v)	Supelcosil C_8	UV 254 nm	serum and tissues	solid-phase extraction.	Greene et al. (1987)
7 mM sodium octane sulphonate: acetonitrile:methanol (600:250:100 by vol.) made up to 1 L with water	Supelco LC-18 5 μm	UV 230 nm	serum	protein precipitation[a]	Rosenthal et al. (1986)
5 mM potassium dihydrogen phosphate: methanol (22:78 v/v)	Nucleosil C_{18} 5 μm	UV 229 nm	serum	solid-phase extraction.	Bauchet et al. (1987)
Acetonitrile:tetrahydrofuran: 0.2% triethylamine phosphate pH 3.2 (5.5:0.8:93.7 by volume)	Ultrasphere ODS 5 μm	UV 254 nm	BAL fluid	extraction procedure.	Lamer et al. (1993)

[a] With 3,4,5-trimethoxyphenylacetonitrile as internal standard.

Table IV. Some published HPLC methods for the assay of teicoplanin

Mobile phase	Stationary phase	Detection	Matrix	Comments	Reference
Acetonitrile: 0.05 M phosphate buffer pH 4 (27.5:72.5 by volume)	Microbondapak C_{18} Z-Module	UV 210 nm	serum	protein precipitation with acetonitrile (internal standard = mephensen).	Levy et al. (1987)
Gradient elution with phosphate buffer:acetonitrile	Nucleosil C_{18} 5 μm	UV 240 nm	serum in renal failure	high specificity extraction on D-Ala-D-Ala-Sepharose; no internal standard.	Falcoz et al. (1987)
Gradient elution with phosphate buffer:acetonitrile	a 5 μm C_{18} packing	UV 214 nm	serum	solid-phase extraction (internal standard = piperacillin).	Jehl et al. (1988)
Methanol:butanol: 0.01 M tetrabutyl-ammonium phosphate pH 7.5 (10:1:10 by volume)	Novapak C_{18}	fluorescence (excitation: 390 nm; emission: 490 nm)	serum	protein precipitation with acetonitrile then fluorescamine derivatization.	Joos & Luthy (1987)
Gradient elution with phosphate buffer:acetonitrile	Erbasil C_{18}	UV 254 nm	urine	extraction procedure, components separated.	Bernareggi et al. (1992)
0.1 M phosphate buffer pH 4: acetonitrile (75:25 v/v)	Hypersil 5 ODS	UV 210 nm	serum	protein precipitation with acetonitrile then dilution (1:1) with water.	White & Shi (1990)

detail. The first method, adapted from method no. AM/WN/0293 of the Merrell Dow Research Institute (now part of Hoechst Marion Roussel), has been used in the author's laboratory and been found to correlate well with the microbiological and immunological methods described above (unpublished results). The second method is a simple procedure developed (White & Shi, 1990) and once used routinely at Southmead Hospital for therapeutic monitoring. Quantification is based on measurement of the major peak (A2-2) only.

Gradient elution and separation of all teicoplanin components (Merrell Dow analytical method no: AM/WN/0293)

Sample preparation. Add 400 μL of serum sample to 400 μL of 0.025 M phosphate buffer pH 4.5, containing 30 mg/L orcinol as internal standard. Mix gently and then precipitate proteins by adding 3 mL of Grade S acetonitrile (Rathburn Chemicals Ltd, Walkerburn, UK). Separate by centrifugation at 1500**g** for 10 min. Remove supernatant to capped glass tubes and add 4 mL of dichloromethane. Mix and centrifuge at 1500**g** for 10 min. Remove aqueous phase for injection.

Chromatographic separation. Separation of individual teicoplanin components is achieved using a Hypersil 5 μm ODS column (4.6 mm × 250 mm), maintained at 40°C, with gradient elution of the mobile phase produced utilizing a Spectra Physics SP8700 pump (Thermo Separation Products, Stoke, UK) at a flow rate of 1.6 mL/min. The eluent for pump resevoir A comprises a mixture of 0.025 M sodium dihydrogen phosphate pH 6.0 and Grade S acetonitrile (87.5:12.5 v/v) and that for pump B comprises 0.025 M sodium dihydrogen phosphate pH 6.0 and Grade S acetonitrile (30:70 v/v).

The gradient used is formed as follows:

Time (min)	Eluent A(%)	Eluent B(%)
0	99	1
30	70	30
33	1	99
35	99	1
40	99	1

Timed events for chromatographic separation and raw data collection are controlled using a Spectra Physics SP4290 integrator and Spectra Station Software.

Quantification. 50 μL of samples and calibrators are injected with detection of the separated teicoplanin components at 210 mm.

This method shows a linear relationship between concentration and the sum of the peak heights of teicoplanin components A2-1, A2-2, A2-3, A2-4 and A2-5, over the range 4–64 mg/L. The retention times for the internal standard (orcinol) and the teicoplanin components are of the order of 6.5 and 21–28 min, respectively.

Simple isocratic assay for teicoplanin (based on the method of White & Shi, 1990)

Sample preparation. Mix serum sample with an equal volume of acetonitrile and centrifuge for approximately 1 min. Mix 100 μL of supernatant with 100 μL water. Note: failure to perform the second dilution may impair the chromatography and cause loss of efficiency. Inject 20–50 μL.

Chromatographic separation. Separation is achieved with a mobile phase of 0.1 M phosphate buffer pH 4: acetonitrile (75:25 v/v) adjusted to pH 3.8–4.0 after the addition of the acetonitrile, pumped at 2 mL/min over a Hypersil 5 μm ODS column (100 mm × 4 mm, or 300 mm × 4 mm). Note: the longer column gives better separation of A2-3 from A2-2.

Quantification. Detection is by UV absorbance at 210 nm and quantification is based on peak height measurement of the major component (A2-2) only. Single point calibration with an aqueous solution of 40 or 50 mg/L teicoplanin may be used for therapeutic monitoring. The response is linear up to 50 mg/L.

Comments. If efficiency declines, the column may be revived by flushing with 500 mL of a solution of 6 g/L anhydrous sodium dihydrogen phosphate. If mobile phase recycling is being used, column life is significantly increased if mobile phase is run to waste when a sample is passing through.

References

Arthur, M. & Courvalin, P. (1993). Genetics and mechanisms of glycopeptide resistance in enterococci. *Antimicrobial Agents and Chemotherapy* 37, 1563–71.

Arthur, M., Reynolds, P. & Courvalin, P. (1996). Glycopeptide resistance in enterococci. *Trends in Microbiology* 4, 401–7.

Assandri, A. & Bernareggi, A. (1987). Binding of teicoplanin to human serum albumin. *European Journal of Clinical Pharmacology* 33, 191–5.

Bauchet, J., Pussard, E. & Garaud, J. J. (1987). Determination of vancomycin in serum and tissues by column liquid chromatography using solid phase extraction. *Journal of Chromatography* 414, 472–6.

Bernareggi, A., Borghi, A., Borgonovi, M., Cavenaghi, L., Ferrari, P., Vekey, K. *et al.* (1992). Teicoplanin metabolism in humans. *Antimicrobial Agents and Chemotherapy* 36, 1744–9.

Billot-Klein, D., Shlaes, D. & Gutmann, L. (1998). Peptidoglycan structure of glycopeptide-resistant enterococci and *Staphylococcus haemolyticus*. In *Bacterial Resistance to Glycopeptides* (Brun-Buisson, C., Eliopoulos, G. M. & Leclercq, R., Eds), pp. 28–38. Médecine-Sciences, Flammarion, Paris.

Borghi, A., Coronelli, C., Faniuolo, L., Allievi, G., Pallanza, R. & Gallo, G. G. (1984). Teichomycins, new antibiotics from

Actinoplanes teichomyceticus nov. sp. IV. Separation and characterization of the components of teichomycin (teicoplanin). *Journal of Antibiotics* 37, 615–20.

Brown, N., Ho, D. H. W., Fong, K.-L. L., Bogerd, L., Maksymiuk, A., Boliver, R. *et al.* (1983). Effects of hepatic function on vancomycin clinical pharmacology. *Antimicrobial Agents and Chemotherapy* 23, 603–9.

Campoli-Richards, D. M., Brogden, R. N. & Faulds, D. (1990). Teicoplanin. A review of its antibacterial activity, pharmacokinetic properties and therapeutic potential. *Drugs* 40, 449–86.

Comité de l'Antibiogramme. (1996). Report of the Comité de l'Antibiogramme de la Société Française de Microbiologie (CA-SFM). Zone sizes and MIC breakpoints for non-fastidious organisms. *Clinical Microbiology and Infectious Diseases* 2, *Suppl. 1*, S46–9.

Evers, S. & Courvalin, P. (1996). Regulation of VanB-type vancomycin resistance gene expression by the VanS$_B$-VanR$_B$ two-component regulatory system in *Enterococcus faecalis* V583. *Journal of Bacteriology* 178, 1302–9.

Falcoz, C., Ferry, N., Pozet, N., Cuisinaud, G., Zech, P. Y. & Sassard, J. (1987). Pharmacokinetics of teicoplanin in renal failure. *Antimicrobial Agents and Chemotherapy* 31, 1255–62.

Farber, B. F. & Moellering, R. C. (1983). Retrospective study of the toxicity of preparations of vancomycin from 1974 to 1981. *Antimicrobial Agents and Chemotherapy* 23, 138–41.

Felmingham, D. (1993). Towards the ideal glycopeptide. *Journal of Antimicrobial Chemotherapy* 32, 663–6.

Felmingham, D., Solomonides, K., O'Hare, M. D., Wilson, A. P. R. & Grüneberg, R. N. (1987). The effect of medium and inoculum on the activity of vancomycin and teicoplanin against coagulase-negative staphylococci. *Journal of Antimicrobial Chemotherapy* 20, 609–10.

Georgopoulos, A., Czejka, M. J., Starzengruber, N., Jager, W. & Lackner, H. (1989). High-performance liquid chromatographic determination of teicoplanin in plasma: comparison with a microbiological assay. *Journal of Chromatography* 494, 340–6.

Geraci, J. E. (1977). Vancomycin. *Mayo Clinic Proceedings* 52, 631–4.

Greenberg, K. N., Saeed, A. M. K., Kennedy, D. J. & McMillian, R. (1986). Instability of vancomycin in Infusaid drug pump model 100. *Antimicrobial Agents and Chemotherapy* 31, 610–1.

Greene, S. V., Abdalla, T. & Morgan, S. L. (1987). High-performance liquid chromatographic analysis of vancomycin in plasma, bone, atrial appendage tissue and pericardial fluid. *Journal of Chromatography* 417, 121–8.

Hiramatsu, K., Hanaki, H., Ino, T., Yabuta, K., Oguri, T. & Tenover, F. C. (1997). Methicillin-resistant *Staphylococcus aureus* clinical strain with reduced vancomycin susceptibility. *Journal of Antimicrobial Chemotherapy* 40, 135–146.

Jehl, F., Gallion, C., Thierry, R. C. & Monteil, H. (1985). Determination of vancomycin in human serum by high-pressure liquid chromatography. *Antimicrobial Agents and Chemotherapy* 27, 503–7.

Jehl, F., Monteil, H. & Tarral, A. (1988). HPLC quantitation of the six main components of teicoplanin in biological fluids. *Journal of Antimicrobial Chemotherapy* 21, *Suppl. A*, 53–9.

Joos, B. & Luthy, R. (1987). Determination of teicoplanin concentrations in serum by high-pressure liquid chromatography. *Antimicrobial Agents and Chemotherapy* 31, 1222–4.

Krogstad, D. J., Moellering, R. C. & Greenblatt, D. J. (1980). Single-dose kinetics of intravenous vancomycin. *Journal of Clinical Pharmacology* 20, 197–201.

Kureishi, A., Jewesson, P. J., Rubinger, M., Cole, C. D., Reece, D. E., Phillips, G. L. *et al.* (1991). Double-blind comparison of teicoplanin versus vancomycin in febrile neutropenic patients receiving concomitant tobramycin and piperacillin: effect on cyclosporin A-associated nephrotoxicity. *Antimicrobial Agents and Chemotherapy* **35**, 2246–52.

Lamer, C., de Beco, V., Soler, P., Calvat, S., Fagon, J. Y., Dombret, M. C. *et al.* (1993). Analysis of vancomycin entry into pulmonary lining fluid by bronchoalveolar lavage in critically ill patients. *Antimicrobial Agents and Chemotherapy* **37**, 281–6.

Levy, J., Truong, B. L., Goignau, H., Van Laethem, Y., Butzler, J. P. & Bourdoux, P. (1987). High pressure liquid chromatographic quantitation of teicoplanin in human serum. *Journal of Antimicrobial Chemotherapy* **19**, 533–9.

Mastin, S. H., Buck, R., L. & Mueggler, P. A. (1993). Performance of a fluorescence polarization immunoassay for teicoplanin in serum. *Diagnostic Microbiology and Infectious Disease* **16**, 17–24.

McCormick, M. H., Stark, W. M., Pittenger, G. E., Pittenger, R. C. & McGuire, J. M. (1956). Vancomycin, a new antibiotic. 1 Chemical and biological properties. In *Antibiotics Annual 1955–1956* (Welch, H. & Marti-Ibanez, F., Eds), pp. 606–11. Medical Encyclopedia Inc., New York.

McMullin, C. M., White, L. O., MacGowan, A. P., Holt, H. A., Lovering, A. M. & Reeves, D. S. (1994). Assay of serum teicoplanin concentrations in clinical specimens: a comparison of isocratic high performance liquid chromatography with polarisation fluoroimmunoassay and bioassay. *Journal of Antimicrobial Chemotherapy* **34**, 425–9.

Morse, G. D., Falolino, D. F., Apicella, M. A. & Walshe, J. J. (1987a). Comparative study of intraperitoneal and intravenous vancomycin pharmacokinetics during continuous ambulatory peritoneal dialysis. *Antimicrobial Agents and Chemotherapy* **31**, 173–7.

Morse, G. D., Nairn, D. K., Bertino, J. S. & Walshe, J. J. (1987b). Overestimation of vancomycin concentrations utilizing fluorescence polarization immunoassay in patients on peritoneal dialysis. *Therapeutic Drug Monitoring* **9**, 212–5.

Navarro, F. & Courvalin, P. (1994). Analysis of genes encoding D-alanine–D-alanine ligase-related enzymes in *Enterococcus casseliflavus* and *Enterococcus flavescens*. *Antimicrobial Agents and Chemotherapy* **38**, 1788–93.

National Committee for Clinical Laboratory Standards. (1998). *Performance Standards for Antimicrobial Susceptibility Testing. Eighth Informational Supplement M100-S8*. NCCLS, Wayne, PA.

O'Hare, M. D., Felmingham, D. & Grüneberg, R. N. (1989). The bactericidal activity of vancomycin and teicoplanin against methicillin-resistant strains of coagulase-negative *Staphylococcus* spp. *Journal of Antimicrobial Chemotherapy* **23**, 800–2.

Olssen-Liljequist, B., Larsson, P., Walder, M. & Maiorner, H. (1997). Antimicrobial susceptibility testing in Sweden. Methodology for susceptibility testing. *Scandinavian Journal of Infectious Diseases* **105**, Suppl., 13–23.

Parenti, F., Beretta, G., Berti, M. & Arioli, V. (1978). Teichomycins, new antibiotics from *Actinoplanes teichomyceticus* nov. sp. 1. Description of the producer strain, fermentation studies and biological properties. *Journal of Antibiotics* **31**, 276–83.

Patton, K. R., Beg, A., Felmingham, D., Ridgway, G. L. & Grüneberg, R. N. (1987). Determination of teicoplanin concentration in serum using a bioassay technique. *Drugs under Experimental and Clinical Research* **13**, 547–50.

Perichon, B., Reynolds, P. & Courvalin, P. (1997). VanD-type glycopeptide-resistant *Enterococcus faecium* BM4339. *Antimicrobial Agents and Chemotherapy* **41**, 2016–8.

Pfeiffer, R. R. (1981). Structural features of vancomycin. *Reviews of Infectious Diseases* **3**, Suppl., 205–9.

Quintiliani, R., Evers, S. & Courvalin, P. (1993). The *vanB* gene confers various levels of self-transferable resistance to vancomycin in enterococci. *Journal of Infectious Diseases* **167**, 1220–3.

Reynolds, P. E. (1985). Inhibitors of bacterial cell wall synthesis. In *The Scientific Basis of Antimicrobial Chemotherapy* (Greenwood, D. & O'Grady, F., Eds), pp. 13–40. Cambridge University Press.

Riva, E., Ferry, N., Cometti, A., Cuisinand, G., Gallo, G. G. & Sassard, J. (1987). Determination of teicoplanin in human plasma and urine by affinity and reversed-phase high-performance liquid chromatography. *Journal of Chromatography* **421**, 99–110.

Rosenthal, A. F., Sarfati, I. & A'Zary, E. (1986). Simplified liquid-chromatographic determination of vancomycin. *Clinical Chemistry* **32**, 1016–9.

Rotschafer, J. C., Crossley, K., Zaske, D. E., Mead, K., Sawchuk, R. J. & Solem, L. D. (1982). Pharmacokinetics of vancomycin: observations in 28 patients and dosage recommendations. *Antimicrobial Agents and Chemotherapy* **22**, 391–4.

Rowland, M. (1990). Clinical pharmacokinetics of teicoplanin. *Clinical Pharmacokinetics* **18**, 184–209.

Rybak, M. J., Albrecht, L. M., Berman, J. R., Warbasse, L. H. & Svensson, C. K. (1990a). Vancomycin pharmacokinetics in burn patients and intravenous drug abusers. *Antimicrobial Agents and Chemotherapy* **34**, 792–5.

Rybak, M. J., Albrecht, L. M., Boike, S. C. & Chandrasekar, P. H. (1990b). Nephrotoxicity of vancomycin, alone and with an aminoglycoside. *Journal of Antimicrobial Chemotherapy* **25**, 679–87.

Rybak, M. J., Lerner, S. A., Levine, D. P., Albrecht, L. M., McNeil, P. L., Thompson, G. A. *et al.* (1991). Teicoplanin pharmacokinetics in intravenous drug abusers being treated for bacterial endocarditis. *Antimicrobial Agents and Chemotherapy* **35**, 696–700.

Schaad, U. B., McCracken, G. H. & Nelson, J. D. (1980). Clinical pharmacology and efficacy of vancomycin in pediatric patients. *Journal of Pediatrics* **96**, 119–26.

Stahl, J. P., Croize, J., Wolff, M., Garaud, J. J., Leclercq, P., Vachon, F. *et al.* (1987). Poor penetration of teicoplanin into cerebrospinal fluid in patients with bacterial meningitis. *Journal of Antimicrobial Chemotherapy* **20**, 141–2.

Uhl, J. R. & Anhalt, J. P. (1979). High performance liquid chromatographic assay of vancomycin in serum. *Therapeutic Drug Monitoring* **1**, 75–83.

Whitby, M., Edwards, R., Aston, E. & Finch, R. G. (1987). Pharmacokinetics of single dose intravenous vancomycin in CAPD peritonitis. *Journal of Antimicrobial Chemotherapy* **19**, 351–7.

White, L. O. & Reeves, D. S. (1991). Problems with the assay of teicoplanin caused by differential binding of the A2 components to plastics and to glass. In *Program and Abstracts of the Thirty-First Interscience Conference on Antimicrobial Agents and Chemotherapy, Chicago, IL, 1991*. Abstract 1312, p. 317. American Society for Microbiology, Washington, DC.

White, L. O., Edwards, R., Holt, H. A., Lovering, A. M., Finch, R. G. & Reeves, D. S. (1988). The in-vitro degradation at 37°C of vancomycin in serum, CAPD fluid and phosphate-buffered saline. *Journal of Antimicrobial Chemotherapy* **22**, 739–45.

White, L. O., McMullin, C., Davis, A. J., MacGowan, A. P., Harding, I. & Reeves, D. S. (1996). The quality of serum teicoplanin assays: an experimental European EQA distribution. *Journal of Antimicrobial Chemotherapy* **38**, 701–6.

White, L. O., Holt, H. A., Reeves, D. S. & MacGowan, A. P. (1997). Evaluation of Innofluor fluorescence polarization immunoassay kits for the determination of serum concentrations of gentamicin, tobramycin, amikacin and vancomycin. *Journal of Antimicrobial Chemotherapy* **39**, 355–61.

White, L. O. & Shi, Y. G. (1990). A simple, straightforward HPLC method for monitoring teicoplanin in serum. In *Infection 1990 Programme, Abstracts and Participants*, Abstract 27, Association of Medical Microbiologists, London.

Wilcox, M. H., Winstanley, T. G. & Spencer, R. C. (1994). Binding of teicoplanin and vancomycin to polymer surfaces. *Journal of Antimicrobial Chemotherapy* **33**, 431–41.

Wilson, A. P. R., Grüneberg, R. N. & Neu, H. (1994). A critical review of the dosage of teicoplanin in Europe and the USA. *International Journal of Antimicrobial Agents* **4**, *Suppl. 1*, S1–30.

Woodley, D. W. & Hall, W. H. (1961). The treatment of severe staphylococcal infections with vancomycin. *Annals of Internal Medicine* **55**, 235–49.

Working Party. (1996). Supplementary report of the Working Party of the British Society for Antimicrobial Chemotherapy. *Journal of Antimicrobial Chemotherapy* **38**, 1103–5.

Chapter 11

The 4-quinolones

Les O. White[a], Caroline M. Tobin[a], Andrew M. Lovering[a] and Jenny M. Andrews[b]

[a]*Department of Medical Microbiology, Southmead Hospital, Bristol BS10 5NB;* [b]*Department of Medical Microbiology, City Hospital NHS Trust, Birmingham B18 7QH, UK*

Introduction

A new antibacterial agent, nalidixic acid (Figure 11.1b) was reported in 1962 and subsequently marketed in 1963 for the treatment of urinary tract infections. Oxolinic acid (Figure 11.1c), which was chemically related and somewhat more active, appeared in 1967 and a third related compound, cinoxacin (Figure 11.1a), was described in 1973. Thus began the history of a family of chemically related antibacterial compounds which are now generally referred to as the 4-quinolones.

The early 4-quinolones were orally active anti-Gram-negative compounds with relatively poor activity which limited them to the treatment of urinary tract infections. Chemical modifications have led to improved compounds. Fluorinated derivatives included flumequine (Figure 11.2d), which showed activity against *Pseudomonas aeruginosa*. The addition of a 7-piperazinyl group gave rise to the modern fluoroquinolones such as ciprofloxacin, norfloxacin, ofloxacin and others (Figure 11.2), with much improved antibacterial activity and pharmacokinetic properties. These fluoroquinolones have a very wide spectrum, which includes pseudomonads and Gram-positive organisms, and are highly active, especially against Enterobacteriaceae. Favourable pharmacokinetic properties and high activity make some of them them suitable for treating both urinary and systemic infections. In addition they are frequently active against strains which have acquired resistance to other antimicrobials, such as aminoglycosides, β-lactams, sulphonamides, trimethoprim and chloramphenicol. More recently developed agents show further improvments in antimicrobial activity, especially against Gram-positive and anareobic bacteria, and improved pharmacokinetics. Most of these are multi-halogenated (Figure 11.3) but grepafloxacin (Figure 11.2e), a dimethyl derivative of ciprofloxacin, and moxifloxacin (Figure 11.2f) are not.

Quinolones are rapidly bactericidal and act by inhibiting the alpha subunit of bacterial DNA gyrase (Crumplin *et al.*, 1984; Smith, 1985). They have been the subject of numerous symposia, proceedings and monographs (Andriole, 1988; Fernandes, 1989; Wilson & Grüneberg, 1997).

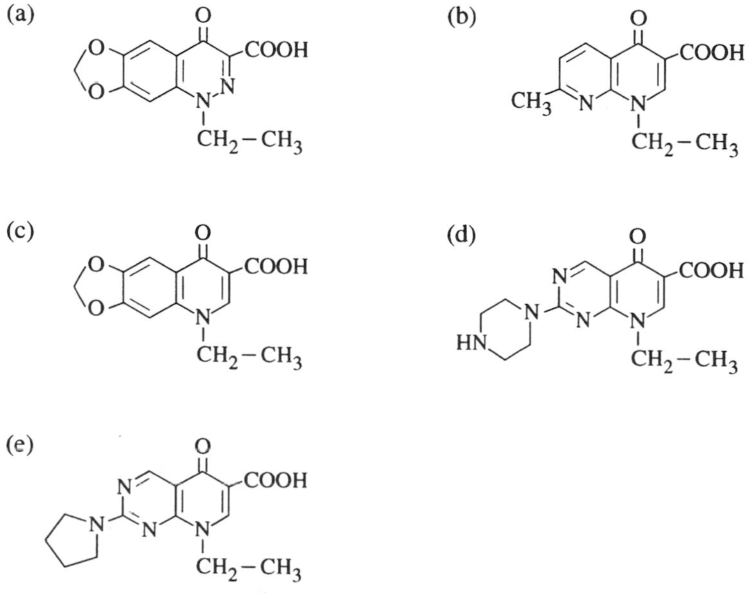

Figure 11.1. Some non-halogenated 4-quinolones: (a) cinoxacin, (b) nalidixic acid, (c) oxolinic acid, (d) pipemidic acid, (e) piromidic acid.

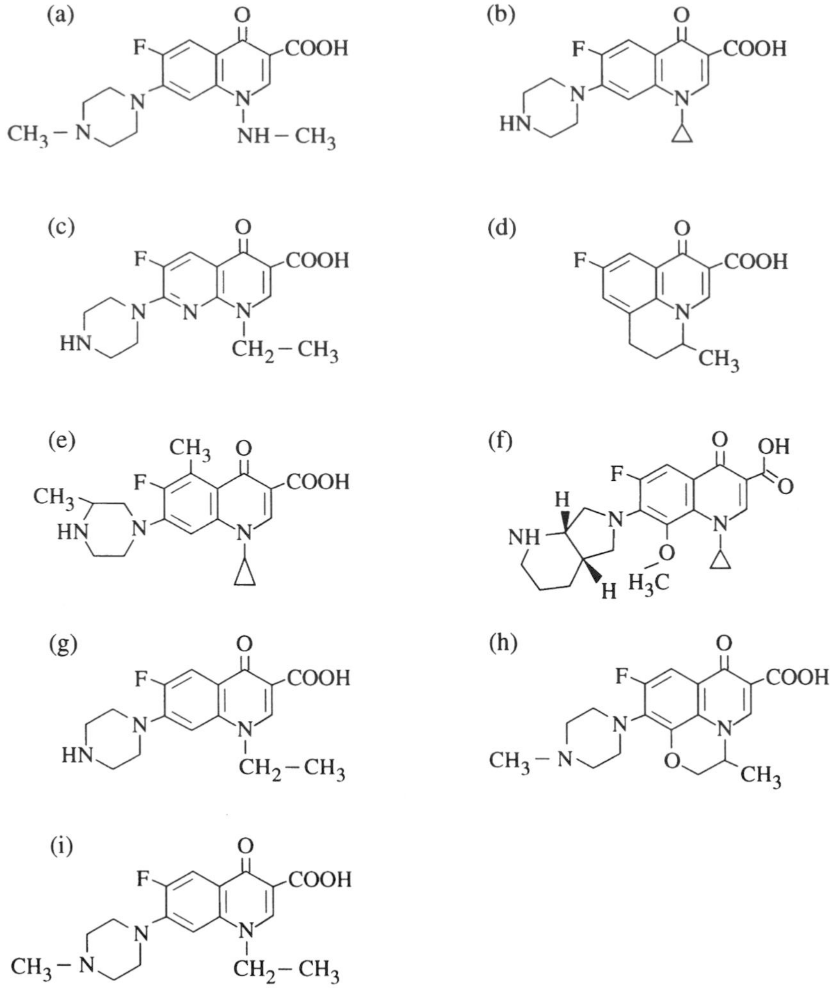

Figure 11.2. Some 6-fluoro-4-quinolones: (a) amifloxacin, (b) ciprofloxacin, (c) enoxacin, (d) flumequine, (e) grepafloxacin, (f) moxifloxacin, (g) norfloxacin, (h) ofloxacin (levofloxacin is L-isomeric form), (i) pefloxacin.

Chemistry and nomenclature

In general, the 4-quinolones are acidic or amphoteric, crystalline compounds, relatively thermostable in solution but often degraded by strong sunlight. 4-Quinolones are soluble in water or in dilute alkali. Dosage forms may be hydrochloride, or esters such as mesylate or lactate. Trovafloxacin is relatively water-insoluble and is formulated as L-Ala-L-Ala-trovafloxacin (alatrofloxacin) for parenteral administration (Cutler *et al.*, 1997).

All the 4-quinolones have the basic structure shown in Figure 11.4. The 4-oxygen and the 3-carboxyl group are essential for antimicrobial activity. The quinolones may be subdivided, depending on whether or not they are N-substituted in the 2, 6 and 8 positions, as follows:

- true quinolones or *quinolines*, e.g. oxolinic acid (Figure 11.1c) and most fluoroquinolones;
- *pyridopyrimidines*, e.g. pipemidic acid (Figure 11.1d);
- *naphthyridines*, e.g. nalidixic acid (Figure 11.1b), enoxa-

cin (Figure 11.2c), tosufloxacin and trovafloxacin (Figure 11.3f and g); and

- *cinolines*, e.g. cinoxacin (Figure 11.1a).

An alternative nomenclature scheme, as suggested by Smith (1985), is also shown in Figure 11.4.

Fluorination of the carbon atom at position 6 increases the antibacterial activity two- to 30-fold compared with the unfluorinated analogue. In addition, 7-substitutions are important in determining increased antimicrobial activity. Many compounds have a 7-piperazine ring and this may or may not be methylated. All compounds have an ethyl group (or some chemical alternative with similar spatial configuration) on 1-N and this appears essential for activity (Schentag & Domagala, 1985). Substitutions at 5-C add anti-Gram-positive activity. Conformation affects antimicrobial activity. Ofloxacin (Figure 11.2h) and flumequine (Figure 11.2d) have a methyl group in the oxazine ring resulting in an asymmetric centre at this position. Levofloxacin, the L(−) isomer of the two isomers which

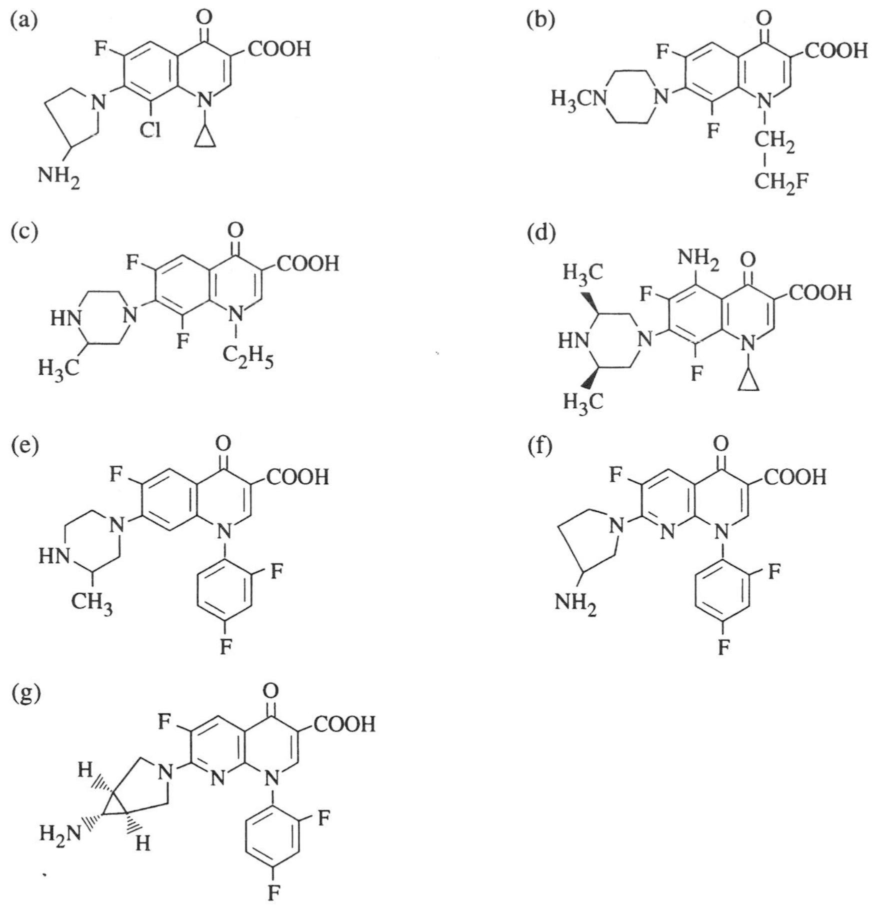

Figure 11.3. Some multi-halogenated 4-quinolones: (a) clinafloxacin, (b) fleroxacin, (c) lomefloxacin, (d) sparfloxacin, (e) temafloxacin, (f) tosufloxacin, (g) trovafloxacin.

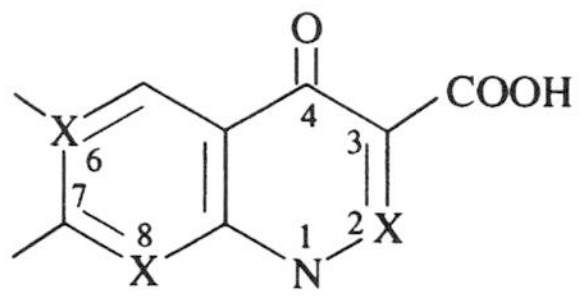

Name	2-X	6-X	8-X	alternative name
Quinoline	C	C	C	4-quinolone
Pyridopyrimidine	C	N	N	6-8 diaza-4-quinolone
Naphthyridine	C	C	N	8-aza-4-quinolone
Cinoline	N	C	C	2-aza-4-quinolone

Figure 11.4. General chemical structure and nomenclature of 4-quinolones.

make up ofloxacin (Hayakawa *et al.*, 1986), is more active than ofloxacin.

Dosage forms

Nalidixic acid is available for oral administration only, as granules containing nalidixic acid and inactive ingredients, as a raspberry-flavoured suspension, or as 500 mg tablets. Norfloxacin is available as 400 mg tablets and cinoxacin as 500 mg capsules. Ciprofloxacin is available as a solution for parenteral administration, tablets or eye drops. The parenteral solution contains ciprofloxacin lactate 254 mg (equivalent to ciprofloxacin 200 mg). The tablets contain ciprofloxacin hydrochloride monohydrate 116.4, 291, 582 or 873 mg (equivalent to ciprofloxacin 100, 250, 500 or 750 mg respectively) together with inert ingredients. The aqueous eye drops contain 0.35% (w/v) ciprofloxacin hydrochloride in 0.006% (w/v) benzalkonium chloride. Ofloxacin is available as 200 or 400 mg tablets and a parenteral infusion containing 110 or 220 mg ofloxacin hydrochloride (equivalent to 100 and 200 mg ofloxacin respectively).

At the time of writing, other agents—including clinafloxacin, grepafloxacin, levofloxacin and sparfloxacin—were not available in the UK.

Pharmacokinetics

There are many reviews of quinolone pharmacokinetics (Hooper & Wolfson, 1985; Janknegt, 1986; Wise *et al.*, 1986; Paton & Reeves, 1988, Brown & Reeves, 1997). The

Table I. Physico-chemical and pharmacokinetic properties of 4-quinolones

Drug	Mol. wt	pK_a	Dosage forms	Elimination half-life (h)	Protein binding (%)	Metabolites	Urinary excretion
Naphthyridines							
clinafloxacin	365.78		hydrochloride	5.2	50	glucuronide	>50%, mostly unchanged
enoxacin	320.32	6.2, 8.8	3/2 hydrate	3–6	35	oxo	60%, 10–15% as oxo
nalidixic acid	232.2	6.7		1.5	93	glucuronide, hydroxy	>80%, mostly metabolites
tosufloxacin	404.33		tosilate	6–7	37		30–35%
trovafloxacin	416.36	5.87, 8.09	L-Ala-L-Ala mesylate	9–13	70	glucuronide, N-acetyl	<10%
Quinolines							
ciprofloxacin	331.3	6.0, 8.8	lactate or hydrochloride	3–4	20–40	sulpho, oxo, formyl, desethyl	50–70%, c. 10% metabolites
fleroxacin	369	5.5, 8.0		9–12	23	desmethyl, N-oxide	>50%, with <11% as metabolites
grepafloxacin hydrochloride	395.86		hydrochloride sesquihydrate	8–15	50	glucuronide, sulphate open-ring	c. 10%
levofloxacin	361.37	7.9	hemihydrate	3–7	30	N-oxide, desmethyl	>75%, <10% as metabolites
lomefloxacin	351.37			7–8	15	glucuronide	>60%, 5–10% as metabolite
moxifloxacin hydrochloride	437.9			10–12	50		c. 20%
norfloxacin	319.3	6.2–6.4, 8.7–8.9		3–4	15	formyl, oxo, desethyl, etc.	35%, with <10% as metabolites
ofloxacin	361.37	7.9	hydrochloride	3–7	30	N-oxide, desmethyl	>75%, <10% as metabolites
pefloxacin	333.37		mesylate dihydrate	8–13	20–30	as norfloxacin plus desmethyl	>60%, mostly as metabolites
sparfloxacin	380.38			15–20	37	glucuronide	<10% mostly as metabolite

Table II. Effect of Mg^{2+} or Al^{3+} containing antacids on absorption of ciprofloxacin and ofloxacin

	Peak serum concentration (mg/L)		
	without antacid	with antacid	% difference
Ofloxacin	2.2	0.5[a]	77
Ciprofloxacin	1.7	0.1[a]	94

[a] These levels would be inadequate for many systemic infections.

Table III. Typical peak serum concentrations for oral fluoroquinolones and nalidixic acid

Drug	Dose (mg)	Serum concentration (mg/L)
Ciprofloxacin	1000	5.5
	500	2.5
Clinafloxacin	200	2.5
Enoxacin	200	1.0
	400	3.8
Grepafloxacin	400	1.5
Nalidixic acid	660	2.8 (2.6)[a]
Norfloxacin	1600	3.9
	400	1.6
Ofloxacin	200	2.2
	400	5.6
Pefloxacin	400	3.8
Trovafloxacin	300	4.4

[a]Hydroxy metabolite.
From Groeneveld & Brouwers (1986), Chang *et al.* (1988), Kucers *et al.* (1997), and Data Sheets.

pharmacokinetic properties of some compounds are summarized in Table I. They are orally absorbed (some also have parenteral formulations), have large volumes of distribution, and are, with the exception of some of the newer agents, eliminated mainly by renal excretion and, to a greater or lesser extent depending on the compound, metabolic conversion. Within this generalization there are quite large differences between the different compounds.

Oral absorption

This is usually rapid (1–2 h) with bioavailability varying from 62% (nalidixic acid) to >80% (ofloxacin) (Bergan, 1986). Some other medications impair absorption and may result in sub-therapeutic serum levels (Rubinstein & Segev, 1987). Antacids containing aluminium or magnesium have the greatest effect (Table II) and are thought to chelate the drugs. Bioavailability may be reduced by as much as 90% as shown by reduced serum levels and urinary recovery (Rubinstein & Segev, 1987). Absorption of ofloxacin in patients with renal failure may be variable and prolonged even in the absence of antacid therapy (Dorfler *et al.*, 1987) and this may be true for other quinolones. Typical peak serum concentrations after oral dosing are shown in Table III.

Elimination

In normal subjects serum half-lives vary from short (1.5 h, nalidixic acid) to very long (40 h, rufloxacin) but there is no apparent correlation between half-life and protein binding —which is relatively low for most fluoroquinolones (Table I). The majority of the dose of most quinolones is excreted in the urine in normal subjects either unchanged or as metabolites. Notable exceptions are grepafloxacin, moxifloxacin, sparfloxacin and trovafloxacin which are extensively excreted via the gut. The degree of metabolism varies greatly from compound to compound. Even with those drugs which are metabolized to a small degree, information on metabolites is important in the context of assays (White, 1987) because metabolites may:

- be microbiologically active and interfere with bioassays
- interfere with other types of assay (e.g. HPLC)
- accumulate in certain fluids or disease states.

Metabolites may be microbiologically active but usually less so than the parent compound. Oxo and formyl metabolites show little activity and N-oxides and glucuronides are essentially inactive. Desmethyl ofloxacin has somewhat less activity than ofloxacin but desmethyl pefloxacin (which is chemically identical to norfloxacin) has similar activity to the parent compound. Oxo-enoxacin is as active as enoxacin against some strains of Enterobacteriaceae. N-Acetyl trovafloxacin has around 10% the activity of the parent compound. It may be necessary to assay metabolite concentrations for pharmacokinetic studies.

It is interesting to note that the importance of metabolites was appreciated with the ancestral quinolone, nalidixic acid, almost all of which is excreted in the in the metabolized form (Figure 11.5). Hydroxy-nalidixic acid is microbiologically active and plays an important role in the efficacy of nalidixic acid. Metabolism of the fluoroquinolones differs somewhat from that of nalidixic acid in

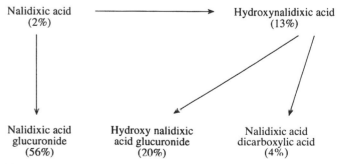

Figure 11.5. Nalidixic acid and metabolites in urine. Only nalidixic acid and hydroxy nalidixic acid have antimicrobial activity. The percentage recovery of a 1 g dose is shown in brackets.

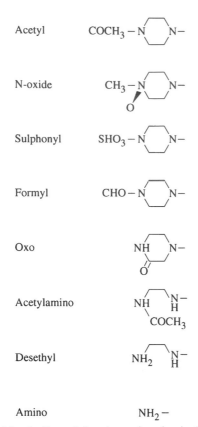

Acetyl	COCH₃ – N⌐⌐⌐N–
N-oxide	CH₃ – N⌐⌐⌐N–
Sulphonyl	SHO₃ – N⌐⌐⌐N–
Formyl	CHO – N⌐⌐⌐N–
Oxo	NH⌐⌐⌐N–
Acetylamino	NH⌐⌐⌐N–H
Desethyl	NH₂⌐⌐⌐N–H
Amino	NH₂ –

Figure 11.6. Metabolism of the piperazine ring in 4-quinolones.

as much as glucuronide formation is relatively minor (some newer agents are again exceptions) and most metabolites are modifications of the 7-piperazine ring (which is not present in nalidixic acid). Types of metabolite are shown in Figure 11.6. *N*-Methylated quinolones such as pefloxacin and ofloxacin undergo demethylation and also form *N*-oxides. The other types of metabolite are produced from non-methylated compounds. Urinary recovery of drug and metabolites is summarized in Tables I and IV.

Toxicity

The overall incidence of side effects of quinolone therapy is 5–23%; side effects are usually mild to moderate, rarely necessitating discontinuation of therapy (Smith, 1987). The most likely reason for monitoring therapy is to ensure that adequate levels are being achieved in patients where the pharmacokinetics are difficult to predict (e.g. in those receiving oral therapy, or with renal impairment, or neonates) and where less susceptible organisms, such as staphylococci, pseudomonads, mycobacteria or legionellae, are being treated.

Efficacy of quinolone treatment has been related to pharmacodynamic predictors such as C_{max}/MIC or AUC/MIC and assays may be performed to determine these parameters in individual patients. Since routine assay of some quinolones in patients may not be feasible (due to a lack of assay facilities) a recent study (Preston *et al.*, 1998) used a population pharmacokinetics approach to attempt to predict C_{max}/MIC ratio in patients but was unable to do so sufficiently reliably for clinical use.

For ciprofloxacin an arbitrary maximum for peak serum concentrations of 8–10 mg/L may be taken as an upper limit since there is unlikely to be any clinical benefit from levels above this and their potential for toxicity is unknown. Adequate trough levels would depend on the MIC of the pathogen, but for most sensitive organisms trough ciprofloxacin levels of 0.1–1.0 mg/L would probably be acceptable.

Gastrointestinal upset

This is the most common side effect and is usually mild to moderate. It is no reason to modify dosage since faecal concentrations are very variable and high, even after iv therapy. Faecal levels can be over 100-fold greater than blood levels.

Table IV. Recovery (% of dose) of fluoroquinolones and metabolites in urine

	Ciprofloxacin		Enoxacin		Norfloxacin	Ofloxacin		Pefloxacin	
	iv	po	iv	po	po	iv	po	iv	po
Parent	62	45	46	44	27	77	74	5	9
Formyl	Tr	Tr			<1				
N-oxide						1	1	18	23
Oxo	6	6	12	7–14	5				<1
Sulphonyl	3	4							
Desmethyl	NA	NA	NA	NA	NA	3	3	9	20
Desethyl	1	1			<2				
Acetylamino					<1				
Amino					<1				

NA, not applicable.
From White (1987) and Lode *et al* (1987).
Tr, traces.

Effects on joints

Studies in animals showed that quinolones impaired the development of cartilage in the joints. For this reason the drugs are not used in children and pregnant women, unless the benefits outweigh the risks.

Phototoxicity

Nalidixic acid shows phototoxicity and this applies to a greater or lesser extent to the fluoroquinolones in a dose-related manner, but no information linking blood levels with phototoxicity is currently available. Lomefloxacin is associated with considerably greater phototoxicity than ciprofloxacin, norfloxacin or ofloxacin (Echols & Oliver, 1993; Scheife et al., 1993) A high incidence of phototoxicity is a problem with many of the multi-halogenated compounds. Quinolone therapy should be stopped if signs of phototoxicity develop. Patients taking quinolones should avoid prolonged exposure to bright sunlight. Photosensitivity is a common side effect in adult cystic fibrosis patients receiving ciprofloxacin (Burdge et al., 1995).

Phototoxicity and photolability may be related, the former being elicited by stable and/or unstable toxic photoproducts. Free radical oxygen may be such an unstable photoproduct (Wagai & Tewara, 1992). Often photodegradation is associated with loss of antimicrobial activity and numerous photoproducts may be formed (McMullin et al., 1992, 1996). Recently photodegredation of sparfloxacin was shown to be associated with loss of a fluorine atom and the cyclopropyl group (Engler et al., 1998)

Neurotoxicity

Some quinolones (notably enoxacin) alter the pharmacokinetics of theophylline and this can induce convulsions. All quinolones inhibit the γ-aminobutyric acid (GABA) receptors in the brain in a dose-dependent manner. High blood concentrations might therefore carry a risk of neurotoxicity and are reason enough to avoid excessive accumulation. Quinolone therapy should be stopped at the first sign of neurotoxicity.

Prolongation of the electrocardiographic QTc interval

This has been extensively studied with regard to sparfloxacin (Jaillon et al., 1996). In dogs the maximum QTc prolongation was observed to coincide with C_{max}. In healthy volunteers doses of 300 mg and above were associated with a moderate prologation of the QTc interval from baseline of 5–8%, and in a Phase I trial 200 mg and 300 mg caused increases of 2–3% and 4–8% respectively. Further data from Phase III trials confirmed a mean increase of around 3%; only ten patients showed an increase to $\geq$500 ms and none experienced arrhythmia.

Microbiological assay methods for quinolones

The literature contains references to many microbiological assay techniques for assaying quinolones using a variety of indicator organisms including Escherichia coli, Bacillus subtilis, Klebsiella pneumoniae and Morganella morganii. However, the possible effect of metabolites is not always stated or considered. Sometimes 'total antimicrobial activity' is measured (this being the overall result obtained with a mixture of parent drug and active metabolites in unknown proportions using pure parent drug in the calibrators). Although this may be of some value in monitoring therapy it is meaningless when applied to any organism other than the assay indicator organism since it depends on the relative MICs of the compounds present. Some investigators have used two indicator organisms, for example B. subtilis for high concentrations (urine) and K. pneumoniae for low concentrations (serum). This is a potentially misleading practice since interfering metabolites may represent a larger proportion of total drug when parent drug concentrations are low and metabolites may be more active against one of the indicator organisms. Since all quinolones produce at least some active metabolites, the results of a microbiological assay will only be accurate if a metabolite-insensitive indicator strain is used. Luckily, for the majority of fluoroquinolones, the metabolites have low activity and are found in small amounts in blood relative to the parent compound, so acceptably accurate assays can be performed on blood samples.

With all quinolones not only the indicator organism but also the composition and pH of the medium will affect metabolite interference. To allow inter-laboratory comparisons to be made a degree of standardization is needed. Some methods are described at the end of this chapter. If a microbiological assay is chosen for therapeutic monitoring then interference from co-administered drugs must be eliminated. Table V summarizes some data regarding microbiological assays for quinolones.

A microbiological assay for ciprofloxacin and other fluoroquinolones

The following method is recommended for assay of ciprofloxacin in serum and urine; it could readily be adapted for assay of other quinolones (see below).

Medium. Isosensitest agar (Oxoid CM 471) prepared according to manufacturer's instructions.

Indicator organism. E. coli 1004 (from Bayer AG, Wuppertal, Germany) stored on nutrient agar (slope) at room temperature. Before use, subculture to check purity then inoculate into digest or infusion broth and incubate overnight at 37°C.

Calibrators. Spike pooled human serum (for serum samples) or phosphate buffer pH 7.2 (for urine samples)

Table V. Bioassays for 4-quinolones

Quinolone	Medium	Indicator strain	Calibrator concentrations (mg/L)	Incubation temperature (°C)	Sample application	LLS (mg/L)	Internal control concentrations (mg/L)
Ciprofloxacin	Isosensitest	*E. coli* 4004	0.05–0.8	30	6 mm well	0.01	0.6, 0.08
Enoxacin	Antibiotic no. 1	*E. coli* I174	0.12–2	30	6 mm well	0.1	1.5, 0.2
Lomefloxacin	Isosensitest	*E. coli* 4004	0.12–2	30	5 mm well	0.12	1.5, 0.2
Norfloxacin	Antibiotic no. 1	*E. coli* NCTC 10418	0.12–2	37	7 mm well	0.1	1.5, 0.2
Pefloxacin	Isosensitest	*K. pneumoniae* 13	2–32	30	6 mm disc	1	20, 3
Rufloxacin	Isosensitest	*E. coli* 4004	0.5–8	30	7 mm well	0.5	6, 0.8
Sparfloxacin	Isosensitest	*E. coli* 4004	0.062–1	30	5 mm well	0.06	0.75, 0.1

LLS, lower limit of sensitivity.

with ciprofloxacin 0.05, 0.1, 0.2, 0.4 and 0.8 mg/L. Dilute samples to fall in this concentration range using the appropriate diluent. Also prepare controls of approximately 0.6 and 0.1 mg/L.

Procedure:

1. Pour the molten agar into plates (100 mL for a Nunc disposable plate or 200 mL for a Mast reusable plate).
2. Allow to set then dry the surface.
3. Dilute the overnight broth culture of *E. coli* 4004 one in 500 in sterile distilled water.
4. Flood the plate, drain off excess fluid and re-dry the surface (this should be done as soon as possible to ensure the organism does not begin to grow).
5. Store at 4°C until needed (up to 8 h).
6. Apply the calibrators and appropriately diluted tests or controls to the agar surface on Whatman 6 mm blotting paper discs.
7. Incubate overnight at 30°C. Measure the zones using a magnifying zone reader.
8. Use a regression procedure to calculate the calibration curve.

The lower limit of sensitivity is 0.01 mg/L and typical CVs should be 7.8% or better.

In a blind interlaboratory comparison the correlation with HPLC assay results (34 samples) was excellent ($r^2 = 1.00$; slope, 1.18; y intercept, –0.059).

A microbiological assay for other quinolones

The method above may be adapted to assay other quinolones as described below:

Ofloxacin. Calibrators: 0.5, 1, 2, 4 and 8 mg/L. Lower limit of sensitivity: 0.25 mg/L. Typical CV: 8.5%. This assay shows a good correlation with HPLC ($r^2 = 0.983$; slope, 1.14; y intercept, 0.187).

Enoxacin in serum. Calibrators: 0.5, 1, 2, 4 and 8 mg/L. Indicator organism: *K. pneumoniae* K13 (Warner-Lambert, Eastleigh, UK). Wells (5 mm) are cut in the agar and filled with sample to give a convex liquid surface (N.B. adding a fixed volume by pipette is not recommended). Lower limit of sensitivity: 0.25 mg/L. Typical CV: 8.9%. This assay shows a good correlation with HPLC for serum samples ($r^2 = 0.99$; slope, 0.925; y intercept, 0.055).

When urine samples are assayed by this method the oxo metabolite causes an overestimation of enoxacin concentration (approx. 30% in a 0–4 h urine, 16% in a 4–16 h urine and 5% in a 24–48 h collection). Microbiological assay is therefore not recommended for urine samples.

Pefloxacin in serum. Calibrators: 2, 4, 8, 16 and 32 mg/L. Indicator organism: *K. pneumoniae* K13 (Warner-Lambert). Lower limit of sensitivity: 0.25 mg/L. Typical

CV: 6.3%. This assay shows a good correlation with HPLC ($r^2 = 0.985$; slope, 1.035; y intercept, 0.11).

Urines cannot be assayed because of pefloxacin's extensive metabolism to desmethyl-pefloxacin (norfloxacin) and oxo-pefloxacin. Urines must be assayed by HPLC.

Norfloxacin in serum. Medium: 5.2 g nutrient broth (Oxoid CM1) plus 4.0 g agar (Difco Bacto Agar) adjusted to pH 6.7 with 1 N HCl. Calibrators: 0.1, 0.2, 0.4, 0.8 and 1.6 mg/L. Indicator organism: *K. pneumoniae* BRL 1003. Samples are placed in 7 mm wells in the agar and filled to give a convex liquid surface. Lower limit of sensitivity: 0.25 mg/L. Typical CV: 7.5%. Microbiological assay is unsuitable for urines because of the activity of oxo-norfloxacin.

Microbiological assay of nalidixic acid

This assay is designed to measure 'total biological activity' by using indicator organisms for which the MICs for nalidixic acid and hydroxy-nalidixic acid are similar. Different procedures are required for serum and urine (see comments above).

Urine. The test organism is *E. coli* ATCC 15764 maintained on Difco tryptose phosphate agar slants by monthly transfer. Two or three tubes of tryptose phosphate broth are subcultured from the slant and incubated at 37°C for 18 h. This inoculum is incorporated into Difco Penassay agar containing 2,3,5-triphenyltetrazolium chloride (TTC, 200 g/L). Calibrators (highest concentration 60 mg/L) are made in pooled human urine and samples and calibrators are applied on paper discs. Incubation is for 16–18 h at 37°C.

Serum. For serum samples the test organism is *K. pneumoniae* ATCC 10872 maintained and inoculated as above into Difco Penassay agar containing only 100 mg/L TTC and 8 M sodium citrate. Calibrators (highest concentration 30 mg/L) are made in serum and the procedure is as described for urine except that the recommended incubation temperature is 30°C.

Non-microbiological assay methods for quinolones

Spectrophotometric and spectrofluorimetric assays were developed for some ancestral quinolones (e.g. nalidixic acid) but such methods have been superseded by HPLC. Nevertheless these methods illustrate some useful points. A spectrophotometric assay for nalidixic acid (see below) exploits the fact that nalidixic acid and its hydroxy metabolite absorb strongly at 336 nm but not at 370 nm. The net value of ($A_{336} - A_{370}$), where A is absorbance and the subscript figure indicates the wavelength in nanometres, gives a relatively specific estimation of nalidixic acid + hydroxy-nalidixic acid. The spectrofluorimetric assay

described below exploits natural fluorescence and this can be utilized to design very specific and sensitive HPLC methods. The results of a fluorimetric assay for flumequine compared well with plate assay and HPLC determinations (Harrison *et al.*, 1984). Fluorimetric assays for the newer fluoroquinolones have been published (Chapman & Georgopapadakou, 1989; Pascual *et al.*, 1989) and have been used to measure uptake into cells. They seem simple straightforward and sensitive but have poor specificity and require blank (quinolone-free) samples; this would prove a problem in therapeutic monitoring. Some fluorescence data for fluoroquinolones are shown in Table VI.

Early work on HPLC methods suggested that reversed phase HPLC might not be suitable for the quinolones. Shargel *et al.* (1973) used ion exchange chromatography on ZIPAX-SAX for nalidixic acid and its hydroxy metabolite, and norfloxacin and its metabolites were also separated by anion exchange on 10 μm Vydac (Boppana & Swanson, 1982). The latter authors had compared anion exchange and reversed-phase HPLC and concluded that the latter technique was preferable. However, subsequent work with various C_{18} columns has shown that reversed-phase HPLC is ideally suited to the assay of fluoroquinolones. Tables VII and VIII summarize some published HPLC methods for quinolones and mixtures of quinolones. It is interesting to note that there are numerous published methods for the earlier compounds (ciprofloxacin and ofloxacin, for example) whereas for the more recently developed compounds there are considerably fewer published methods. This may reflect the tightening of regulatory controls related to drug development.

The natural fluorescence of ciprofloxacin, ofloxacin, norfloxacin, grepafloxacin, moxifloxacin and others has been used in highly specific, highly sensitive HPLC methods. Non-fluorescent or poorly fluorescent compounds are detected by UV absorption.

Unfortunately not all published HPLC methods have been tested for metabolite interference. This will occur if a

Table VI. Fluorescence excitation and emission data for quinolones

Quinolone	Excitation (nm)	Emission (nm)
Ciprofloxacin	287	445
Difloxacin	280	452
Flumequine	245	365
Nalidixic acid	246	354
Norfloxacin	281	440
Ofloxacin	292	496
Oxolinic acid	230	305
Pefloxacin	277	442

Data (except those for ciprofloxacin) from Chapman & Georgopapadakou (1989).

Table VII. Some published HPLC methods for assay of 4-quinolones and their metabolites in biological materials

Quinolone	Mobile phase	Column	Detection	Matrix	Comments	Reference
Amifloxacin	Water + ACN/TBAP/NaCl, pH 2.3	μbondapak phenyl	UV 273 nm	urine	direct injection	Johnson & Benzinger (1985)
Amifloxacin	ACN/TBAP/NaCl pH 2.3	μbondapak phenyl	UV 280 nm	plasma/urine	chloroform extraction; N-desmethyl metabolite assayed	McCoy et al. (1985)
Amifloxacin	not stated	not stated	UV	urine/plasma	column-switching	Stroshane et al. (1990)
Ciprofloxacin	MeOH/water + cetrimide pH 7.5	C$_8$ Z-Module	UV 274 nm	serum ultrafiltrate	direct injection	Reeves et al. (1984)
Ciprofloxacin	TBAB/ACN pH 2	Ultrasphere ODS	UV 254 nm	body fluids	methylene chloride/ phosphoric acid extraction	Brogard et al. (1985)
Ciprofloxacin	PA/TBAH/ACN pH 3	Spherisorb 5 ODS II (heated to 50°C)	fluorescence ex. 277 nm	body fluids	samples diluted with PA or HCl	Gau et al. (1985)
Ciprofloxacin	not stated	Nucleosil 5 C$_{18}$	fluorescence ex. 275 nm em. 414 nm	urine	direct injection; metabolites assayed (but incorrectly quantified)	Hoffken et al. (1985)
Ciprofloxacin	ACN/TBAB/PA pH 2	Ultrasphere C$_{18}$ 5 μm	UV 254 nm	biological fluids	methylene chloride/ phosphoric acid extraction	Jehl et al. (1985)
Ciprofloxacin	ACN/TBAH/PA pH 3	Spherisorb 5 ODS	fluorescence ex. 278 nm em. 456 nm	serum/urine	protein precipitation with TCA	Joos et al. (1985)
Ciprofloxacin	ACN/MeOH + PB/TBAH/PA pH 3	μbondapak C$_{18}$	fluorescence ex. 270 nm em. 440 nm	serum/urine	protein precipitation with PCA	Nix et al. (1985)
Ciprofloxacin	MeOH/PB pH 3.5	μbondapak C$_{18}$	fluorescence ex. 445 nm em. 277 nm	body fluids	protein precipitation with ACN	Weber et al. (1985)
Ciprofloxacin	PA/TBAH/ACN pH 3	μbondapak C$_{18}$	fluorescence ex. 425 nm em. 254 nm	animal model		Bamberger et al. (1986)
Ciprofloxacin	MeOH/THF/PB pH 3	μbondapak C$_{18}$	fluorescence	serum/urine		Drusano et al. (1986)
Ciprofloxacin	MeOH/PB/HDTA pH 7.4	Nova Pak	UV 313 nm	body fluids	protein precipitation with ACN + methylene chloride extraction; LLS = 0.06 mg/L	LeBel et al. (1986)
Ciprofloxacin Ciprofloxacin	ACN/TBAS PB/TBAH + ACN/MeOH	μbondapak C$_{18}$	fluorescence ex. 278 nm	serum urine	desethyl metabolite assayed dichloromethane extraction	Awni et al. (1987)
Ciprofloxacin	PA/TBAH/ACN pH 3	Versapak C$_{18}$ 10 μm	fluorescence ex. 274 nm em. 418 nm	body fluids	sera diluted with HCl	Dudley et al. (1987)

Compound	Mobile phase/pH	Column	Detection	Matrix	Method/notes	Reference
Ciprofloxacin	ACN/PP pH 2.5	MicroPak MCH 10 μm	fluorescence ex. 330 nm em. 440 nm	serum	protein precipitation with ACN and chloroform metabolites assayed	Myers & Blumer (1987)
Ciprofloxacin	ACN/TBABS	Nucleosil C$_{18}$ 5 μm (heated to 40°C)	fluorescence ex. 278 nm em. 445 nm	body fluids and tissues	all metabolites assayed using a specially constructed post-column photothermal derivatization device	Scholl et al. (1987)
Ciprofloxacin	MeOH/PIC A + PA pH 3	Lichrosorb RP18 5 μm	fluorescence ex. 276 nm em. 441 nm	serum/CSF	proteins precipitated with acetone	Wolff et al. (1987)
Ciprofloxacin	not stated	C$_{18}$	fluorescence ex. 280 nm	biological	metabolites	Parry et al. (1988)
Ciprofloxacin	TBAHS/PA/ACN	Novapak C$_{18}$	fluorescence em. 445 nm	biological samples	protein precipitation with MeOH; metabolites assayed	Kees et al. (1989)
Ciprofloxacin	PA/TBAH/ACN	μBondapak C$_{18}$	fluorescence ex. 280 nm em. 456 nm	saliva/ nasal secretions	same method as Joos et al. (1985) but protein precipitation step omitted	Piercy et al. (1989)
Ciprofloxacin	ACN/PB pH 3	not stated	not stated	serum	extraction with methylene chloride and isopropyl alcohol	Polk et al. (1989)
Ciprofloxacin	PA/ACN/TBAH pH 3	Spherisorb 5 ODS II (heated to 50°C)	fluorescence ex. 310 nm em. 445 nm	body fluids	protein precipitation with methanol	White et al. (1989)
Ciprofloxacin	ACN/PB pH 6.5	Lichrosorb RP-2 5 μm	UV 244 nm	serum	extraction procedure	Anadon et al. (1990)
Ciprofloxacin	MeOH/PP/SLS pH 2.5	Wakosil 5 C$_{18}$ (heated to 40°C)	fluorescence ex. 277 nm em. 445 nm	rat brain and CSF	extraction with choloroform–2–propanol	Katagiri et al. (1990)
Ciprofloxacin		Chemcosorb 5 ODS-H		biological fluids	extraction	Naora et al. (1990)
Ciprofloxacin	ACN/CB pH 4.8	Ultrabase RP8	fluorescence ex. 340 nm em. 440 nm	serum/ bronchial mucosa	extraction; LLS = 25 ng/mL	Fabre et al. (1991)
Ciprofloxacin	ACN/PB/TPAB pH 2.5	LiChrospher RP18	fluorescence ex. 400 nm em. 462 nm	neutrophils	extraction; LLS = 0.01 mg/L	Garraffo et al. (1991)
Ciprofloxacin	PP:NaP/MeOH + HDTA pH 7.4	μbondapak	fluorescence ex. 280 nm em. 425 nm	serum	extraction with chloroform–isopropanol LLS = 0.03 mg/L	Paradis et al. (1992)
Clinafloxacin	ACN/CA/AP/TBAH	Partisil 5 ODS-3	UV 340 nm	serum	protein precipitation with ACN/PA LLS = 0.075 mg/L	Kaatz et al. (1992)
CP-74667	ACN/PB/TBAH pH 3	Novapak C$_{18}$	fluorescence ex. 338 nm em. 425 nm	mouse serum/ urine/tissue	sample preparation by solid phase extraction method metabolites assayed; LLS = 0.02 mg/L (serum), 0.05 mg/L (urine), or 0.1 mg/kg (tissue)	Girard et al. (1992)

Table VII. *Continued*

Quinolone	Mobile phase	Column	Detection	Matrix	Comments	Reference
Difloxacin	ACN/PA	μbondapak C$_{18}$	fluorescence ex. 281 nm em. 418 nm	body fluids	direct injection	Fernandes et al. (1986)
Difloxacin	PB/ACN	Adsorbosphere C$_{18}$ 7 μm	fluorescence ex. 280 nm em. 389 nm UV 280 nm	serum	metabolites assayed sample preparation using ultrafiltration	Granneman et al. (1986)
Enoxacin	MeOH/CA/ACN	Bio-Sil ODS 5S	UV 340 nm	plasma	methylene choride; oxo metabolite assayed	Tseui et al. (1984)
Enoxacin	PA/TBAH/ACN pH 3	Techsil 10 C$_{18}$ (heated to 50°C)	UV 340 nm	bronchial secretions	samples diluted with mobile phase	Fong et al. (1987)
Enoxacin	CA/ACN + MeOH	μbondapak C$_{18}$	UV 278 nm	body fluids	methylene chloride extraction method oxo metabolite assayed; LLS = 0.1 mg/L (enoxacin), 0.02 mg/L (oxo-enoxacin.)	Wise et al. (1987a)
Enoxacin	MeOH/CA/ACN + TEA for urine	Bio-Sil ODS-5S	UV 340 nm	plasma/ urine	direct injection (urine) or methylene chloride extraction (plasma)	Chang et al. (1988)
Enoxacin	CA/MeOH	μbondapak C$_{18}$	UV 254 nm	serum/ prostate	extraction with dichloromethane	White et al. (1988a)
Enoxacin	ACN/CA/TBAH/AP	Partisil ODS 3 5μm	UV 340 nm	serum/ urine	protein precipitation with ACN and PCA	Grasela et al. (1989)
Enoxacin	MeOH/PB	C$_{18}$	UV 340 nm	lung tissue	extraction with methylene chloride	Newsom et al. (1989)
Enoxacin		Cosmosil 5 C$_{18}$	UV 274 nm	rat plasma	protein precipitation with MeOH; LLS = 0.05 mg/L	Hasegawa et al. (1990)
Enoxacin	ACN/CA/AP/TBAH	Partisil 5 ODS	UV 340 nm	serum	protein precipitation with ACN/PCA	Marchbanks et al. (1990)
Enoxacin	CA/AA/ACN	Cosmosil 5 C$_{18}$	fluorescence ex. 330 nm em. 400 nm	plasma	protein precipitation with ACN and dilution with mobile phase	Nadai et al. (1990)
Enoxacin	CA/ACN/TBAH/AP	Lichrospher 100RP C$_{18}$	UV 340 nm	serum/ dialysate	protein with ACN/PCA	Van Der Auwera et al. (1990)
Enrofloxacin	PA/ACN	PLRP-S polymer adsorbent column	UV 289nm fluorescence ex. 278 nm em. 440 nm	fish serum/ tissue		Rogstad et al. (1991)
Fleroxacin	ACN/PA	Nucleosil 5 CN	UV 287 nm	animal body fluids	extraction with chloroform	Kusajima et al. (1986)
Fleroxacin	TBAS/MeOH	Toyo Soda ODS-120-T	fluorescence ex. 290 nm em. 450 nm	body fluids	extraction with choroform and isopropyl alcohol; desmethyl and N-oxide metabolite assayed	Weidekamm et al. (1987)

Compound	Mobile phase	Column	Detection	Matrix	Comments	Reference
Fleroxacin	PDP/TBAH	μbondapak C$_{18}$	fluorescence ex. 290 nm em. 470 nm	serum/urine		Awni et al. (1988)
Fleroxacin	TBAS/MeOH	Toya Soda TSK Gel ODS 20T	fluorescence ex. 290 nm em. 450 and UV 290 nm	plasma urine	extraction method with SDS dichloromethane and isopropanol; N-desmethyl and N-oxide metabolites assayed	Dell et al. (1988)
Fleroxacin	PP/HSA/MeOH/PA	Nova-Pak C$_{18}$	not stated	body fluids	protein precipitation with PCA; N-oxide and desmethyl metabolites assayed	Griggs et al. (1988)
Fleroxacin	CA/ACN/TBAH/AP	Sperisorb ODS-II	fluorescence ex. 290 nm em. 460 nm	serum	protein precipitation with ACN/PCA; N-oxide and desmethyl metabolites assayed	Sorgel et al. (1988)
Fleroxacin	TBAH/MeOH	Toyo Soda ODS-120 T	fluorescence ex. 290 nm em. 450 nm	body fluids	protein precipitation with TCA; N-oxide and desmethyl metabolites assayed	Weidekamm et al. (1988)
Fleroxacin	PB/NaOH	Nova-Pak C$_{18}$	fluorescence em. 425 nm	body fluids	extraction	Panneton et al. (1988)
Fleroxacin	TBAHS/MeOH (serum) or THF/TEAP/OSA (bile); pH 2.8	Toyo-Soda ODS-120 T	fluorescence ex. 290 nm em. 450 nm	serum/urine/ bile	protein precipitation with TCA	Hayton et al. (1990)
Fleroxacin	MeOH/TBAHS	TSK gel ODS-120T	fluorescence ex. 290 nm	serum/urine	SDS/chloroform–isopropanol extraction metabolites assayed in urine	Shiba et al. (1990)
Fleroxacin	MeOH/TBAHS	Toyo-Soda ODS-120T	fluorescence ex. 290 nm em. 450 nm	plasma/ urine	extraction; metabolites assayed	Blouin et al. (1992)
Fleroxacin	PDP/HSA/ MeOH/PA	μbondapak C$_{18}$ (Z-Module)	fluorescence ex. 288 nm em. 475 nm	body fluids	protein precipitation with ACN	Wise et al. (1987b)
Fleroxacin	MeOH/TBAHS	Toyo-Soda ODS-120T	fluorescence ex. 290 nm em. 450 nm	plasma/ breast milk/ urine	protein precipitation with TCA(plasma) or dichloromethane-isopropanol extraction procedure (breast milk)	Dan et al. (1993)
Flumequine	SS/3A pH 9	Ion exchange Zipax SAX	UV 254 nm	biological fluids	chloroform extraction procedure	Harrison et al. (1984)
Flumequine	ACN/PA (gradient) pH 3.5	RP8	fluorescence ex. 320 nm em. 380 nm	biological fluids and meat	chloroform extraction procedure	Ellerbroek & Bruhn (1989)
Flumequine Grepafloxacin	DMF/ACN/PA phosphate buffer/MeOH pH 3	Cp-Sper-C$_8$ Nova-Pak C$_{18}$	UV 320 nm fluorescence ex. 288 nm em. 475 nm	serum biological samples	acyl glucuronide assayed direct injection or dilution	Vree et al. (1992) Woodcock et al. (1994)
Lomefloxacin	ACN/CA/AA	Nucleosil C$_{18}$	fluorescence ex. 280 nm em. 455 nm	plasma/urine	chloroform/isoamyl alcohol extraction procedure	Morrison et al. (1988)

Table VII. *Continued*

Quinolone	Mobile phase	Column	Detection	Matrix	Comments	Reference
Lomefloxacin	PB/ACN/SHS pH 2.5	TSK gel ODS-80T	radioactivity counter	serum	glucuronide assayed	Saito *et al.* (1988)
Lomefloxacin	PA/TBAH/ACN pH 3	Spherisorb 5 ODS II (heated to 40°C)	fluorescence ex. 288 nm em. 475 nm	serum/urine/ tissue fluid	samples diluted with HCl	Stone *et al.* (1988)
Lomefloxacin	ACN/CA/AA	μbondapak C_{18}	fluorescence	serum		Okezaki *et al.* (1989)
Lomefloxacin	not stated	Nucleosil C_{18}	fluorescence	plasma/urine	chloroform/isoamyl alcohol extraction procedure	Gros & Carbon (1990)
Lomefloxacin	ACN/AA/CA	μbondapak C_{18}	fluorescence ex. 280 nm em. 455 nm	plasma	extraction	Hooper *et al.* (1990)
Lomefloxacin	ACN/AA/CA	Nucleosil C_{18}	fluorescence ex. 280 nm em. 445 nm	serum/ prostate	chloroform extraction procedure LLS = 0.01 mg/L	Kovarik *et al.* (1990)
Lomefloxacin	not stated	Novapak C_{18}	fluorescence ex. 280 nm em. 455 nm	plasma	LLS = 0.01 mg/L	LeBel *et al.* (1990)
Lomefloxacin	ACN/PA/TBAH pH 3	Sperisorb 5 ODS II	fluorescence ex. 310 nm em. 445 nm	serum	protein precipitation with MeOH modification of the ciprofloxacin assay of Gau *et al.* (1985)	Cowling *et al.* (1991)
Lomefloxacin	ACN/Aa/CA	Nucleosil C_{18}	fluorescence ex. 280 nm em. 418 nm	serum	chloroform–isoamylalcohol extraction procedure	Healy *et al.* (1991)
Lomefloxacin	not stated	Nucleosil C_{18}	fluorescence ex. 280 nm em. 455 nm	serum/urine/ saliva/ faeces	chloroform–isoamylalcohol extraction procedure; LLS = 0.01 mg/L (serum, saliva and urine); 0.05 mg/L (faeces)	Leigh *et al.* (1991)
Miloxacin	CA/SNB pH 5	ion exchange Zipax SAX	UV 330 nm	serum/urine	ethyl acetate extraction procedure; metabolites assayed	Yoshitake *et al.* (1980)
Moxifloxacin	ACN/TBAP	Techsphere 5C_8 at 50°C	fluorescence ex. 290 nm em. 500 nm	serum	protein precipitation with MeOH	Sunderland *et al.* (1997)
Nalidixic acid	SS/BA	ion exchange Zipax SAX	not stated	plasma	chloroform extraction hydroxy metabolite assayed	Shargel *et al.* (1973)
Norfloxacin	ACN/PB pH 7	ion exchange Vydac column	UV 273 nm	serum/ urine	methylene chloride extraction procedure metabolites assayed	Boppana & Swanson (1982)
Norfloxacin	ACN/PB	Lichrosorb RP8	UV 282 nm	serum/ urine	solid phase extraction (Baker 10 C_{18})	Eandi *et al.* (1983)

162

Drug	Mobile phase	Column	Detection	Matrix	Notes	Reference
Norfloxacin	DCM/ACN/MeOH Amm/water	Lichrosorb Si 100	fluorescence ex. 300 nm em. 420 nm	serum/plasma/urine	protein precipitation with TCA and dichloromethane	Shimada et al. (1983)
Norfloxacin	ACN/PB pH 7	ion exchange Vydac + AXGU column	UV 273 nm	body fluids	methylene chloride extraction procedure; metabolites assayed	Forchetti et al. (1984)
Norfloxacin	ACN/MeOH/PB pH 2.5	Partisil PXS C_8	fluorescence ex. 300 nm em. 420 nm	serum/tissues	solid phase extraction (Carbopak B); LLS = 0.001 mg/L	Lagana et al. (1988)
Norfloxacin	ACN/TBAH pH 3	Spherisorb 5 ODS II	fluorescence ex. 310 nm em. 445 nm	serum/urine	samples diluted; metabolites assayed	MacGowan et al. (1988)
Norfloxacin	gradient	Apex II C_{18}	not stated	urine	solid phase extraction procedure (Supelco LC-18)	Davis et al. (1989)
Norfloxacin	ACN/CA/AP/TBAH	Partisil Rac II	UV 280 nm	plasma/urine	protein precipitation with PCA/ACN	Nix et al. (1990)
Ofloxacin	ACN/PB pH 7	ion exchange Vydac	UV 280 nm	body fluids	extraction procedure	Carlucci et al. (1986)
Ofloxacin	Hx/DCE/EtOH	Sumipax OA-4200	not stated		method for isomer synthesis	Hayakawa et al. (1986)
Ofloxacin	TBAP/ACN pH 2	Nucleosil $C_{:8}$	fluorescence ex. 295 nm em. 418 nm	body fluids	N-oxide and desmethyl metabolites assayed	Lode et al. (1987)
Ofloxacin	CA/MeOH	µbondapak C_{18}	fluorescence ex. 310 nm em. 467 nm	serum	N-oxide and desmethyl metabolites assayed	White et al. (1987)
Ofloxacin	PA/TEAH/ACN pH 1.85	Nucleosil 120-5 C_{18}	fluorescence ex. 298 nm em. 458 nm	biological fluids	for isomer assay	Lehr & Damm (1988)
Ofloxacin	PP,TBAP/ACN	µbondapak C_{18}	fluorescence ex. 290 nm em. 500 nm	plasma/urine	dichloromethane extraction, urine diluted with water	Mignot et al. (1988)
Ofloxacin	ACN/PA	Hypersil 5 ODS	fluorescence ex. 370 nm em. 400–700 nm		chloroform extraction	Notarianni & Jones (1988)
Ofloxacin	CA/MeOH	µbondapak	fluorescence ex. 310 nm em. 489 nm	serum	sera diluted with HCl	White et al. (1988b)
Ofloxacin	PA/TMAC/DMF/ACN	Spherisorb 5 ODS	UV 294 nm	lung tissue		Wijnands et al. (1988)
Ofloxacin	ACN/PB/TBAC pH 2.2	Nucleosil C_{18}	fluorescence ex. 295 nm em. 475 nm	serum/CSF	methylene chloride extraction procedure, direct injection of CSF LLS = 0.5 mg/L	Bitar et al. (1989)

Table VII. *Continued*

Quinolone	Mobile phase	Column	Detection	Matrix	Comments	Reference
Ofloxacin	ACN/Pic A/PP/FA pH 2.5	Lichrosorb RP 18	fluorescence ex. 290 nm em. 500 nm	serum/CSF	methylene chloride extraction	Pioget et al. (1989)
Ofloxacin			UV	plasma/blister fluid	methylene chloride extraction procedure	Sanchez-Navarro et al. (1990)
Ofloxacin	ACN/PA/SLS	Ultrasphere ODS	fluorescence ex. 358 nm em. 495 nm	tears		Richman et al. (1990)
Ofloxacin	ACN/TEA/PA pH 2.6	NovaPak C$_{18}$	fluorescence ex. 295 nm em. 505 nm	serum/saliva/blister fluid	protein precipitation with PCA	Warlich et al. (1990)
Ofloxacin	PP/HSA/MeOH/PA pH 3	µbondapak C$_{18}$	UV 313 nm	serum	dicholormethane extraction	Flor et al. (1990)
Ofloxacin	TBA/PP/ACN	Licrosorb RP 18	UV	serum	LLS = 0.04 mg/L	Martin et al. (1991)
Ofloxacin	CA/MeOH	µbondapak C$_{18}$	fluorescence ex. 313 nm em. 495 nm	serum/bronchial tissue	serum sample and sonicated bronchial tissue diluted with HCl, LLS = 0.06 mg/L protein precipitated with PCA metabolites assayed	Davey et al. (1991)
Ofloxacin	MeOH/ACN/CA	Licrosorb RP 18	UV 340 nm	serum/	LLS = 0.1 mg/L	Yuk et al. (1991)
Ofloxacin	PP/HSA/MeOH/PA, pH 3	µbondapak C$_{18}$	UV 313 nm	plasma	dicholormethane extraction, LLS = 0.01 mg/L	Guay et al. (1992)
Ofloxacin	ACN/TBAB/PA pH 2	C$_{18}$	fluorescence ex. 277 nm em. 445 nm	serum/heart tissue/fat/bone marrow	dicholormethane extraction (serum and tissues)	Mertes et al. (1992)
Oxolinic acid	MeOH/PB	NovaPak C$_{18}$	UV 258 nm	fish serum	solid phase extraction (SepPak Accell) LLS = 1 µg/L	Hustvedt et al. (1989)
Oxolinic acid	ACN/MeOH/OA	C$_{18}$	fluorescence ex. 327 nm em. 369 nm	fish muscle	ethyl acetate extraction	Carignan et al. (1991)
Oxolinic acid	ACN/MeOH/OA	Partsil ODS-3	fluorescence ex. 327 nm em. 369 nm	fish muscle	ethyl acetate extraction	Larocque et al. (1991)
Pefloxacin	gradient	Lichrosorb RP 18	UV 270 nm	body fluids	chloroform extraction; metabolites assayed	Montay et al. (1983)
Pefloxacin	gradient	Lichrosorb RP 18	UV 270 nm	urine	XAD2 resin or chloroform extraction procedure metabolites assayed	Montay et al. (1984)

Drug	Mobile phase	Column	Detection	Matrix	Comments	Reference
Pefloxacin	ACN/SAT/CA/TEA	Nucleosil C$_{18}$ 10 μm	fluorescence ex. 330 nm em. 440 nm	plasma/tissue	chloroform/isopentanol extraction, metabolites assayed	Montay & Tassel (1985)
Pefloxacin	ACN/CA pH 3	cation exchange	fluorescence ex. 279 nm em. 440 nm	plasma/dialysate	protein precipitation with methanol, metabolites assayed	Rose et al. (1990)
Pefloxacin	ACN/SAT/CA/TEA/FA	Nucleosil C$_8$	UV 280 nm	rabbit serum aqueous and vitreous humour	extraction procedure; metabolites assayed	Cochereau-Massin et al. (1991)
Pefloxacin	ACN/CA/SAT/TEA	Novapak C$_{18}$	fluorescence ex. 330 nm em. 418 nm	plasma/dialysate	protein precipitation with ACN, metabolites assayed LLS = 0.03 mg/L	Schmit et al. (1991)
Pefloxacin	ACN/CA/SA/TEA pH 4.8	Ultrabase	fluorescence ex. 340 nm em. 440 nm	plasma/gynaecological tissue	extraction procedure LLS = 25 μg/L	Bouvet et al. (1992)
Pipemidic acid	SDPD/MeOH/ACN	μbondapak C$_{18}$	UV 280 nm	serum/urine	urine diluted with mobile phase, sera diluted with acetonitrile	Smethurst & Mann (1983)
Pipemidic acid	MeOH/PB pH 8.2	μbondapak C$_{18}$	not stated	biological fluids	protein precipitation with PCA and chloroform extraction procedure	Klinge et al. (1984)
Pipemidic acid	PCA/MeOH/CF	Lichrosorb Si60	UV 265 nm	plasma	acidification with HCl followed by chloroform extraction procedure	Ito et al. (1985)
Pipemidic acid	ACA/ACN	C$_{18}$ 5 μm	fluorescence ex. 277 nm em. 440 nm	biological fluids	protein precipitation with methanol	Fukuhara & Matsuki (1987)
Rosoxacin	ACN/PA	ion-exchange Partisil-PXS 10/25 PAC	UV 280 nm	plasma/urine	chloroform extraction N-oxide metabolite assayed	Kullberg et al. (1979)
Rufloxacin	PB/ACN pH 7	anion exchange Vydac	UV 296 nm	serum/urine	extraction procedure LLS = 0.1 mg/L (serum) or 0.05 mg/L (urine)	Carlucci & Palumbo (1991)
Rufloxacin	PA/ACN/THF/TEA, pH 5	PRP1 column	UV 300 nm	plasma/urine	protein precipitation with PCA (plasma), or dichloromethane extraction (urine)	Imbimbo et al. (1991)
Rufloxacin	PA/ACN/THF/TEA, pH 5	Ultrasphere 5	fluorescence ex. 360 nm em. 470 nm	plasma/urine	protein precipitation with PCA or dichloromethane extraction procedure (urine), metabolites assayed LLS = 0.05 mg/L (plasma and urine)	Kisicki et al. (1992)
Rufloxacin	PA/ACN/TEA pF 5.6	Hamiton PRP-1	fluorescence ex. 350 nm em. 510 nm	biological fluids	extraction procedure metabolites assayed	Lombardi et al. (1992)
Sparfloxacin	ACN/CA/TBAH/AP	Partisil ODS-3	UV 380 nm	guinea pig serum/tissue	protein precipitation with ACN/PCA	Edelstein et al. (1990)
Sparfloxacin	SA/TBAHS pH 5	Cosmosil 5 C$_{18}$ (heated to 35°C)	UV 278 nm	urine	(theophylline also assayed using the same method)	Takagi et al. (1991)

Table VII. *Continued*

Quinolone	Mobile phase	Column	Detection	Matrix	Comments	Reference
Sparfloxacin	ACN/PA/NaOH pH 3.82	Nucleosil 100 5SA	fluorescence ex. 295 nm em. 525 nm	serum/urine		Borner et al. (1992)
Sparfloxacin	MEOH/ACN/AA	AsahiPAK OD1	UV 346 nm	serum/urine	extraction procedure; glucuronide also assayed	Fillastre et al. (1994)
Temafloxacin	ACN/PA/ SDPD/SDS/AHAA	Adsorbosphere HS C$_{18}$	fluorescence ex. 280 nm em. 389 nm	serum/urine/ dialysate	extraction procedure; LLS = 0.01 mg/L	Granneman et al. (1991a)
Temafloxacin	ACN/PA/ SDPD/SDS/AHAA	Adsorbosphere HS C$_{18}$	UV and fluorescence ex. 280 nm em. 389 nm	plasma/urine	protein removal by ultrafiltration	Granneman et al. (1991b)
Temafloxacin	ACN/PA/ SDPD/SDS/AHHA	Adsorbosphere HS C$_{18}$	fluorescence ex. 280 nm em. 460 nm	plasma/ urine	extraction procedure; LLS = 0.01 mg/L (plasma) or 1 mg/L (urine)	Granneman et al. (1992)
Tosufloxacin see Table VIII						
Trovafloxacin	PB ACN/TBAP/ dibutylamine phosphate	C$_{18}$	UV 275 nm	biological samples	solid-phase extraction; LLS = 0.1 mg/L	Teng et al. (1996)

AA, ammonium acetate; ACA, acetic acid; ACN, acetonitrile; AHAA, acetylhydroxamic acid; Amm, ammonia; AP, ammonium perchlorate; BA, boric acid; CA, citric acid; CB, citrate buffer; CF, chloroform; DCE, dichloroethane; DCM, dichloromethane; DDS, dodecanesulphonate; DMF, dimethylformamide; EtOH, ethanol; em, emission; ex, excitation; FA, formic acid; HCl, hydrochloric acid; HDTA, hexadecyltrimethylammonmium; HSA, heptane sulphonic acid; Hx, hexane; MeOH, methanol; NaC, sodium carbonate; NaCl, sodium chloride; SDPD. sodium dihydrogen phosphate; dihydrate; SHS, sodium hydrogen sulphate; SLS, sodium lauryl sulphate; NaOH, sodium hydroxide; NaP, sodium phosphate; NaS, sodium sulphate; OSA, octane sulphonic acid; OX, oxalic acid; PA, phosphoric acid; PB, phosphate buffer; PCA, perchloric acid; PDP, potassium dihydrogen phosphate; Pic A, Waters paired-ion reagent; PP, potassium phosphate; SAT, sodium acetate trihydrate; SB, sulphate buffer; SDS, sodium dodecyl sulphate; SNB, sodium nitrate buffer; SPS, sodium propane-2-sulphonate; SS, sodium sulphate; TB, tetraborate buffer; TBAB, tetrabutylammonium bromide; TBABS, tetrabutylammonium bisulphate; TBAH, tetrabutylammonium hydroxide; TBAHS, tetrabutylammonium hydrogen sulphate; TBAI, tetrabutylammonium iodide; TBAP, tetrabutylammonium phosphate; TBAS, tetrabutylammonium sulphate; TCA, trichloracetic acid; TEA, triethylamine; TEAP, triethylamine phosphate buffer; THF, tetrahydrofuran; TMAC, tetramethylammonium chloride; TPAB, tetraphenylammonium bromide.

Table VIII. Some published HPLC methods for assay of mixtures of 4-quinolones and their metabolites in biological materials

Mixture	Mobile phase	Column	Detection	Matrix	Comments	Reference
Cinoxacin, nalidixic acid, oxolinic acid	TB/SB pH 9.2	ion exchange Zipax SAX	UV 254 nm	dosage form	order of elution: oxolinic acid; nalidixic acid; cinoxacin.	Sondack & Koch (1977)
Cinoxacin, nalidixic acid (hydroxy metabolite); oxolinic acid; pipemidic acid	ACN/oxalic acid	Nova-Pak C_{18}	UV 265 nm fluorescence ex. 260/270 nm em. 380/440 nm	urine		Duran Meras et al. (1997)
Norfloxacin, pefloxacin	ACN/SAT/CA/TEA	Nucleosil C_{18}	fluorescence ex. 330 nm em. 440 nm	plasma/tissue	chloroform/isopentanol extraction procedure; metabolites assayed order of elution: norfloxacin; pefloxacin; N-oxide; oxo-norfloxacin	Montay & Tassel (1985)
Ciprofloxacin, norfloxacin, pefloxacin, ofloxacin	PA/TBAI	Nucleosil C_{18}	UV 278 nm or UV 294 nm	serum/tissue	dichloromethane extraction procedure	Groeneveld & Brouwers (1986)
Ciprofloxacin, norfloxacin	PA/TBAH/ACN/ pH 3	μbondapak C_{18}	fluorescence ex. 278 nm em. 456 nm		protein precipitation with methanol	Morton et al. (1986)
Nalidixic acid, oxolinic acid, piromidic acid	ACN/MeOH/CA or TBAB	Nucleosil C_8	UV 254 nm	fish	extraction prcedure	Horie et al. (1987)
Nalidixic acid, oxolinic acid, piromidic acid	ACN/PB	Kaseisorb LC ODS 300-5	fluorescence ex. 325 nm em. 365 nm or UV 280 nm[a]	fish tissue	methanol/phosphoric acid extraction procedure	Horie et al. (1987)
Ciprofloxacin, enoxacin, ofloxacin, norfloxacin, pefloxacin, lomefloxacin	MeOH/ACN/CA or ACN/CA[b]	Lichrosorb RP18	UV 340 nm UV 275 nm	clinical specimens	protein precipitation metabolites assayed[b]	Chan et al. (1989)
A-64730, fleroxacin, temafloxacin	TBAB/PB/ACN	Ultrasphere	fluorescence ex. 265–277 nm em. 433–450 nm	biological fluids	two-step extraction procedure, LLS = 2.5–20 μg/L	Koechlin et al. (1989)
Ciprofloxacin, enoxacin, fleroxacin, lomefloxacin, norfloxacin, ofloxacin, pefloxacin	PDP/HSA/PA	μbcndapak C_{18} or Novapak C_{18}	fluorescence or UV	serum	metabolites assayed	Griggs & Wise (1989)
Enrofloxacin, ciprofloxacin	ACN/MeOH/DDS/ OSA/PA/TEA	Lichrosorb RP18	fluorescence ex. 280 nm em. 360 nm	canine serum/ tissue	protein precipitation with acetonitrile and ultrafiltration; LLS = 4 μg/L (enrofloxacin), 2 μg/L (ciprofloxacin)	Tyczkowska et al. (1989)
Clinafloxacin, fleroxacin, levofloxacin, sparfloxacin, tosufloxacin	MeOH/ACN/CA	Radial Pak 4 μm C_{18}	UV 275 nm or 349 nm (fleroxacin)	serum	protein precipitation LLS = 0.1 mg/L	Lyon et al. (1994)

AA, ammonium acetate; ACA, acetic acid; ACN, acetonitrile; AHAA, acetylhydroxamic acid; Amm, ammonia; AP, ammonium perchlorate; BA, boric acid; CA, citric acid; CB, citrate buffer; CF, chloroform; DCE, dichloroethane; DCM, dichloromethane; DDS, dodecanesulphonate; DMF, dimethylformamide; EtOH, ethanol; ex, excitation; em, emission; FA, formic acid; HCl, hydrochloric acid; HDTA, hexadecyltrimethylammonium; HSA, heptane sulphonic acid; Hx, hexane; MeOH, methanol; NaC, sodium chloride; SDPD, sodium dihydrogen phosphate; dihydrate; SHS, sodium hydrogen sulphate; SLS, sodium lauryl sulphate; NaOH, sodium hydroxide; NaP, sodium phosphate; NaS, sodium sulphate; OSA, octane sulphonic acid; OX, oxalic acid; PA, phosphoric acid; PB, phosphate buffer; PCA, perchloric acid; PDP, potassium dihydrogen phosphate; Pic A, Waters paired-ion reagent; PP, potassium phosphate; SAT, sodium acetate trihydrate; SB, sulphate buffer; SDS, sodium dodecyl sulphate; SNB, sodium nitrate buffer; SPS, sodium propane-2-sulphonate; SS, sodium sulphate; TB, tetraborate buffer; TBAB, tetrabutylammonium bromide; TBABS, tetrabutylammonium bisulphate; TBAH, tetrabutylammonium hydroxide; TBAHS, tetrabutylammonium hydrogen sulphate; TBAI, tetrabutylammonium iodide; TBAP, tetrabutylammonium phosphate; TBAS, tetrabutylammonium sulphate; TCA, trichloracetic acid; TEA, triethylamine; TEAP, triethylamine phosphate buffer; THF, tetrahydrofuran; TMAC, tetramethylammonium chloride; TPAB, tetraphenylammonium bromide. [a]For piromidic acid. [b]For norfloxacin and pefloxacin.

metabolite gives a detector response and co-elutes with either parent drug (causing over-estimation) or internal standard (causing under-estimation). With fluorescence detection only fluorescent metabolites will be detected. All the metabolites of ofloxacin are fluorescent, whereas most of the metabolites of ciprofloxacin (with the exception of desethyl-ciprofloxacin, which is considerably more fluorescent than ciprofloxacin) have poor fluorescent emission (Table IX).

Assaying metabolites in addition to the parent drug presents a bigger problem than assaying the parent alone. Since there may be more than one metabolite, they must be resolved from the parent and from one another, and they may be present in very low concentration. With non-fluorescent metabolites, obtaining sufficient sensitivity to assay serum levels may be onerous if UV absorbance is the only detection method available. Scholl *et al.* (1987) devised a post-column photothermal derivatization technique to render the metabolites of ciprofloxacin fluorescent, enabling them to be readily assayed in low concentrations. The technique could be adapted for other quinolones but requires highly specialized custom-built post-column derivatization equipment.

HPLC with fluorescence detection can be recommended for monitoring ciprofloxacin, ofloxacin, norfloxacin or pefloxacin. The technique is rapid, sensitive and highly specific. For enoxacin solvent extraction and HPLC with UV detection is recommended. There is currently no clinical indication for assaying metabolites. The interested reader should consult the following articles: ciprofloxacin metabolites (Borner *et al.*, 1986; Krol *et al.*, 1986; Scholl *et al.*, 1987); norfloxacin metabolites (Boppana & Swanson, 1982); ofloxacin metabolites (Lode *et al.*, 1987; White *et al.*, 1987; Korner *et al.*, 1994; Nau *et al.*, 1994); enoxacin (Tsuei *et al.*, 1984); pefloxacin (Montay *et al.*, 1983; Montay & Tassel, 1985).

Table IX. Relative fluorescence data for quinolone metabolites

Metabolite	% Fluorescence relative to parent drug
Desethyl-ciprofloxacin	1800
Sulpho-ciprofloxacin	0.64
Oxo-ciprofloxacin	24
Formyl-ciprofloxacin	9
Desmethyl ofloxacin	85
Ofloxacin *N*-oxide	107

For each metabolite the fluorescence of the parent compound was taken as 100%.
Data for ofloxacin metabolites from White *et al.* (1987) (emission 467 nm); data for ciprofloxacin metabolites from Scholl *et al.* (1987) (emission 445 nm).

Spectrophotometric assay of nalidixic acid (Andrews & Nicol, 1978)

This assay is based on the fact that in aqueous solution nalidixic acid and its hydroxy metabolite both absorb UV strongly at 336 nm but hardly at all at 370 nm. The difference in absorbance at these wavelengths gives a highly specific measure of the total amount of the two compounds.

Procedure for serum or urine. Note: values for urine are given in brackets.

1. Mix 2 (1) mL of sample with 4 (5) mL of 0.1 M HCl.
2. Add 50 (50) mL toluene and shake well for 10–15 (2–3) min.
3. Centrifuge 1000**g** for 20 min.
4. Remove the aqueous phase and transfer 40 (40) mL of the toluene to a graduated cylinder.
5. Add 4 mL of 0.4 M borate buffer pH 9.2; shake for 2–3 min.
6. Remove the toluene layer by aspiration, transfer the aqueous phase to a spectrophotometer cuvette and read the absorbance at 336 nm and 370 nm.

The original method quotes the net absorbance of a 1 mg/L solution of nalidixic acid in borate buffer as 0.049 and suggests calculating the results for the unknowns from this after: (i) deducting the net absorbance of a drug-free blank; (ii) allowing for only 83% recovery from the extraction; and (iii) allowing for any dilutions made. We, however, would recommend the processing of calibrators in the usual manner.

HPLC assay of ciprofloxacin, norfloxacin or lomefloxacin

This assay is essentially that of Gau *et al.* (1985) modified slightly (White *et al.*, 1985). Note that it is not suitable for the assay of metabolites since many are not fluorescent (see below). None of the metabolites of ciprofloxacin, norfloxacin or lomefloxacin interfere. This assay has been used for over 10 years for monitoring ciprofloxacin in patients (White *et al.*, 1989).

Column. Spherisorb ODS II 5 μm particle size (4 mm × 25 cm) maintained at 50°C.

Mobile phase. (a) acetonitrile and (b) 0.025 M phosphoric acid in deionized or distilled water adjusted to pH 3.0 with tetrabutylammonium hydroxide solution (40% w/w, Sigma), 50:1000 v/v (a:b). Flow rate: 1 or 2 mL/min.

Detection. Fluorescence detection, with excitation at 287–310 nm and emission at 445 nm.

Sample preparation. In the original method sera are mixed with an equal volume of 0.16 M HCl, centrifuged and the supernatant injected. It was found that if plasma was inadvertently handled the same way the column rapidly developed high back-pressure and became unusable (L. O. White, unpublished observations). Because of this the procedure has been replaced by protein precipitation with methanol since it eliminates the problem associated with plasma.

Mix samples with an equal volume of methanol, allow to stand at room temperature for 10–15 min, then centrifuge and inject 20 μL of the clear supernatant.

Additional information. Liquefy sputum samples by adding dithiothreitol powder to give a final concentration of approximately 0.05% and then treat the samples like sera. Dilute bile samples 1/20 with mobile phase.

Since quinolones bind strongly to faeces the following extraction procedure is recommended (Scholl *et al.*, 1987). Mix 100 mg of faeces with 500 μL of dichloromethane: propan-2-ol:0.33 M phosphoric acid (1:5:4 by volume) and shake for 60 min. Inject 20 μL of the supernatant after centrifugation. Failure to use an extraction solution results in poor recovery from faeces (White *et al.*, 1989)

Run time. The retention time for quinolones is less than 10 min but the run time between samples should be at least 20 min because of the possibility of late eluting peaks. Bile and CAPD fluids are a particular problem and run times of up to 3 h may be required to elute all the late-running peaks.

Performance. Single point calibration with 5 mg/L is suitable for monitoring. The lower limit of sensitivity is approximately 0.05 mg/L but this is very dependent on the signal-to-noise ratio of the fluorescence detector.

HPLC assay of ofloxacin and its metabolites

This is the method of White *et al.* (1987). It is is preferred to that of Gau *et al.* (1985) for ofloxacin assay since ofloxacin co-elutes with desmethyl ofloxacin in the latter. The assay should also be suitable for levofloxacin.

Column. Microbondapak C$_{18}$ (Waters) (a radially compressed column was used in the original work but a steel column should perform satisfactorily). As an alternative Techsphere C$_{18}$ (HPLC Technology, Macclesfield, UK) may be used).

Mobile phase. 0.1 M citric acid:methanol (75:25). Flow rate: 2 mL/min.

Detection. Fluorescence detection, excitation at 310 nm and emission at 467 nm or 489 nm; the latter gives approximately twice the sensitivity of the former.

Sample preparation. Mix sera with an equal volume of 0.16 M HCl, centrifuge and inject from the supernatant. Note that no protein precipitation occurs.

Run time. Typical retention times are as follows: desmethyl ofloxacin, 321 s; ofloxacin, 438 s; ofloxacin *N*-oxide, 624 s.

Performance. Multi- or single-point calibration (5 or 10 mg/L) may be used. Calibrators may be prepared in serum. The lower limit of sensitivity is approximately 0.1 mg/L but this is very dependent on the signal-to-noise ratio of the fluorescence detector.

HPLC assay of enoxacin and oxo-enoxacin

This is based on the method of White *et al.* (1988a). Enoxacin is only poorly fluorescent so UV detection coupled with an extraction procedure is used. An internal standard and multi-point calibration are recommended for optimum accuracy and reproducibility. The assay is a simplified version of the assay of Tsuei *et al.* (1984) in which enoxacin and the metabolite were converted to their ethyl carbamate derivatives during the extraction. In the assay described here the drugs are not converted to ethyl carbamates.

Column. Microbondapak C$_{18}$ (a radially compressed column was used in the original work).

Mobile phase. 0.1 M citric acid:methanol (15–25% methanol depending on the column). Flow rate 2 mL/min.

Detection. UV absorption at 254 nm.

Sample preparation. Mix serum (0.5 mL) in a glass stoppered tube with 125 μL of internal standard (50 mg/L ciprofloxacin), 125 μL 0.2 M phosphate buffer pH 7.5 and 6.0 mL dichloromethane. After gentle mixing for 10 min remove 3 mL of the organic layer and dry under nitrogen at 30°C. Reconstitute the residue in 75 μL mobile phase and inject 30 μL.

The method can also be used for tissue samples, they should be chopped and digested with an equal weight of 2.5 mg of collagenase plus 0.1 mg of trypsin in 10 mL phosphate buffer for 3 h at 37°C and then extracted as described above.

Run time. Typical retention times are as follows: enoxacin, 204 s; ciprofloxacin, 228 s; oxo-enoxacin, 888 s. **Note**: In the method of Tsuei *et al.* (1984) oxo-enoxacin eluted *before* enoxacin.

Performance. Calibrators (0, 1, 2, 4, 6 and 8 mg/L in serum) should be processed in exactly the same way as the tests. Lower limit of sensitivity: approximately 0.2 mg/L.

References

Anadon, A., Martinez-Larranaga, M. R., Fernandez, M. C., Diaz, M. J. & Bringas, P. (1990). Effect of ciprofloxacin on antipyrine pharmacokinetics and metabolism in rats. *Antimicrobial Agents and Chemotherapy* **34**, 2148–51.

Andrews, R. S. & Nicol, C. G. (1978). In *Laboratory Methods in Antimicrobial Chemotherapy*, (Reeves, D. S., Phillips, I., Williams, J. D., Wise, R., Eds), pp. 254–64. Churchill Livingstone, Edinburgh.

Andriole, V. T. (1988). *The Quinolones.* Academic Press, London.

Awni, W. M., Clarkson, J. & Guay, D. R. P. (1987). Determination of ciprofloxacin and its 7-ethylenediamine metabolite in human serum and urine by high-performance liquid chromatography. *Journal of Chromatography* **419**, 414–20.

Awni, W. M., Maloney, J. A. & Heim-Duthoy, K. L. (1988). Liquid-chromatographic determination of fleroxacin in serum and urine. *Clinical Chemistry* **34**, 2330–2.

Bamberger, D. M., Peterson, L. R., Gerding, D. M., Moody, J. A. & Fasching, C. E. (1986). Ciprofloxacin, azlocillin, ceftizoxime and amikacin alone and in combination against Gram-negative bacilli in an infected chamber model. *Journal of Antimicrobial Chemotherapy* **18**, 51–63.

Bergan, T. (1986). Quinolones. *Antimicrobial Agents Annual* **1**, 164–78.

Bitar, N., Claes, R. & Van der Auwera, P. (1989). Concentrations of ofloxacin in serum and cerebrospinal fluid of patients without meningitis receiving the drug intravenously and orally. *Antimicrobial Agents and Chemotherapy* **33**, 1686–90.

Blouin, R. A., Hamelin, B. A., Smith, D. A., Foster, T. S., John, W. J. & Welker, H. A. (1992). Fleroxacin pharmacokinetics in patients with liver cirrhosis. *Antimicrobial Agents and Chemotherapy* **36**, 632–8.

Boppana, V. K. & Swanson, B. N. (1982). Determination of norfloxacin, a new nalidixic acid analog, in human serum and urine by high-performance liquid chromatography. *Antimicrobial Agents and Chemotherapy* **21**, 808–10.

Borner, K., Lode, H., Hoffken, G., Prinzing, C., Glatzel, P. & Wiley, R. (1986). Liquid chromatographic determination of ciprofloxacin and some metabolites in human body fluids. *Journal of Clinical Chemistry and Clinical Biochemistry* **24**, 325–31.

Borner, K., Borner, E. & Lode, H. (1992). Determination of sparfloxacin in serum and urine by high-performance liquid chromatography. *Journal of Chromatography* **579**, 285–9.

Bouvet, O., Bressolle, F., Courtieu, C. & Galtier, M. (1992). Penetration of pefloxacin into gynaecological tissues. *Journal of Antimicrobial Chemotherapy* **29**, 579–87.

Brogard, J. M., Jehl, F., Monteil, H., Adloff, M., Blickle, J. F. & Levy, P. (1985). Comparison of high-pressure liquid chromatography and microbiological assay for the determination of biliary elimination of ciprofloxacin in humans. *Antimicrobial Agents and Chemotherapy* **28**, 311–4.

Brown, E. M. & Reeves, D. S. (1997). Quinolones. In *Antibiotic and Chemotherapy*, 7th edn (O'Grady, F., Lambert, H. P., Fince, R. G. & Greenwood, D., Eds), pp. 419–52. Churchill Livingstone, Edinburgh.

Burdge, D. R., Nakielna, E. M. & Rabin, H. R. (1995). Photosensitivity associated with ciprofloxacin use in adult patients with cystic fibrosis. *Antimicrobial Agents and Chemotherapy* **39**, 793.

Carignan, G., Larocque, L. & Sved, S. (1991). Assay of oxolinic acid residues in salmon muscle by liquid chromatography with fluorescence detection: interlaboratory study. *Journal – Association of Official Analytical Chemists* **74**, 906–9.

Carlucci, G. & Palumbo, G. (1991). Analytical procedure for the determination of rufloxacin, a new pyridobenzothiazine, in human serum and urine by high-performance liquid chromatography. *Journal of Chromatography (Biomedical Applications)* **564**, 346–51.

Carlucci, G., Guadagni, S. & Palumbo, G. (1986). Determination of ofloxacin, a new oxazine derivative in human serum, urine and bile, by high-performance liquid chromatography. *Journal of Liquid Chromatography* **9**, 2539–47.

Chan, C. Y., Lam, A. W. & French, G. L. (1989). Rapid high performance liquid chromatographic assay of fluoroquinolones in clinical specimens. *Journal of Antimicrobial Chemotherapy* **23**, 597–604.

Chang, T., Black, A., Dunky, A., Wolf, R., Sedman, A., Latts, J. *et al.* (1988). Pharmacokinetics of intravenous and oral enoxacin in healthy volunteers. *Journal of Antimicrobial Chemotherapy* **21**, Suppl. B, 49–56.

Chapman, J. S. & Georgopapadakou, N. H. (1989). Fluorometric assay for fleroxacin uptake by bacterial cells. *Antimicrobial Agents and Chemotherapy* **33**, 27–9.

Cochereau-Massin, I., Bauchet, J., Faurisson, F., Vallois, J. M., Lacombe, P. & Pocidalo, J. J. (1991). Ocular kinetics of pefloxacin after intramuscular administration in albino and pigmented rabbits. *Antimicrobial Agents and Chemotherapy* **35**, 1112–5.

Cowling, P., Rogers, S., McMullin, C. M., White, L. O., Lovering, A. M., MacGowan, A. P. *et al.* (1991). The pharmacokinetics of lomefloxacin in elderly patients with urinary tract infection following daily dosing with 400 mg. *Journal of Antimicrobial Chemotherapy* **28**, 101–7.

Crumplin, G. C., Kenwright, M. & Hirst, T. (1984). Investigations into the mechanism of action of the antibacterial agent norfloxacin. *Journal of Antimicrobial Chemotherapy* **13**, Suppl. B, 9–23.

Cutler, N. R., Vincent, J., Jhee, S. S., Teng, R., Wardle, T., Lucas, G. *et al.* (1997). Penetration of trovafloxacin into cerebrospinal fluid in humans following intravenous infusion of alatrofloxacin. *Antimicrobial Agents and Chemotherapy* **41**, 1298–300.

Dan, M., Weidekamm, E., Sagiv, R., Portmann, R. & Zakut, H. (1993). Penetration of fleroxacin into breast milk and pharmacokinetics in lactating women. *Antimicrobial Agents and Chemotherapy* **37**, 293–6.

Davey, P. G., Precious, E. & Winter, J. (1991). Bronchial penetration of ofloxacin after single and multiple oral dosage. *Journal of Antimicrobial Chemotherapy* **27**, 335–41.

Davis, R. L., Kelly, H. W., Quenzer, R. W., Standefer, J., Steinberg, B. & Gallegos, J. (1989). Effect of norfloxacin on theophylline metabolism. *Antimicrobial Agents and Chemotherapy* **33**, 212–4.

Dell, D., Partos, C. & Portman, R. (1988). The determination of a new trifluorinated quinolone, fleroxacin, its *N*-demethyl, and *N*-oxide metabolites in plasma and urine by high performance liquid chromatography with fluorescence detection. *Journal of Liquid Chromatography* **11**, 1299–312.

Dorfler, A., Schulz, W., Burkhardt, F. & Zichner, M. (1987). Pharmacokinetics of ofloxacin in patients on haemodialysis treatment. *Drugs* **34**, Suppl. 1, 62–70.

Drusano, G. L., Plaisance, K. I., Forrest, A. & Standiford, H. C. (1986). Dose ranging study and constant infusion evaluation of ciprofloxacin. *Antimicrobial Agents and Chemotherapy* **30**, 440–3.

Dudley, M. N., Ericson, J. & Zinner, S. H. (1987). Effect of dose on serum pharmacokinetics of intravenous ciprofloxacin with identification and characterization of extravascular compartments using noncompartmental and compartmental pharmacokinetic models. *Antimicrobial Agents and Chemotherapy* **31**, 1782–6.

Duran Meras, I., Galeano Diaz, T., Rodriguez Caceres, M. I. & Salinas Lopez, F. (1997). Determination of the chemotherapeutic quinolonic and cinolonic derivatives in urine by high-performance liquid chromatography with ultraviolet and fluorescence detection in series. *Journal of Chromatography A* **787**, 119–27.

Eandi, M., Viano, I., Di Nola, F., Leone, L. & Genazzani, E. (1983). Pharmacokinetics of norfloxacin in healthy volunteers and patients with renal and hepatic damage. *European Journal of Clinical Microbiology* **2**, 253–9.

Echols, R. M. & Oliver, M. K. (1993). Ciprofloxacin safety relative to temafloxacin and lomefloxacin. In *Recent Advances in Chemotherapy, Proceedings of the Eighteenth International Congress of*

Chemotherapy, Stockholm, Sweden, pp. 349–50. American Society for Microbiology, Washington, DC.

Edelstein, P. H., Edelstein, M. A. C., Weidenfeld, J. & Dorr, M. B. (1990). *In vitro* activity of sparfloxacin (CI-978; AT-4140) for clinical Legionella isolates, pharmacokineticcs in guinea pigs, and use to treat guinea pigs with *L. pneumophila* pneumonia. *Antimicrobial Agents and Chemotherapy* **34**, 2122–7.

Ellerbroek, L. & Bruhn, M. (1989). Determination of flumequine in biological fluids and meat by high-performance liquid chromatography. *Journal of Chromatography* **495**, 314–7.

Engler, M., Rüsing, G., Sörgel, F. & Holzgrabe, U. (1998). Defluorinated sparfloxacin as a new photoproduct identified by liquid chromatography coupled with UV detection and tandem mass spectrometry. *Antimicrobial Agents and Chemotherapy* **42**, 1151–9.

Fabre, D., Bressolle, F., Gomeni, R., Arich, C., Lemesle, F., Beziau, H. & Galtier, M. (1991). Steady-state pharmacokinetics of ciprofloxacin in plasma from patients with nosocomial pneumonia: penetration of the bronchial mucosa. *Antimicrobial Agents and Chemotherapy* **35**, 2521–5.

Fernandes, P. B. (1989). *International Telesymposium on Quinolones.* J. R. Prous, Barcelona.

Fernandes, P. B., Chu, D. T. W., Bower, R. R., Jarvis, K. P., Ramer, N. R. & Shipkowitz, N. (1986). In-vivo evaluation of A-56619 (difloxacin) and A-56620: new aryl-fluoroquinolones. *Antimicrobial Agents and Chemotherapy* **29**, 201–8.

Fillastre, J. P., Montay, G., Bruno, R., Etienne, I., Dhib, M., Vivier, N. *et al.* (1994). Pharmacokinetics of sparfloxacin in patients with renal impairment. *Antimicrobial Agents and Chemotherapy* **38**, 733–7.

Flor, S., Guay, D. R. P., Opsahl, J. A., Tack, K. & Matzke, G. R. (1990). Effects of magnesium–aluminium hydroxide and calcium carbonate antacids on bioavailability of ofloxacin. *Antimicrobial Agents and Chemotherapy* **34**, 2436–8.

Fong, I. W., Vandenbroucke, A. & Simbul, M. (1987). Penetration of enoxacin into bronchial secretions. *Antimicrobial Agents and Chemotherapy* **31**, 748–51.

Forchetti, C., Flammini, D., Carlucci, G., Cavicchio, G., Vaggi, L. & Bologna, M. (1984). High-performance liquid chromatographic procedure for the quantitation of norfloxacin in urine, serum and tissues. *Journal of Chromatography* **309**, 177–82.

Fukuhara, K. & Matsuki, Y. (1987). Simple method for the determination of pipemidic acid in biological fluids by high-performance liquid chromatography. *Journal of Chromatography* **416**, 409–13.

Garraffo, R., Jambou, D., Chichmanian, R. M., Ravoire, S. & Lapalus, P. (1991). In vitro and in vivo ciprofloxacin pharmacokinetics in human neutrophils. *Antimicrobial Agents and Chemotherapy* **35**, 2215–8.

Gau, W., Ploschke, H. J., Schmidt, K. & Weber, B. (1985). Determination of ciprofloxacin in biological fluids by high performance liquid chromatography. *Journal of Liquid Chromatography* **8**, 485–97.

Girard, D., Gootz, T. D. & McGuirk, P. R. (1992). Pharmacokinetic studies of CP-74,667, a new quinolone, in laboratory animals. *Antimicrobial Agents and Chemotherapy* **36**, 1671–6.

Granneman, G. R., Snyder, K. M. & Shu, V. S. (1986). Difloxacin metabolism and pharmacokinetics in humans after single oral doses. *Antimicrobial Agents and Chemotherapy* **30**, 689–93.

Granneman, G. R., Braeckman, R., Kraut, J., Shupien, S. & Craft, J. C. (1991a). Temafloxacin pharmacokinetics in subjects with normal and impaired renal function. *Antimicrobial Agents and Chemotherapy* **35**, 2345–51.

Granneman, G. R., Carpentier, P., Morrison, P. J. & Pernet, A. G. (1991b). Pharmacokinetics of temafloxacin in humans after single oral doses. *Antimicrobial Agents and Chemotherapy* **35**, 436–41.

Granneman, G. R., Carpentier, P., Morrison, P. J. & Pernet, A. G. (1992). Pharmacokinetics of temafloxacin in humans after multiple oral doses. *Antimicrobial Agents and Chemotherapy* **36**, 378–86.

Grasela, T. H., Schentag, J. J., Sedman, A. J., Wilton, J. H., Thomas, D. J., Schultz, R. W. *et al.* (1989). Inhibition of enoxacin absorption by antacids or ranitidine. *Antimicrobial Agents and Chemotherapy* **33**, 615–7.

Griggs, D. J. & Wise, R. (1989). A simple isocratic high-pressure liquid chromatographic assay of quinolones in serum. *Journal of Antimicrobial Chemotherapy* **24**, 437–45.

Griggs, D. J., Wise, R., Kirkpatrick, B. & Ashby, J. P. (1988). The metabolism and pharmacokinetics of fleroxacin in healthy subjects. *Journal of Antimicrobial Chemotherapy* **22**, *Suppl. D*, 191–4.

Groeneveld, A. J. N. & Brouwers, J. R. B. J. (1986). Quantitative determination of ofloxacin, ciprofloxacin, norfloxacin and pefloxacin in serum by high pressure liquid chromatography. *Pharmaceutisch Weekblad, Scientific Edition* **8**, 79–84.

Gros, I. & Carbon, C. (1990). Pharmacokinetics of lomefloxacin in healthy volunteers: comparison of 400 milligrams once daily and 200 milligrams twice daily given orally for 5 days. *Antimicrobial Agents and Chemotherapy* **34**, 150–2.

Guay, D. R. P., Opsahl, J. A., McMahon, F. G., Vargas, R., Matzke, G. R. & Flor, S. (1992). Safety and pharmacokinetics of multiple doses of intravenous ofloxacin in healthy volunteers. *Antimicrobial Agents and Chemotherapy* **36**, 308–12.

Harrison, L. I., Schuppan, D., Rohlfing, S. R., Hansen, A. R., Hansen, C. S., Funk, M. L. *et al.* (1984). Determination of flumequine and a hydroxy metabolite in biological fluids by high-pressure liquid chromatographic, fluorometric, and microbiological methods. *Antimicrobial Agents and Chemotherapy* **25**, 301–5.

Hasegawa, T., Nadai, M., Kuzuya, T., Muraoka, I., Apichartpichean, R., Takagi, K. *et al.* (1990). The possible mechanism of interaction between xanthines and quinolone. *Journal of Pharmacy and Pharmacology* **42**, 767–72.

Hayakawa, I., Atarashi, S., Yokohama, S., Imamura, M., Sakano, K. I. & Furukawa, M. (1986). Synthesis and antibacterial activities of optically active ofloxacin. *Antimicrobial Agents and Chemotherapy* **29**, 163–4.

Hayton, W. L., Vlahov, V., Bacracheva, N., Viachki, I., Portmann, R., Muirhead, G. *et al.* (1990). Pharmacokinetics and biliary concentrations of fleroxacin in cholecystectomized patients. *Antimicrobial Agents and Chemotherapy* **34**, 2375–80.

Healy, D. P., Schoenle, J. R., Stotka, J. & Polk, R. E. (1991). Lack of interaction between lomefloxacin and caffeine in normal volunteers. *Antimicrobial Agents and Chemotherapy* **35**, 660–4.

Hoffken, G., Lode, H., Prinzing, C., Borner, K. & Koeppe, P. (1985). Pharmacokinetics of ciprofloxacin after oral and parenteral administration. *Antimicrobial Agents and Chemotherapy* **27**, 375–9.

Hooper, D. C. & Wolfson, J. S. (1985). The fluoroquinolones: pharmacology, clinical uses, and toxicities in humans. *Antimicrobial Agents and Chemotherapy* **28**, 716–21.

Hooper, W. D., Dickinson, R. G. & Eadie, M. J. (1990). Effect of food on absorption of lomefloxacin. *Antimicrobial Agents and Chemotherapy* **34**, 1797–9.

Horie, M., Saito, K., Hoshino, Y., Nose, N., Mochizuki, E. & Nakazawa, H. (1987). Simultaneous determination of nalidixic acid, oxolinic acid and piromidic acid in fish by high-performance liquid chromatography with fluorescence and UV detection. *Journal of Chromatography* **402**, 301–8.

Hustvedt, S. O., Salte, R. & Benjaminsen, T. (1989). Rapid high-performance liquid chromatographic method for the determination of oxolinic acid in fish serum employing solid-phase extraction. *Journal of Chromatography* **494**, 335–9.

Imbimbo, B. P., Broccali, G., Cesana, M., Crema, F. & Attardo-Parrinello, G. (1991). Inter- and intra-subject variabilities in the pharmacokinetics of rufloxacin after single oral administration to healthy volunteers. *Antimicrobial Agents and Chemotherapy* **35**, 390–3.

Ito, H., Inoue, M., Morikawa, M., Tsuboi, M. & Oka, K. (1985). Determination of pipemidic acid in plasma by normal-phase high-pressure liquid chromatography. *Antimicrobial Agents and Chemotherapy* **28**, 192–4.

Jaillon, P., Morganroth, J., Brumpt, I., Talbot, G. & the Sparfloxacin Safety Group. (1996). Overview of electrocardiographic and cardiovascular safety data for sparfloxacin. *Journal of Antimicrobial Chemotherapy* **37**, Suppl. A, 161–7.

Janknegt, R. (1986). Fluorinated quinolones. A review of their mode of action, antimicrobial activity, pharmacokinetics and clinical efficacy. *Pharmaceutisch Weekblad, Scientific Edition* **8**, 1–21.

Jehl, F., Gallion, C., Debs, J., Brogard, J. M., Monteil, H. & Minck, R. (1985). High-performance liquid chromatographic method for the determination of ciprofloxacin in biological fluids. *Journal of Chromatography* **3**, 347–57.

Johnson, J. A. & Benziger, D. P. (1985). Metabolism and disposition of amifloxacin in laboratory animals. *Antimicrobial Agents and Chemotherapy* **27**, 774–81.

Joos, B., Ledergerber, B., Flepp, M., Bettex, J. D., Luthy, R. & Siegenthaler, W. (1985). Comparison of high-pressure liquid chromatography and bioassay for determination of ciprofloxacin in serum and urine. *Antimicrobial Agents and Chemotherapy* **27**, 353–6.

Kaatz, G. W., Seo, S. M., Lamp, K. C., Bailey, E. M. & Rybak, M. J. (1992). CI-960, a new fluoroquinolone, for therapy of experimental ciprofloxacin-susceptible and -resistant *Staphylococcus aureus* endocarditis. *Antimicrobial Agents and Chemotherapy* **35**, 1192–7.

Katagiri, Y., Naora, K., Ichikawa, N., Hayashibara, M. & Iwamoto, K. (1990). High-performance liquid chromatographic determination of ciprofloxacin in rat brain and cerebrospinal fluid. *Chemical and Pharmaceutical Bulletin* **38**, 2884–6.

Kees, F., Naber, K. G., Meyer, G. P. & Grobecker, H. (1989). Pharmacokinetics of ciprofloxacin in elderly patients. *Arzneimittel-Forschung* **39**, 523–7.

Kisicki, J. C., Griess, R. S., Ott, C. L., Cohen, G. M., McCormack, R. J., Troetel, W. M. *et al.* (1992). Multiple-dose pharmacokinetics and safety of rufloxacin in normal volunteers. *Antimicrobial Agents and Chemotherapy* **36**, 1296–301.

Klinge, E., Mannisto, P. T., Mantyla, R., Mattila, J. & Hanninen, U. (1984). Single- and multiple-dose pharmacokinetics of pipemidic acid in normal human volunteers. *Antimicrobial Agents and Chemotherapy* **26**, 69–73.

Koechlin, C., Jehl, F., Linger, L. & Monteil, H. (1989). High-performance liquid chromatography for the determination of three new fluoroquinolones, fleroxacin, temafloxacin and A-64730, in biological fluids. *Journal of Chromatography* **491**, 379–87.

Korner, R. J., McMullin, C. M., Bowker, K. A., White, L. O., Reeves, D. S. & MacGowan, A. P. (1994). The serum concentrations of desmethyl ofloxacin and ofloxacin N-oxide in seriously ill patients and their possible contributions to the antibacterial activity of ofloxacin. *Journal of Antimicrobial Chemotherapy* **34**, 300–3.

Kovarik, J. M., de Hond, J. A. P. M., Hoepelman, L. M., Boon, T. &

Verhoef, J. (1990). Intraprostatic distribution of lomefloxacin following multiple-dose administration. *Antimicrobial Agents and Chemotherapy* **34**, 2398–401.

Krol, G. J., Noe, A. J. & Beerman, D. (1986). Liquid chromatographic analysis of ciprofloxacin metabolites in body fluids. *Journal of Liquid Chromatography* **9**, 2897–919.

Kucers, A., Crowe, S., Grayson, M. L. & Hoy, J. (1997). *The Use of Antibiotics,* 5th edn. Butterworth-Heinemann, Oxford.

Kullberg, M. P., Koss, R., O'Neil, S. & Edelson, J. (1979). High-performance liquid chromatographic analysis of rosoxacin and its N-oxide metabolite in plasma and urine. *Journal of Chromatography* **173**, 155–63.

Kusajima, H., Ishikawa, N., Machida, M., Uchida, H. & Irikura, T. (1986). Pharmacokinetics of a new quinolone, AM-833, in mice, rats, rabbits, dogs and monkeys. *Antimicrobial Agents and Chemotherapy* **30**, 304–9.

Lagana, A., Curini, R., D'Ascenzo, G., Marino, A. & Rotari, M. (1987). High-performance liquid chromatographic determination of norfloxacin in human tissues and plasma with fluorescence detection. *Journal of Chromatography* **417**, 135–42.

Larocque, L., Schnurr, M., Sved, S. & Weninger, A. (1991). Determination of oxolinic acid residues in salmon muscle tissue by liquid chromatography with fluorescence detection. *Journal of the Association of Official Analytical Chemists* **74**, 608–1.

LeBel, M. (1988). Ciprofloxacin: chemistry, mechanism of action, pharmacokinetics, clinical trials and adverse reactions. *Pharmacotherapy* **8**, 3–33.

LeBel, M., Vallee, F. & St-Laurent, M. (1990). Influence of lomefloxacin on the pharmacokinetics of theophylline. *Antimicrobial Agents and Chemotherapy* **34**, 1254–6.

Lehr, K. H. & Damm, P. (1988). Quantification of the enantiomers of ofloxacin in biological fluids by high-performance liquid chromatography. *Journal of Chromatography* **425**, 153–61.

Leigh, D. A., Harris, C., Tait, S., Walsh, B. & Hancock, P. (1991). Pharmacokinetic study of lomefloxacin and its effect on the faecal flora of volunteers. *Journal of Antimicrobial Chemotherapy* **27**, 655–62.

Lode, H., Hoffken, G., Olschewski, P., Sievers, B., Kirch, A., Borner, K. & Koeppe, P. (1987). Pharmacokinetics of ofloxacin after parenteral and oral administration. *Antimicrobial Agents and Chemotherapy* **31**, 1338–42.

Lombardi, F., Ardemagni, R., Colzani, V. & Visconti, M. (1992). High-performance liquid chromatographic determination of rufloxacin and its main active metabolite in biological fluids. *Journal of Chromatography* **576**, 129–34.

Lyon, D. J., Cheung, S. W., Chan, C. Y. & Cheng, A. F. B. (1994). Rapid HPLC assay of clinafloxacin, fleroxacin, levofloxacin, sparfloxacin and tosufloxacin. *Journal of Antimicrobial Chemotherapy* **34**, 446–8.

MacGowan, A. P., Greig, M. A., Clarke, E. A., White, L. O. & Reeves, D. S. (1988). The pharmacokinetics of norfloxacin in the aged. *Journal of Antimicrobial Chemotherapy* **22**, 721–7.

Marchbanks, C. R., Mikolich, D. J., Mayer, K. H., Zinner, S. H. & Dudley, M. N. (1990). Pharmacokinetics and bioavailability of intravenous-to-oral enoxacin in elderly patients with complicated urinary tract infections. *Antimicrobial Agents and Chemotherapy* **34**, 1966–72.

Martin, C., Lambert, D., Bruguerolle, B., Saux, P., Freney, J., Fleurette, J. *et al.* (1991). Ofloxacin pharmacokinetics in mechanically ventilated patients. *Antimicrobial Agents and Chemotherapy* **35**, 1582–5.

McCoy, L. F., Crawmer, B. P. & Benziger, D. P. (1985). Analysis of amifloxacin in plasma and urine by high-pressure liquid chromatography and intravenous pharmacokinetics in rhesus monkeys. *Antimicrobial Agents and Chemotherapy* **27**, 769–73.

McMullin, C. M., Lovering, A. M., Lewis, R. J. & White, L. O. (1992). A comparative photodegradation study of the fluoroquinolones ciprofloxacin and norfloxacin. In *Program and Abstracts of the Thirty-Second Interscience Conference on Antimicrobial Agents and Chemotherapy, Anaheim, CA, 1992.* Abstract 788, p. 242. American Society for Microbiology, Washington, DC.

McMullin, C. M., White, L. O., Reeves, D. S., Lovering, A. M. & Lewis, R. J. (1996). Photodegradation of fluoroquinolones using a continuous-flow photochemical reaction unit. *Journal of Antimicrobial Chemotherapy* **37**, 392–4

Mertes, P. M., Jehl, F., Burtin, P., Dopff, C., Pinelli, G., Villemot, J. P. *et al.* (1992). Penetration of ofloxacin into heart valves, myocardium, mediastinal fat, and sternal bone marrow in humans. *Antimicrobial Agents and Chemotherapy* **36**, 2493–6.

Mignot, A., Lefebvre, M. A. & Fourtillan, J. B. (1988). High-performance liquid chromatographic determination of ofloxacin in plasma and urine. *Journal of Chromatography* **430**, 192–7.

Montay, G. & Tassel, J. P. (1985). Improved high-performance liquid chromatographic determination of pefloxacin and its metabolite norfloxacin in human plasma and tissue. *Journal of Chromatography* **339**, 214–8.

Montay, G., Blain, Y., Roquet, F. & Le Hir, A. (1983). High-performance liquid chromatography of pefloxacin and its main active metabolites in biological fluids. *Journal of Chromatography* **272**, 359–65.

Montay, G., Goueffon, Y. & Roquet, F. (1984). Absorption, distribution, metabolic fate, and elimination of pefloxacin mesylate in mice, rats, dogs, monkeys, and humans. *Antimicrobial Agents and Chemotherapy* **25**, 463–72.

Morrison, P. J., Mant, T. G., Norman, G. T., Robinson, J. & Kunka, R. L. (1988). Pharmacokinetics and tolerance of lomefloxacin after sequentially increasing oral doses. *Antimicrobial Agents and Chemotherapy* **32**, 1503–7.

Morton, S. J., Shull, V. H. & Dick, J. D. (1986). Determination of norfloxacin and ciprofloxacin concentrations in serum and urine by high-pressure liquid chromatography. *Antimicrobial Agents and Chemotherapy* **30**, 325–7.

Myers, C. M. & Blumer, J. L. (1987). High-performance liquid chromatography of ciprofloxacin and its metabolites in serum, urine and sputum. *Journal of Chromatography* **422**, 153–64.

Nadai, M., Hasegawa, T., Kuzuya, T., Muraoka, I., Takagi, K. & Yoshizumi, H. (1990). Effects of enoxacin on renal and metabolic clearance of theophylline in rats. *Antimicrobial Agents and Chemotherapy* **34**, 1739–43.

Naora, K., Katagiri, Y., Ichikawa, N., Hayashibara, M. & Iwamoto, K. (1990). Simultaneous high-performance liquid chromatographic determination of ciprofloxacin, fenbufen and felbinac in rat plasma. *Journal of Chromatography* **530**, 186–91.

Nau, R., Kinzig, M., Dreyhaupt, T., Kolenda, H., Sorgel, F. & Prange, H. W. (1994). Kinetics of ofloxacin and its metabolites in cerebrospinal fluid after a single intravenous infusion of 400 milligrams of ofloxacin. *Antimicrobial Agents and Chemotherapy* **38**, 1849–53.

Newsom, S. W. B., Eden, C. G., Wells, F. C. & Meredith, P. (1989). Penetration of enoxacin into lung tissue *Journal of Antimicrobial Chemotherapy* **23**, 113–5.

Nix, D. E., De Vito, J. M. & Schentag, J. J. (1985). Liquid chromatographic determination of ciprofloxacin in serum and urine. *Clinical Chemistry* **31**, 684–6.

Nix, D. E., Wilton, J. H., Ronald, B., Distlerath, L., Williams, V. C. & Norman, A. (1990). Inhibition of norfloxacin absorption by antacids. *Antimicrobial Agents and Chemotherapy* **34**, 432–5.

Notarianni, L. J. & Jones, R. W. (1988). Method for the deter-

mination of ofloxacin, a quinolone carboxylic acid antimicrobial, by high-performance liquid chromatography. *Journal of Chromatography* **431**, 461–4.

Okezaki, E., Terasaki, T., Nakamura, M., Nagata, O., Kato, H. & Tsuji, A. (1989). Serum protein binding of lomefloxacin, a new antimicrobial agent, and its related quinolones. *Journal of Pharmaceutical Sciences* **78**, 504–7.

Panneton, A. C., Bergeron, M. G. & LeBel, M. (1988). Pharmacokinetics and tissue penetration of fleroxacin in serum and urine. *Clinical Chemistry* **34**, 2330–2.

Paradis, D., Vallee, F., Allard, S., Bisson, C., Daviau, N., Drapeau, C. *et al.* (1992). Comparative study of pharmacokinetics and serum bactericidal activities of cefpirome, ceftazidime, ceftriaxone, imipenem and ciprofloxacin. *Antimicrobial Agents and Chemotherapy* **36**, 2085–92.

Parry, M. F., Smego, D. A. & Digiovanni, M. A. (1988). Hepatobiliary kinetics and excretion of ciprofloxacin. *Antimicrobial Agents and Chemotherapy* **32**, 982–5.

Pascual, A., Garcia, I. & Perea, E. J. (1989). Fluorometric measurement of ofloxacin uptake by human polymorphonuclear leukocytes. *Antimicrobial Agents and Chemotherapy* **33**, 653–6.

Paton, J. H. & Reeves, D. S. (1988). Fluoroquinolone antibiotics. Microbiology, pharmacokinetics and clinical use. *Drugs* **36**, 193–228.

Piercy, E. A., Bawdon, R. E. & Mackowiak, P. A. (1989). Penetration of ciprofloxacin into saliva and nasal secretions and effect of the drug on the oropharyngeal flora of ill subjects. *Antimicrobial Agents and Chemotherapy* **33**, 1645–6.

Pioget, J., Wolff, M., Singlas, E., Laisne, M., Clair, B., Regnier, B. *et al.* (1989). Diffusion of ofloxacin into cerebrospinal fluid of patients with purulent meningitis or ventriculitis. *Antimicrobial Agents and Chemotherapy* **33**, 933–6.

Polk, R. E., Healy, D. P., Sahai, J., Drwal, L. & Racht, E. (1989). Effect of ferrous sulfate and multivitamins with zinc on absorption of ciprofloxacin in normal volunteers. *Antimicrobial Agents and Chemotherapy* **33**, 1841–4.

Preston, S. L., Drusano, G. L., Berman, A. L., Fowler, C. L., Chow, A. T., Dornseif, B. *et al.* (1998). Levofloxacin population pharmacokinetics and creation of a demographic model for prediction of individual drug clearance in patients with serious community-acquired infection. *Antimicrobial Agents and Chemotherapy* **42**, 1098–104.

Reeves, D. S., Bywater, M. J., Holt, H. A., and White, L. O. (1984). In-vitro studies with ciprofloxacin, a new 4-quinolone compound. *Journal of Antimicrobial Chemotherapy* **13**, 333–46.

Richman, J., Zolezio, H. & Tang-Liu, D. (1990). Comparison of ofloxacin, gentamicin, and tobramycin concentrations in tears and in vitro MICs for 90% of test organisms. *Antimicrobial Agents and Chemotherapy* **34**, 1602–4.

Rogstad, A., Hormazabal, V. & Yndestad, M. (1991). Extraction and high-performance liquid chromatographic determination of enrofloxacin in fish serum and tissues. *Journal of Liquid Chromatography* **14**, 521–31.

Rose, T. F., Bremner, D. A., Collins, J., Ellis-Pegler, R., Isaacs, R., Richardson, R. *et al.* (1990). Plasma and dialysate levels of pefloxacin and its metabolites in CAPD patients with peritonitis. *Journal of Antimicrobial Chemotherapy* **25**, 657–64.

Rubinstein, E. & Segev, S. (1987). Drug interactions of ciprofloxacin with other non-antibiotic agents. *American Journal of Medicine* **82**, Suppl. 4A, 119–23.

Saito, A., Nagata, O., Takahara, Y., Okezaki, E., Yamada, T. & Ito, Y. (1988). Enterohepatic circulation of NY-198, a new difluorinated quinolone, in rats. *Antimicrobial Agents and Chemotherapy* **32**, 156–7.

Sanchez Navarro, A., Martinez Lanao, J., Sanchez Recio, M. M., Dominguez-Gil Hurle, A., Tabernero Romo, J. M., Gomez Sanchez, J. C. *et al.* (1990). Effect of renal impairment on distribution of ofloxacin. *Antimicrobial Agents and Chemotherapy* **34**, 455–9.

Scheife, R. T., Cramer, W. R. & Decker, E. L. (1993). Photosensitizing potential of ofloxacin. *International Journal of Dermatology* **32**, 413–6.

Schentag, J. J. & Domagala, J. M. (1985). Structure–activity relationships with the quinolone antibiotics. *Research and Clinical Forums* **7**, 9–13.

Schmit, J. L., Hary, L., Bou, P., Renaud, H., Westeel, P. F., Andrejak, M. *et al.* (1991). Pharmacokinetics of single-dose intravenous, oral, and intraperitoneal pefloxacin in patients on chronic ambulatory peritoneal dialysis. *Antimicrobial Agents and Chemotherapy* **35**, 1492–4.

Scholl, H., Schmidt, K. & Weber, B. (1987). Sensitive and selective determination of picogram amounts of ciprofloxacin and its metabolites in biological samples using high-performance liquid chromatography and photothermal post-column derivatization. *Journal of Chromatography* **416**, 321–30.

Shargel, L., Koss, R. F., Crain, A. V. R. & Boyle, V. J. (1973). Nalidixic acid and hydroxynalidixic acid analysis in human plasma and urine by liquid chromatography. *Journal of Pharmaceutical Sciences* **62**, 1452–4.

Shiba, K., Saito, A., Shimada, J., Hori, S., Kaji, M., Miyahara, T. *et al.* (1990). Renal handling of fleroxacin in rabbits, dogs and humans. *Antimicrobial Agents and Chemotherapy* **34**, 58–64.

Shimada, J., Yamaji, T., Ueda, Y., Uchida, H., Kusajima, H. & Irikura, T. (1983). Mechanism of renal excretion of AM-715, a new quinolonecarboxylic acid derivative, in rabbits, dogs, and humans. *Antimicrobial Agents and Chemotherapy* **23**, 1–7.

Smethurst, A. M. & Mann, W. C. (1983). Determination by high-performance liquid chromatography of pipemidic acid in human serum and urine. *Journal of Chromatography* **274**, 421–7.

Smith, C. R. (1987). The adverse effects of fluoroquinolones. *Journal of Antimicrobial Chemotherapy* **19**, 709–11.

Smith, J. T. (1985). The 4-quinolone antibacterials. In *The Scientific Basis of Antimicrobial Chemotherapy*: 38th Symposium of the Society for General Microbiology. (Greenwood, D. & O'Grady, F., Eds), pp. 69–94. Cambridge University Press, London.

Sondack, D. L. & Koch, W. L. (1977). Analysis of urinary tract antibacterial agents in pharmaceutical dosage form by high-performance liquid chromatography. *Journal of Chromatography* **132**, 352–5.

Sörgel, F., Seelmann, R., Naber, K., Metz, R. & Muth, P. (1988). Metabolism of fleroxacin in man. *Journal of Antimicrobial Chemotherapy* **22**, *Suppl. D*, 169–78.

Stone, J. W., Andrews, J. M., Ashby, J. P., Griggs, D. & Wise, R. (1988). Pharmacokinetics and tissue penetration of orally administered lomefloxacin. *Antimicrobial Agents and Chemotherapy* **32**, 1508–10.

Stroshane, R. M., Silverman, M. H., Sauerschell, R., Brown, R. R., Boddy, A. W. & Cook, J. A. (1990). Preliminary study of the pharmacokinetics of oral amifloxacin in elderly subjects. *Antimicrobial Agents and Chemotherapy* **34**, 751–4.

Sunderland, J., McMullin, C. M., White, L. O., MacGowan, A. P. & Reeves, D. S. (1997). An isocratic high performance liquid chromatography assay (HPLC) for the new 8-methoxyquinolone BAY 1208039. In *Program and Abstracts of the Thirty-Seventh Interscience Conference on Antimicrobial Agents and Chemotherapy, Toronto 1997*. Abstract F-150, p. 171. American Society for Microbiology, Washington, DC.

Takagi, K., Yamaki, K., Nadai, M., Kuzuya, T. & Hasegawa, T.

(1991). Effect of a new quinolone, sparfloxacin, on the pharmacokinetics of theophylline in asthmatic patients. *Antimicrobial Agents and Chemotherapy* **35**, 1137–41.

Teng, R., Tensfeldt, T. G., Liston, T. E. & Foulds, G. (1996). Determination of trovafloxacin, a new quinolone antibiotic, in biological samples by reversed-phase high-performance liquid chromatography. *Journal of Chromatography B: Biomedical Applications* **675**, 53–9.

Tsuei, S. E., Darragh, A. S. & Brick, I. (1984). Pharmacokinetics and tolerance of enoxacin in healthy volunteers administered at a dosage of 400 mg twice daily for 14 days. *Journal of Antimicrobial Chemotherapy* **14**, *Suppl. C*, 71–4.

Tyczkowska, K., Hedeen, K. M., Aucoin, D. P. & Aronson, A. L. (1989). High-performance liquid chromatographic method for the simultaneous determination of enrofloxacin and its primary metabolite ciprofloxacin in canine serum and prostatic tissue. *Journal of Chromatography* **493**, 337–46.

Van der Auwera, P., Stolear, J. C. & Dudley, M. N. (1990). Pharmacokinetics of enoxacin and its oxo metabolite following intravenous administration to patients with different degrees of renal impairment. *Antimicrobial Agents and Chemotherapy* **34**, 1491–7.

Vree, T. B., van Ewijk-Beneken Kolmer, E. W. J. & Nouws, J. F. M. (1992). Direct-gradient high-performance liquid chromatographic analysis and preliminary pharmacokinetics of flumequine and flumequine acyl glucuronide in humans: effect of probenecid. *Journal of Chromatography* **579**, 131–41.

Wagai, N. & Tawara, K. (1992). Possible reasons for differences in phototoxic potential of a 5 quinolone antibacterial agents: generation of toxic oxygen. *Free Radical Research Communications* **17**, 387–98.

Warlich, R., Korting, H. C., Schafer-Korting, M. & Mutschler, E. (1990). Multiple-dose pharmacokinetics of ofloxacin in serum, saliva, and skin blister fluid of healthy volunteers. *Antimicrobial Agents and Chemotherapy* **34**, 78–81.

Weber, A., Chaffin, D., Smith, A. & Opheim, K. E. (1985). Quantitation of ciprofloxacin in body fluids by high-pressure liquid chromatography. *Antimicrobial Agents and Chemotherapy* **27**, 531–4.

Weidekamm, E., Portmann, R., Suter, K., Partos, C., Dell, D. & Lucker, P. W. (1987). Single- and multiple-dose pharmacokinetics of fleroxacin, a trifluorinated quinolone, in humans. *Antimicrobial Agents and Chemotherapy* **31**, 1909–14.

Weidekamm, E., Portmann, R., Partos, C. & Dell, D. (1988). Single and multiple dose pharmacokinetics of fleroxacin. *Journal of Antimicrobial Chemotherapy* **22**, *Suppl. D*, 145–54.

White, L. O. (1987). Metabolism of 4-quinolones. *Quinolones Bulletin* **3**, 1–4.

White, L. O., Bowyer, H. M., Lovering, A. M. & Reeves, D. S. (1985). Assay of ciprofloxacin in body fluids by HPLC with fluorimetric detection. Second European Congress of Clinical Microbiology, Brighton. Abstract 22/16.

White, L. O., MacGowan, A. P., Lovering, A. M., Reeves, D. S. & Mackay, I. G. (1987). A preliminary report on the pharmacokinetics of ofloxacin, desmethyl ofloxacin and ofloxacin *N*-oxide in patients with chronic renal failure. *Drugs* **34**, *Suppl. 1*, 56–61.

White, L. O., Bowyer, H. M., McMullin, C. H. & Desai, K. (1988a). Assay of enoxacin in human serum and prostatic tissue by HPLC. *Journal of Antimicrobial Chemotherapy* **21**, 512–3.

White, L. O., MacGowan, A. P., Mackay, I. G. & Reeves, D. S. (1988b). The pharmacokinetics of ofloxacin, desmethyl ofloxacin and ofloxacin *N*-oxide in haemodialysis patients with end-stage renal failure. *Journal of Antimicrobial Chemotherapy* **22**, *Suppl. C*, 65–72.

White, L. O., McMullin, C. M., Bowyer, H. M. & Lovering, A. M. (1989). The therapeutic monitoring of 4-quinolone antibacterials with special reference to ciprofloxacin. In *International Telesymposium on Quinolones* (Fernandes, P. B., Ed.), pp. 351–67. J. R. Prous, Barcelona.

Wijnands, W. J. A., Vree, T. B., Baars, A. M., Hafkenscheid, J. C. M., Kohler, B. E. M. & van Herwaarden, C. L. A. (1988). The penetration of ofloxacin into lung tissue. *Journal of Antimicrobial Chemotherapy* 22, Suppl. C, 85–89.

Wilson, A. P. R. & Grüneberg, R. N. (1997). *Ciprofloxacin: 10 Years of Clinical Experience.* Maxim Medical, Oxford.

Wise, R., Lister, D., McNulty, C. A. M, Griggs, D. & Andrews, J. M. (1986). The comparative pharmacokinetics of five quinolones. *Journal of Antimicrobial Chemotherapy* 18, Suppl. D, 71–81.

Wise, R., Baker, S. L., Misra, M. & Griggs, D. (1987a). The pharmacokinetics of enoxacin in elderly patients. *Journal of Antimicrobial Chemotherapy* 19, 343–50.

Wise, R., Kirkpatrick, B, Ashby, J. & Griggs, D. J. (1987b). Pharmacokinetics and tissue penetration of Ro 23-6240, a new trifluoroquinolone. *Antimicrobial Agents and Chemotherapy* 31, 161–3.

Wolff, M., Boutron, L., Singlas, E., Clair, B., Decazes, J. M. & Regnier, B. (1987). Penetration of ciprofloxacin into cerebrospinal fluid of patients with bacterial meningitis. *Antimicrobial Agents and Chemotherapy* 31, 899–902.

Woodcock, J. M., Andrews, J. M., Honeybourne, D. & Wise, R. (1994). Determination of OPC-17116, a new fluoroquinolone, in human alveolar macrophages and other biological matrices by high performance liquid chromatography (HPLC). *FEMS Microbiology Letters* 119, 315–20.

Yoshitake, A., Kawahara, K., Shono, F., Umeda, I., Izawa, A. & Komatsu, T. (1980). Determination of miloxacin and metabolites in human serum and urine by high-pressure liquid chromatography. *Antimicrobial Agents and Chemotherapy* 18, 45–9.

Yuk, J. H., Nightingale, C. H., Quintiliani, R. & Sweeney, K. R. (1991). Bioavailability and pharmacokinetics of ofloxacin in healthy volunteers. *Antimicrobial Agents and Chemotherapy* 35, 384–6.

Chapter 12

Chloramphenicol

Richard Wise

Department of Medical Microbiology, City Hospital NHS Trust, Dudley Road, Birmingham B18 7QH, UK

Introduction

Chloramphenicol (Figure 12.1) was the first wide-spectrum antibiotic. Although originally isolated from *Streptomyces venezuelae* it is now synthesized chemically. It has a molecular weight of 323.1 and a pK_a of 5.5.

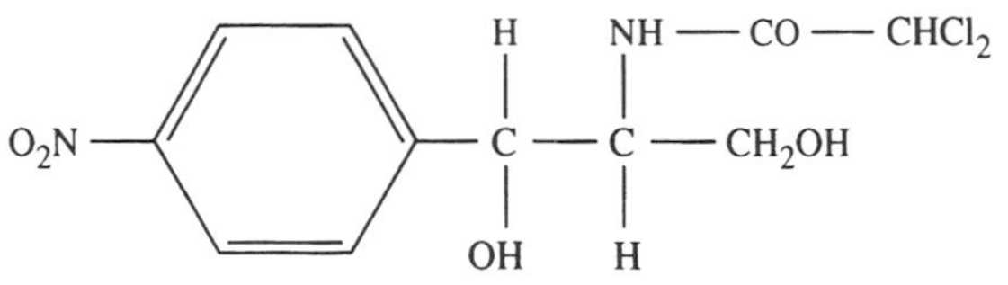

Figure 12.1 Structure of chloramphenicol.

Although not now widely used, chloramphenicol has an important role to play in clinical practice. Its toxicity is well known and, certainly in neonates, high concentrations may well be associated with a greater incidence of toxicity. This antimicrobial has had certain definite clinical indications, for example for treating typhoid fever. The emergence of ampicillin-resistant strains of *Haemophilus influenzae* may lead to chloramphenicol being used as an alternative to, say, the fluoroquinolones in the treatment of severe infections with this organism and hence there may be an increasing need to assay. The widest (but now greatly diminishing) use of chloramphenicol in the UK was in the treatment of confirmed or suspected neonatal sepsis.

Pharmacokinetics

In adults chloramphenicol is usually given as free drug or the palmitate prodrug by mouth. The dose is usually 0.5 g every 6 h. A 1 g dose will give serum levels of about 6–11 mg/L (Lindberg *et al.*, 1966). The bioavailability of the free drug and the palmitate is high, but that of the iv formulation, chloramphenicol succinate, is highly variable, 6–82% being lost in the urine before the prodrug is hydrolysed, rather slowly *in vivo*, to the parent chloramphenicol. The succinate is present in the commercially available product as the 3-monosuccinate and 1-monosuccinate isomers in a ratio of 4:1; this ratio is maintained in the blood.

The dose is usually 25 mg/kg/day in newborns and 50 mg/kg/day in infants (1–4 weeks), both in divided doses. In neonates and children it is generally accepted that the therapeutic range is between 15 and 25 mg/L (Mulhall *et al.*, 1983a, b). The serum concentration in infants receiving the same dose varies widely, emphasizing the need to assay the agent in every baby receiving the agent.

As it is lipophilic, chloramphenicol is well distributed, with a volume of distribution of about 1 L/kg. The serum protein binding is around 50%.

Only 5–10% of chloramphenicol appears in the urine as unchanged compound, some 80% being metabolized to the 1- and 3-glucuronides. A number of other minor metabolites have also been described, including the glycol alcohol, the oxamic acid, the aldehyde, the amine, nitroso derivatives and chloramphenicol base (Figure 12.2). None of these have useful antimicrobial activity nor have they been clearly linked with toxicity (Holt *et al.*, 1995).

When chloramphenicol is given intravenously, it is important to obtain samples for assay at the correct times.

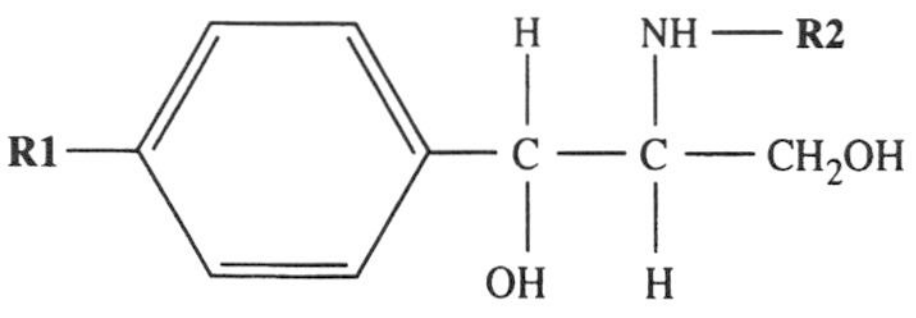

R1	R2	Compound
$-NO_2$	$-COCHCl_2$	chloramphenicol
$-NO_2$	$-COH_2COH$	glycol alcohol
$-NO_2$	$-COCOOH$	oxamic acid
$-NO_2$	$-COCHO$	aldehyde
$-NO_2$	$-H$	base
$-NH_2$	$-COCHCl_2$	amine
$-NO$	$-COCHCl_2$	nitro-chloramphenicol

Figure 12.2 Chloramphenicol metabolites.

The pre-dose sample should be taken just before the next dose is given, but the post-dose sample should be delayed for 1–2 h as in-vivo hydrolysis to free chloramphenicol is slow.

Assay methods

Chloramphenicol can be assayed by a variety of procedures, but it should be remembered that inactive metabolites may interfere with non-microbiological assays; this is particularly important in urine assays. In renal failure there will be preferential accumulation of the glucuronide which is thought to be as toxic as the parent compound, so it may be necessary to assay both components.

Chemical methods

Colorimetric determinations have been described (Bessman & Stevens, 1950; Grove & Randall, 1955), as has a turbidimetric procedure (Grove & Randall, 1955). The colorimetric method should not now be employed as it detects the succinate as well as the parent compound.

Enzymatic methods

A method utilizing an acetyltransferase procedure, similar to that which used to be employed for gentamicin assay, has been developed and could be used in those laboratories which already use procedures involving radiochemicals (Daigneault & Guitard, 1976). A modification which does not require radiochemicals has been described (Hammond et al., 1987; Morris et al., 1988) in which the chloramphenicol reacts with the chloramphenicol acetyltransferase in the presence of acetyl coenzyme A. The coenzyme A formed is then used to initiate a second reaction whereby oxoglutamate dehydrogenase is used to convert oxoglutaric acid to succinyl coenzyme A and concomitant conversion of NAD to NADH, the latter compound then reacting to convert a colourless tetrazolium compound to its coloured form. The inactive prodrug forms of chloramphenicol are not detected by this assay system and of the metabolites only the glycolic acid shows cross-reactivity (of 81%). Thiamphenicol, a chloramphenicol analogue, is also detected. Grossly haemolysed serum, or serum with high bilirubin levels, interfere with this assay.

Microbiological assay

The microbiological assay usually employs *Sarcinea lutea* as an indicator strain for a cup plate method (Grove & Randall, 1955). We have found that a modification using wells and a different indicator strain (*Escherichia coli* Sch 12655) is accurate and reproducible, the 95% confidence limits being better than ±20% when performed correctly (see *Chapter 4*). It is suggested that this be used in clinical microbiological laboratories. However, like all microbiological methods it is slow and may suffer from interference from other agents and, since chloramphenicol is often co-administered with other antimicrobials, it is suggested that one of the non-microbiological methods described below be used in preference.

High performance liquid chromatography assays

Those laboratories which possess high performance liquid chromatography (HPLC) equipment would probably take advantage of the more rapid methods available. The advantage of HPLC is that, if need be, the metabolites can also be measured (Holt et al., 1995) as well as the parent drug and prodrug. Assay specificity can be improved by the simultaneous use of detection at two different wavelengths (see below).

Immunoassay

An EMIT method has been developed utilizing a monoclonal antibody which is specific and reliable (Dalbey et al., 1985; White et al., 1988). However, the calibration curve is somewhat unstable (Andrews et al., 1988) compared with other EMIT assays and it is suggested that high and low controls be included with each sample run. The assay comes as a 50-assay kit and can be run on all analysers normally used for EMIT assays (see *Chapter 5*). Although in terms of specificity this assay is superior to other methods, the reduced usage of chloramphenicol and the consequent drop in demand for clinical assays makes this assay very expensive to run.

Changes in the numbers of laboratories performing chloramphenicol assays and the methods used based on UK NEQASEQA returns are shown in Table I. Numbers have declined and so has the use of the EMIT assay. Currently HPLC remains the most popular assay method despite its limitations.

Useful information

Chloramphenicol in solution or as dry powder is very stable, but is light-sensitive and therefore should be stored in the dark. The potency of the powder is usually 100%. Powder can be obtained from Parke Davis & Co. Ltd (Pontypool, UK). Its solubility is 2.5 mg/mL in water and 400 mg/mL in ethanol.

Dosage forms are: free chloramphenicol, oral prodrug, the palmitate ester and chloramphenicol succinate, the parenteral formulation. Both succinate and palmitate are microbiologically inactive. Chloramphenicol 'base', as described in certain chemical catalogues, should not be used to prepare calibrators or controls as it has no microbiological activity.

Examples of assay methods

Microbiological plate assay

The method suggested below is a modification of the procedure recommended by Grove & Randall (1955).

Preparation of calibrators. The equivalent of 100 mg of chloramphenicol is accurately weighed out from stock powder of known potency. This is placed in a 100 mL

Chloramphenicol

Table I. Changes in the popularity of methods (expressed as percentage of laboratories using each method) used to perform clinical chloramphenicol assays as shown by UK NEQAS returns 1990–1997

Year	No. of participants	Percentage of laboratories using each method of assay			
		EMIT	HPLC	Gram-negative bioassay	Gram-positive bioassay
1990	36	53	36	8	3
1991	31	48	45	7	0
1992	30	47	47	6	0
1993	24	42	54	4	0
1994	15	27	66	7	0
1995	14	21	72	7	0
1996	13	31	61	8	0
1997	13	23	69	8	0

volumetric flask and 2–3 mL of ethanol is added and gently warmed to assist dissolving the material. The volume is then made up to 100 mL with phosphate-buffered saline, pH 7.2. This stock solution can be kept deep-frozen for several weeks. From the stock solution calibrators are prepared in horse serum or pooled human serum at 40, 20, 10, 5 and 2.5 mg/L. On the rare occasions that urine concentrations of chloramphenicol need to be measured, the calibrators are prepared in phosphate-buffered saline.

Medium. It is recommended that 'Oxoid' Antibiotic Medium No. 1, pH 6.6 (CM 327) be used. This should be dispensed in 200 mL amounts.

Indicator strain. E. coli Sch 12655 (obtained from Schering Laboratories) is employed. This is grown in nutrient broth overnight and before use is diluted, 1 mL in 100 mL of sterile water. This is used to flood the plate and excess fluid drained off. The plate is then dried for 30 min in a 37°C incubator.

Procedure. The calibrators and samples are applied in triplicate, following a random pattern. The plates are incubated at 37°C overnight and read, preferably with an optical zone reader. A curve approximating to a straight line should be obtained. The lower limit of sensitivity of the assay is about 2.5 mg/L.

Table II. Two methods of HPLC assay for chloramphenicol

	Method A[a]	Method B[b]
Column	μBondapak C_{18} 100 × 8 mm, 10 μm particle size, used with a Radial Compression Module (Waters, Watford, UK)	Hypersil 5 ODS 150 × 4 mm, 5 μm particle size (HPLC Technology, Macclesfield, UK)
Mobile phase	35% methanol, 1% glacial acetic acid, 64% water (v/v) pumped at 3 mL/min	35% methanol, 1% phosphoric acid, 64% water (v/v) pumped at 2 mL/min
Detection	UV 276 and 254 nm (dual wavelength)	UV 214 and 254 nm (dual wavelength)
Sample preparation	Add an equal volume of 7% (v/v) perchloric acid (usually 200 μL serum and 200 μL 7% perchloric acid) and centrifuge at 11,600**g** for 5 min	Mix serum sample (100 μL) an equal volume of acetonitrile and centrifuge at 11,600**g** for 1 min
Retention time	Approximately 5 min	Approximately 5–10 min
Calibration	Single point calibration with 30 mg/L calibrator is satisfactory; a 15 mg/L internal control is also run	Single point calibration with 30 or 50 mg/L aqueous calibrator is satisfactory; an internal control of 10–50 mg/L is also run

[a]The succinate, being more polar, will elute later and will not interfere with this assay. If it is necessary to measure the glucuronide then it is suggested that the method of Aravind *et al.* (1982) be employed.
[b]This method also uses dual-wavelength detection to improve assay specificity but requires only fixed-wavelength detectors at 214 and 254 nm.

Problems encountered. The major difficulty is that chloramphenicol is often prescribed in combination with other drugs such as penicillin, sulphonamide or aminoglycoside. The first two of these can be inactivated by β-lactamase or *para*-aminobenzoic acid respectively. The presence of an aminoglycoside raises certain problems, which can be overcome by using ion-exchange resin or phosphocellulose paper to remove it. The use of aminoglycoside-resistant Gram-negative indicators can, if care is not taken in their choice, lead to poor zone edges and inferior sensitivity.

HPLC assay

A number of methods have been described (Koup *et al.*, 1978; Thies & Fischer, 1978) which, like the two given here, do not measure the glucuronide. Holt *et al.* (1995) describe methods for assaying the metabolites, although this would not be done routinely in clinical studies.

Since high serum concentrations of chloramphenicol in neonates are potentially life-threatening, it is important that clinical assay results have a high degree of reliability. Falsely elevated results could result in therapy being stopped or the dosage being reduced unnecessarily. Dual wavelength detection improves assay reliability. The chromatograms are obtained at two wavelengths and the results are calculated from both. The two sets of results should agree. If they are widely discrepant (>10% difference) there may be an interfering peak co-eluting with the chloramphenicol. In such a situation the assay result should be the lowest result and this should be reported as '$<x$ mg/L'. Two suggested methods are given in Table II.

References

Andrews, J. M., Ashby, J. P. & Wise, R. (1988). Chloramphenicol assay by EMIT. *Journal of Antimicrobial Chemotherapy* **21**, 682–3.

Aravind, M. K., Miceli, J. N., Done, A. K. & Kauffman, R. E. (1982). Determination of chloramphenicol-glucuronide in urine by high-performance liquid chromatography. *Journal of Chromatography* **232**, 461–4.

Bessman, S. P. & Stevens, B. S. (1950). A colorimetric method for the determination of chloromycetin in serum or plasma. *Journal of Laboratory and Clinical Medicine* **35**, 129–35.

Daigneault, R. & Guitard, M. (1976). An enzymatic assay for chloramphenicol with partially purified chloramphenicol acetyltransferase. *Journal of Infectious Diseases* **133**, 515–22.

Dalbey, M., Gano, C., Izatsu, A., Collins, C., Jaklitsch, A., Hu, M. *et al.* (1985). Quantitative chloramphenicol determination by homogenous enzyme Immunoassay. *Clinical Chemistry* **31**, 933.

Grove, D. C. & Randall, W. A. (1955). *Assay Methods of Antibiotics, a Laboratory Manual*, pp. 66–75. Medical Encyclopaedia Inc., New York.

Hammond, P. M., Miller, J., Price, C. P. & Morris, H. C. (1987). Enzyme-based chloramphenicol estimation. *Lancet ii*, 449.

Holt, D. E., Hurley, R. & Harvey, D. (1995). A reappraisal of chloramphenicol metabolism; detection and quantification of metabolites in the sera of children. *Journal of Antimicrobial Chemotherapy* **35**, 115–27.

Koup, J. R., Brodsky, B., Lau, A. & Beam, T. R. (1978). High-performance liquid chromatographic assay of chloramphenicol in serum. *Antimicrobial Agents and Chemotherapy* **14**, 439–43.

Lindberg, A. A., Nilsson, L. H., Bucht, H. & Kallings, L. O. (1966). Concentration of chloramphenicol in urine and blood in relation to renal function. *British Medical Journal ii*, 724–8.

Morris, H. C., Miller, J., Campbell, R. S., Hammond, P. M., Berry, D. J. & Price, C. P. (1988). A rapid, enzymatic method for the determination of chloramphenicol in serum. *Journal of Antimicrobial Chemotherapy* **22**, 935–44.

Mulhall, A., de Louvois, J. & Hurley, R. (1983a). Chloramphenicol toxicity in neonates: its incidence and prevention. *British Medical Journal* **287**, 1424–7.

Mulhall, A., de Louvois, J. & Hurley, R. (1983b). The pharmacokinetics of chloramphenicol in the neonate and young infant. *Journal of Antimicrobial Chemotherapy* **12**, 629–39.

Thies, R. L. & Fischer, L. J. (1978). High-performance liquid-chromatographic assay for chloramphenicol in biological fluids. *Clinical Chemistry*, **24**, 778–81.

White, L. O., Bywater, M. J., Lovering, A. M., Holt, H. A. & Reeves, D. S. (1988). Chloramphenicol assay by EMIT. *Journal of Antimicrobial Chemotherapy* **21**, 683–4.

Chapter 13

Macrolides and azalides

Jocelyn M. Woodcock and Jenny M. Andrews

Department of Medical Microbiology, City Hospital NHS Trust, Birmingham B18 7QH, UK

Introduction

The group of antibiotics known as the macrolides (a contraction of the term 'macrocyclic lactonide') consists of several sub-groups, namely the 14-, 15- and 16-membered macrolides. The 14-membered macrolide sub-group includes erythromycin, clarithromycin and dirithromycin and is so called because of the characteristic 14-membered macrocyclic lactone ring structure (Figures 13.1 and 13.2). Similarly, the 15- and 16-membered macrolides have 15- and 16-membered rings respectively.

The insertion of a methyl-substituted nitrogen into erythromycin has produced a new class of macrolide, the azalides. Azithromycin (Figure 13.3), a 15-membered macrolide, is the prototype of this class. Recently the ketolides, a new class of semi-synthetic macrolides with the cladinose moiety of the 14-membered lactone ring being replaced by a keto group at position 3 have been introduced. Examples showing improved antimicrobial activity over other macrolides include the investigational compounds RU 64004 and RU 66647 (Biedenbach & Jones, 1996).

Macrolide antibiotics have been used clinically for more than three decades in the treatment of respiratory and soft tissue infections against susceptible strains of staphylococci, streptococci and *Haemophilus influenzae* and as an alternative to penicillin in patients with penicillin hypersensitivity. Erythromycin has been the most widely used

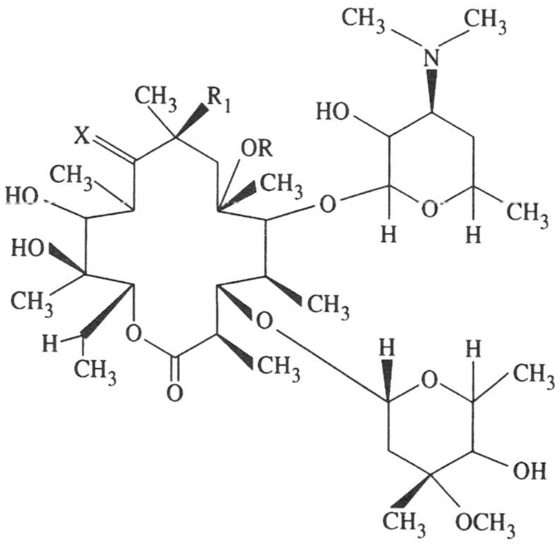

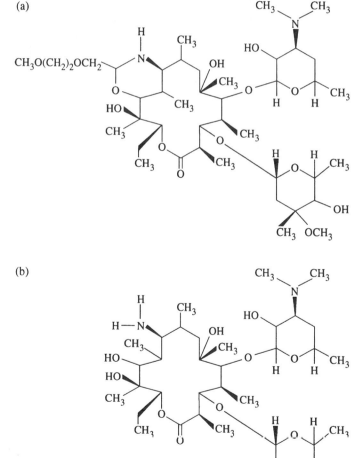

Figure 13.1 Structure of 14-membered macrolides, erythromycin A, clarithromycin and roxithromycin.

	X	R	R$_1$
Erythromycin A	O	H	H
Clarithromycin	O	CH$_3$	O
Roxithromycin	NOCH$_2$O(CH$_2$)$_2$OCH$_3$	H	H

Figure 13.2 Structure of (a) dirithromycin and (b) erythromycycline.

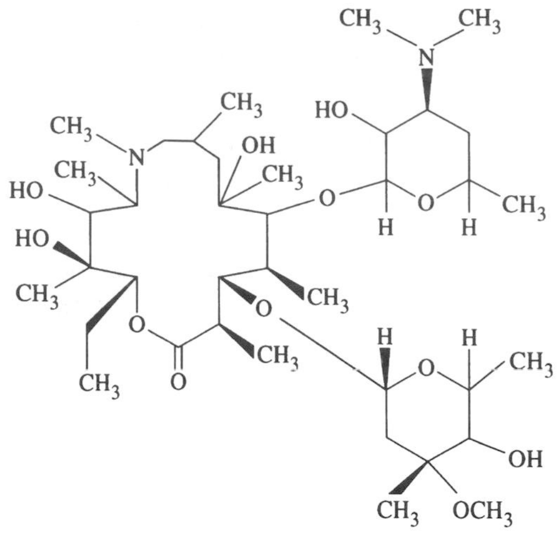

Figure 13.3 Structure of azithromycin.

natural product is a mixture of four slightly different macrolides designated erythromycins A, B, C and D although erythromycin A (Figure 13.1; mol. wt 733.9, pK_a 8.8) is the major component and has the greatest activity. Degradation of erythromycin in acidic conditions yields products that have no significant antimicrobial activity, but which may play a role in the gastrointestinal side-effects of erythromycin therapy (Kirst & Sides, 1989). Protection of erythromycin base from decomposition by gastric acid during oral administration can be achieved in several ways, Firstly, more stable forms such as the salts or esters of erythromycin, e.g. estolate (mol. wt 1056.4), stearate (mol. wt 1018.4), acistrate (mol. wt 1061) can be used; secondly, enteric coatings can be employed; and thirdly, the propionate, a combination of two salts of propionyl erythromycin (the N-acetylcysteinate and mercaptosuccinate) can be administered so that antibiotic and mucolytic properties are combined in a single agent. Although erythromycin is usually administered orally, erythromycin ethyl succinate (mol. wt 862.1) can be administered intramuscularly and erythromycin lactobionate (mol. wt 1092.2) intravenously. The active ingredient in all the above forms of erythromycin is erythromycin base.

The gastric absorption of erythromycin is affected by food, so the drug should be administered at least 1 h before meals. Peak serum concentrations (approximately 2 h after the dose) following a 250 mg dose are variable depending on preparation and mode of administration, ranging from 1.4, 0.4 and 0.3 mg/L for the estolate, stearate and base respectively (Griffith & Black, 1962). Controversy exists regarding the high levels observed for the estolate compared with those obtained for other preparations. Both ester and base are present in blood after oral administration and there is a continuation of transformation of erythromycin ester to erythromycin base. Croteau et al. (1987) showed that, in vitro, erythromycin estolate is less stable than erythromycin succinate to physicochemical hydrolysis, which may explain the observed differences.

The extent to which erythromycin is protein-bound has been determined by many methods including equilibrium dialysis, ultrafiltration and ultracentrifugation which give

agent of this class, particularly more recently, with the recognition of the importance of intracellular pathogens such as *Legionella* spp., *Mycoplasma* spp. and *Chlamydia* spp. in lower respiratory tract infections since macrolides concentrate in tissues and cells. The limitations of erythromycin, namely its variable activity against *H. influenzae*, gastro-intestinal irritation (mainly in adults) and frequency of administration, have led to the development of newer macrolides which possess an improved spectrum of activity and better pharmacokinetic properties. A summary of pharmacokinetic properties is shown in Table I.

This chapter gives information on the most frequently used macrolides, but the principles of the assays (both HPLC and microbiological) discussed could be adapted for other macrolides.

Erythromycin

Introduction

Erythromycin was isolated in 1952 from a strain of *Streptomyces erythreus* (Abbott Laboratories, 1966). The

Table I. Macrolide pharmacokinetic parameters (fasting volunteers)

Macrolide	Dose (mg)	t_{max} (h)	C_{max} (mg/L)	$t_{1/2}$ (h)	AUC (mg·h/L)	References
Erythromycin stearate	500	1.2 ± 0.6	2.1 ± 1.2	1.6 ± 0.4	7.3 ± 3.9	Kavi *et al.* (1988)
Clarithromycin	400	1.9 ± 0.7	1.1 ± 0.2	3.6 ± 0.1		Prokocimer & Hardy (1991)
14-Hydroxyclarithromycin		2.1 ± 0.6	0.8 ± 0.2	3.9		Prokocimer & Hardy (1991)
Dirithromycin/ Erythromycylamine	500	3.9 ± 0.4	0.29 ± 0.21	20–50	0.86 ± 0.59	Sides *et al.* (1993)
Azithromycin	500	1.7	0.4	11–14	3.4	Foulds *et al.* (1990)
Roxithromycin	150	1.0 ± 0.6	9.2 ± 2.3	13.2 ± 2.9	114.4 ± 27.9	Wise *et al.* (1987)
Spiramycin	2000	3.3 ± 0.5	3.1 ± 1.9	3.8 ± 3.0	8.5 ± 3.0	Kavi *et al.* (1988)

values of approximately 70%, 80% and 40% respectively (Lorian, 1991). The reasons for these differences have not yet been elucidated. The plasma elimination half-life of erythromycin is 1.2–2.6 h. Excretion of erythromycin is partly in urine (2.5% oral administration, 15% parenteral administration), but the majority of the drug is excreted in an active form in bile. The major side-effects associated with erythromycin are gastro-intestinal and hepatic disturbance. Concomitant administration of erythromycin may interfere with the clearance of drugs such as theophylline, antipyrine and carbamezepine (Bartolucci et al., 1984).

Assay of erythromycin—general information

Erythromycin base is soluble in ethanol, slightly soluble in water and unstable in acidic conditions. Laboratory reference powder (available from Lilly or Abbott Laboratories) should be prepared by dissolving a known weight in a small amount of absolute ethanol then adding distilled water to the volume required. Erythromycin is unstable in acidic conditions; neutral solutions remain stable for several weeks at 4°C, but some loss will occur at room temperature. In powder form, erythromycin is stable if stored in ampoules at 4°C (Ryden, 1978).

HPLC assays of erythromycin

Several HPLC assay methods for the determination of erythromycin have been published (Tsuji, 1978; Tsuji & Goetz, 1978; Chen & Chiou, 1983; Duthu, 1984; Stubbs et al., 1985; Croteau et al., 1987). However, methods employing electrochemical detection (ECD) (Chen & Chiou, 1983; Duthu, 1984; Croteau et al., 1987), as opposed to UV absorbance or fluorescence detection, are more suitable for determining levels in serum and other biological matrices, because of the high sensitivity which can be achieved.

Of the three HPLC-ECD methods quoted above, the most sensitive is that of Chen & Chiou (1983) with lower limits of detection for serum of 0.005–0.03 mg/L depending upon the volume of sample used. Although the method by Croteau et al., 1987 is less sensitive (lower limit 0.25 mg/L for 2 mL serum sample), it has been used to quantify concentrations of erythromycin estolate and ethyl succinate and not just to separate them from the erythromycin base. The main difference between these two methods is the type of electrochemical detector used; a coulometric detector was employed by Chen & Chiou (1983), whereas Croteau et al. used an amperometric detector (see Chapter 5).

HPLC assay method of Chen & Chiou (1983)

Column. C$_{18}$ MicroBondapak, 300 × 3.9 mm, 10 μm (Waters).

Detection. An Environmental Sciences Association coulometric detector (ESA 5100A) in the oxidative screen mode, the first and second electrode potentials are set to +0.7 and +0.9 V respectively.

Mobile phase. Acetonitrile:methanol: 0.2 M sodium acetate buffer pH 6.7 (40:5:55) pumped at 1 mL/min. The pH of the acetate buffer is adjusted to 6.7 with 0.2 M acetic acid.

Serum sample preparation. The following are pipetted into glass tubes with caps lined with aluminium foil: plasma sample (200 μL); internal standard solution comprising erythromycin B, 0.01–0.1 mg/mL (10 μL); saturated sodium carbonate solution (20 μL); ethyl ether (1000 μL). The mixture is vortexed for 20 s. Following centrifugation at 800**g** for 5 min, 750 μL of the diethyl ether layer is transferred to a clean tube and evaporated at room temperature in a vortex evaporator at reduced pressure for 10–15 min. Just before HPLC analysis the residue is reconstituted in 50 μL of mobile phase, vortexed for 10 s and centrifuged. A 20 μL volume is injected on to the column.

Urine sample preparation. The following are mixed: internal standard (erythromycin B, 0.2 mg/mL) (10 μL); urine sample (100 μL); acetonitrile (200 μL). The mixture is vortexed for 10 s and then centrifuged at 800**g** for 2 min. A 20 μL volume is injected into the chromatograph.

Retention times: Erythromycin A, 9.6 min; erythromycin B, 14.22 min; erythromycin ethylsuccinate, 19.5 min; erythromycin estolate, 22.86 min. There is good resolution between these compounds.

Assay characteristics. The assay was linear ($r^2 = 0.9999$) for erythromycin A in plasma, over the range 0.25–5 mg/L. Reproducibility at 0.5 and 2.5 mg/L ($n = 6$) gave coefficient of variation (CV) values of 1.5% and 2.3% respectively. Day-to-day reproducibility was 3.9% and within-day precision usually between 1.5 and 3.6%. Recovery of erythromycin A from plasma was virtually 100%. Specificity has been tested against a number of drugs (including oleandomycin and rifampicin), none of which interfered, and also against blank human serum which produced no interfering peaks.

For urine the assay is linear ($r^2 = 0.9999$) over the range 2.5–50 mg/L.

Comments. The original reference did not include details of the type of analytical cell used with the electrochemical detector, or about whether or not a guard cell was employed. One assumes that a 5010 analytical cell and a 5020 guard cell (set at a potential of about +0.95 V) should be used.

Microbiological assay of erythromycin

As mentioned previously, erythromycin is often administered as an ester which is hydrolysed *in vivo* to erythromycin base. If ester is still present in the samples, hydrolysis

Table II. Microbiological assay of macrolides (methods based on the microbiological assay method described in *Chapter 4*)

Antimicrobial	Plasma calibrator range (mg/L)	Well size (mm)	Incubation temperature (°C)	LLS	CV%	References
Erythromycin	0.03–0.5	7	30	0.01	9.39	Kavi *et al.* (1988)
Dirithromycin (erythromycycline)	0.06–1	5	37	0.06	8.88	Unpublished data
Azithromycin	0.01–0.64		37	0.01	9.44	Baldwin *et al.* (1990)
Roxithromycin	0.5–8	7	37 tor 1 h	0.5	8.9	Wise *et al.* (1987)
Spiramycin	0.06–1	7	30	0.06	9.7	Kavi *et al.* (1988)

LLS, Lower limit of sensitivity.
In all of these assays the medium used was anti I (CM327) adjusted to pH 8.5; the surface of the plate was flooded with a suspension of *M. lutea* NCTC 8340/ATCC 9341 with an optical density of 0.05 at 630 nm.

of the ester continues during the assay procedure, which is undertaken at pH 8. Unexpectedly high levels have been observed when concentrations obtained with the estolate have been compared with those of the stearate and the base. The explanation for these spurious results is the continued hydrolysis of the estolate in the assay plate (Croteau *et al.*, 1987). To ensure that hydrolysis is complete, plasma samples diluted with an equal volume of pH 8 buffer can be incubated for 2 h before assay against erythromycin base calibrators. Results of protein binding have been variable so it would seem prudent to prepare calibrators in pooled human serum. Table II outlines a plate assay method suitable for the assay of erythromycin in clinical samples.

Other published assay methods for erythromycin

Determination of erythromycin and its esters in plasma has been achieved by the use of fast atom bombardment mass spectrometry (Taskinen & Ohoila, 1988) but this specialized technique is beyond the scope of this book since few laboratories are likely to have the facilities for this type of assay.

Clarithromycin

Introduction

Clarithromycin (6-*O*-methylerythromycin A; mol. wt 747.96) is structurally similar to erythromycin (Figure 13.1), but is far more stable in acidic conditions. Eight metabolites of clarithromycin have been isolated from human urine and faeces, but only 14-hydroxyclarithromycin is present in significant quantities in plasma (Peters & Clissold, 1992). 14-Hydroxyclarithromycin is formed by hydroxylation at position 14 of the lactone ring. This major metabolite has identical in-vitro activity to that of the parent drug and it is claimed by some to enhance (either additively or synergically) the antibacterial activity of the

parent compound (Neu, 1991). Unlike erythromycin, clarithromycin is stable to gastric acid and is rapidly absorbed after oral administration, with a bioavailability of 52–55% (Langtry & Brogden, 1997). Absorption is not affected by food and the drug is well tolerated, with little gastro-intestinal disturbance. Peak serum concentrations (1–2 h after dose) of 1–2.5 mg/L have been achieved following a 250 or 500 mg dose (Neu, 1991). The plasma elimination half-life of clarithromycin is 3.5–5 h and the apparent half-life of the 14-hydroxymetabolite is 7 h. In contrast with erythromycin, a significant proportion of the dose (20–40%) is excreted in urine. The extent to which clarithromycin binds to protein has been reported as 42–50% at clarithromycin concentrations between 0.25 and 5 mg/L and 65–70% at concentrations between 0.45 and 4.5 mg/L.

A decrease in protein binding has been observed with an increase in concentration above 1 mg/L (Peters & Clissold, 1992).

Assays for clarithromycin—general information

Clarithromycin laboratory reference powder is available from Abbott Laboratories, but supplies of the metabolite 14-hydroxyclarithromycin are limited and may prove difficult to obtain. Clarithromycin is extremely stable and does not degrade under normal storage conditions. Its solubility in water, 0.01 M aqueous HCl and methanol is 0.07, 7.6 and 9.0 mg/mL, respectively.

HPLC assay method of Chu et al. (1991)

Column. C_8 Nucleosil, 250 × 4.6 mm, 5 μm.

Detection. An ESA 5100A coulometric detector with screening and working electrodes set at +0.5 V and +0.78 ± 0.4 V respectively.

Mobile phase. Acetonitrile:methanol:water (39:9:52, v/v) containing 0.04 M sodium dihydrogen phosphate, pH adjusted to 6.8 with sodium hydroxide, pumped at 1.2–1.4 mL/min.

Sample preparation (plasma). The following are mixed together: 750 ng (e.g. 75 µL of a 10 mg/L solution in a 1:1 (v/v) mixture of acetonitrile and water) of erythromycin A 9-*O*-methyl-oxime (Abbott 41036), the internal standard; plasma sample (500 µL); 0.1 M sodium carbonate solution (200 µL); a 1:1 (v/v) mixture of ethyl acetate and hexane (3000 µL). The mixture is then stirred vigorously for 1 min. Following centrifugation at 800**g** for 5 min, the organic layer is transferred to a clean tube and the solvent evaporated to dryness at 45°C under a stream of air. The residue is reconstituted in 200–400 µL of a 1:1 (v/v) mixture of acetonitrile and water. Between 20 and 80 µL are injected on to the column.

Sample preparation (urine). This is similar to that for plasma samples. The following are mixed together: 3 µg of internal standard (e.g. 300 µL of a 10 mg/L solution in 50% aqueous acetonitrile); urine sample (200 µL); 0.1 M sodium carbonate solution (100 µL); 3000–4000 µL of a 1:1 (v/v) mixture of ethyl acetate:hexane. The tubes are mixed and centrifuged as described above. The organic layer is transferred to a clean tube and the solvent evaporated to dryness. Residues are dissolved in 800–1200 µL of 50% aqueous acetonitrile.

Retention times. 14-Hydroxyclarithromycin, 10 min; clarithromycin, 20 min; internal standard, 27 min.

Assay characteristics. Clarithromycin is well resolved from the internal standard, 14-hydroxyclarithromycin and other metabolites (e.g. 14(*S*)-hydroxyclarithromycin and M7, an anhydro derivative). A late eluting peak appeared on the chromatogram after about 60 min, so that the injection intervals had to be adjusted to avoid interference from this peak. Normal pooled human plasma and urine and pre-dose plasma and urine samples did not cause interference.

The assay is linear over the range 0.1–2.0 mg/L in plasma for both clarithromycin and 14-hydroxyclarithromycin, with correlation coefficients of 0.9999 and 0.9989, respectively. For urine over the range 1.0–100.0 mg/L correlation coefficients of 0.9995 for clarithromycin and 0.9998 for 14-hydroxyclarithromycin were obtained.

Recovery from plasma was >85% for clarithromycin, 14-hydroxyclarithromycin and the internal standard. From urine, at 10 mg/L and 40 mg/L recoveries were 97.3% and 95.3% for clarithromycin and 104.8% and 103.1% for 14-hydroxyclarithromycin. Lower limits of detection were 0.03 mg/L for both parent and metabolite in plasma (0.5 mL samples); 0.5 mg/L in urine (0.2 mL sample).

Comments. Although not stated in the publication, a model 5020 guard cell set at +0.83 ± 0.04 V would be suitable for use with the stated electrochemical detector settings and it is most likely that the type of analytical cell employed was a model 5010.

During sample extraction care should be taken not to transfer any aqueous phase along with the organic solution, because 14-hydroxyclarithromycin breaks down when heated at 45°C in the presence of sodium carbonate.

This method has been employed in the authors' laboratory for the assay of clarithromycin in human serum with some minor modifications: a 150 × 4.6 mm, 5 µm, C_8 Nucleosil column and a flow rate of 1.5 mL/min were employed; the residue from the extraction procedure was reconstituted in 100% acetonitrile. Retention times of approximately 6, 9 and 12 min were observed for 14-hydroxyclarithromycin, clarithromycin and the internal standard, respectively, and all three compounds were well resolved.

Microbiological assay of clarithromycin

As mentioned previously, clarithromycin is metabolized in humans and the major metabolite, 14-hydroxyclarithromycin, has equivalent microbiological activity to that of the parent drug. As yet no organisms have been found which can differentiate between the activity of clarithromycin and 14-hydroxyclarithromycin, so microbiological assays are not considered suitable. An attempt to utilize microbiological assays has been made by some workers who have used a conversion factor to estimate the amount of parent drug in a sample (Fish *et al.*, 1993). However, this process assumes a fixed ratio of parent drug to metabolite which may not be the case at all sites in the body. In other animal species, clarithromycin may not be metabolized to the active major metabolite and therefore a microbiological assay based on those used for other macrolides may be appropriate (Veber *et al.*, 1993).

Dirithromycin

Introduction

Dirithromycin (mol. wt 835.1, pK_a 5.2 and 9.3) is a 14-membered ring macrolide and is the C9-oxazine derivative of erythromycylamine (Figure 13.2). It is non-enzymically hydrolysed *in vivo* and *in vitro* to the microbiologically active erythromycylamine (Figure 13.2). Both dirithromycin and erythromycylamine are found in plasma, tissues and cells (Kirst & Sides, 1989; Bergogne-Berezin, 1993). However, in a ^{14}C-radiolabelling study it was shown by Sides *et al.* (1993) that 60–90% of the administered dose was hydrolysed to erythromycylamine in 35 min of infusion and that by 1.5 h conversion to erythromycylamine was virtually complete. Dirithromycin may therefore be considered a pro-drug of erythromycylamine.

Dirithromycin and erythromycylamine have equivalent antibacterial activity, which is similar to that of erythromycin (Barry *et al.*, 1993). Plasma concentrations and AUCs are low because it penetrates rapidly into tissue ($C_{max} = 0.3$ mg/L, AUC = 1.0 g·h/L following 500 mg dose).

However, unlike erythromycin, the plasma elimination half-life is long ($t_{1/2}$ = 20–50 h) allowing a once-daily dosing schedule (Sides *et al.*, 1993). Only 17–25% of the dose is excreted in urine as erythromycylamine, the primary route of excretion being faecal/hepatic. Plasma protein binding of 15–30% as determined by ultracentrifugation is modest compared with erythromycin (Sides *et al.*, 1993).

Dirithromycin/erythromycylamine assay

Laboratory reference powders for both dirithromycin and erythromycylamine are available from Lilly Laboratories. Solutions can be prepared by dissolving a known weight of powder in a few drops of absolute ethanol and then making the solution up to the volume required with water. As mentioned previously, dirithromycin spontaneously converts to erythromycylamine. Microbiological assays are therefore normally performed using erythromycylamine calibrators.

HPLC assay of Whitaker & Lindstrom (1988)

Column. An Analytichem Sepralyte 2DP, 250 × 4.6 mm, 5 μm column maintained at 4°C.

Detection. An ESA 5100A coulometric detector with a 5020 guard cell (potential set at +1.0 V) and a 5011 analytical cell with first and second electrode potentials set at +0.7 and +0.9 V, respectively.

Mobile phase. Acetonitrile:methanol:water (60:10:30) with a final ammonium acetate buffer concentration of 50 mM and an apparent pH of 7.5 pumped at 1 mL/min.

Sample preparation. The following are mixed: plasma sample (500 μL); pH 10 buffer (1000 μL); 250 ng (in 10 μL methanol) of LY023907, the internal standard. The solution is mixed for 10 s. Dichloromethane (5000 μL) is added, the samples mixed for 10–15 s, and then centrifuged at 1000**g** for 10 min. The dichloromethane layer is removed to a clean tube and the extraction repeated using a further 5000 μL of dichloromethane. The two dichloromethane extracts are combined, evaporated to dryness under nitrogen and then reconstituted in 100 μL acetonitrile. Between 40 and 60 μL are injected on to the column.

Retention times: LY023907 (the internal standard), 6 min; dirithromycin, 8 min; erythromycylamine, 21 min.

Assay characteristics. The assay was linear over the range 0.02–0.5 mg/L, correlation coefficients were 0.999 and 0.996 for dirithromycin and erythromycylamine, respectively. Intra-assay CVs were 2–11% for dirithromycin and 3–12% for erythromycylamine depending on the concentration assayed (the assay was less reproducible at lower concentrations). Recovery from plasma at 0.05, 0.2 and 0.35 mg/L was 102–111% for dirithromycin and 91–106% for erythromycylamine.

Erythromycin was well resolved from erythromycylamine and dirithromycin, but roxithromycin had a retention time similar to that of dirithromycin. Specificity was not tested against other antimicrobial agents. Blank dog plasma did not produce any interfering peaks.

Microbiological assay of dirithromycin/erythromycylamine

As mentioned above, it is suggested that erythromycylamine powder is used for the preparation of calibrators and controls; however, dirithromycin laboratory reference powder has been used to assay concentrations of drug in tissue (Bergogne-Berezin, 1993). A good correlation between HPLC and microbiological assays has been observed (Bergogne-Berezin, 1993) and no other metabolites of dirithromycin have been detected in serum (Sides *et al.*, 1993). Details of a suggested microbiological assay are given in Table II.

Azithromycin

Introduction

Azithromycin (mol. wt 749.0, although the powder is usually the dihydrate form which has a mol. wt of 785.03, pK_a 8.48) is the prototype of a new class of antimicrobial agents, the azalides (Bright *et al.*, 1988). It is structurally different from erythromycin because it has a 15-membered lactone ring with a methyl-substituted nitrogen at position 9a (Figure 13.3). This structural difference gives azithromycin superior stability in acidic conditions. A metabolite of azithromycin, 9a-*N*-desmethylazithromycin, with some antimicrobial activity has been identified (Shepard, 1991), although its concentration in plasma, human urine and bile is low and it does not contribute significantly to antimicrobial activity (G. Foulds, personal communication). Azithromycin has an enhanced microbiological activity compared with erythromcyin, significantly so in its activity against Gram-negative organisms (Retsema *et al.*, 1987). Concentrations of azithromycin in tissues and cells are greatly in excess of simultaneous serum concentrations and also persist for days after the cessation of dosing (Baldwin *et al.*, 1990; Foulds *et al.*, 1990). Following a 500 mg dose a mean peak serum concentration of 0.4 mg/L at a t_{max} of 1.7 h with a $t_{1/2}$ of 11–14 have been observed (Foulds *et al.*, 1990). Serum protein binding of azithromycin has been determined by equilibrium dialysis and found to be concentration-dependent, being approximately 50% at concentrations of 0.02 and 0.05 mg/L and declining to 12% at concentration of 0.5 mg/L (Foulds *et al.*, 1990). Azithromycin is well tolerated, producing little gastro-intestinal disturbance.

Assays for azithromycin—general information

Azithromycin is highly soluble (>100 mg/mL) in ethanol, methanol, acetonitrile and dimethylsulphoxide; its solubility at pH 6.9 in pure water is 1.1 mg/mL, giving a final pH of 7.8. Azithromycin is stable under mild alkaline conditions but undergoes slow hydrolysis of the cladinose sugar to form desosaminylazithromycin under acidic conditions. This breakdown product has no in-vitro antimicrobial activity against organisms sensitive to erythromycin and azithromycin (Feise & Steffen, 1990).

Azithromycin in human serum or tissue stored at –20°C is stable for at least 2 and 3 months, respectively (Shepard, 1991). Serum samples stored at –70°C are stable for 8 months and thawing and refreezing does not have any adverse effect. Azithromycin appears to be stable in human urine at –20°C for at least 4 months (G. Foulds, personal communication) and this stability is not impaired if the urine is heated at 56°C for 1 h, a procedure which may be necessary if samples are obtained from patients infected with hepatitis B or HIV.

The HPLC assay of Shepard *et al.* (1991) has two separate methods: a coulometric method for serum and an amperometric method for serum and tissues. Both are described below.

Coulometric HPLC assay method of Shepard et al. (1991)

Column. Chromegabond alkylphenyl 50 × 4.6 mm, 5 μm with a guard column of glass beads, 21 × 3.0 mm, 40 μm (ES Industries, Marlton, NJ, USA).

Detection An ESA 5100A Coulochem electrochemical detector (Environmental Sciences Association) with a 5020 guard cell (set at +1.0 V) and a 5010 analytical cell, operated in the oxidative screen mode with applied potentials of the screen and detector electrodes of +0.7 and +0.8 V.

Mobile phase. 0.02 M ammonium acetate:0.02 M sodium perchlorate:acetonitrile:methanol (22:23:45:10, by volume), with the apparent pH adjusted to 6.8–7.2 with concentrated acetic acid, pumped at 1 mL/min.

Serum sample preparation. Two ranges of calibrators are used: 0.02–0.25 and 0.2–2.0 mg/L. Serum sample (500 μL), internal standard solution (0.25 or 2.5 μg of the 9a-*N*-*n*-propyl analogue of azithromycin in acetonitrile for the low and high range calibrators, respectively) (25 μL) and 0.03 M potassium carbonate (1000 μL) are placed in a tube and vortex mixed. (For an assay range of 0.01–0.1 mg/L, 1000 μL aliquots of serum and 0.25 μg of internal standard can be used). Methyl-*tert*-butyl (5000 μL) ether is added, the samples are vortexed for 30 s and centrifuged at 2000**g** for 10 min. The ether layer is transferred to a clean tube and evaporated to dryness at 37°C under reduced pressure in a vortex evaporator. The residue is reconstituted in 300 μL of mobile phase by vortex mixing for 30 s; 300 μL hexane is added, the sample is mixed and the phases are separated by centrifugation. The hexane layer is discarded and 100 μL of the aqueous layer is injected on to the column.

Amperometric HPLC assay method of Shepard et al. (1991)

Column. Chromegabond RP-1 alumina, 150 × 4.6 mm, 5 μm with a guard column of glass beads, 21 × 3 mm, 40 μm (ES Industries, Marlton, NJ, USA).

Detection. A BAS LC-4B amperometric electrochemical detector (Bioanalytical Systems) with a glassy carbon electrode set at an oxidation potential of +0.8 V relative to a silver/silver chloride reference electrode.

Mobile phase. Potassium phosphate (0.05 M):acetonitrile (70:30, v/v), adjusted to an apparent pH of 11.0 with 1.0 M potassium hydroxide, and pumped at 1 mL/min.

Serum sample preparation. Two ranges of calibrators are used: 0.01–0.2 and 0.2–2.0 mg/L. Serum sample (500 μL), internal standard solution (0.05 or 0.5 μg of the 9a-*N*-propargyl analogue of azithromycin in acetonitrile for the low and high range, respectively) (25 μL) and 0.06 M potassium carbonate (500 μL) are placed in a tube and vortexed. Five millilitres of methyl-*tert*-butyl ether is added, the samples are vortex mixed and centrifuged at 2000**g** for 10 min. The ether layer is transferred to a clean tube. For samples in the low concentration range, 500 μL of 0.015 M citric acid pH 3.1 is added to the ether layer, and this is vortexed for 30 s and centrifuged at 2000**g** for 5 min. The ether layer is removed and then the aqueous layer is alkalinized with 1 mL of 0.06 M potassium carbonate and then vortex mixed for 1 min. Further extraction with 5 mL of methyl-*tert*-butyl ether is then performed and this ether layer is transferred to a clean tube. This final ether layer for both high and low range samples is evaporated to dryness at 37°C under reduced pressure in a vortex evaporator. The residue is reconstituted in 300 μL (for the low range) or 500 μL (for the high range) of 1:1 (v/v) acetonitrile:water, with vortex mixing and washed with one volume of hexane. The phases are separated by centrifugation and the hexane layer discarded. A 60 μL volume of the aqueous phase is injected on to the column.

Tissue sample preparation. Calibrators covering four concentration ranges are employed, namely 0.1–2, 1–10, 10–100 and 100–1000 mg/kg. An accurately weighed portion (approximately 1 g) of each tissue sample is placed in a tube, and (assuming 1 mg is equivalent to 1 μL) nine volumes of acetonitrile and 50–200 μL of internal standard solution in methanol (in amounts of 1, 5, 50 or 500 μg depending on which calibrator range is used) is added. The sample is homogenized with a tissue homogenizer for 10 s and centrifuged at 2000**g** for 10 min. Five hundred microlitres of supernatant is pipetted into a clean tube and evaporated to dryness at 50°C under reduced pressure in a vortex evaporator. The residue is reconstituted in 500 μL of 0.06 M potassium carbonate and extracted with 5 mL methyl-*tert*-butyl ether as described for serum samples. The ether layer is transferred to a clean tube, evaporated to dryness and reconstituted in 1 mL (300 μL only for the lowest concentration range) of 1:1 (v/v) acetonitrile:water with vortex mixing. Washing with one volume of hexane is carried out as for serum samples, and 20–60 μL of aqueous phase is injected.

Comments on both coulometric and amperometric assays for azithromycin

Retention times. Coulometric assay: azithromycin, 9 min; internal standard 13 min. Amperometric assay: azithromycin, 8 min; internal standard, 10 min.

Specificity. Possible azithromycin metabolites (3'-N-desmethylazithromycin, 9a-N-desmethylazithromycin, descladinose azithromycin and azithromycin-3'-N-oxide) eluted earlier than azithromycin and therefore did not interfere.

Linearity. For both assays correlation coefficients of >0.997 were obtained for each range of calibrators.

Recoveries. Recoveries were 88% from rat serum (0.1 mg/L), 86% from human serum (0.5 mg/L), 85% from rat liver (10 mg/kg) and 84% from rat kidney (50 mg/kg).

Sensitivity limits. These were 0.01 mg/L in serum (both coulometric and amperometric assays) and 0.1 mg/kg in tissues.

Precision. Intra-assay CVs ranged from 1% to 8% for both assays (serum and tissues). Inter-assay CVs ranged from 4% to 11%.

Additional information. The coulometric assay was only used for serum samples, but it may be possible to use it for tissue samples. Several concentration ranges were used for both assays, as larger ranges seemed to deviate from linearity.

Microbiological assay of azithromycin

While a microbiologically active metabolite of azithromycin is known (9a-N-desmethylazithromycin), its concentration in plasma, human urine and bile is low and it does not contribute significantly to antimicrobial activity. In validation studies, undertaken with Pfizer Laboratories, comparing an HPLC-ECD assay and a microbiological assay an r^2 value of 0.910 was obtained (the authors' unpublished results). A methodological problem which might be encountered is the measurement of azithromycin in fluids which may contain phagocytes, since the drug accumulates in these cells. If cells loaded with azithromycin are disrupted into the fluid for assay, as may be the case with sputum samples, then falsely elevated concentrations of the drug may be observed. Details of a suggested microbiological assay method are given in Table II.

Roxithromycin

Introduction

Roxithromycin (an erythromycin A, 9-[O-{(2-methoxyethoxy)-methyl}oxime] derivative; mol. wt 837.1) is similar in structure to erythromycin (see Figure 13.1) but, unlike erythromycin, its oxime derivatives are more stable under acid conditions. Its spectrum of antibacterial acitivity is similar to that of erythromycin. Serum concentrations are considerably higher than those found with erythromycin (Wise *et al.*, 1987): at steady state following twice-daily dosing of 150 mg, mean peak serum concentrations of 9.2 mg/L at 1 h were achieved. The plasma elimination half-life of roxithromycin is longer than that of erythromycin, being 13.2 h (Wise *et al.*, 1987). Absorption is not significantly affected by food. Roxithromycin is approximately 86% protein bound in human serum. The protein fraction in human serum to which roxithromycin is preferentially bound is the acute-phase protein, α_1 acid glycoprotein. This is not found in all animal species, so wide variation in protein binding has been observed depending on animal species (Andrews *et al.*, 1987). Although generally well tolerated, gastrointestinal disturbances, headache, dizziness, rash, reversible liver function tests (LFTs) and glucose, increased eosinophils and platelets have been reported. The half-life of simultaneously administered theophylline is increased (by approximately 10%), but there is no effect on carbamazepine pharmacokinetics.

Roxithromycin assay—general considerations

Roxithromycin laboratory reference powder may be obtained from Roussell UCLAF (Romainville, Paris, France). This powder may be solublized in methanol or ethanol to obtain a stock solution from which to prepare calibrators or controls.

Roxithromycin (9 mg/L) appears to be stable in plasma when stored at either 4, 20 or 37°C over 48 h (Demotes-Mainaird *et al.*, 1989). Because of variations in protein binding between species, the use of animal serum in the place of human serum is not recommended.

HPLC assays of roxithromycin

A number of HPLC-ECD assays have been reported which may be suitable for determining roxithromycin in serum (Grgurinovich & Matthews, 1988; Whitaker & Lindstrom, 1988) and in serum and urine (Demotes-Mainaird *et al.*, 1989). The method of Grgurinovich & Matthews (1989) is described below.

HPLC assay method of Grgurinovich & Matthews (1989)
Column. Methyl, 250 × 4.6 mm, 5 μm (ICI Australia, Instrument Division, Adelaide, Australia).

Detection. An ESA 5100A Coulochem electrochemical detector with a model 5010 cell, potentials set at +0.6 and +0.75 V for cells 1 and 2 respectively.

Mobile phase. 42% acetonitrile, 58% water containing 15 mM diammonium hydrogen phosphate adjusted to pH 7.0 with concentrated phosphoric acid, pumped at 2 mL/min.

Serum sample preparation. Serum sample (1000 μL), internal standard (25 mg/L erythromycin in aqueous solution) (100 μL), 1 M sodium carbonate solution (50 μL), and diethyl ether:isopentane (3:2) (6000 μL) are placed in a tube fitted with PTFE-lined screw caps. The tubes are mixed thoroughly by inversion and centrifuged for 5 min at 1500**g**. The lower aqueous phase is frozen by immersion in a dry ice–acetone bath and the upper, organic layer is transferred to a clean tube. The solvent is removed by vortex evaporation and the residue is reconstituted in 200 μL of a methanol:15 mM phosphate buffer (1:1) mixture (pH 6.5). One millilitre of hexane:octanol (19:1) is then added, the tubes are vortex-mixed, and the aqueous and organic layers separated by centrifugation for 5 min at 1500**g**. The lower, aqueous layer is transferred to a clean tube and 100 μL is injected.

Retention times. Erythromycin, 8 min; roxithromycin, 12 min.

Linearity. A typical linear regression equation was $y = 0.324x - 0.001$ ($r^2 = 0.999$).

Lower limit of sensitivity. This was 0.01 mg/L.

Recovery. Over the range 0.1–5 mg/L ($n = 10$), the mean recovery was 88.9 ± 3.5% for roxithromycin and 85.6 ± 3.8% for erythromycin.

Precision. Intra-assay CVs of 4.21% ($n = 10$) and 1.57% ($n = 10$) were found for 0.1 and 5.0 mg/L roxithromycin, respectively. Inter-assay CVs were 6.47% ($n = 24$) and 2.99% ($n = 24$) for 0.25 and 5.0 mg/L roxithromycin, respectively.

Specificity. Specificity was tested against a number of different drugs; some caused interference to the roxithromycin peak (for example, erythromycin estolate, amitriptylline) and others to the internal standard peak (for example, erythromycin ethylsuccinate, oleandomycin). Specificity was not tested against potential metabolites but no extra peaks which could be attributed to metabolites or degradation products were observed during analysis of samples taken for a pharmacokinetic study.

Microbiological assay of roxithromycin

Roxithromycin is not extensively metabolized, so correlation between microbiological and HPLC assay results is good (Puri & Lassman, 1987). Although *Bacillus subtilis* has been used to measure concentrations of roxithromycin in plasma samples, it is inhibited by antibiotic-free pooled human plasma and so is not recommended for the assay of roxithromycin (Andrews *et al.*, 1987). As mentioned above, roxithromycin is approximately 86% bound to human serum protein, and it preferentially binds the acute phase protein, α_1 acid glycoprotein. This protein is not found in all animal species so it is not advisable to attempt to substitute serum from other animal species for the preparation of calibrators or control. Details of a microbiological assay method are given in Table II.

Spiramycin

Introduction

Spiramycin is a 16-membered ring macrolide derived from *Streptomyces ambofaciens* (Pinnert-Sindico, 1954) which has a similar spectrum of antibacterial activity to that of erythromycin although MICs are generally higher (Chabbert, 1955). It is a complex of three major components which differ in the grouping at the 4-position of the hexadecadiene ring (Figure 13.4). The components are: spiramycin I (–OH, mol. wt 843); spiramycin II (*O*-acetyl, mol. wt 885) and spiramycin III (*O*-propionyl, mol. wt 899). The pK_a is 8.0. The proportions of these three compenents in commercial spiramycin are 63% (I), 24% (II) and 13% (II). Of the three major components only spiramycin I is detected in plasma (Dow *et al.*, 1985). Co-administration of spiramycin with food significantly reduces bioavailability (50%) and delays the time of the peak concentration from 4 to 6 h (Frydman *et al.*, 1988). Following a single 2 g dose of spiramycin, a mean peak concentration of 3.1 mg/L was achieved at 3.3 h; the plasma elimination half-life was 3.8 h (Kavi *et al.*, 1988). Spiramycin is not bound to plasma proteins to a significant

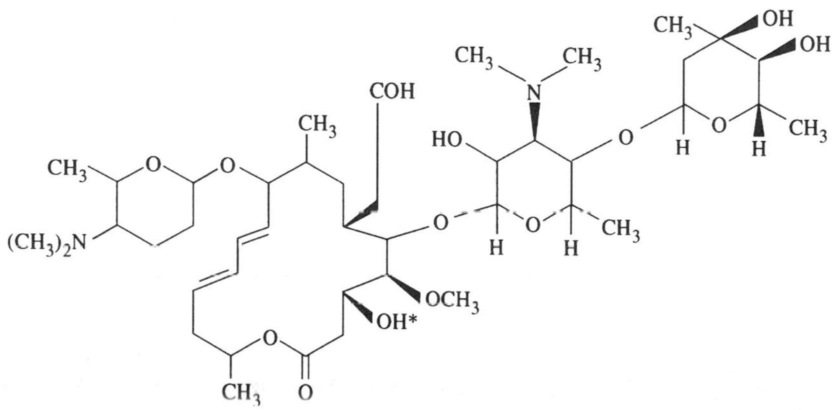

Figure 13.4 Structure of spiramycin I. * is $OCOCH_3$ in spiramycin II and $OCOCH_2CH_3$ in spiramycin III.

extent (approximately 17% to albumin and 6% to α_1 acid glycoprotein; Frydman *et al.*, 1988). Spiramycin is generally well tolerated, the commonest adverse reactions being gastrointestinal disturbance. However, rashes resulting from contact or industrial exposure have been reported.

HPLC assays of spiramycin

Allen *et al.* (1988) and Harf *et al.* (1988) report HPLC assay methods for determining spiramycin in tissues and alveolar macrophages. Both of these methods are modified versions of the assay method reported by Dow *et al.* (1985) which employs a column-switching technique and UV detection. This method is detailed below.

HPLC assay method of Dow *et al.* (1985)

Column. Nucleosil 5 C_8, 150 × 4.6 mm, 5 μm (Macherey-Nagel, Duren, Germany) with a pre-column of Perisorb RP-18, 50 × 7 mm, 30–40 μm (Merck, Darmstadt, Germany).

Detection. UV at 230 nm.

Column switching device. A Rheodyne 7010 six-port valve fitted with a Model 7001 pneumatic actuator and a solenoid valve for remote actuation. Remote-controlled column switching is carried out by a programmable external events time-relay connected to a computing integrator.

Sample application/column switching procedure. Acetonitrile (1250 μL of a 4% solution) followed by 250 μL of 4% acetonitrile containing 0.75 mg/L spiramycin II (the internal standard) are added to 1000 μL of plasma sample; 1000 μL of this mixture is then injected on to the pre-column (equilibrated with 4% aqueous acetonitrile, pumped at 2.5 mL/min). Four hundred seconds after injection, the pneumatic valve is switched automatically so that the mobile phase pumped by a second pump (26% acetonitrile in 2% perchloric acid (v/v) pumped at 1.8 mL/min) washes the pre-column, effectively back-flushing the compounds (from the sample) retained on the top of the pre-column on to the analytical column. Eighty seconds later the pneumatic valve is switched again, so that the second pump delivers mobile phase directly to the analytical column and the first pump again delivers its mobile phase to the pre-column, thus washing and re-equilibrating the pre-column whilst separation and elution continue on the analytical column. The valve was kept in this position throughout analysis (40 min) so that the pre-column is ready for the injection of the next sample.

Assay characteristics. The recovery of spiramycin I from plasma was 71.96% over the range 0.1–1 mg/L. A linear regression coefficient of 0.9965 was obtained from a five-point calibration curve (0.05–0.5 mg/L). The CV did not exceed 12% for six replicates at 0.05, 0.1, 0.2, 0.3 and 0.5 mg/L concentrations. The lower limit of the assay was 0.05 mg/L. Good agreement was found between this HPLC method and a microbiological method.

Additional information. The absorbance maximum for spiramycin in the eluent was 230 nm. Dilution of human plasma samples with 4% acetonitrile appeared to stabilize the spiramycin. Spiramycin II was used as the internal standard because its chemical structure is similar to that of spiramycin I. Commercial preparations of spiramycin contain a small percentage of spiramycin II, but it would appear that this amount is either too small to interfere or that spiramycin II is rapidly converted to spiramycin I *in vivo*.

Microbiological assay of spiramycin

Microbiological assays using *Micrococcus lutea* as indicator organism have been used to study the pharmacokinetics of spiramycin in humans (Kavi *et al.*, 1988; Kitzis *et al.*, 1988; Frydman *et al.*, 1988). In a study comparing HPLC with the microbiological assay method, good agreement was observed at concentrations of >1 mg/L; at concentrations of <1 mg/L microbiological assay results were approximately five times higher (Harf *et al.*, 1988). In a study measuring concentrations of spiramycin in plasma by HPLC, levels were two to three times higher than those previously reported (Allen *et al.*, 1988). Although methodological differences could not be excluded, the authors attributed the differences to multiple dosing. It would appear that methodological problems may exist with the assay of spiramycin. It would therefore seem essential that any assay developed be validated externally before clinical samples are analysed. A suggested microbiological assay method is outlined in Table II.

References

Abbott Laboratories. (1966). *Erythromycin. A Review of Its Properties and Clinical Status.* Abbott Laboratories, Scientific Division, North Chicago, IL.

Allen, H. H., Khalil, M. W., Vachon, D. & Glasier, M. A. (1988). Spiramycin concentrations in female pelvic tissues, determined by HPLC: a preliminary report. *Journal of Antimicrobial Chemotherapy* **22**, Suppl. B, 111–6.

Andrews, J. M., Ashby, J. P. & Wise, R. (1987). Factors affecting the in-vitro activity of roxithromycin. *Journal of Antimicrobial Chemotherapy* **20**, Suppl. B, 31–7.

Baldwin, D. R., Wise, R., Andrews, J. M., Ashby, J. P. & Honeybourne, D. (1990). Azithromycin concentrations at the sites of pulmonary infection. *European Respiratory Journal* **3**, 886–90.

Barry, A. L., Pfaller, M. A. & Fuchs, P. C. (1993). Dirithromycin disc susceptibility tests: interpretative criteria and quality control parameters. *Journal of Antimicrobial Chemotherapy* **31**, Suppl. C, 27–37.

Bartolucci, L., Gradol, C., Vincenzi, V., Ispadre, M. & Valori, C. (1984). Macrolide antibiotics and serum theophylline levels in relation to the severity of respiratory impairment: a comparison between the effects of erythromycin and josamycin. *Chemiterapia* **3**, 286–90.

Bergogne-Berezin, E. (1993). Tissue distribution of dirithromycin: comparison with erythromycin. *Journal of Antimicrobial Chemotherapy* **31**, Suppl. C, 77–87.

Biedenbach, D. J. & Jones, R. N. (1996). In vitro evaluation of new ketolides RU-64004 and RU-66647 against macrolide-resistant Gram-positive pathogens. In *Program and Abstracts of the Thirty-sixth Interscience Conference on Antimicrobial Agents and Chemotherapy, New Orleans, Louisiana, 1996.* Abstract F221, p. 138. American Society for Microbiology, Washington, DC.

Bright, G. M., Nagel, A. A., Bordner, J., Desai, K. A., Dibrino, J. N., Nowakowska, J. *et al.* (1988). Synthesis, in vitro and in vivo activity of novel 9-deoxo-9a-aza-9a-homoerythromycin A derivatives: a new class of macrolide antibiotics, the azalides. *Journal of Antibiotics* **41**, 1029–47.

Chabbert, Y. A. (1955). Etudes *in vitro* sur la spiramycine. Activité, resistance, antibiogramme, concentrations humorales. *Annales de l'Institut Pasteur* **89**, 434–46.

Chen, M. L. & Chiou, W. L. (1983). Analysis of erythromycin in biological fluids by high-performance liquid chromatography with electrochemical detection. *Journal of Chromatography* **278**, 91–100.

Chu, S. Y., Sennello, L. T. & Sonders, R. C. (1991). Simultaneous determination of clarithromycin and 14(R)-hydroxyclarithromycin in plasma and urine by high-performance liquid chromatography with electrochemical detection. *Journal of Chromatography* **571**, 199–208.

Croteau, D., Vallee, F., Bergeron, M. G. & Lebel, M. (1987). High-performance liquid chromatographic assay of erythromycin and its esters using electrochemical detection. *Journal of Chromatography* **419**, 205–12.

Demotes-Mainaird, F. M., Vincon, G. A., Jarry, C. H. & Albin, H. C. (1989). Micro-method for the determination of roxithromycin in human plasma and urine by high-performance liquid chromatography using electrochemical detection. *Journal of Chromatography* **490**, 115–23.

Dow, J., Lemar, M., Frydman, A. & Gaillot, J. (1985). Automated high-performance liquid chromatographic determination of spiramycin by direct injection of plasma, using column switching for sample clean-up. *Journal of Chromatography* **344**, 275–83.

Duthu, G. S. (1984). Assay of erythromycin from human serum by high-performance liquid chromatography with electrochemical detection. *Journal of Liquid Chromatography* **7**, 1023–32.

Fiese, E. F. & Steffen, S. H. (1990). Comparison of the acid stability of azithromycin and erythromycin A. *Journal of Antimicrobial Chemotherapy* **25**, Suppl. A, 39–47.

Fish, D. N., Gotfried, M. H., Danziger, L. H. & Rodvold, K. A. (1993). Penetration of clarithromycin into lung tissues at steady-state. In *Programme and Abstracts of the 18th International Congress of Chemotherapy, Stockholm.* June 27–July 2. Abstract 358.

Foulds, G., Shepard, R. M. & Johnson, R. B. (1990). The pharmacokinetics of azithromycin in human serum and tissues. *Journal of Antimicrobial Chemotherapy* **25**, Suppl. A, 73–82.

Frydman, A. M., Le Roux, Y., Desnottes, J. F., Kaplan, P., Djebbar, F., Cournot, A. *et al.* (1988). Pharmacokinetics of spiramycin in man. *Journal of Antimicrobial Chemotherapy* **22**, Suppl B, 93–103.

Grgurinovich, N. & Matthews, A. (1988). Analysis of erythromycin and roxithromycin in plasma or serum by high-performance liquid chromatography using electrochemical detection. *Journal of Chromatography* **433**, 298–304.

Griffith, R. S. & Black, H. R. (1962). A comparison of blood levels after oral administration of erythromycin and erythromycin estolate. *Antibiotic Chemotherapy* **12**, 398–403.

Harf, R., Panteix, G., Desnottes, J. F., Diallo, N. & Leclercq, M. (1988). Spiramycin uptake by alveolar macrophages. *Journal of Antimicrobial Chemotherapy* **22**, Suppl. B, 135–40.

Kavi, J., Webberley, J. M., Andrews, J. M. & Wise, R. (1988). A comparison of the pharmacokinetics and tissue penetration of spiramycin and erythromycin. *Journal of Antimicrobial Chemotherapy* **22**, Suppl. B, 105–10.

Kirst, H. A. & Sides, G. D. (1989). New directions for macrolide antibiotics: structural modifications and in vitro activity. *Antimicrobial Agents and Chemotherapy* **33**, 1413–8.

Kitzis, M., Desnottes, J. F., Brunel, D., Giudicelli, A., Jacotot, F. & Andreassian, B. (1988). Spiramycin concentrations in lung tissue. *Journal of Antimicrobial Chemotherapy* **22**, Suppl. B, 123–6.

Langtry, H. D. & Brogden, R. N. (1997). Clarithromycin, a review of its efficacy in the treatment of respiratory tract infections in immunocompetent patients. *Drugs* **53**, 973–1004.

Lorian, V. (Ed.) (1991). *Antibiotics in Laboratory Medicine*, 3rd edn. Williams & Wilkins, Baltimore, MD.

Neu, H. C. (1991). The development of macrolides: clarithromycin in perspective. *Journal of Antimicrobial Chemotherapy* **27**, Suppl. A, 1–9.

Peters, D. H. & Clissold, S. P. (1992). Clarithromycin: a review of its antimicrobial activity, pharmacokinetic properties and therapeutic potential. *Drugs* **44**, 117–64.

Pinnert-Sindico, S. (1954). Une nouvelle espèce de *Streptomyces* productrice d'antibiotiques: *Streptomyces ambofaciens* n.sp. Caractères culturaux. *Annales de l'Institut Pasteur* **87**, 702–7.

Prokocimer, P. & Hardy, D. (1991). In *Journées de l'Hôpital Claude-Bernard. Nouveaux Macrolides, Nouvelles Quinolones.* (Conland, J. P., Pocidalo, J. J., Vachan, F. & Vilde, J. L., Eds), pp. 123–44. Arnette, Paris.

Puri, S. K. & Lassman, H. B. (1987). Roxithromycin: a pharmacokinetic review of a macrolide. *Journal of Antimicrobial Chemotherapy* **20**, Suppl. B, 89–100.

Reeves, D. S., Phillips, I., Williams, J. D. & Wise, R. (Eds) (1978). *Laboratory Methods in Antimicrobial Chemotherapy.* Churchill Livingstone, Edinburgh.

Retsema, J., Girard, A., Schelkley, W., Manousos, M., Anderson, M., Bright, G. *et al.* (1987). Spectrum and mode of action of azithromycin (CP-62,993), a new 15-membered-ring macrolide with improved potency against gram-negative organisms. *Antimicrobial Agents and Chemotherapy* **31**, 1939–47.

Ryden, R. (1978). Erythromycin. In *Laboratory Methods in Antimicrobial Chemotherapy* (Reeves, D. S., Phillips, I., Williams, J. D. & Wise, R., Eds), pp. 208–11. Churchill Livingstone, Edinburgh.

Shepard, R. M., Duthu, G. S., Ferraina, R. A. & Mullins, M. A. (1991). High performance liquid chromatographic assay with electrochemical detection for azithromycin in serum and tissues. *Journal of Chromatography* **565**, 321–37.

Sides, G. D., Cerimele, B. J., Black, H. R., Busch, U. & DeSante, K. A. (1993). Pharmacokinetics of dirithromycin. *Journal of Antimicrobial Chemotherapy* **31**, Suppl. C, 65–75.

Stubbs, C., Haigh, J. M. & Kanfer, I. (1985). Determination of erythromycin in serum and urine by high-performance liquid chromatography with ultraviolet detection. *Journal of Pharmaceutical Sciences* **74**, 1126–8.

Taskinen, J. & Ohoila, P. (1988). Hydrolysis of 2'-esters of erythromycin. *Journal of Antimicrobial Chemotherapy* **21**, Suppl. D, 1–8.

Tsuji, K. (1978). Fluorimetric determination of erythromycin and erythromycin ethylsuccinate in serum by a high-performance liquid chromatographic post-column, on-stream derivatization and extraction method. *Journal of Chromatography* **158**, 337–48.

Tsuji, K. & Goetz, J. F. (1978). High-performance liquid chromatographic determination of erythromycin. *Journal of Chromatography* **147**, 359–67.

Veber, B., Vallée, E., Desmonts, J. M., Pocidalo, J. J. & Azoulay-Dupuis, E. (1993). Correlation between macrolide lung pharmacokinetics and therapeutic efficacy in a mouse model of pneumococcal pneumonia. *Journal of Antimicrobial Chemotherapy* **32**, 473–82.

Whitaker, G. W. & Lindstrom, T. D. (1988). Determination of dirithromycin, LY281389 and other macrolide antibiotics by HPLC with electrochemical detection. *Journal of Liquid Chromatography* **11**, 3011–20.

Wise, R., Kirkpatrick, B., Ashby, J. & Andrews, J. M. (1987). Pharmacokinetics and tissue penetration of roxithromycin after multiple dosing. *Antimicrobial Agents and Chemotherapy* **31**, 1051–3.

Chapter 14

Lincosamides: clindamycin and lincomycin

Andrew M. Lovering

Department of Medical Microbiology, Southmead Hospital, Bristol BS10 5NB, UK

Physico-chemical properties

Lincomycin (Figure 14.1) and clindamycin (Figure 14.2) are the only members of the lincosamide class of antibiotics to be used clinically in Europe. They exhibit good activity against Gram-positive organisms, notably staphylococci and streptococci, and anaerobic organisms, but have no useful activity against enterobacteria. Lincomycin was first described in 1963 from the fermentation broth of a soil streptomycete, *Streptomyces lincolnensis*, whilst clindamycin was synthesized in 1967 by the 7(*S*)-chloro substitution of the 7(*R*)-hydroxyl group of lincomycin (Birkenmeyer & Kagen, 1970). Both antibiotics are supplied for clinical use as the hydrochloride salts, with clindamycin additionally available as the phosphate and

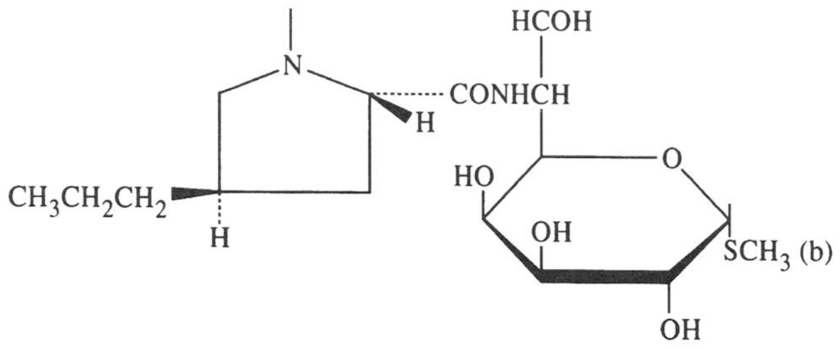

Figure 14.1 Structure of lincomycin and its major metabolites. (a) is H in desmethyl lincomycin and (b) is $SOCH_3$ in lincomycin sulphoxide.

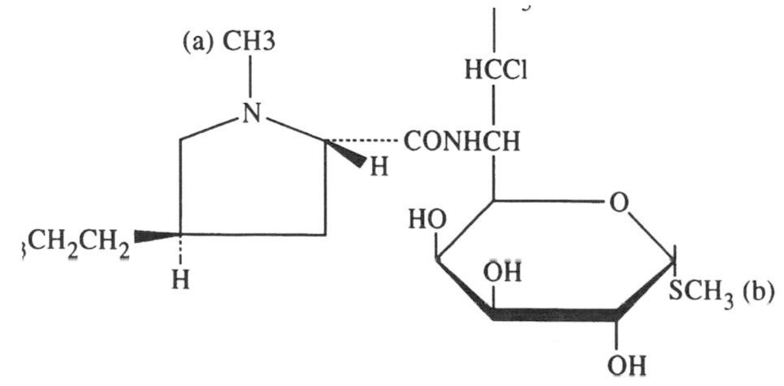

Figure 14.2 Structure of clindamycin and its major metabolites. (a) is H in desmethyl clindamycin and (b) is $SOCH_3$ in clindamycin sulphoxide.

the palmitate esters (Table I). Although both lincomycin and clindamycin are relatively soluble in water, a marked pH effect on solubility is observed. The solubility of clindamycin hydrochloride in water decreases from >160 g/L below pH 6 to <3 g/L at pH 7, while the solubility of the phosphate ester increases from 25 g/L at pH 4 to >200 g/L at pH 7 (Upjohn, unpublished data). The pK_a of both antibiotics is close to neutral (Table I). In the crystalline state both antibiotics are extremely stable and, although stability studies have shown clindamycin to be stable for >20 days at 70°C, it is recommended that dry drug powder is stored desiccated at 4°C. Aqueous solutions of ≤1 mg/mL of the antibiotics are stable for more than a month at 4°C, while serum calibrators are stable for at least 56 days at –20°C, similar stability being reported for bone samples (La Follette *et al.*, 1988). An evaluation of the effect of HIV deactivation procedures (56–58°C for 35–45 min) on plasma samples containing clindamycin (concentration of 75, 200 or 500 μg/L), found no statistically significant loss of potency (Gatti *et al.*, 1993).

Dosage forms available

For clinical use lincomycin is available in capsules (as the hydrochloride salt) or syrup (as the palmitate ester) for

Table I. Molecular weights and pK_a values of clindamycin and lincomycin formulations

	Molecular weight	pK_a
Lincomycin		
base	406.5	7.6
hydrochloride	443	–
Clindamycin		
base	425	7.7
hydrochloride	461.5	7.5
hydrochloride monohydrate	479.5	7.5
phosphate	504	5.8, 8.1
palmitate	700	–

oral administration, or as a solution (as the hydrochloride salt) for either iv or im administration. However, the use of lincomycin is declining and it is no longer available in a number of countries. For oral administration clindamycin is supplied in capsules as the hydrochloride salt or in a syrup as the palmitate ester. The palmitate undergoes rapid hydrolysis in the gut with little or none of the intact ester detectable in plasma (Brodasky & Sun, 1984). For iv and im administration clindamycin is supplied as the phosphate ester. Clindamycin is also available as a topical lotion consisting of a 1% preparation of the hydrochloride salt; there have been reports of up to 7.5% systemic absorption following application of this lotion to the whole face (Eller et al., 1989). For the preparation of laboratory calibrators, free clindamycin and lincomycin base should be obtained from the manufacturers (Upjohn). Alternatively, the hydrochloride salts (Table I) may be used after correction for potency. Clindamycin hydrochloride has a typical potency of 89%; whilst lincomycin hydrochloride has a typical potency of 92%. Neither the palmitate nor the phosphate esters are suitable for preparing calibrators as they do not readily yield free clindamycin.

Pharmacokinetics

The pharmacokinetic profiles of both lincomycin and clindamycin are complex because of their extensive metabolism to microbiologically active compounds and because clindamycin is administered as the phosphate ester, an inactive pro-drug. Unfortunately the reliability of many of the existing pharmacokinetic data on these compounds is somewhat suspect because of the extensive use of microbiological assay, which has the potential of interference from the active metabolites (especially the demethyl metabolites).

Both clindamycin and lincomycin are rapidly absorbed ($t_{1/2}$ of absorption < 10 min) following oral administration, both with bioavailability of >90%. Typically, peak serum concentrations are observed at 45–60 min for clindamycin and 2–3 h for lincomycin following an oral dose, although peak levels may be delayed if the antibiotics are given with food. Peak concentrations of 3–4 mg/L have been reported after a 300 mg oral dose of clindamycin, while concentrations of 2–7 mg/L have been reported after a 500 mg oral dose of lincomycin (Metzler et al., 1973). Following iv administration, peak serum concentrations of 5.4 mg/L have been found after a 300 mg dose of clindamycin, with concentrations of 10–16 mg/L being reported after a 600 mg dose of clindamycin and concentrations of 21 mg/L after a 600 mg dose of lincomycin (DeHaan et al., 1973; Metzler et al., 1973; Flaherty et al., 1988; Plaisance et al., 1989).

Following parenteral administration of clindamycin phosphate, although hydrolysis to free clindamycin is relatively rapid in most subjects ($t_{1/2}$ = 10 min), appreci-

able concentrations of the phosphate ester may persist for some time following a dose (Plaisance et al., 1989). Furthermore, the rate of hydrolysis of the phosphate ester is concentration-dependent, almost ceasing at low concentration, and at 8 h following an im dose the majority of the drug in serum is present as the phosphate (DeHaan et al., 1973). In contrast, lincomycin is administered parenterally as the hydrochloride salt which rapidly dissociates. Bioavailability of about 75% has been reported for im dosing with clindamycin phosphate (DeHaan et al., 1973) and 53% after oral dosing with clindamycin hydrochloride (Gatti et al., 1993).

Clindamycin and lincomycin have large volumes of distribution of 40–75 L (0.79–1.10 L/kg), and distribute widely both as the free drugs and, in the case of clindamycin, also as the phosphate ester (Plaisance et al., 1989; Gatti et al., 1993). They distribute into most human body tissues and fluids, with particularly good penetration into bone (80–110% of the serum concentrations) (Berger et al., 1978). However, extensive plasma protein binding of >90% for clindamycin and 70–80% for lincomycin may substantially decrease this apparent antimicrobial activity. Neither agent penetrates adequately into cerebrospinal fluid, even in the presence of bacterial meningitis (Garrod et al., 1981); however, it has been suggested that damage to the blood–brain barrier in patients with HIV infection may allow clindamycin to reach effective levels in the brain (Hofflin & Remington, 1987). The main routes of elimination of these antibiotics are by metabolism and the faecal (61% of an iv dose) route, with plasma $t_{1/2}$ values of 2–3 h for clindamycin and 4–6 h for lincomycin (Metzler et al., 1973; Flaherty et al., 1988; Plaisance et al., 1989; Gatti et al., 1993). Both antibiotics are extensively metabolized in the liver, with at least six metabolites identified for clindamycin (Table II), a number of which retain antimicrobial

Table II. The metabolites of clindamycin and lincomycin and their relative activities against *Micrococcus lutea*

Compound	Activity (% of clindamycin activity)
Clindamycin	100
N-demethyl clindamycin	140
clindamycin sulphoxide	15
hydroxylated clindamycin	13
N-demethyl clindamycin sulphoxide	3
clindamycose	3
5'-oxo-clindamycin	<0.5
5'-oxo-clindamycin sulphoxide	<0.5
Lincomycin	25
N-demethyl lincomycin	1
lincomycin sulphoxide	0.5

Data from Kucers & Bennett (1979) and Osono & Umezawa (1985).

activity. The relative proportions of the metabolites found in urine differ from those found in faeces, with a greater proportion of the oxidized metabolites found in faeces. Urinary excretion of clindamycin, following oral administration, has been reported as: clindamycin, 27%; clindamycin sulphoxide, 35%; N-demethyl clindamycin, 6%; N-demethyl clindamycin sulphoxide, 2% (Daniels et al., 1976). Other workers have reported urinary recoveries to be <10% of the dose, although it is not clear whether this is unchanged clindamycin (Leigh, 1981).

Although the pharmacokinetics of clindamycin and lincomycin do not appear to be altered in the aged, in children of <1 year, and particularly in neonates, prolonged half-lives of up to 8 h have been reported (Bell et al., 1984). Half-life is also longer in patients with severe liver damage. In contrast, quite severe renal impairment appears to have only a limited effect on the pharmacokinetics, and neither of the drugs is significantly removed by CAPD or haemodialysis. Relatively few side-effects, other than those affecting the gastrointestinal tract, have been reported. Faecal concentrations of 300–500 mg/kg are common. Up to 10% of patients treated develop diarrhoea, with a much smaller percentage going on to develop pseudomembranous colitis.

Assay considerations

Instances where there is a direct clinical need to monitor serum concentrations of either clindamycin or lincomycin are rare. The pharmacokinetics of both drugs are affected little by changes in renal function, elimination of parent drug by the renal route is generally considered to be <10%, and no dosage reduction is required in patients with changing renal function unless the degree of renal impairment is severe (Leigh, 1981). As both drugs are primarily cleared by hepatic metabolism, a case could be made for monitoring serum levels in patients with severe hepatic impairment. However, in practice these agents are rarely given in such patients. Clindamycin has also been used in the treatment of *Toxoplasma gondii* and *Pneumocystis carinii* infections in patients with HIV infection (Leport et al., 1989; Toma et al., 1993; Safrin et al., 1996). As the pharmacokinetics of clindamycin in patients with HIV infection differs significantly from that seen in healthy volunteers (Gatti et al., 1993), a case could be made for monitoring serum levels in this patient group. Serum clindamycin concentrations in patients with HIV infection are significantly higher than those in healthy volunteers and, although increased protein binding has been found, the reasons for the elevated concentrations are largely unknown (Flaherty et al., 1996).

Since instances where there is a need to monitor either clindamycin or lincomycin are so rare, clinical assay methodology for these compounds is poorly developed. Consequently laboratories attempting to set up such methods, for whatever reason, should expect a significant method work-up time. A number of workers have reported microbiological assay of these agents (Hanaka et al., 1962; Metzler et al., 1973; Daniels et al., 1976; Hextall et al., 1994), but none of these are specific for the parent compound in the presence of its metabolites. Since studies using methodologies specific for clindamycin (Flaherty et al., 1988; Plaisance et al., 1989; Gatti et al., 1993) have found that pharmacokinetic parameters differ significantly from those reported by workers using microbiological assay, we would question the validity of using microbiological assay for these compounds, particularly in the case of urine samples. Although the results from microbiological assay may be of adequate quality for clinical use, data generated for comparative or research purposes may be potentially misleading.

Although the assay of clindamycin has been reported using TLC (Onderdonk et al., 1981), HPLC (Brown, 1978; Landis et al., 1980; Munson & Kubiak, 1985; La Follette et al., 1988; Sarkar et al., 1991) and GLC (Plaisance et al., 1989; Gatti et al., 1993), a number of these methods have been developed for pharmaceutical preparations and are not relevant to clinical practice. Likewise, although a radioimmunoassay has been described for clindamycin, it is not available commercially and, again, is of limited relevance to clinical practice (Gilbertson & Styrd, 1976). To the best of our knowledge, there are no similar methods available for the assay of lincomycin in serum. However, owing to its very close chemical similarity with clindamycin, it is probable that assays developed for clindamycin will be applicable to lincomycin. All of the methods reported for the assay of clindamycin in serum are technically demanding and are probably only of relevance to those centres which are highly specialized and experienced in the use of GLC or HPLC.

As a result of the relatively slow in-vivo hydrolysis of clindamycin phosphate, blood samples from patients receiving this formulation require special treatment to prevent in-vitro degradation giving falsely elevated clindamycin levels. Samples should be held on ice, centrifuged as soon as possible and the serum stored at –20°C to prevent hydrolysis of the inactive ester by serum alkaline phosphatase. The addition of phosphate buffer pH 8 to give a final concentration of 0.3–0.5 M further retards hydrolysis of the ester and dilution of samples 1:1 in this buffer should be carried out if facilities for the rapid freezing of samples are not available. If samples are diluted in stabilizing buffer, the calibrators should be made up at double strength and also diluted in buffer because of the high protein binding of these compounds. However, evaluations have shown that dilution of calibrators is not required in GLC assay (Gatti et al., 1993). Should it be necessary to determine the concentration of the phosphate ester, this can be achieved by determining both the free clindamycin and total clindamycin (free plus phosphate ester) concentrations and subtracting one from the other. If a sample collection protocol using stabilizing buffer is in use, separate samples (without buffer) will need to be collected. The clindamycin phosphate is hydrolysed by adding 0.25 mL of a solution of diethanolamine and magnesium chloride buffer (pH 9.8) containing 100 U of alkaline phosphatase to 1 mL of sample and incubation at 37°C for 3 h (Plaisance *et al.*, 1989). Internal controls containing known concentrations of clindamycin, clindamycin phosphate and a mixture of clindamycin phosphate and clindamycin should be treated in an identical way.

Assay methods

Microbiological assay

The definitive method for assaying clindamycin and lincomycin is that described by the United States Code of Federal Regulations (1980) which uses a small plate design, with base and seed layers and application of antibiotic solutions using cylinders. However, in the UK it is normal practice for microbiological assay to use large assay plates and a single-layer medium. As a result the CFR method may be of limited relevance to most laboratories.

Cylinder-plate assay

Indicator strain. *Micrococcus lutea* ATCC 9341 is inoculated on to an agar slope and incubated for 24 h at 32–35°C. The agar slope is composed of bacteriological peptone 6.0 g, pancreatic digest of casein 4.0 g, yeast extract 3.0 g, beef extract 1.5 g, dextrose 1.0 g, and agar 15.0 g, with distilled water to 1000 mL; the pH is adjusted to pH 6.5–6.6. Media of this composition can be obtained from a number of manufacturers as 'Antibiotic Assay Medium No. 1'). After incubation the growth is washed from the slope with 3 mL of sterile saline and spread over a large agar surface, such as a Roux bottle, containing 250 mL of the same medium. After incubation at 32–35°C for 24 h the growth is harvested by washing with 50 mL of saline, diluted a further 1 in 40 in saline and stored at 4°C for up to 2 weeks.

Procedure. Assay plates are prepared by pouring 21 mL of medium (same formula as used for preparing the indicator strain, but with the pH adjusted to 7.8–8.0) into Petri dishes (90 mm diameter). A further 4 mL of this medium, seeded with 60 μL of the indicator strain suspension, is then added as a seed layer. Calibrators (0.64, 0.80, 1.00, 1.25 and 1.56 mg/L) and samples are added to the surface of the plates using cylinders (10 × 6 mm). Following incubation at 36–37°C for 18 h, zones of inhibition are read. The procedures described in *Chapter 4* should then be followed to determine the results.

Large-plate assay

This is a more practical method for laboratories accustomed to using large assay plate procedures based upon that reported by Hanaka *et al.* (1962). One hundred millilitres of molten Antibiotic Assay Medium 1 (see above), but with the pH adjusted to 7.8–8.0, are seeded with 1 mL of a suspension of *M. lutea* ATCC 9341 containing approximately 10^9 cfu/L. This indicator suspension may be prepared as described under the cylinder-plate assay. The seeded agar is poured into a level assay plate (250 mm × 250 mm), to give an agar depth of about 2 mm, and is left to set. Wells approximately 9 mm in diameter are cut in the agar, typically using a no. 5 cork borer, and filled with calibrators or samples. After overnight incubation at 37°C, zones of inhibition are measured. For the assay of serum samples, calibrators are prepared in antibiotic-free human serum at a concentration of 0.325, 0.65, 1.25, 2.5, 5.0 and 10.0 mg/L. For the assay of samples with a low protein content, such as bone or urine, calibrators are prepared at the same concentrations in 0.1 M pH 8.0 phosphate buffer. Using similar methodology a lower limit of sensitivity of 0.06 mg/L and a coefficient of variation of 8.7% have been reported (Hextall *et al.*, 1994).

Considerable variation has been observed in the dose–response of serum concentrations between patients and it is occasionally necessary to dilute patient samples (1:1 in antibiotic-free serum) to bring concentrations to within the calibrator range. Likewise it is recommended that urine samples are diluted 1:5 in 0.1 M phosphate buffer pH 8 before assay. For samples such as bone, tissue, ascitic fluid, bile and sweat, calibrators are prepared in 0.1 M phosphate buffer pH 8, and the samples are either diluted or homogenized with an appropriate volume of the same buffer. However, because of the high protein binding of lincomycin and clindamycin, the possibility of incomplete extraction must be considered and, where appropriate blank material is available, evaluated.

HPLC assays

The HPLC method for assaying clindamycin in serum (Flaherty *et al.*, 1988; La Follette *et al.*, 1988) is based upon ion-pair chromatography on a reversed-phase packing. Chromatography is performed on a Nova-Pak C_{18} column, using a mobile phase composed of acetonitrile:water:phosphoric acid:tetramethylammonium chloride (30:70:0.2:0.075), with the pH adjusted to 6.7, and pumped at a flow rate of 2 mL/min. Detection is by UV absorbance at 198 nm. Sample preparation consists of mixing 200 μL of sample with 500 μL of acetonitrile (containing the internal standard triazolam), centrifugation and then concentration of the resultant supernatant to 200 μL by evaporation under nitrogen. Using a 30 μL injection of the concentrated supernatant, the authors were able to detect clindamycin concentrations as low as 0.17 mg/L. The method is not suitable for detecting clindamycin metabolites.

Principally as a result of the low wavelength used, the authors experienced problems with oxygen quenching and thermal-drift instability. These were partially overcome by passing a constant stream of purified nitrogen over the detector's monochromator, by bubbling pure helium through the mobile phase reservoir, by insulating the column and by using overnight runs when 'activity in the laboratory and demand on circuitry were minimal'. In view of these difficulties we would have reservations about recommending the method for laboratories other than those highly specialized and experienced in the use of HPLC.

GLC assays

In the methods reported by Plaisance *et al.* (1989) and Gatti *et al.* (1993), plasma clindamycin concentrations were determined by the use of a capillary column gas–liquid chromatograph. Sample preparation involves liquid–liquid partitioning with back-extraction, using ethyl acetate at a basic pH followed by water at an acidic pH and then chloroform at a basic pH. Samples are then derivatized with heptofluorobutyric anhydride. The final clindamycin derivative is injected on to a GLC equipped with an electron capture detector. A structural analogue of clindamycin (U-33232E, Upjohn) is used as an internal standard. Helium carrier gas (flow rate 2 mL/min) and nitrogen (flow rate 35–40 mL/min) are used. Injector and detector temperatures are 225°C and 350°C respectively. A two-step temperature gradient of 165–200°C and then 200–250°C is used. Calibrators are prepared in plasma over the concentration range 0.025–0.75 mg/L and patient samples are diluted so that assayed concentrations fall within this range; an evaluation established that such dilution did not affect the calculated concentrations. The limit of detection is 0.025 mg/L. The clindamycin metabolites do not interfere with the assay of clindamycin.

Unfortunately, fuller methodological details were not given and any laboratory attempting to implement this method would encounter a number of difficulties and should expect a significant work-up time. Furthermore, without fuller details it is difficult to assess the technical difficulty involved with the assay, but as with the HPLC procedures, we would have reservations about recommending the method to laboratories other than those highly specialized and experienced in the use of GLC.

References

Bell, M. J., Shackelford, P., Smith, R. B. & Schroeder, K. (1984). Pharmacokinetics of clindamycin phosphate in the first year of life. *Journal of Pediatrics* **105**, 482–6.

Berger, S. A., Barza, M., Haher, J., McFarland, J. J., Louie, S. & Kane, A. (1978). Penetration of clindamycin into decubitis ulcers. *Antimicrobial Agents and Chemotherapy* **14**, 498–9.

Birkenmeyer, R. D. & Kagen, F. (1970). Lincomycin XI. Synthesis and structure of clindamycin. A potent antibacterial agent. *Journal of Medical Chemistry* **13**, 616–9.

Brodasky, T. F. & Sun, F. (1974). Determination of clindamycin 2-palmitate in clinical human serum samples. *Journal of Pharmaceutical Sciences* **63**, 360–5.

Brown, L. W. (1978). High pressure liquid chromatographic assays for clindamycin, clindamycin phosphate and clindamycin palmitate. *Journal of Pharmaceutical Sciences* **67**, 1254–7.

Daniels, E. G., Stafford, J. E. & Van Eyk, R. L. (1976). Technical Report 7261-75-7261-015. Upjohn.

DeHaan, R. M., Metzler, C. M., Schallenberg, D. & Van den Bosch, W. D. (1973). Pharmacokinetic studies of clindamycin phosphate. *Journal of Clinical Pharmacology* **13**, 190–209.

Eller, M. G., Smith, R. B. & Phillips, J. R. (1989). Absorption kinetics of topical clindamycin preparations. *Biopharmacology and Drug Disposition* **10**, 505–12.

Flaherty, J. F., Rodondi, L. C., Guglielmo, B. J., Fleishaker, J. C., Townsend, R. J. & Gambertoglio, J. G. (1988). Comparative pharmacokinetics and serum inhibitory activity of clindamycin in different dosing regimes. *Antimicrobial Agents and Chemotherapy* **32**, 1825–9.

Flaherty, J. F., Gatti, G., White, J., Bubp, J., Borin, M. & Gambertoglio, J. G. (1996). Protein binding of clindamycin in sera of patients with AIDS. *Antimicrobial Agents and Chemotherapy* **40**, 1134–8.

Garrod, L. P., Lambert, H. P. & O'Grady, F. (1981). *Antibiotics and Chemotherapy*, 5th edn, p. 183. Churchill Livingstone, Edinburgh.

Gatti, G., Flaherty, J., Bubp, J., White, J., Borin, M. & Gambertoglio, J. (1993). Comparative study of bioavailabilities and pharmacokinetics of clindamycin in healthy volunteers and patients with AIDS. *Antimicrobial Agents and Chemotherapy* **37**, 1137–43.

Gilbertson, T. J. & Styrd, R. P. (1976). Radioimmunoassay for clindamycin. *Clinical Chemistry* **22**, 828–31.

Hanaka, L. J., Mason, D. J., Burch, M. R. & Treick, R. W. (1962). Lincomycin a new antibiotic III. Microbial assay. *Antimicrobial Agents and Chemotherapy*, 565–9.

Hextall, A., Radley, S., Andrews, J. M., Boyd, E. J. S., Dent, J. C. & Donovan, I. (1994). Mucosal concentrations and excretion of clindamycin by the human stomach. *Journal of Antimicrobial Chemotherapy* **33**, 595–602.

Hofflin, J. M. & Remington, J. S. (1987). Clindamycin in a murine model of toxoplasmic encephalitis. *Antimicrobial Agents and Chemotherapy* **31**, 492–6.

Kucers, A. & Bennett, N. M. (1979) Lincomycin and clindamycin. 3rd Edn. In *The Use of Antibiotics*, pp. 470–95. Heinemann, London.

La Follette, G., Gambertoglio, J., White, J. A., Knuth, D. W. & Lin, E. T. (1988). Determination of clindamycin in plasma or serum by high-performance liquid chromatography with ultraviolet determination. *Journal of Chromatography—Biomedical Applications* **431**, 379–88.

Landis, J. B., Grant, M. E. & Nelson, S. A. (1980). Determination of clindamycin in pharmaceuticals by high performance liquid chromatography using ion-pair formation. *Journal of Chromatography* **202**, 99–106.

Leigh, D. A. (1981). Antibacterial activity and pharmacokinetics of clindamycin. *Journal of Antimicrobial Chemotherapy* **7**, Suppl. A, 3–9.

Leport, C., Bastuji-Garin, C., Perronne, C., Salmon, D., Marche, C. & Bricaire, F. (1989). An open study of the pyrimethamine-clindamycin combination in AIDS patients with brain toxoplasmosis. *Journal of Infectious Diseases* **160**, 557–8.

Metzler, C. M., De Hann, R., Schellenberg, D. & Van den Bosch, W. D. (1973). Clindamycin dose–bioavailability relationships. *Journal of Pharmaceutical Sciences* **62**, 591–8.

Munson, J. W. & Kubiak, E. J. (1985). A high performance liquid chromatographic assay for clindamycin phosphate and its principal degradation product in bulk drug and formulations. *Journal of Pharmaceutical and Biomedical Analysis* **3**, 523–33.

Onderdonk, A. B., Brodasky, T. F. & Bannister, B. (1981).

Comparative effects of clindamycin and clindamycin metabolites on the hamster model of antibiotic-associated colitis. *Journal of Antimicrobial Chemotherapy* **8**, 383–93.

Osono, T. & Umezawa, H. (1985). Pharmacokinetics of macrolides, lincosamides and streptogramins. *Journal of Antimicrobial Chemotherapy* **16**, *Suppl. A*, 151–66.

Plaisance, K. I., Drusano, G. L., Forrest, A., Townsend, R. J. & Standiford, H. C. (1989). Pharmacokinetic evaluation of two dosage regimens of clindamycin phosphate. *Antimicrobial Agents and Chemotherapy* **33**, 618–20.

Safrin, S., Finkelstein, D. M., Feinberg, J., Frame, P., Simpson, G., Wu, A. *et al.* (1996). Comparison of three regimens for treatment of mild to moderate *Pneumocystis carinii* pneumonia in patients with AIDS. A double blind randomised trial of oral trimethoprim–sulphamethoxazole, dapone–trimethoprim, and clindamycin–primaquine. ACTG Study Group. *Annals of Internal Medicine* **124**, 792–802.

Sarkar, M. A., Rogers, E., Reinhard, M., Wells, B. & Karnes, H. T. (1991). Stability of clindamycin phosphate, ranitidine hydrochloride and piperacillin sodium in polyolefin containers. *American Journal of Hospital Pharmacy* **10**, 2184–6.

Toma, E., Fournier, S., Dumont, M., Bouldec, P. & Deschamps, H. (1993). Clindamycin/primaquine versus trimethoprim–sulphamethoxazole as primary therapy for *Pneumocystis carinii* pneumonia in AIDS: a randomised, double-blind pilot trial. *Clinical Infectious Diseases* **17**, 178–84.

United States Code of Federal Regulations. (1980). The Office of the Federal Register, National Archives and Records Administration, Washington, DC.

Chapter 15

Fusidic acid

Nicholas M. Brown

Clinical Microbiology and Public Health Laboratory, Addenbrooke's Hospital, Cambridge CB2 2QW, UK

Introduction

Fusidic acid is a tetracyclic triterpenoid (steroid-like) antibiotic with a chemical structure based on the cyclopentanoperhydrophenanthrene ring (Figure 15.1) (Godtfredsen & Vangedal, 1962). It was originally isolated from the fermentation broth of *Fusidium coccineum* (Godtfredsen *et al.*, 1962a). There are several other triterpenoid antibiotics, of which helvolic acid and cephalosporin P1 are best known, but none are as active as fusidic acid, and, as yet, none have proved to be clinically useful (von Daehne *et al.*, 1979). The chemical structure of fusidic acid is difficult to represent in two dimensions due to the unique arrangement of the four carbon rings, and is unlike any other triterpenoid antibiotic (Godtfredsen *et al.*, 1965). It is this that may confer the unusual properties of the antibiotic. Several derivatives and modifications of fusidic acid have been studied in an attempt to produce alternative compounds with improved antibacterial activity or enhanced pharmacokinetic properties, but none has been as active as fusidic acid itself (von Daehne *et al.*, 1979).

Chemical properties

Fusidic acid is a colourless crystalline compound with a molecular weight of 517 and a melting point of 192–193°C

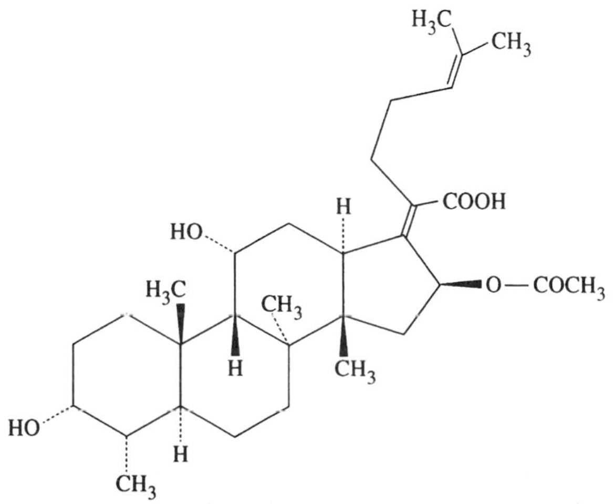

Figure 15.1 Structure of fusidic acid.

(Godtfredsen *et al.*, 1962b). It is an unsaturated carboxylic acid with a pK_a of 5.35, and is only sparingly soluble in water, although it will dissolve in alcohols, acetone and chloroform (Godtfredsen *et al.*, 1962a). The chemical structure is very similar to that of bile salts, and fusidic acid shares many of their physiochemical properties (Carey *et al.*, 1975). It is lipid-soluble, as shown *in vitro* by a tendency to diffuse from an aqueous phase to oil when the two are mixed (Stewart, 1964), and forms micelles when exposed to lipids, including the phospholipids of cell membranes. These characteristics may explain the good tissue penetration of fusidic acid and also its hepatic metabolism and excretion in bile.

For clinical use, the water-soluble salt sodium fusidate (mol. wt 539) is used in both iv and oral preparations. The 250 mg film-coated tablet is equivalent to 240 mg fusidic acid, although dosage regimens are usually expressed in terms of sodium fusidate. A flavoured oral suspension of fusidic acid as a hemihydrate (molecular weight 526) is also available (5 mL containing 250 mg, therapeutically equivalent to 175 mg sodium fusidate), and has replaced the older oily suspension (Masterson & Goldman, 1970). Preparations in a cream, ointment or gel are available for use topically, and a viscous 1% aqueous formulation is available for topical application to the eye.

For laboratory use the parenteral sodium salt may be used (500 mg, which is equivalent to 482 mg fusidic acid). The diethanolamine salt (mol. wt 622) is no longer available.

Pharmacokinetics

The pharmacokinetics of fusidic acid have been reviewed recently (Reeves, 1987). Current knowledge is far from complete, and is based on scattered information published over the last 30 years. Many early reports give data in a particular clinical situation from one or few patients and are unsatisfactory. In many cases these studies have yet to be repeated.

Following oral administration, considerable intersubject variation in serum concentrations can be found. Absorption may be reduced if fusidic acid is taken with food or oral chelating agents. Mean C_{max} values obtained following a single 500 mg dose of various oral formulations are

30–34 mg/L (Godtfredsen et al., 1962b; Wise et al., 1977; MacGowan et al., 1989; Taburet et al., 1990). However, this similarity in mean values disguises the wide range of concentrations found. Adsorption of the oral fusidic acid suspension was approximately 70% of the oral sodium fusidate tablet (Wise et al., 1977). The t_{max} following oral administration was 2–3 h, and $t_{1/2}$ values of 5.6 h (Wise et al., 1977), 8.7–9.5 h (MacGowan et al., 1989) and 16 h (Taburet et al., 1990) have been reported. There was a trend to increasing $t_{1/2}$ with repeat dosing. Evidence of accumulation can be seen from rising trough concentrations. With a standard oral dose of 500 mg tds trough serum concentrations were 11–41 (mean 21) mg/L after three doses, and 30–144 (mean 71) mg/L after 12 doses (Godtfredsen et al., 1962b). In eight of the ten subjects studied, steady-state concentrations had not been reached at this time. It has been postulated that this may be the result of non-linear kinetics of hepatic clearance, perhaps because of saturation of one or more metabolic pathways involved in hepatic elimination (Reeves, 1987; Taburet et al., 1990). The unpublished observations of Menday & Atkins, quoted by Vaillant et al. (1992), suggest that, with the standard oral dosing regimen of 500 mg tds, steady-state pharmacokinetics are reached on the third day. Godtfredsen et al. (1962b) found no accumulation with a 500 mg bd regimen. Likewise no accumulation was seen with a dosage of 250 mg tds (Stewart, 1964).

Intravenous administration of the now discontinued diethanolamine fusidate 580 mg preparation (equivalent to 500 mg sodium fusidate) as a 2 h infusion gave a C_{max} of 40 mg/L (Wise et al., 1977). Intravenous infusion of sodium fusidate 500 mg gave a mean C_{max} of 52.4 mg/L (Taburet et al., 1990). As with oral administration, accumulation was seen, and after nine doses the mean C_{max} was 123 mg/L.

Fusidic acid is extensively metabolized and excreted by the liver. Less than 2% of active drug is found in faeces. Fusidic acid concentrations of <1 mg/L were found in urine after a dosing regimen of 500 mg tds for 4 days (Godtfredsen et al., 1962b). Also, serum concentrations of only 7.3 and 5.2 mg/L were found in a completely anephric patient who had had a bilateral nephrectomy and was awaiting renal transplantation (Lautenbach et al., 1975). Serum concentrations of fusidic acid in renal failure are not altered by haemodialysis (Hobby et al., 1970; Brown et al., 1997). However, very little is known of the pharmacokinetics in patients on continuous ambulatory peritoneal dialysis (CAPD). Brown et al. (1997) reported fusidic acid concentrations of 0–2.3 mg/L in the dialysis fluid of patients without peritonitis receiving 500 mg orally tds for 3 days. Higher concentrations would be expected in the presence of peritonitis. In an animal model, fusidic acid concentrations in peritoneal dialysis fluid of 0.13 mg/L were found 1 h after a single dose, and 0.63 mg/L 1 h after multiple doses, representing 6.4–24.5% of the corresponding serum concentrations (Rowe et al., 1992). Intra-peritoneal administration has been used in human patients

therapeutically (Taylor et al., 1981) although this practice is not recommended because of the incompatibility of sodium fusidate and some peritoneal dialysis solutions (data on file, Leo Labs Ltd).

Fusidic acid is metabolized by the liver and excreted in the bile. Stewart (1964) reported that ligation of the bile duct in the rat increased liver fusidic acid concentrations. Several different metabolites have been identified from bile (Godtfredsen & Vangedal, 1966) the major ones being the glucuronide, dicarboxylic, hydroxyl, desacetyl and 3-keto- derivatives. However, little detail is known of the metabolism and excretion of fusidic acid and its metabolites by the liver or whether metabolites accumulate in renal or hepatic failure. Brown et al. (1997) found no accumulation of metabolites in patients with end-stage renal disease. The study of fusidic acid pharmacokinetics in hepatic impairment is difficult because of the association between fusidic acid and reversible jaundice, therefore contraindicating its use in such patients. It has been suggested that factors such as hypoalbuminaemia and reduced protein-binding in liver impairment may be as important as reduced metabolism and excretion by the liver (Peter et al., 1993).

Tissue distribution

The major factor influencing tissue distribution of fusidic acid is the high reversible protein-binding to serum albumin of >97% (Güttler & Tybring, 1971; Henry et al., 1981; Bannatyne & Cheung, 1982). It would be expected that only unbound drug would be able to diffuse freely between different compartments, but tissue fusidic acid concentrations are, in general, higher than would be predicted from the protein content, suggesting that other factors contribute also (Table). The lipid solubility of fusidic acid is probably the most important of these (Stewart, 1964). Intracellular penetration of fusidic acid has been shown in HeLa cells and leucocytes (Brown & Percival, 1978) and inflammatory exudate (Raeburn, 1971). In the inflammatory exudate produced by a full thickness burn (the burn crust) fusidic acid concentrations were considerably higher than serum concentrations (Sorensen et al., 1966). However, as a result of high protein-binding, the clinical significance of these tissue concentrations is not clear, as the unbound (microbiologically active) fraction is very small. Activity of fusidic acid in vitro is reduced by the addition of protein in the form of serum (Godtfredsen et al., 1962b). Concentrations in tissue with very low protein content, for example CSF or aqueous humour, are very low. However, this is all unbound and therefore presumably microbiologically active drug.

Assay methods

Routine assay of fusidic acid concentrations in patients is unnecessary since high serum concentrations have not

Table. Tissue distribution of fusidic acid

Tissue	Fusidic acid concentration (mg/L or mg/kg)	Reference
Cortical bone	0.2–16.7	Chater *et al.* (1972)
Sequestrum	0–6.9	
Soft tissue	3–17.7	
Infected bone	1.7–14.9	Hierholzer *et al.* (1974)
Healthy cortical bone	17.9	
Healthy cancellous bone	31.9	
Synovial fluid		
osteoarthritis	7–70	Deodhar *et al.* (1972)
rheumatoid arthritis	16–107	
Sputum	0.1–1.6	Saggers & Lawson (1968)
Pus	4.0–29.0	Güttler & Tybring (1971)
Cerebral abscess	6.2–>10	de Louvois *et al.* (1977)
Aqueous humour		
normal	0.8–2.0	Chadwick & Jackson (1969)
inflamed	12.8	Williamson *et al.* (1970)
Vitreous humour, inflamed	28.8	
CSF	<1	Godtfredson *et al.* (1962)
		Mindermann *et al.* (1993)

been associated with toxicity. However, because of the great variability in absorption, occasionally it may be useful to monitor therapy in patients with potentially serious infection, patients failing to respond to therapy, or patients with infection at a site with poor antibiotic penetration, for example the central nervous system. The most widely used method for the assay of fusidic acid has been the agar plate diffusion method, adapted from that originally produced by Leo Laboratories (Bywater, 1978). This method has the advantage of relative simplicity and widespread availability, but can be time-consuming and is unable to differentiate between fusidic acid and its many metabolites, some of which have antibacterial activity (Reeves, 1987). Specificity problems may arise when assaying samples containing a mixture of antibiotics but these may be sometimes overcome by the choice of indicator organism. Other assay techniques have been described and include those based on calorimetry (Presser *et al.*, 1966) and first-derivative spectrophotometry (Hassan *et al.*, 1987). Both of these methods are relatively complex, and neither has been validated for use with clinical specimens. The preferred method of assay is HPLC, due to its inherent good reproducibility and specificity. Several methods are available (Hikal *et al.*, 1982; Rahmen & Hoffman, 1988; Sorensen, 1988), but that of Sorensen (1988) has the advantage that it has been used in several laboratories and found to be simple and reliable. It does not require a complex extraction procedure, as do some other methods. Many of the metabolites of fusidic acid are clearly separated from the parent drug, fusidic acid. The assay has

been validated for a variety of clinical specimens, but, at the time of writing, not tissue or bone samples.

For laboratory use, sodium fusidate (500 mg of which is equivalent to 482 mg fusidic acid) and the HPLC internal standard, 24,25-dihydrofusidic acid may be obtained from Leo Laboratories. The dry powder of sodium fusidate is stable for 3 years when stored below 25°C.

HPLC of fusidic acid (based on that of Sorensen (1988))

HPLC conditions. The mobile phase of acetonitrile: methanol: 0.01 M phosphoric acid (50:20:30) is pumped at a flow rate of 3 mL/min through a 250 mm × 4.6 mm LiChrosorb RP18 column (HPLC Technology Ltd, Macclesfield, UK). Detection is by UV absorbance at 235 nm. When metabolites are being investigated it may be beneficial to use a reduced flow rate of 1 mL/min. The retention time of fusidic acid at 3 ml/min is approximately 7 min. The lower limit of detection of the assay is 1 mg/L. Recovery of fusidic acid from water, serum and peritoneal dialysis fluid is effectively 100% (this includes both free and protein-bound drug), and peak height versus fusidic acid concentration is linear over a range of 0–250 mg/L. No interference has been demonstrated with other commonly used antibiotics.

Sample and calibrator preparation. If the internal standard is to be used, samples are prepared by mixing 100 μL of sample with 100 μL of internal standard in acetonitrile

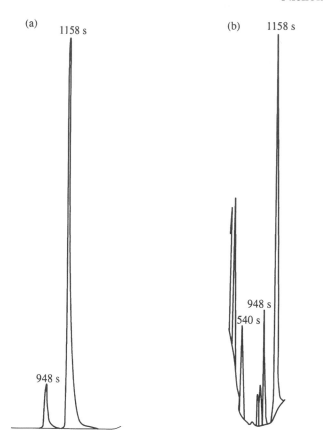

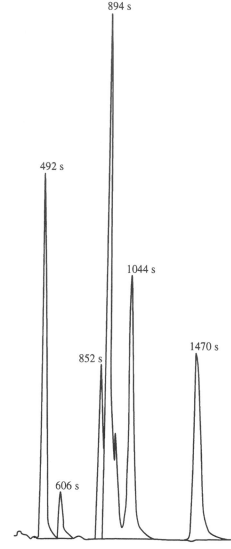

Figure 15.2 Chromatograms of fusidic acid in (a) water and (b) the serum of a patient. Fusidic acid has a retention time of 1158 s. The peaks at 540 and 948 s are unidentified.

and 100 μL of acetonitrile. The mixture is centrifuged and 75 μL of the supernatant is injected. Calibrators are treated in a similar fashion. It may be preferable to omit the internal standard because of the possibility of interference with the assay, particularly when fusidic acid metabolites are being investigated. If no internal standard is used, the sample is prepared by adding 100 μL acetonitrile to 100 μL sample, centrifuging, and then adding a further 100 μL acetonitrile to 100 μL of the supernatant. The mixture is centrifuged again and 100 μL of the resultant supernatant is injected. Calibrators are treated in a similar fashion. This procedure minimizes the risk of further protein precipitation in the HPLC system due to the high organic content of the mobile phase. Quantification is by comparison of peak heights of the tests and calibrators.

Stability. Samples and calibrators may be kept at room temperature for up to 4 days. After this time a reduction of peak height may be demonstrated. This occurs rapidly at temperatures above 22°C (N. M. Brown, unpublished observations). Samples are stable if stored at 4°C for 2 weeks, and may be kept at –20°C for at least 6 months.

Agar plate diffusion assay (from Bywater (1978))

Sample and calibrator preparation. Sodium fusidate is dissolved in phosphate buffer pH 7.3. Final dilutions

Figure 15.3 Chromatogram of fusidic acid metabolites and laboratory derivatives: 24,25-dihydrofusidic acid (1470 s), 16-epi-deacetylfusidic acid (1044 s), sodium 16-deacetylfusidate (894 s), 3-ketofusidic acid (852 s) and 27-carboxyfusidic acid (492 and 606 s).

should be made in 25% human serum in phosphate–citrate buffer (see Appendix). Calibrators in the range 0.5–8 mg/L (for the staphylococcus indicator strain) or 0.125–2 mg/L (for the corynebacterium indicator strain) are used. Serum samples are diluted 1:4 with phosphate–citrate buffer and then further diluted in the same 25% human serum in buffer used for calibrators. Ideally the final dilution should be such that the sample fusidic acid concentration lies in the middle of the calibrator range. The pH and protein content of samples and calibrators must be the same, because of the increased activity at a low pH and the high protein binding of fusidic acid. Samples and calibrators may be stored at –20°C for several months, but should not be kept at room temperature for more than 4 days.

Indicator strains. The following indicator strains may be used: *Staphylococcus aureus* (NCTC 6571), which in general is the easiest to use and suitable for most clinical purposes; *Corynebacterium xerosis* NCTC 9755, which is sensitive to fusidic acid and methicillin; *C. xerosis* Leo FF-M (available from Leo Laboratories), which is sensitive to fusidic acid but resistant to methicillin; *C. xerosis* Leo FF-ZN6 (available from Leo Laboratories), which is resistant to fusidic acid.

Organisms may be stored at –70°C, and are prepared for inoculation by 1 in 10 dilution of an 18 h culture in nutrient broth at 37°C.

Assay procedure. The assay medium outlined in the Appendix may be used, and is adjusted to pH 6.0 for increased sensitivity. One millilitre of bacterial suspension is incorporated into 100 mL of cooled molten agar (45°C), which is poured immediately into levelled 25 cm × 25 cm assay plates. After cooling and drying, 7 mm wells are cut using a no. 4 cork borer. Clearer zones of inhibition are obtained if the plate is exposed to UV light for 5 min to kill surface bacteria. Samples and calibrators are applied in triplicate according to a random pattern design. Incubation is at 37°C for 18 h. Results may be calculated from the inhibition zone diameters by a method such as that described by Bennett *et al.* (1966) (see *Chapter 4*).

Bone and other samples. Samples other than serum may be assayed provided the protein content of calibrators is similar to that of the specimens. Weighed bone samples should be cut into small pieces, dried over phosphorus pentoxide for 48 h at 4°C and then crushed in a stainless steel mortar. The same weight of phosphate–citrate buffer pH 6.0 is added, and the mixture is shaken and then left at 4°C for 1–2 h, with further shaking at intervals. The mixture is centrifuged and the supernatant diluted in phosphate–citrate buffer and assayed using calibrators prepared in the same buffer.

References

Bannatyne, R. M. & Cheung, R. (1982). Protein binding of fusidic acid. *Current Therapeutic Research* **31**, 159–61.

Bennett, J. V., Brodie, J. L., Benner, E. J. & Kirby, W. M. M. (1966). Simplified, accurate method for antibiotic assay of clinical specimens. *Applied Microbiology* **14**, 170–7.

Brown, K. N. & Percival, A. (1978). Penetration of antimicrobials into tissue culture cells and leucocytes. *Scandinavian Journal of Infectious Diseases, Suppl. 14*, 251–60.

Brown, N. M., Reeves, D. S. & McMullin, C. M. (1997). The pharmacokinetics and protein binding of fusidic acid in patients with severe renal failure requiring either haemodialysis or continuous ambulatory peritoneal dialysis. *Journal of Antimicrobial Chemotherapy* **39**, 803–9.

Bywater, M. J. (1978). Fusidic acid. In *Laboratory Methods in Antimicrobial Chemotherapy* (Reeves, D. S., Phillips, I., Williams, J. D. & Wise, R., Eds), pp. 219–21. Churchill Livingstone, Edinburgh.

Carey, M. C., Montet, J. C. & Small, D. M. (1975). Surface and solution properties of steroid antibiotics: 3-acetoxylfusidic acid, cephalosporin P1 and helvolic acid. *Biochemistry* **14**, 4896–905.

Chadwick, A. J. & Jackson, B. (1969). Intraocular penetration of the antibiotic fucidin. *British Journal of Ophthalmology* **53**, 26–9.

Chater, E. H., Flynn, J. & Wilson, A. L. (1972). Fucidin levels in osteomyelitis. *Journal of the Irish Medical Association* **65**, 506–8.

de Louvois, J., Gortvai, P. & Hurley, R. (1977). Antibiotic treatment of abscesses of the central nervous system. *British Medical Journal ii*, 985–7.

Deodhar, S. D., Russell, F., Dick, W. C., Nuki, G. & Buchanan, W. W. (1972). Penetration of sodium fusidate (fucidin) in the synovial cavity. *Scandinavian Journal of Rheumatology* **1**, 33–9.

Godtfredsen, W. O. & Vangedal, S. (1962). The structure of fusidic acid. *Tetrahedron* **18**, 1029–48.

Godtfredsen, W. O. & Vangedal, S. (1966). On the metabolism of fusidic acid in man. *Acta Chemica Scandinavica* **20**, 1599–607.

Godtfredsen, W. O., Jahnsen, S., Lorck, H., Roholt, K. & Tybring, L. (1962a). Fusidic acid: a new antibiotic. *Nature* **193**, 987.

Godtfredsen, W. O., Roholt, K. & Tybring, L. (1962b). Fucidin. a new orally active antibiotic. *Lancet i*, 928–31.

Godtfredsen, W. O., Albrethsen, C., von Daehne, W., Tybring, L. & Vangedal, S. (1965). Transformations of fusidic acid and the relationship between structure and antibacterial activity. *Antimicrobial Agents and Chemotherapy* **5**, 132–7.

Güttler, F. & Tybring, L. (1971). Interaction of albumin and fusidic acid. *British Journal of Pharmacology* **43**, 151–60.

Hassan, S. M., Amer, S. M. & Amer, M. M. (1987). Determination of fusidic acid and sodium fusidate in pharmaceutical dosage forms by first-derivative ultraviolet spectrophotometry. *Analyst* **112**, 1459–61.

Henry, J. A., Dunlop, A. W., Mitchell, S. N., Turner, P. & Adams, P. (1981). A model for the pH dependence of drug–protein binding. *Journal of Pharmacy and Pharmacology* **33**, 179–82.

Hierholzer, G., Rehn, J., Knothe, H. & Masterson, J. (1974). Antibiotic therapy of chronic post-traumatic osteomyelitis. *Journal of Bone and Joint Surgery* **56B**, 721–9.

Hikal, A. H., Shibl, A. & El-Hoofy, S. (1982). Determination of sodium fusidate and fusidic acid in dosage forms by high-performance liquid chromatography and a microbiological method. *Journal of Pharmaceutical Sciences* **71**, 1297–8.

Hobby, J. A. E., Beeley, L. & Whitby, J. L. (1970). Fucidin in patients on haemodialysis. *Journal of Clinical Pathology* **23**, 484–6.

Lautenbach, E. E. G., Robinson, R. G. & Koornhof, H. J. (1975). Serum and tissue concentrations of sodium fusidate in patients with chronic osteomyelitis and in normal volunteers. *South African Journal of Surgery* **13**, 21–32.

MacGowan, A. P., Greig, M. A., Andrews, J. M., Reeves, D. S. & Wise, R. (1989). Pharmacokinetics and tolerance of a new film-coated tablet of sodium fusidate administered as a single oral dose to healthy volunteers. *Journal of Antimicrobial Chemotherapy* **23**, 409–15.

Masterson, J. & Goldman, L. (1970). Studies on the absorption and tolerance of a new fucidin suspension. *Irish Journal of Medical Science* **3**, 115–22.

Mindermann, T., Zimmerli, W., Rajacic, Z. & Gratzl, O. (1993). Penetration of fusidic acid into human brain tissue and cerebrospinal fluid. *Acta Neurochirurgica* **121**, 12–4.

Peter, J. D., Jehl, F., Pottecher, T., Dupeyron, J. P. & Monteil, H. (1993). Pharmacokinetics of intravenous fusidic acid in patients with cholestasis. *Antimicrobial Agents and Chemotherapy* **37**, 501–6.

Presser, J. E., Wilkomirsky, F. T. & Brieva, A. J. (1966). Spectrophotometric determination of steroids. *Chemical Abstracts* **64**, 6408–9.

Raeburn, J. A. (1971). A method for studying antibiotic concentrations in inflammatory exudate. *Journal of Clinical Pathology* **24**, 633–5.

Rahman, A. & Hoffman, N. E. (1988) High-performance liquid chromatographic determination of fusidic acid in plasma. *Journal of Chromatography (Biomedical Applications)* **433**, 159–66.

Reeves, D. S. (1987). The pharmacokinetics of fusidic acid. *Journal of Antimicrobial Chemotherapy* **20**, 467–76.

Rowe, L., Findon, G. & Miller, T. E. (1992). An experimental evaluation of the pharmacokinetics of fusidic acid in peritoneal dialysis. *Journal of Medical Microbiology* **36**, 71–7.

Saggers, B. A. & Lawson, D. (1968). *In vivo* penetration of antibiotics into sputum in cystic fibrosis. *Archives of Diseases of Childhood* **43**, 404–9.

Sorensen, B., Sejrsen, P. & Thomsen, M. (1966). Fucidin, pro-staphlin, and penicillin concentration in burn crusts. *Acta Chirgica Scandinavica* **131**, 423–9.

Sorensen, H. (1988). Liquid chromatographic determination of fusidic acid in serum. *Journal of Chromatography (Biomedical Applications)* **430**, 400–5.

Stewart, G. T. (1964). Steroid antibiotics. *Pharmakotherapie* **2**, 136–48.

Taburet, A. M., Guibert, J., Kitzis, M. D., Sorensen, H., Acar, J. F. & Singlas, E. (1990). Pharmacokinetics of sodium fusidate after single and repeated infusions and oral administration of a new formulation. *Journal of Antimicrobial Chemotherapy* **25**, *Suppl. B*, 23–31.

Taylor, R., Schofield, I. S., Ramos, J. M., Bint, A. J. & Ward, M. K. (1981). Ototoxicity of erythromycin in peritoneal dialysis patients. *Lancet ii*, 935–6.

Vaillant, L., Machet, L., Taburet, A. M., Sorensen, H. & Lorette, G. (1992). Levels of fusidic acid in skin blister fluid and serum after repeated administration of two dosages (250 and 500 mg). *British Journal of Dermatology* **126**, 591–5.

von Daehne, W., Godtfredsen, W. O. & Rasmussen, P. R. (1979). Structure–activity relationships in fusidic acid-type antibiotics. *Advances in Applied Microbiology* **25**, 95–146.

Williamson, J., Russell, F., Doig, W. M. & Paterson, R. W. W. (1970). Estimation of sodium fusidate levels in human serum, aqueous humour, and vitreous body. *British Journal of Ophthalmology* **54**, 126–30.

Wise, R., Pippard, M. & Mitchard, M. (1977). The disposition of sodium fusidate in man. *British Journal of Clinical Pharmacology* **4**, 615–9.

Appendix

Medium for microbiological assay: vitamin-free casamino acids (Difco), 15.0 g; yeast extract (Difco), 5.0 g; L-cysteine, 0.05 g; sodium chloride, 2.5 g; dextrose, 1.0 g; agar (Difco), 24.0 g; distilled water to 1000 mL.

Phosphate–citrate buffer pH 6.0: disodium hydrogen phosphate, 19.8 g; citric acid, 6.59 g; distilled water to 1000 mL.

Phosphate buffer pH 7.3: sodium chloride, 8.0 g; potassium chloride, 0.2 g; disodium hydrogen phosphate, 1.15 g; potassium dihydrogen phosphate, 0.2 g; distilled water to 1000 mL.

Chapter 16

Antimycobacterial drugs

Peter J. Jenner

Division of Mycobacterial Research, National Institute for Medical Research, Mill Hill, London NW7 1AA, UK

Introduction

Mycobacterial infections continue to cause widespread disease particularly in developing countries, where tuberculosis and leprosy are common. In the West, after many years of decline, the incidence of tuberculosis is now on the increase. Infections in AIDS patients with opportunistic mycobacteria, such as *Mycobacterium avium*, and the emergence of multiple-drug-resistant strains of *Mycobacterium tuberculosis* (MDRMT) are also causing concern and this has once more focused attention on these largely forgotten diseases. The treatment of mycobacterial infections has seen some important changes during the last ten years, with the introduction of newer drugs, and the re-assessment of the role of older drugs to treat atypical mycobacterial infections, which are often less susceptible to conventional anti-tuberculosis chemotherapy. Drugs such as clofazimine, regarded previously as having little use in tuberculosis therapy, are now being re-evaluated for the treatment of *M. avium*. The established regimens for treating tuberculosis contain combinations of rifampicin, isoniazid, ethambutol, pyrazinamide or streptomycin, while the recommended drugs for the treatment of leprosy are dapsone, rifampicin and clofazimine. Of the newer drugs, much interest is centred on the fluoroquinolones, such as ciprofloxacin, ofloxacin, pefloxacin and sparfloxacin; as more potent derivatives are likely to be developed, the fluoroquinolones are likely to become important in the treatment of mycobacterial infections. Newer rifamycins, such as rifapentine and rifabutin, may also prove useful in treating both tuberculosis and leprosy, and in dealing with atypical mycobacterial infection in AIDS patients. Since the fluoroquinolones and streptomycin are covered elsewhere, this chapter will deal with assays of the older anti-mycobacterial drugs and the newer rifamycins.

Assays of patients' samples

Many of the assay methods for the antimycobacterial drugs are neither simple nor routine, and it is unlikely that hospital laboratories will regularly have to measure serum levels of these drugs. Requests for measurements of any of the antimycobacterial drugs will need to be assessed to consider whether the value of the information obtained is worth the cost of obtaining it. Many of the drugs used for treating tuberculosis are bactericidal and their therapeutic efficacy is not directly related to their plasma concentration or to the time for which blood levels exceed their MIC. Incidences of poor adsorption and thus inadequate serum concentrations are exceedingly rare, at least in non-AIDS patients, and by far the most common reason for treatment failure is poor compliance. A significant number of the requests for assays of antimycobacterial drugs may be made in order to assess compliance. Fortunately, simple and quick urine tests for the commonly used drugs are available and should give the information required; details of these are given later. More sophisticated assay methods, such as HPLC, GLC, fluorimetric, spectrophotometric and immunological assays, may have been developed for measuring serum concentrations and can be used where appropriate; for example, for research investigations of pharmacokinetics in patients, for measurements where sample volume is low, such as in neonates, or for determination of CSF levels, since tubercular meningitis, although uncommon, can be a particularly life-threatening form of the disease. In tuberculosis patients also undergoing renal dialysis there may be requests for blood levels to be monitored. However, the dosage of most anti-tuberculosis drugs (streptomycin and possibly cycloserine being exceptions) need not be reduced in renal patients (Ellard, 1993). Only when inappropriate dosage regimens causing toxic side-effects have been prescribed should it be appropriate to investigate blood levels. In patients with clinically significant hepatic impairment, the potential hepatotoxicity of isoniazid, rifampicin and pyrazinamide means that monitoring serum levels of these drugs could be helpful, although the exact relationship between blood levels and liver damage is unclear. AIDS patients may need to be monitored more closely, since cases of abnormal pharmacokinetics can occur, particularly impaired oral adsorption.

Methods for estimating the concentration of antimycobacterial drugs in body fluids

Almost all of the drugs used to treat mycobacterial infections were discovered 40 or more years ago, with a consequence that many different methods have been used to determine their concentrations in body fluids. A number of the older assays are still valuable, but recent years have seen an increase in chromatographic methods, particularly HPLC, for the antimycobacterial drugs. In general, HPLC methods are often regarded as the most reliable and sensitive of the techniques for determining drug concentrations in biological fluids, but satisfactory HPLC assays have yet to be developed for some of the antimycobacterial drugs. Moreover, whilst HPLC equipment is widely found in hospital laboratories in developed countries, in many areas of the world where tuberculosis and leprosy are major health problems such equipment may not be available and alternative methods of assay must be used.

Of the less commonly used antimycobacterial drugs, sensitive HPLC assay methods have been published for thiacetazone (Jenner, 1983) and for the thioamides, ethionamide and prothionamide (Jenner & Ellard, 1981; Jenner et al., 1984; Peloquin et al., 1991).

Since combined chemotherapy is the exclusive form of treatment of mycobacterial diseases, it is important that any analytical method be specific, with no interference from other co-administered antimycobacterials. The recommended methods described below have this attribute. In the case of HPLC methods, retention times of other drugs or metabolites may not always be reported, and this could lead to spurious peaks being misidentified. It is therefore important to analyse appropriate blank serum samples and aqueous drug solutions, and to identify correctly the peaks in the chromatogram so as to avoid misinterpretation of the results. The concentration ranges used to construct suitable calibration curves together with estimates of assay reproducibility have been included where available. However, it is essential that methods are properly validated before clinical samples are assayed.

Isoniazid

Most anti-tuberculosis regimens include isoniazid (mol. wt 137.1; Figure 16.1a). Isoniazid powder can be obtained commercially (Aldrich Chemicals, Poole, UK) and can be recrystallized from butanol (melting point 171–173°C). The dry powder is stable at 25°C and absorbs insignificant amounts of moisture, but is best stored in airtight containers, protected from light. Isoniazid is a weak base (pK_a 1.8, 3.5, 9.5); it is soluble in water, ethanol or methanol and a stock solution of 1 g/L can be conveniently

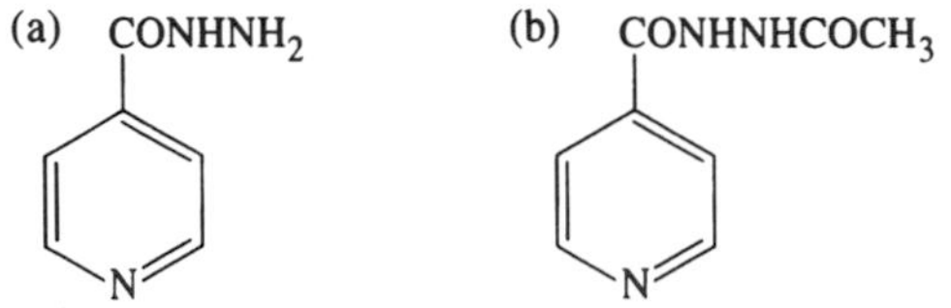

prepared. Aqueous solutions can be sterilized at 120°C for 30 min. There is contention over the stability of the drug in frozen plasma samples and as a precaution it is probably wise to deproteinize serum and plasma samples within 6 h of sampling or to extract the drug as soon as possible. Isoniazid is available clinically as tablets or as capsules, and in combined formulation with rifampicin, ethambutol or rifampicin plus pyrazinamide for oral dosage. It is also available as a parenteral preparation of 50 mg in 2 mL of fluid for iv or im administration.

Pharmacokinetics

Isoniazid is well absorbed and widely distributed throughout the body water, with peak plasma concentrations normally occurring 1–3 h post-dose (Table). The drug appears to penetrate readily through cell membranes, and to pass rapidly into the CSF, even in the absence of inflammation of the meninges. The major route of elimination of isoniazid is by metabolism to acetylisoniazid (Figure 16.1b), with only a small proportion of the dose being excreted unchanged in the urine. The extent to which individuals acetylate isoniazid varies considerably; the distribution of half-lives in any population is bimodal, with subjects characterized as either rapid or slow acetylators. This polymorphic acetylation is genetically determined, and different racial populations have different proportions of rapid and slow acetylators. Typical peak plasma concentrations are given in the Table, and peak urinary concentrations can be as high as 100–200 mg/L following a 300 mg dose (peak urinary excretion rates can be 10–20 mg/h), necessitating considerable dilution of urine samples before analysis. Absorption can be affected by food, resulting in lower than normal peak plasma concentrations. Isoniazid is not thought to be significantly bound to serum proteins, although a more recent publication suggests binding may be about 20% (Woo et al., 1996).

It is not necessary to reduce the dosage of isoniazid in patients with renal insufficiency. Its half-life in slow acetylator patients with renal failure is increased by about 40%. In patients with liver disease plasma isoniazid concentrations may be slightly elevated but at normal therapeutic doses these concentrations should not be toxic. As a precaution blood levels could be monitored but this should not be necessary for patients with less severe disease. Isoniazid, like all antituberculosis drugs, exhibits toxic side-effects only rarely (Girling, 1982). Asymptomatic hepatitis, manifest in transient increases in serum liver enzyme concentrations, can occur, but clinical hepatitis is only found in less than about 1% of patients. Other uncommon chronic toxic manifestations include cutaneous and generalized hypersensitivity, and anorexia. Neurological toxicity such as peripheral neuropathy, results from isoniazid's inhibition of the action of coenzymes produced from pyridoxine, and is related to plasma levels of the

Figure 16.1 Structure of (a) isoniazid and (b) its metabolite acetylisoniazid.

Table. Pharmacokinetic properties and MICs of anti-mycobacterial drugs

Drug	Dose (mg)	Peak (mg/L)	Half-life	Protein binding[a]	MIC[b] (mg/L) for *M. tuberculosis*
Isoniazid	300	5[c]	3 h[c]	0–20	0.2
		4[d]	1.3 h[d]		
Rifampicin	600	10	3 h	85	0.2
Rifabutin	300	0.5	16 h	10	0.04
Rifapentine	300	12	15 h	?	0.08
Ethambutol	1200	3	4 h	20	1.5
Pyrazinamide	2000	40	9 h	0–40	20
Cycloserine	250	10+	4–30 h	20	5–20
Dapsone	100	2	27 h	75	NA[e]
Clofazimine	100	0.4	70 days (?)	?	NA[f]

The dose and peak plasma concentrations are appropriate for a 60 kg subject.
?, Uncertain; NA, not active against *M. tuberculosis* at conventional dosage.
[a]Percent bound to serum proteins.
[b]MICs determined for *M. tuberculosis* in liquid medium without Tween 80.
[c]Slow acetylator.
[d]Rapid acetylator.
[e]MIC against *M. leprae* 0.01 mg/L, using the mouse footpad method.
[f]MIC against *M. leprae* 0.1–1 mg/L, using the mouse footpad method.

drug, and is therefore more common in slow acetylators but can be treated with pyridoxine. Isoniazid is safe in pregnancy, and other side-effects are very rare.

Interactions between isoniazid and a number of drugs have been reported, but in some cases the evidence is conflicting. There are, however, some well-documented examples of significant drug interactions that may affect either the serum concentrations of isoniazid, or of the co-administered drug. In particular, the half-life of isoniazid is increased when taken with procainamide, phenyramidol and chlorpromazine, and decreased when given with sodium salicylate and ethanol. Antacids such as aluminium hydroxide may lead to reduced isoniazid adsorption. Isoniazid inhibits the hepatic metabolism of phenytoin and, particularly in slow acetylators, this can lead to toxic serum phenytoin concentrations. Isoniazid may also increase the serum concentrations of primidone diazepam and carbomazepine, although not necessarily to toxic levels.

Assay of isoniazid in biological samples

A number of assays have been developed for isoniazid, from simple colorimetric assays to more complex chromatographic ones. Colorimetric methods will probably give reasonable results if peak blood concentrations are to be determined, while HPLC measurements are useful if a more sensitive assay is required. The HPLC method detailed below is relatively simple; the alternative assays of Kubo *et al.* (1990); El-Yazigi & Yusuf (1991) and Walubo *et al.* (1991) are more complex, all requiring either electrochemical or fluorimetric detection, and either pre- or post-column derivatization.

Determination of isoniazid by HPLC

The method of Hutchings *et al.* (1983), which also measures the metabolite acetylisoniazid, is recommended. A reversed-phase nitrile (CN) column is used (250 × 4.5 mm internal diameter, Spherisorb nitrile; Anachem, Luton, UK) as it gives very good resolution of the peaks. The mobile phase is 0.01 mol/L phosphoric acid in acetonitrile:water (20:80 v/v) pumped at 2 mL/min and detection is by UV absorbance at 266 nm.

Procedure

1. To 1 mL of plasma or 1 mL of diluted of urine (diluted 1 in 11), add 50 μL of a 40 mg/L aqueous solution of the internal standard iproniazid (Sigma Chemicals, Poole, UK).

2. After adding 500 μL of 0.5 mol/L pH 7.4 phosphate buffer saturated with sodium chloride, extract the mixture for 10 min with 10 mL chloroform/butan-1-ol (70:30 v/v).

3. Filter the solution through phase-separating paper, and extract the organic phase with 500 μL of 0.05 mol/L phosphoric acid; after centrifugation, inject 20 μL of the aqueous phase.

Calibration. Prepare calibrators by adding known amounts of isoniazid and, if required, acetylisoniazid to plasma, blood or urine, with isoniazid concentrations covering the range 0.1–15 mg/L for plasma and blood, and 2–100 mg/L for urine.

Comments. Isoniazid elutes with a retention time of approximately 4 min, with the peaks of acetylisoniazid at approximately 2.8 min and that of the internal standard at

5 min. CVs of around 5% should be expected. Minor modification of these conditions to suit individual requirements should not affect the results significantly and with a detection limit of 0.02 mg/L the method should be sufficiently sensitive for monitoring plasma concentrations for at least 10 h post-dose in both acetylator phenotypes following therapeutic doses.

Fluorimetric assay for isoniazid

A number of assays are available based on a method originally described by Scott & Wright (1967) which used salicylaldehyde to form a highly fluorescent hydrazone derivative. That of Ellard *et al.* (1972) used a lengthy extraction procedure to separate isoniazid from its metabolites, enabling quantification of both metabolites and parent drug. Such an extraction method is very useful for detailed studies on isoniazid metabolism.

A simpler fluorimetric method which can be used for isoniazid is that of Olson *et al.* (1977) and is described below.

Procedure

1. Prepare a stock solution of the derivatizing reagent, salicylaldehyde, by dissolving 0.1 mL in 3 mL of absolute ethanol, adding 13.67 g of sodium acetate trihydrate and 2.1 mL of 10 mol/L sodium hydroxide, and diluting the solution to 250 mL with water.

2. Prepare a solution of 0.38 g of sodium metabisulphite in 250 mL of 0.4 mol/L sodium acetate trihydrate, 0.167 mol/L sodium hydroxide (metabisulphite reagent).

3. Prepare a solution of 10% aqueous ascorbic acid with the pH adjusted to 5.7 ± 0.05 (ascorbic acid reagent).

4. Precipitate serum or plasma proteins by mixing 0.2 mL of serum or heparinized plasma sample (not EDTA-treated) with 1 mL of 10% (w/v) aqueous trichloro-acetic acid (TCA).

5. Leave to stand at room temperature for 10 min, then centrifuge at 500**g** for 15 min. Pipette 400 μL of the TCA supernatant into a test tube and add 500 μL of the derivatizing reagent; adjust to pH 4.0 ± 0.1 using good quality, narrow range pH paper.

6. After 15 min at room temperature the hydrazone derivative will have formed; add 1 mL metabisulphite reagent and mix well.

7. Reduce the derivative by adding 0.1 mL ascorbic acid reagent and heating the mixture to 50°C for 10 min.

8. Extract the derivative into 1.5 mL of isobutanol (2-methyl-1-propanol), place the test tubes on ice for 15 min, and separate the phases by centrifugation at 500**g** for 10 min.

9. Keep the samples on ice and measure the fluorescence in the upper organic phase as soon as possible using 385 nm for excitation and 462 nm for emission.

Comment. Failure to measure the fluorescence quickly can result in inaccurate measurements. For this reason, if a large number of samples are to be measured, they are best done in small batches.

Calibration. Isoniazid concentrations are calculated by reference to a calibration curve prepared using serum calibrators covering the range 0.05–4 mg/L. Outside this concentration range the relationship between concentration and fluorescence intensity is not linear, and sera thought to contain >4 mg/L should be suitably diluted. Replicate errors of around 5% can be expected. The limit of sensitivity is approximately 0.05 mg/L.

Spectrophotometric assay for isoniazid

A rapid, but rather insensitive assay for isoniazid is that of Dymond & Russell (1970), which is described below.

Procedure

1. Freshly prepare an aqueous 0.5% solution of 2,4,6-trinitrobenzene sulphonic acid (picrylsulphonic acid, Sigma).

2. Pipette plasma or urine (1 mL) into a stoppered centrifuge tube together with 1 mL of 1 mol/L dipotassium hydrogen phosphate and 1.5 mL of 2,4,6-trinitrobenzene sulphonic acid solution.

3. Add 3 mL of methyl-isobutyl-ketone, mix and leave in the dark for 15 min.

4. Shake the two phases thoroughly, centrifuge and measure the optical density of the upper phase at 480 nm.

Calibration. The isoniazid concentration is calculated by comparison with suitable calibrators in the range 1–10 mg/L. The lower limit of sensitivity is approximately 1 mg/L.

Rifampicin

Rifampicin (Figure 16.2a) is a synthetic derivative of a natural antibiotic, rifamycin B, and is a first-line anti-tuberculosis drug. Rifampicin powder for use as an analytical standard can be obtained from Sigma. The dry, brick-red powder is fairly stable at 25°C but is best stored below 15°C in an airtight container, protected from light. Aqueous solutions lose potency over 24 h without the addition of a reducing agent such as ascorbic acid. Rifampicin is a zwitterion, with a molecular weight of 822.9 and pK_as of 1.7 (associated with the 4-hydroxy group) and 7.9 (due to the 3-piperizine nitrogen). It is freely soluble in ethanol, methanol and chloroform, but its solubility in aqueous solution at neutral pH is about 3 g/L. Solutions of rifampicin may be sterilized by membrane filtration, but this is rarely necessary provided sterile diluents are used. Clinically, rifampicin is available for oral administration as capsules, syrup or tablets, and in combined formulation with isoniazid or isoniazid plus pyrazinamide.

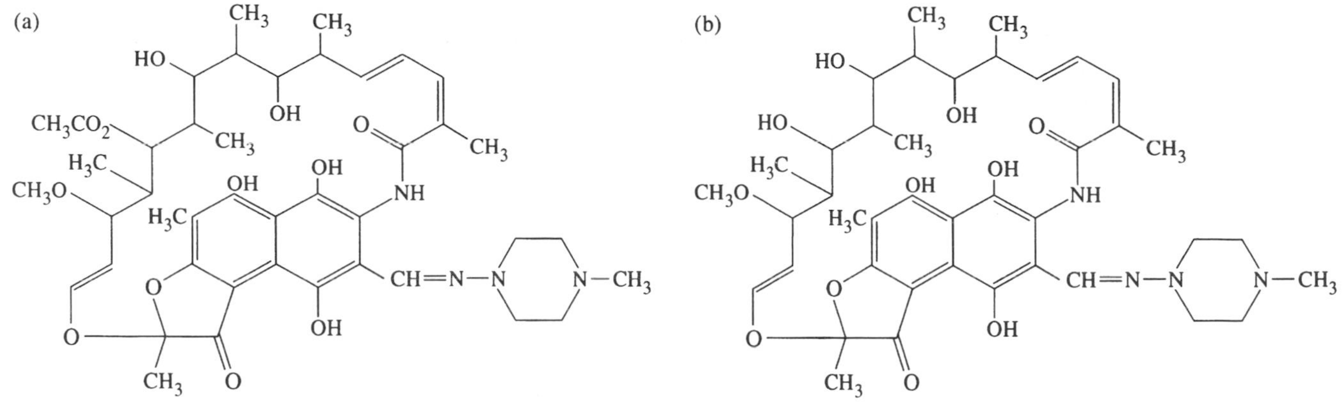

(a)

(b)

Figure 16.2 Structure of (a) rifampicin and (b) its principal metabolite 25-desacetyl rifampicin.

Pharmacokinetics

Rifampicin should be administered before meals rather than afterwards, since absorption is sometimes impaired by food. When given on an empty stomach, peak plasma concentrations normally occur 1–3 h later. Rifampicin is widely distributed in body fluids and tissues, and readily penetrates through cell membranes, but passes slowly and poorly into the CSF, where concentrations are only about 8% or less of the concomitant serum concentrations, although higher levels are found while the meninges remain inflamed (Ellard *et al.*, 1993).

Rifampicin is extensively metabolized by the liver and is a potent inducer of hepatic microsomal oxidases. Thus, on repeated dosage, it will induce its own metabolism, so reducing its elimination half-life from about 3.5 h to 2 h. Following a single 600 mg dose, peak plasma concentrations average about 10 mg/L (Table), while urinary concentrations can reach 500 mg/L. Only about 18% of the dose is excreted in the urine unchanged and the major metabolite, 25-*O*-desacetyl rifampicin (Figure 16.2b), has significant antimicrobial activity, which can contribute to estimates of rifampicin in microbiological assays. Rifampicin is also hydrolysed to 3-formyl rifampicin. Rifampicin dosage should not be reduced in patients with renal failure. Patients with liver disease have somewhat elevated rifampicin serum concentrations and increased elimination half-lives, but liver damage appears not to affect the metabolism of the drug. The increased rifampicin concentrations may lead to increased bilirubin levels and the risk of hyperbilirubinaemia in some cases. Although it is probably not necessary to measure rifampicin levels in the majority of cases, it is advisable to monitor liver function in patients with chronic liver disease since rifampicin is itself occasionally hepatotoxic. Other side-effects are rare, and are associated with intermittent and/or daily administration. Those occurring with intermittent rifampicin regimens include 'flu' syndrome, acute haemolytic anaemia and renal failure, while those sometimes observed with either daily or intermittent treatment include hypersensitivity, gastrointestinal reactions, asymptomatic hepatitis and thrombocytopenia.

The induction of hepatic microsomal enzymes by rifampicin causes significant reduction in the serum half-life and clinical efficacy of a number of drugs. These drug interactions include narcotics and analgesics (morphine, methadone and phenobarbitone), digoxin, corticosteroids, coumarin anticoagulants and oral contraceptives.

Assay of rifampicin in patient samples

Microbiological assays for rifampicin all suffer from interference from active metabolites such as desacetyl rifampicin. HPLC assays are more specific but much less sensitive than some microbiological assays. Two HPLC methods are described below, although the alternative reversed phase procedure of Guillaumont *et al.* (1982), or the normal phase method of Lecaillon *et al.* (1978) may also be worth consideration.

One of the problems associated with rifampicin assay is its adsorption to plastic and glass. This adsorption should be minimized by avoiding the use of small volumes of sample in large containers and the pre-filling of pipettes to saturate any binding sites.

HPLC method of Ishii & Ogata (1988) for rifampicin and its metabolites

This method measures rifampicin, 25-desacetyl rifampicin, 3-formyl rifampicin and other metabolites. Blood samples (3 mL) should be treated with 10 mg ascorbic acid to prevent oxidation of rifampicin and its metabolites.

The column is Nucleosil C_{18} 250 × 4.0 mm (Fisons, Loughborough, UK) operating at 40°C (the temperature is probably not critical but some increase in retention time is to be expected at ambient temperature) and the mobile phase is acetonitrile:0.1 mol/L potassium phosphate pH 4.0 (38:62 v/v) pumped at a flow rate of 1.2 mL/min. Detection is by UV absorption at 340 nm.

Procedure

1. Mix plasma sample (0.5 mL) with 2 mL of 0.5 mol/L phosphate buffer pH 7.2, 100 μL of a 20 mg/L solution of papaverine hydrochloride (Sigma), and extract with 7 mL of chloroform.

2. Shake well, centrifuge, and discard the upper aqueous layer.

3. Evaporate the lower organic phase to dryness under nitrogen at 50°C.

4. Dissolve the residue in 300 μL acetonitrile/2-propanol (1:1 v/v) and inject 20 μL.

Comments. Rifampicin elutes after 6.7 min, desacetyl rifampicin after 3.5 min and the internal standard after 4.5 min, with other metabolites appearing at approximately 5 and 15 min, although plasma concentrations of the minor metabolites are normally below the lower limit of detection.

Calibration. A calibration curve is constructed with serum calibrators covering the range 0.5–10 mg/L. CVs of 2% for five replicates of known concentration can be expected. Detection limits are 0.1 mg/L for rifampicin, and 0.06 mg/L for the desacetyl metabolite.

HPLC method of Y. G. Shi and L. O. White (unpublished) for rifampicin

The column is Hypersil 5 ODS (100 × 4 mm) and the mobile phase 0.2 M sodium phosphate buffer pH 6.0: acetonitrile (70:30 v/v) pumped at 2 mL/min. Detection is by UV absorption at 254 nm.

Procedure. Mix serum sample, calibrator and control (100–200 μL) with an equal volume of acetonitrile, stand for 2 min, and centrifuge. Inject 20–50 μL of supernatant.
Calibration. Single-point calibration with an aqueous solution of 5 mg/L rifampicin is suitable for therapeutic monitoring. Reproducibility is 5% or better but care must be taken to avoid loss of rifampicin by binding to glass and plastics.

Microbiological assay of rifampicin

A conventional agar diffusion assay may be used to measure rifampicin in serum or CSF (Dickinson *et al.*, 1974). *Staphylococcus aureus* NCTC 10702, which is sensitive to rifampicin but highly resistant to streptomycin, kanamycin, viomycin, capreomycin and cycloserine, is used as the indicator strain.

Procedure

1. Prepare seeded assay medium by adding 10 μL of an overnight broth culture of *S. aureus* NCTC 10702 to 100 mL of Bacto antibiotic medium no. 1 (Difco 0263), modified by the addition of 3 mL of 1 mol/L KH_2PO_4, and add 10 mL to 9 cm diameter plates.

2. Cut four 8 mm diameter wells into each plate, using one plate for each of the unknowns to be measured.

3. Prepare two-fold diluted calibrators in pooled normal serum from 1.28 to 0.02 mg/L. These calibrators should be prepared in at least triplicate (enough for six unknown samples).

4. Pipette three of the calibrator solutions chosen at random and a test sample into the four wells of a plate and continue with all the calibrators and samples.

5. After overnight incubation at 37°C, measure the diameters of the zones of inhibition (including the well) with Vernier callipers or a zone reader.

Calibration. The relationship between the zone diameters and log rifampicin concentrations may be linear, the line being fitted to the points by linear regression methods. Alternatively, use a curve-fitting procedure as outlined in *Chapter 4*.

Comments. The method is much less precise than HPLC assays: the estimated precision is about ± 25% for samples falling near the middle of the calibration curve, and about ± 50% near the extremities of the line. The main metabolite of rifampicin, desacetyl rifampicin, is also active against *S. aureus* NCTC 10702 but, in order to give zones of the same diameter as those produced by rifampicin, 4.5 times as much metabolite is required. Samples containing more than 1.28 mg/L should be suitably diluted (1:40 say) in human serum and re-assayed.

Rifabutin and rifapentine

Other new rifamycin derivatives are currently under investigation and of these, rifapentine (Figure 16.3a, DL 473, Merrell-Dow/Lepetit) and rifabutin (Figure 16.3b, LM 427 (ansamycin), Farmitalia Carlo Erba) appear to be promising for the treatment of mycobacterial disease. Rifabutin is a broad-spectrum antibiotic that in mice is about six times more active against *M. tuberculosis* than rifampicin. Rifapentine has an antibacterial spectrum similar to that of rifampicin, but is two to ten times more active against *M. tuberculosis*. In several studies rifapentine and, more so, rifabutin has been found to be of value in the treatment of *M. avium–intracellulare* which frequently cause infection in patients with AIDS. Both drugs show higher activity than rifampicin against *Mycobacterium leprae* in the mouse footpad model.

Rifabutin (mol. wt 847.1) is a violet crystalline powder, highly soluble in chloroform, soluble in methanol, but only slightly soluble in water. Rifapentine (mol. wt 877) is a red-orange crystalline powder and is soluble in methanol. Samples of both drugs may be available from the manufacturers.

Both rifabutin and rifapentine have half-lives four to five times longer than that of rifampicin (Table). Peak plasma concentrations of rifabutin are rather low (0.5 mg/L after 300 mg), but tissue levels are five to ten times higher. Peak plasma concentrations of rifapentine are about 12 mg/L.

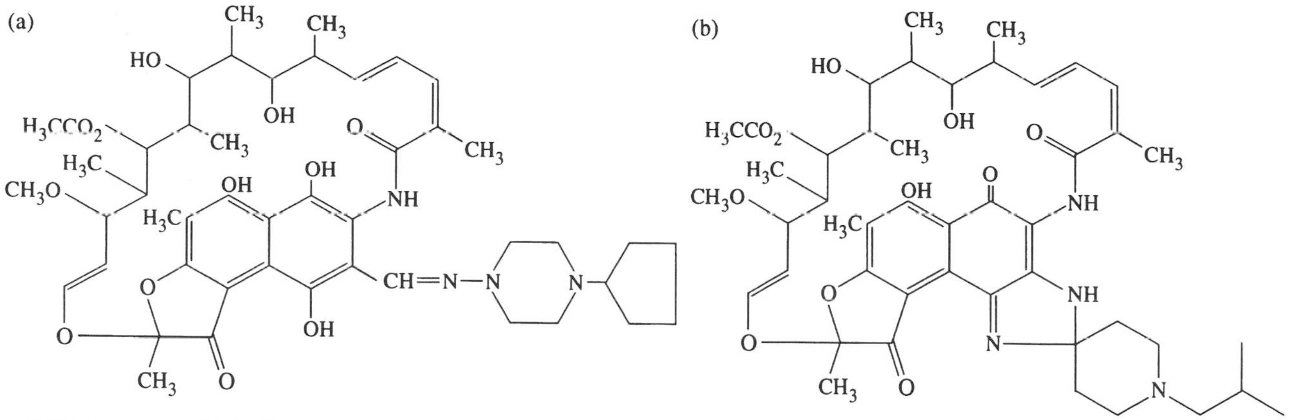

Figure 16.3 Structure of (a) rifapentine (DL473) (b) and rifabutin (LM 427).

Rifapentine exhibits higher and rifabutin exhibits lower plasma protein binding than rifampicin. Like rifampicin, the new rifamycins are potent inducers of P450 enzymes and can alter the pharmacokinetics of themselves and co-administered drugs.

Both drugs can be assayed with the agar diffusion method as described above for rifampicin, using *Micrococcus lutea* ATCC 9341 as the indicator strain. This organism is also sensitive to the desacetyl metabolite of rifabutin; but an assay employing *S. aureus* as indicator organism would give better specificity.

Relatively few pharmacokinetic studies using HPLC methods have been reported for the newer rifamycins but two new HPLC methods are outlined below.

Assay methods

HPLC assay for rifapentine (He *et al.*, 1996)
This method is suitable for measuring rifapentine in serum. To avoid oxidation of rifapentine, blood (2mL) is sampled into a heparinized tube with 0.04 mL of ascorbic acid (10 g/L) before serum is obtained.

A Spherisorb 5 μm C_{18} column (250 × 4.6 mm, Waters) is used, and operated using a column oven at 28°C, although it is likely that similar results will be obtained at ambient temperature. The mobile phase is 0.01 mol/L pH 5.5 NaH_2PO_4/methanol (32:68 v/v) at 1 mL/min. UV detection is at 336 nm, and rifampicin is used as an internal standard.

Procedure

1. Aliquot 1 mL samples of serum and add 0.4 mL of 1 mol/L pH 4.0 phosphate buffer, 40 μL of ascorbic acid (10 g/L) and 0.05 mL of rifampicin (100 mg/L, made by dissolving 10 mg in 1 mL ethanol and diluting to 100 mL with water).

2. Precipitate the serum proteins with 4 mL ethyl acetate by vortexing for 5 min and centrifuge at 1800**g** for 10 min.

3. Transfer 3 mL of the upper aqueous phase to a new

tube and evaporate at 100°C under a stream of nitrogen.

4. Dissolve the residue in 0.5 mL methanol and, after vortexing for 1 min, inject 20 μL.

Comments. The retention time for rifapentine is about 7 min, and that for rifampicin 4.3 min.

Calibration. Prepare calibration curves using blank serum spiked with rifapentine in the concentration range 0.5–30 mg/L, and calculate the peak area rations of rifapentine/rifampicin. CVs of <10% can be expected, and the limit of detection is 5 ng.

HPLC assay of rifabutin in serum (Lau *et al.*, 1996).
This method measures both rifabutin and its main metabolite, 25-desacetyl-rifabutin, using a reversed phase Zorbax RX C_8 5 μm 250 × 4.6 mm column (manufactured by Rockland Technologies; supplied by a variety of suppliers) with a Brownlee RP-8 pre-column (15 × 3.2 mm, 7 μm). The mobile phase is 0.05 mol/L KH_2PO_4 and 0.05 mol/L sodium acetate adjusted to pH 4.0 with acetic acid and acetonitrile (53:47 v/v) and pumped at a flow rate of 1 mL/min. The UV detector is set at 275 nm. Sulindac (Sigma) is used as an internal standard.

Procedure

1. Make up stock solutions of rifabutin (100 mg/L) in methanol, and Sulindac (100 mg/L) in water, and store at 5°C for at least 2 months if protected from light.

2. Dilute the internal standard to a working solution concentration of 2 mg/L when required.

3. Take 1 mL of plasma and add 25 μL of the internal standard, mix and centrifuge at 630 rpm for 5 min.

4. Condition C_8 Bond Elut extraction columns with 1 mL of methanol, followed by 1 mL of water.

5. Load the plasma mixture on to the column and wash with water.

6. Elute the analytes with 1 mL of methanol and dry down under nitrogen.

7. Reconstitute with 250 μL of mobile phase and inject 100 μL aliquots.

Comments. The retention time of rifabutin is about 11 min and that of Sulindac 7 min. The 25-desacetyl metabolite elutes after 6 min.

Calibration. Prepare calibrator in serum over the range 20–400 ng/mL and calculate peak height ratios. Typical CVs of <9% should be obtained. The lower limit of detection is 5 ng/mL.

Ethambutol

Ethambutol (Figure 16.4) is another first-line anti-tuberculosis agent. It is a strong base (mol. wt 204.3, pK_a 6.3 and 9.5) and may be purchased from Sigma as the dihydrochloride (mol. wt 277.2), a stable white, crystalline substance which should be stored in a tightly closed container. The dihydrochloride is very water-soluble and is also readily dissolved in ethanol. Aqueous solutions can be heated to 120°C for 10 min without significant decomposition. Clinically the drug is available for oral use as powder or tablets containing the hydrochloride, and in a tablet combined with isoniazid.

Pharmacokinetics

About 80% of an oral dose of ethambutol is absorbed from the gastrointestinal tract, and food is not thought to impair adsorption. The drug is widely distributed throughout most of the body, and is concentrated within erythrocytes so that red cell concentrations are two to three times higher than plasma concentrations. However, ethambutol penetrates slowly and poorly the blood–brain barrier and the CSF, although penetration is improved when the meninges are inflamed. Peak plasma concentrations are achieved approximately 2 h post-dose. Typical blood concentrations and other pharmacokinetic details are shown in the Table. The principal route of elimination of ethambutol is renal excretion of the unchanged drug, about 70% being excreted unmetabolized. Much of the remainder is excreted as inactive dialdehyde and dicarboxylic acid metabolites. Despite this dependence on urinary elimination, only a modest increase in elimination half-life has been reported in patients with renal impairment.

Ethambutol may produce ocular toxicity in the form of retrobulbar neuritis and since both the efficacy and ocular toxicity are dosage-dependent, so a significant dosage reduction in tuberculosis patients with renal failure to reduce potential ocular toxicity is likely to result in therapeutically inadequate plasma concentrations. Peritoneal dialysis and haemodialysis remove ethambutol so that larger intermittent doses can be given. However, there remains a significant risk of over-dosage that warrants therapeutic drug monitoring. A review has concluded that ethambutol is not recommended for treatment of patients in renal failure (Ellard, 1993).

Ethambutol is well tolerated, but it should not be used to treat children, in whom retrobulbar neuritis is difficult to diagnose. Ocular toxicity is reversible if the drug is stopped, but can result in permanent blindness if administration is continued, and is related to dosage and duration of treatment. Hyperuricaemia is common, but not normally clinically significant and other side-effects such as peripheral neuropathy are rare.

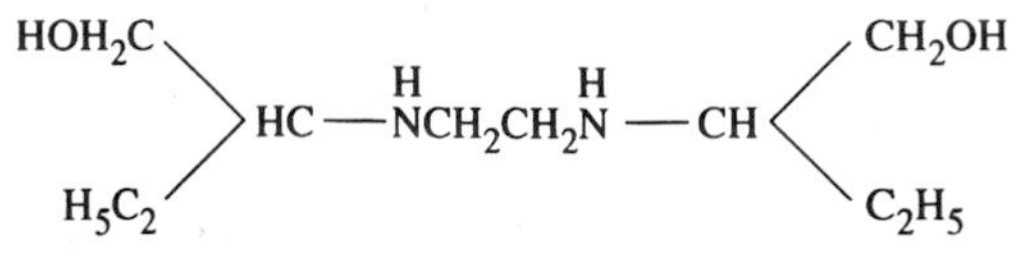

Assays for ethambutol

Microbiological assays using *Mycobacterium smegmatis* were employed when the drug was first introduced, but have been criticized for having poor sensitivity and reproducibility. Until recently the only satisfactory assays were rather complex GLC ones, and are recommended if the facilities are available. An HPLC method is now available which should enable more convenient measurements of the drug.

GLC determination of ethambutol

The GLC method of Lee & Benet (1976) is sensitive and reliable although perhaps rather tedious to set up for the occasional sample. Since the method was first reported, a number of minor modifications have been incorporated to improve sensitivity (Lee & Benet, 1978; Lee & Wang, 1980, Varughese *et al.*, 1986).

The instrumentation required comprises a gas–liquid chromatograph equipped with a ^{63}Ni electron capture detector; an Ni column, 3 feet long and 1/8 inch internal diameter containing OV-101 (5%) on Gas-chrom Q, 100/120 mesh, with a carrier gas (nitrogen) flow of 20 mL/min; and oven temperature of 155°C, injection port temperature 210°C, and detector block temperature 240°C (Varughese *et al.*, 1986).

Procedure

1. Add an appropriate amount of internal standard, the methyl analogue of ethambutol ((dextro-2'-ethylenediimino)-di 1-propanol) to 100–200 μL plasma or diluted urine and extract for 10 min under alkaline conditions with 5–8 mL of spectroquality chloroform. (Note that there is no readily obtainable supply of the internal standard mentioned above. An alternative internal standard, decanediol (Aldrich Chemicals) has been used (Lee & Benet, 1978) for measuring urinary concentrations of ethambutol, but it is very insoluble in water; this method is substantially different from that detailed here and much less sensitive.)

Figure 16.4 Structure of ethambutol.

2. Transfer the organic extract to fresh tubes and evaporate to dryness under nitrogen, ensuring that all traces of water are removed. This can be facilitated by dissolving the residues in 1 mL methylene chloride and evaporating to dryness with the azeotropic removal of any residual water.

3. Dissolve the residues in benzene, alkalinized with dilute pyridine (1:4 in benzene) and derivatize with 20 μL of trifluoroacetic anhydride (Sequanal grade).

4. After 2 h at room temperature, remove excess pyridine and derivatizing agent by washing with 0.01 mol/L HCl.

5. Inject an appropriate volume (1–3 μL).

Calibration and comments. The retention times of ethambutol and the internal standard are approximately 4 and 2.5 min, respectively. The assay is calibrated with calibrators covering the concentration range 0.1–1 mg/L. CVs of 8–3.5% can be expected. An earlier version of this method is ten times more sensitive, and is suitable for assays when only a small sample volume is available (Lee & Wang, 1978).

HPLC assay for ethambutol

The assay of Breda *et al.* (1996) is sensitive for plasma samples, but less so for urine. It uses pre-column derivatization, reversed phase chromatography and fluorescence detection. A 250×4.6 mm Spherisorb CN (5 μm) column (Waters), with a Pellicular ODS (37–53 μm) guard column (Whatman) is used, and the mobile phase is acetonitrile, 0.01 mol/L H_3PO_4, adjusted to pH 2.5 with a few drops of 10 mol/L KOH (30:70 v/v), pumped at 1 mL/min. Derivatization is with 4-fluoro-7-nitrobenzo-2-oxo-1,3-diazole (NBD-F, Aldrich) prepared daily by dissolving the reagent (4 mg) in 1 mL acetonitrile and the fluorescent detector is set at 490 nm excitation, 540 nm emission.

Procedure for plasma samples

1. Mix 0.2 mL of plasma with 0.3 mL of 5 mol/L NaOH, add 5 mL of diethyl ether, vortex for 1 min and centrifuge at 1200**g** (all subsequent extractions are performed using these conditions).

2. Transfer the upper ether layer to a fresh tube and re-extract the bottom aqueous layer with a further 5 mL of ether.

3. Combine the two organic phases and back-extract with 0.2 mL of 0.01 mol/L phosphoric acid.

4. Discard the organic phase and add 0.3 mL of 0.2 mol/L borate buffer (pH 7.5), followed by 40 μL of NBD-F solution (4 mg/mL in acetonitrile).

5. Cap the tubes and heat in a multiblock heater at 80°C for 30 min.

6. Stop the reaction with 50 μL of 1 mol/L phosphoric acid followed by rapid cooling in an acetone bath at about −40°C for 1 min.

7. Remove excess derivatizing agent by extracting with 2 mL of ethyl acetate.

8. Discard upper organic layer, add 50 μL of 5 mol/L NaOH and extract with 2 mL ethyl acetate–methanol (9:1 v/v).

9. Dry the upper organic layer under a stream of nitrogen at 37°C, dissolve the residue in 0.25 mL of 0.01 mol/L phosphoric acid (pH 2.5), and inject 0.2 mL.

Comments. Note that since no internal standard is used, volumes are critical, and complete transfer of phases is important.

Procedure for urine samples

1. Dilute 0.1 mL of urine to 20 mL with distilled water, then add 0.3 mL of 0.2 mol/L borate buffer (pH 7.5) and 40 μL of NBD-F solution (4 mg/mL).

2. Treat the mixture in the same way as for plasma samples.

3. Dissolve the residue in 1 mL of 0.01 mol/L phosphoric acid (pH 2.5) and inject 0.2 mL.

Calibration and comment. The ethambutol derivative elutes with a retention time of about 14 min. Calibrators should be prepared over the range 10–1500 ng/mL for plasma and 10–500 μg/mL for urine. An inter-day variation of up to 14% for plasma and 6% for urine samples can be expected. The sensitivity of the method is 10 ng/mL for plasma and 10 μg/mL for urine. Since this method has had little published application, care should be taken to establish the method in individual laboratories.

Pyrazinamide

Pyrazinamide (Figure 16.5) is an anti-tuberculosis drug which is used only in combination with other agents. It has a molecular weight of 123.1 and a pK_a of 0.5. The white powder, which can be purchased from Aldrich, should be stored at room temperature in an airtight container protected from light. It is a very weak base and is readily soluble in water, ethanol and methanol. Clinically it is available for oral administration as tablets in combination with isoniazid plus rifampicin. Pyrazinamide is only active against *M. tuberculosis* at acidic pH values, and although its MIC is high (Table) and serum concentrations after a therapeutic dose only exceed its MIC by a factor of two, it is considered to have a unique sterilizing activity when included in multi-drug regimens.

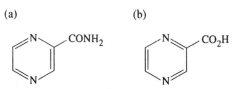

Figure 16.5 Structure of (a) pyrazinamide and (b) its principal metabolite, pyrazinoic acid.

Pharmacokinetics

After rapid absorption, pyrazinamide is distributed throughout the body water, readily penetrates the blood–brain barrier, and is found in the CSF where the levels are similar to the concomitant plasma concentrations in patients with tuberculosis meningitis (Table). Pyrazinamide is extensively metabolized to inactive metabolites, pyrazinoic acid (Figure 16.5b) and 5-hydroxypyrazinoic acid, only about 3% of the dose being excreted unchanged in the urine by glomerular filtration. Urinary concentrations following a 1.5 g dose reach about 50 mg/L; much higher levels of the metabolites are excreted. Pyrazinamide is thought not to be bound to plasma proteins, although a report, using the HPLC method outlined below, put the binding at 40% (Woo *et al.*, 1996).

The adverse effects of pyrazinamide are normally mild, and include cutaneous hypersensitivity, anorexia and nausea. More serious side-effects are hepatitis and arthralgia. As with isoniazid, transient asymptomatic increases in serum hepatic enzyme levels can occur, but these are normally without clinical significance. More overt hepatitis is very rare and occurs in <1% of patients. Arthralgia is much more common, and is thought to be due to the inhibition of the renal secretion of uric acid by the metabolite pyrazinoic acid. Arthralgia is less common in intermittent regimens; it is usually self-limiting and can be treated symptomatically with analgesics.

Determination of pyrazinamide by HPLC

Laboratories are recommended the method of Chan *et al.* (1986), which can be used for both plasma and CSF samples. To overcome possible loss of pyrazinamide by adsorption on to glass, it is recommended that glassware is treated with a suitable silanizing agent.

A reversed phase C_{18} column (Hibar, LiChrocart RP-8; Merck) is used with a mobile phase of acetonitrile: 0.01 mol/L pH 3.5 phosphate buffer (1:9 v/v) pumped at a flow rate of 1.5 mL/min. Detection is by UV absorbance at 215 nm, which in the absence of a suitable variable wavelength detector could be changed to around 270 nm near another absorption maximum, but with decreased sensitivity.

Procedure

1. Mix 8 μL of a 12.5 mg/L solution of the internal standard paracetamol with 200 μL of plasma sample and 200 μL of methanol, followed by 1 mL of 1 mol/L potassium dihydrogen phosphate in 0.2% ascorbic acid.

2. Adjust the pH to 4.2. Extract this acidic solution twice with dichloromethane/diethylether (2:3 v/v) and, after centrifugation for 10 min at 2500g to break the emulsion, evaporate the organic supernatant to dryness at 45°C.

3. Dissolve the residue in 50 μL methanol and inject.

Calibration and comments. Pyrazinamide and the internal standard have retention times of 2.9 and 3.9 min, respectively. Calculate the pyrazinamide concentrations from a calibration curve covering the range 0–40 mg/L. The CV should be around 4.5% at 20 mg/L. The method will measure down to at least 2.5 mg/L.

Colorimetric assay for pyrazinamide

Because of the high concentrations of pyrazinamide found in serum, a relatively insensitive, but rapid, colorimetric assay (Ellard, 1969) can be used.

Procedure

1. Pipette serum sample (3 mL) into a stoppered centrifuge tube, add 6 mL toluene:butan-1-ol (4:1 v/v) and shake well.

2. Add sodium hydroxide (100 μL of 10 mol/L) and re-shake the tube.

3. Immediately add 4 g of anhydrous sodium sulphate and shake the tube again vigorously.

4. Centrifuge, remove 4 mL of the organic phase and wash with 400 μL 0.05 mol/L sulphuric acid. This removes any traces of an unidentified inhibitor present in stored serum that can affect the formation of the final colour.

5. Mix 3 mL of the organic phase with 3 mL *n*-heptane and 1.6 mL 0.05 mol/L sulphuric acid, shake and centrifuge.

6. Discard the organic phase and take 1 mL of the acid extract (diluted if necessary so that it contains approximately 10 mg/L pyrazinamide) and react with 20 μL of a freshly prepared aqueous solution of sodium nitroprusside (Analar, 2% w/v) and 400 μL of 1 mol/L sodium hydroxide.

7. Exactly 5 min later, read the optical density at 495 nm.

Calibration and comments. The concentration of pyrazinamide is determined by comparison with serum calibrators covering the range 0–50 mg/L pyrazinamide. CVs of around 3–5% can be expected and the method is sensitive to 1 mg/L for serum.

Cycloserine

D-Cycloserine (Figure 16.6) is a second-line anti-tuberculosis drug for use in combination with other agents when primary treatment has failed and the organism is cycloserine-sensitive. Both D and L isomers and the DL

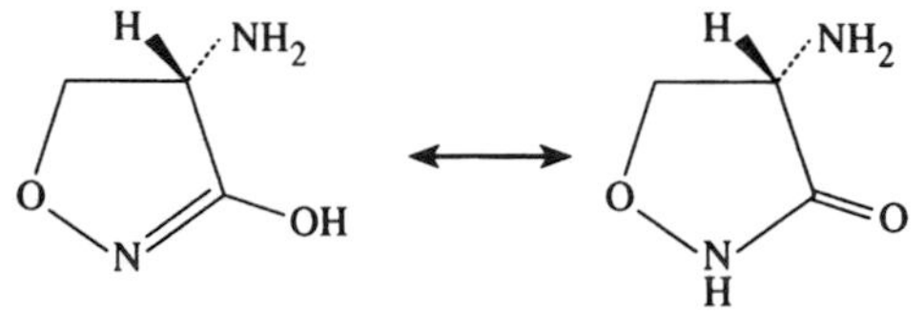

Figure 16.6 The structure of D-cycloserine, showing both *enol* and *keto* configurations.

mixture are available from chemical suppliers but only D-cycloserine (mol. wt 102.1, pK_a 4.4 and 7.4) is used clinically and should be used for preparing calibrators or controls. Reference substance is available from the British Pharmacopoeia Commission Laboratory. D-Cycloserine is a hygroscopic powder, very soluble in water and alcohol, but unstable below pH 7. In solution the drug undergoes *keto–enol* transformation and can also dimerize, the rate being concentration- and pH-dependent. Clinically cycloserine is available as 125 and 250 mg capsules containing cycloserine and inert diluent.

Pharmacokinetics

Cycloserine is administered orally and shows high bio-availability and a large volume of distribution which includes breast milk, CSF, pleural fluids and fetal blood. Peak plasma concentrations occur up to 4 h after oral dosing and serum protein binding is <20%. The drug is eliminated partially by metabolism (30–40%) to unidentified metabolites and partially by glomerular filtration (60–70%). Dosage reduction is therefore indicated in patients with impaired renal function. Pharmacokinetic studies were performed many years ago and the mean elimination half-life is reported as 10 h but may vary from 4 to 30 h. Cycloserine may produce a number of side-effects, the most important being neurotoxicity producing psychotic disturbances. Neurotoxicity is more commonly seen in patients receiving higher dosages (500 mg bd) and may be related to serum concentrations >30 mg/L (Storey & McLean, 1957). It may also be potentiated by ethionamide or alcohol. For this reason the drug should not be used in alcoholics or patients with mental illness or epilepsy. Weekly serum monitoring is recommended for all patients receiving cycloserine but particularly those with renal impairment, or receiving >500 mg per day, or showing suspected toxicity.

Assay of cycloserine in biological fluids

Spectrophotometric, microbiological (Mattila *et al.*, 1969) and HPLC assays for cycloserine in body fluids have been described. For therapeutic monitoring the spectrophotometric or HPLC methods are recommended. The spectrophotometric method described below is simple to perform and can be applied to serum, CSF or urine, but requires a relatively large sample volume. HPLC assays require only small samples and a simple straightforward method which is suitable for therapeutic monitoring is given below together with a more complex method showing greater sensitivity. Although cycloserine is unstable at pH 7 and below, serum calibrators may be prepared and stored at –70°C for at least 3 months and samples may be transported or kept at room temperature without significant loss. A 1 g/L stock solution in 0.1 mol/L sodium hydroxide may be stored for at least 2 weeks at –20°C.

Spectrophotometric assay of cycloserine (based on that of Jones, 1956)

Procedure and calibration

1. Prepare serum or plasma calibrators containing 0, 50 and 75 mg/L D-cycloserine.

2. Pipette 2 mL of each calibrator and each control and patient's sample in 15 mL centrifuge tubes and add 2 mL of aqueous 10% sodium tungstate and 2 mL of 0.16 mol/L sulphuric acid to each tube.

3. Mix the tubes well, allow to stand on the bench for 5 min, and then centrifuge for 15 min.

4. Carefully transfer 3 mL of each clear supernatant to 10 mm spectrophotometer tubes and add 3 mL of 0.1 mol/L sodium hydroxide in an additional similar tube (reagent blank).

5. Add 1 mL of 9 mol/L acetic acid and 1 mL of freshly prepared colour reagent (4 mol/L sodium hydroxide: 4% aqueous sodium nitroprusside ($Na_2Fe(CN)_5NO \cdot 2H_2O$) (50:50 v/v)) to each tube. After 10 min a blue coloration develops.

6. Measure the absorbance at 615 nm against the reagent blank in a suitable spectrophotometer.

7. Determine the concentrations in the patients' samples from the calibration curve of optical density versus calibrator concentration.

Comments. Samples giving results >75 mg/L should be diluted and re-assayed. Reproducibility at around 30 mg/L should be better than ±5% as long as great care is taken in separating clear supernatant and precipitate. This assay may be performed with urine also. The urines should be diluted to give an expected concentration <75 mg/L and, if any are very concentrated, or expected concentrations are low, they should initially be adjusted to pH 7 and decolorized with activated charcoal.

HPLC assay of cycloserine in plasma or urine (Musson *et al.*, 1987)

The stationary phase is Hypersil ODS (240×5 mm, 10 μm) at room temperature and the mobile phase is isopropanol 75 mL, or 65 mL for urine samples, water (800 mL), glacial acetic acid (5 mL) and decanesulphonic acid (500 mg) adjusted to pH 4.4 with potassium hydroxide. Flow rate is 2.3 mL/min. Detection is by fluorescence (excitation 340 nm, emission 455 nm) after post-column derivatization with fluoraldehyde (a commercially available mixture of *ortho*-phthalaldehyde and 2-mercaptoethanol obtained from Pierce, Rockford, IL, USA).

Post-column derivatization requires specialized apparatus and the authors employed a steel column (250×4.5 mm) packed with 50 μm glass beads and heated to 40°C. The fluoraldehyde is pumped at 1.2 mL/min and mixes with the column eluant at a T-junction before passing through the post-column derivatization apparatus and then the detector.

Procedure for serum or plasma samples

1. Mix 1 mL of sample with 250 μL of 0.2 M borate buffer pH 9.75 and 30 μL of internal standard (240 mg/L 6-aminocaproic acid).

2. Ultrafilter through an Amicon ultrafilter by centrifugation for 15 min at 723**g**.

3. Inject 20–25 μL of the ultrafiltrate.

Procedure for urine samples

1. Mix 1 mL of sample with 50 μL of internal standard (3 g/L α-aminobutylhistidine) and 1 mL of 200 g/L sodium carbonate.

2. Add 2 mL of isopropanol, mix by vortexing and centrifuge at 723**g** for 5 min.

3. Inject 25 μL of the top organic layer.

Comments. The retention time of cycloserine is around 5 min. The limits of sensitivity are 0.3 mg/L (plasma) and 2 mg/L (urine). Typical day-to-day CVs are 5–11%.

HPLC assay of cycloserine in serum (D. Ryder & L. O. White, unpublished)
This method is based on the method of Musson *et al.* (1987) but uses UV detection at 210 or 214 nm without post-column derivatization. Single-point calibration with a serum calibrator of 20–30 mg/L cycloserine is suitable for therapeutic monitoring.

The stationary phase is IB-SIL 5 C$_{18}$ BD 100 × 4.6 mm (Phenomenex, Macclesfield, UK) and the mobile phase water:orthophosphoric acid:isopropanol (98:0.5:1.5 by volume) containing 10 g/L octanesulphonic acid and adjusted to pH 4.4–4.5 with sodium hydroxide. Flow rate is 0.7 mL/min.

Procedure

1. Mix sample, serum calibrator and serum control (50–100 μL) with equal volumes of methanol.

2. Allow to stand for 5 min, then centrifuge.

3. Inject 10 μL of the clear supernatant.

Comments. Cycloserine retention time is <10 min but late-eluting peaks with retention times up to 30 min will be encountered. Larger injection volumes and higher pump flow rates should be avoided since they may result in loss of adequate chromatographic separation and poor efficiency. CVs are better than 5%.

Dapsone

Dapsone (Figure 16.7a) is used primarily to treat leprosy, although it is also used in the prophylaxis of malaria, in the treatment of dermatitis herpetiformis and other skin disorders, and in combination with trimethoprim to treat

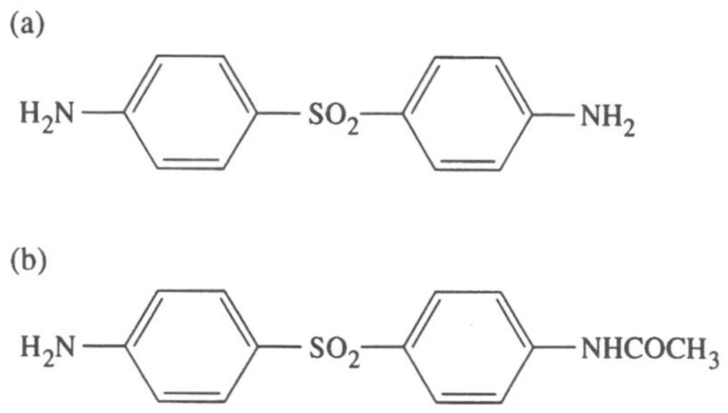

Figure 16.7 Structure of (a) the anti-leprosy drug dapsone and (b) its metabolite monoacetyldapsone.

pneumocystis infections. Dapsone powder (mol. wt 248.3, pK$_a$ 2.3) can be obtained as the 4-aminophenyl sulphone from Aldrich. This is stable when kept in an airtight container, protected from sunlight. Dapsone is a weak base, rather insoluble in water, but soluble in ethanol, methanol and dilute mineral acids. Stock solutions of 1 g/L in ethanol can be prepared and are stable for many months at 4°C.

Pharmacokinetics

Dapsone is well absorbed with peak plasma concentrations occurring approximately 4 h post-dose (Table), and is widely distributed in the tissues. Urinary concentrations can reach 50 mg/L after a 100 mg dose. The drug is extensively metabolized; a major metabolite being monoacetyldapsone (Figure 16.7b) formed by polymorphic acetylation of dapsone. Between 5% and 15% of the dose is excreted unchanged in the urine and none of the metabolites appear to have any antimicrobial activity. Clinically dapsone is available as tablets and in combination with pyrimethamine.

Side-effects are rare, the most common being dose-related haemolysis and methaemoglobinaemia, but this is rarely seen at doses of <200 mg, unless the patient is deficient in glucose-6-phosphate dehydrogenase. Other rare side-effects are hypersensitivity, agranulocytosis and peripheral neuropathy.

Assays for dapsone in biological samples

A number of HPLC assays are available for dapsone, and most are sensitive and specific. Details are given below of the method of Horai & Ishizaki (1985) which is very similar to that of Carr *et al.* (1978) but uses an internal standard that is more easily obtainable. Other HPLC methods that are suitable are those of Edstein (1984) and Peters *et al.* (1981).

HPLC determination of dapsone by the method of Horai & Ishizaki (1985)
The stationary phase is a 5 μm Hibar LiChrosorb RP-18 column (Merck, Darmstadt, Germany), although other similar columns would probably suffice. The mobile phase

is water:acetonitrile:acetic acid (730:250:20 by volume) pumped at 1.3 mL/min. Detection is by UV absorption at 250 nm. At higher wavelengths the absorption of the internal standard is much reduced, although 250 nm is not the absorption maximum of dapsone.

Procedure

1. Mix plasma (500 μL) or urine (100–500 μL) in a 10 mL PTFE-lined screw-capped tube with 50 μL of the internal standard (*m*-aminophenyl sulphone solution, Aldrich), 350 μL of water and 100 μL of 1 mol/L sodium hydroxide.

2. Extract the samples with 3 mL of dichloromethane.

3. Following centrifugation, discard the upper aqueous phase and transfer the organic phase to fresh tubes.

4. Evaporate to dryness under nitrogen. Dissolve the residue in 30–50 μL of mobile phase and inject.

Comments. The retention time of dapsone is 4.5 min, with its mono-acetyl metabolite eluting at 5.3 min and the internal standard at 6.2 min. The limit of sensitivity is approximately 0.01 mg/L, and the day-to-day CVs are 2.5–8%.

Clofazimine

Clofazimine (Figure 16.8) is primarily used to treat leprosy, although some interest has been expressed recently in its use in *M. avium–intracellulare* infections in AIDS patients. The drug (mol. wt 473.4) is not readily obtainable as an analytical standard, although small amounts of the dark brown powder may be available on request from its manufacturer, Ciba. The powder should be stored at room temperature, protected from light in an airtight container. Clofazimine is practically insoluble in water, but stock solutions of 1 g/L may be prepared in ethanol. For clinical use it is available as both tablets and capsules containing a microcrystalline suspension in an oil–wax base.

Clofazimine is adsorbed strongly on to glass, so during laboratory procedures, such as preparation of calibrators or sample preparation, glassware should be silylated or disposable polypropylene vessels used instead.

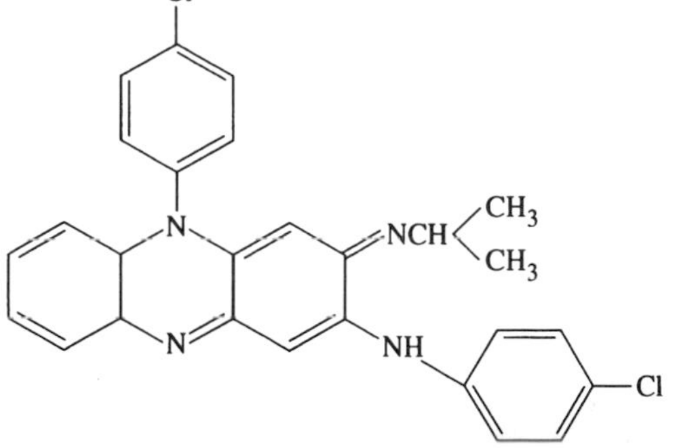

Figure 16.8 Structure of clofazimine (B663).

Pharmacokinetics

Clofazimine is poorly absorbed by the gastro-intestinal tract, and blood concentrations vary greatly between individuals on daily treatment, the mean steady-state peak concentrations being approximately 0.4 mg/L after 600 mg daily. Clofazimine has the unusual property of being retained as crystals in tissues. Little is known about its metabolism; only a very small proportion of the dose is excreted in the urine, but larger amounts are found in faeces. Clofazimine is cleared very slowly from the body: its exact half life, although uncertain, is extremely long, and may be much as 70 days.

When given over an extended period of time, clofazimine can lead to red-brown skin discoloration as a result of the deposition of crystals in the tissues. This is most noticeable in light-skinned patients, but the side-effect is mostly cosmetic. Other side-effects, such as gastrointestinal effects, are uncommon, and dose-related.

HPLC determination of clofazimine

The HPLC method of Krishnan & Abraham (1992) is recommended for determining clofazimine, although it has the drawback that the internal standard used is salicylic acid, which precludes measurements on plasma from patients taking aspirin.

The stationary phase is μ-Bondapak (150 × 3.9 mm, Waters) with a C_{18} guard column. A mobile phase of tetrahydrofuran (THF):0.5% acetic acid:methanol (8:11:1 by volume) plus 2.5 mmol/L hexane sulphonic acid, is pumped at a flow rate of 1.8 mL/min. Detection is by UV absorption at 280 nm. Solid-phase extraction (SPE) is used to extract clofazimine from plasma samples.

Procedure and calibration

1. Prepare stock solutions of clofazimine (10 mg/L) in 50% methanol in water and of the internal standard salicylic acid in the same solvent (34 mg/mL) and store at 5°C.

2. Prepare five different calibrator solutions of clofazimine by diluting the stock solution with 10% methanol in water to cover the range 0.33–6.6 mg/L.

3. Mix five 1 mL drug-free plasma samples with 100 μL of a calibrator solution of clofazimine of known concentration.

4. Prepare a sixth blank sample with 100 μL of 10% methanol in water.

5. Dilute these samples with 5 mL of 0.1 mol/L pH 6.0 phosphate buffer, and leave to stand while the SPE columns are conditioned.

6. Treat the test samples in the same way.

7. Treat SPE columns sequentially with 3 mL methanol, 5 mL water and 5 mL 0.1 mol/L pH 6.0 phosphate buffer.

8. Add the diluted plasma samples and allow to percolate slowly under reduced pressure.

9. Air-dry the column for 2 min and add 50 μL of a solution containing 170 ng salicylic acid.

10. Extract the samples with 4 × 1 mL solution of THF:acetonitrile:methanol (2:2:1 v/v) containing 0.7 mmol/L hexane sulphonic acid.

11. Combine the eluate, evaporate at 37°C under nitrogen and reconstitute in 100 μL mobile phase (see above).

12. Inject 25 μL aliquots.

13. Calculate the peak height ratio of clofazimine to the internal standard.

Comments. Clofazimine elutes after 6.0 min and salicylic acid after 3.3 min. CVs of <5% should be expected; the lower limit of detection is 3 μg/L.

Urine tests for measuring patient compliance

Perhaps the simplest method of determining whether tablets have been ingested is to prescribe combined formulations containing rifampicin plus either isoniazid and/or pyrazinamide and to check the colour of the patient's urine. The red coloration caused by the excretion of rifampicin can usually be seen in urine collected up to 10 h post-dose and will confirm that all components of the formulation have been taken (Fox *et al.*, 1987). However, in the absence of a combined tablet containing rifampicin, or as a confirmation test, a colorimetric assay for isoniazid or a microbiological urine test for rifampicin can be performed.

Urine test for isoniazid

The method of Ellard & Greenfield (1977) which detects the isoniazid metabolites isonicotinic acid and isonicotinyl glycine is recommended. The test is best carried out in 5 mL test tubes.

Procedure

1. Mix 500 μL samples of the test urine together with 200 μL of 4 mol/L acetate buffer pH 5.0.

2. At 15 s intervals, add 100 μL of 10% aqueous potassium cyanide, 100 μL 10% aqueous chloramine-T and 500 μL 1% barbituric acid in acetone:water (1:1 v/v). After 30 min a blue colour appears in positive samples.

Comment. Among smokers this colour may be somewhat masked by the reaction given by nicotine and its metabolites (Peach *et al.*, 1985), but the resulting grey/black to brown colour (the orange/red from nicotine mixed with the blue from the isoniazid metabolites) is easily recognized.

Urine test for rifampicin

An agar diffusion assay similar to the method described earlier in this chapter can be used to determine patient compliance, following dosing with rifampicin (Mitchison *et al.*, 1970). By using both a rifampicin-sensitive indicator organism (*S. Aureus* NCTC 10702) and rifampicin-resistant one (*S. Aureus* NCTC 10703), a specific and sensitive urine test is achieved by comparison of zone diameters.

References

Breda, M., Marrari, P., Pianezzola, E. & Strolin Benedetti, M. (1996). Determination of ethambutol in human plasma and urine by high-performance liquid chromatography with fluorescence detection. *Journal of Chromatography A* **729**, 301–7.

Carr, K., Oates, J. A., Nies, A. S. & Woosley, R. L. (1978). Simultaneous analysis of dapsone and monoacetyldapsone employing high performance liquid chromatography: a rapid method for detection of acetylator phenotype. *British Journal of Clinical Pharmacology* **6**, 421–7.

Chan, K., Wong, C. L. & Lok, S. (1986). High-performance liquid chromatographic determination of pyrazinamide in cerebrospinal fluid and plasma in the rabbit. *Journal of Chromatography* **380**, 367–73.

Dickinson, J. M., Aber, V. R., Allen, B. W., Ellard, G. A. & Mitchison, D. A. (1974). Assay of rifampicin in serum. *Journal of Clinical Pathology* **27**, 457–62.

Dymond, L. C. & Russell, D. W. (1970). Rapid determination of isonicotinic acid hydrazide in whole blood with 2,4,6-trinitrobenzenesulphonic acid. *Clinica Chimica Acta* **27**, 513–20.

Edstein, M. (1984). Quantification of antimalarial drugs. II. Simultaneous measurement of dapsone, monoacetyldapsone and pyrimethamine in human plasma and urine. *Journal of Chromatography* **307**, 426–31.

Ellard, G. A. (1969). Absorption, metabolism and excretion of pyrazinamide in man. *Tubercle* **50**, 144–58.

Ellard, G. A. (1993). Chemotherapy of tuberculosis for patients with renal impairment. *Nephron* **64**, 169–81.

Ellard, G. A. & Greenfield, C. (1977). A sensitive urine-test method for monitoring the ingestion of isoniazid. *Journal of Clinical Pathology* **30**, 84–7.

Ellard, G. A., Gammon, P. T. & Wallace, S. M. (1972). The determination of isoniazid and its metabolites acetylisoniazid, monoacetylhydrazine, diacetylhydrazine, isonicotinic acid and isonicotinylglycine in serum and urine. *Biochemical Journal* **126**, 449–58.

Ellard, G. A., Humphries, M. J. & Allen, B. W. (1993). Cerebrospinal fluid drug concentrations and the treatment of tuberculous meningitis. *American Review of Respiratory Diseases* **148**, 650–5.

El-Yazigi, A. & Yusuf, A. (1991). An expedient liquid chromatographic micromeasurement of isoniazid in plasma by use of electrochemical detection. *Therapeutic Drug Monitoring* **13**, 254–9.

Fox, W., Prabhakar, R., Ganapathy, S. & Somasundaram, P. R. (1987). An enquiry into the attitude of South Indian patients to the coloration of their urine by rifampicin. *Tubercle* **68**, 201–7.

Girling, D. J. (1982). Adverse effects of antituberculosis drugs. *Drugs* **23**, 56–74.

Guillaumont, M., Leclercq, M., Frobert, Y., Guise, B. & Harf, R. (1982). Determination of rifampicin, desacetylrifampicin, isoniazid and acetylisoniazid by high-performance liquid chromatography: application to human serum extracts, polymorphonucleocytes and alveolar macrophages. *Journal of Chromatography* **232**, 369–76.

He, X., Wang, J., Liu, X. & Chen, X. (1996). High-performance liquid chromatography assay of rifapentine in human serum. *Journal of Chromatography* **681**, 412–5.

Horai, Y. & Ishizaki, T. (1985). Rapid and sensitive liquid chromatographic method for the determination of dapsone and monoacetyldapsone in plasma and urine. *Journal of Chromatography* **345**, 447–52.

Hutchings, A., Monie, R. D., Spragg, B. & Routledge, P. A. (1983). High-performance liquid chromatographic analysis of isoniazid and acetylisoniazid in biological fluids. *Journal of Chromatography* **277**, 385–90.

Ishii, M. & Ogata, O. (1988). Determination of rifampicin and its main metabolites in human plasma by high-performance liquid chromatography. *Journal of Chromatography* **426**, 412–6.

Jenner, P. J. (1983). High-performance liquid chromatographic determination of thiacetazone in body fluids. *Journal of Chromatography* **276**, 463–70.

Jenner, P. J. & Ellard, G. A. (1981). High-performance liquid chromatographic determination of ethionamide and prothionamide in body fluids. *Journal of Chromatography* **225**, 245–51.

Jenner, P. J., Ellard, G. A., Gruer, P. J. K. & Aber, V. R. (1984). A comparison of the blood levels and urinary excretion of ethionamide and prothionamide in man. *Journal of Antimicrobial Chemotherapy* **13**, 267–77.

Jones, L. R. (1956). Colorimetric determination of cycloserine, a new antibiotic. *Analytical Chemistry* **28**, 39–41.

Krishnan, T. R. and Abraham, I. (1992). A rapid and sensitive high performance liquid chromatographic analysis of clofazimine in plasma. *International Journal of Leprosy* **60**, 549–55.

Kubo, H., Kinoshita, T., Matsumoto, K. & Nishikawa, T. (1990). Fluorometric determination of isoniazid and its metabolites in urine by high performance liquid chromatography. *Chromatographia* **30**, 69–72.

Lau, Y. Y., Hanson, G. D. & Carel, B. J. (1996). Determination of rifabutin in human plasma by high-performance liquid chromatography with ultraviolet detection. *Journal of Chromatography* **676**, 125–30.

Lecaillon, J. B., Febvre, N., Metayer, J. P. & Souppart, C. (1978). Quantitative assay of rifampicin and three of its metabolites in human plasma, urine and saliva by high-performance liquid chromatography. *Journal of Chromatography* **145**, 319–24.

Lee, C. S. & Benet, L. Z. (1976). Gas–liquid chromatographic determination of ethambutol in plasma and urine of man and monkey. *Journal of Chromatography* **128**, 188–92.

Lee, C. S. & Benet, L. Z. (1978). Micro and macro GLC determination of ethambutol in biological fluids. *Journal of Pharmaceutical Sciences* **67**, 470–3.

Lee, C. S. & Wang L. H. (1980). Improved GLC determination of ethambutol. *Journal of Pharmaceutical Sciences* **69**, 362–3.

Mattila, M. J., Nieminen, E. & Tiitinen, H. (1969). Serum levels, urinary excretion, and side-effects of cycloserine in the presence of isoniazid and p-aminosalicylic acid. *Scandinavian Journal of Respiratory Diseases* **50**, 291–300.

Mitchison, D. A., Allen, B. W. & Miller, A. B. (1970). Detection of rifampicin in urine by a simple microbiological assay. *Tubercle* **51**, 300–4.

Musson, D. G., Maglietto, S. M., Hwang, S. S., Gravellese, D. & Bayne, W. F. (1987). Simultaneous quantification of cycloserine and its prodrug acetylacetonylcycloserine in plasma and urine by high-performance liquid chromatography using ultraviolet absorbance and fluorescence after post-column derivatization. *Journal of Chromatography* **414**, 121–9.

Olson, W. A., Dayton, P. G., Israili, Z. H. & Pruitt, A. W. (1977). Spectrophotofluorometric assay for isoniazid and acetylisoniazid in plasma adapted to paediatric studies. *Clinical Chemistry* **23**, 745–8.

Peach, H., Ellard, G. A., Jenner, P. J. & Morris, R. W. (1985). A simple, inexpensive urine test of smoking. *Thorax* **40**, 351–7.

Peloquin, C. A., James, G. T. & McCarthy, E. (1991). Improved high-performance liquid chromatographic assay for the determination of ethionamide in serum. *Journal of Chromatography* **563**, 472–5.

Peters, J. H., Murray, J. F., Gordon, G. R. & Gelber, R. H. (1981). Dapsone in saliva and plasma of man. *Pharmacology* **22**, 162–71.

Scott, E. M. & Wright, R. C. (1967). Fluorometric determination of isonicotinic acid hydrazide in serum. *Journal of Laboratory and Clinical Medicine* **70**, 355–60.

Storey, P. B. & McLean, R. L. (1957). A current appraisal of cycloserine. *Antibiotics in Medical and Clinical Therapy* **4**, 223–32.

Varughese, A., Brater, D. C., Benet, L. Z. & Lee, C. S. (1986). Ethambutol kinetics in patients with impaired renal function. *American Review of Respiratory Diseases* **134**, 34–8.

Walubo, A., Chan, K. & Wong, C. L. (1991). Simultaneous assay for isoniazid and hydrazine metabolite in plasma and cerebrospinal fluid in the rabbit. *Journal of Chromatography* **567**, 261–6.

Woo, J., Cheung, W., Chan, R., Chan, H. S., Cheng, A. & Chan, K. (1996). *In vitro* protein binding characteristics of isoniazid, rifampicin, and pyrazinamide to whole plasma, albumin and α-1-acid glycoprotein. *Clinical Biochemistry* **29**, 175–7.

Chapter 17

Antifungal drugs

David W. Warnock, Elizabeth M. Johnson and Debra A. Oliver

PHLS Mycology Reference Laboratory, Public Health Laboratory, Bristol BS2 8EL, UK

Introduction

The concentrations in biological fluids of antifungal drugs are measured for two principal reasons: to ensure that adequate concentrations of the drug are attained, and to ensure that excessive concentrations that could cause unwanted side-effects are avoided. A number of assay methods are available. Microbiological methods are often the simplest to perform, but long incubation periods are required. With the judicious selection of assay strain and medium, these methods can often be modified to permit determination of one drug when a second is also present. Non-biological methods have also been developed for the quantification of most antifungal drugs. Spectrophotometric and fluorimetric methods have been described, but as no separation step is performed these methods cannot be relied upon to distinguish the parent compound from metabolites. GLC and HPLC methods have also been described for most antifungal drugs. These methods offer improved reproducibility and specificity over many other methods and permit determination of concentrations of individual drugs in patients receiving combination regimens.

Amphotericin B

Physicochemical properties

Amphotericin B (Figure 17.1) is a polyene antibiotic with a molecular weight of 924.1 and empirical formula of $C_{47}H_{73}NO_{17}$. The molecule consists of a large closed macrolide ring with well demarcated hydrophilic and hydrophobic regions. It is insoluble in water unless complexed with sodium desoxycholate, when concentrations of

50 g/L are attainable. Its pK_a values are 5.7 (carboxyl group) and 10.0 (amino group). It is soluble in dimethyl sulphoxide (DMSO) at concentrations of 10 g/L.

Dosage forms

Amphotericin B is normally supplied for parenteral administration in lyophilized form containing 50 mg (50,000 international units) of amphotericin B together with 41 mg of sodium desoxycholate (which acts as a dispersing agent) and a sodium phosphate buffer. The addition of 10 mL of sterile water gives a clear micellar suspension and this is further diluted with 5% dextrose solution before infusion.

Amphotericin B is now also available for parenteral administration encapsulated in liposomes. The drug is provided in lyophilized form in 50 mg amounts and is first reconstituted in 12 mL of cold sterile water (for injection) before being further diluted with 5% dextrose solution and filter sterilized.

Other new formulations include amphotericin B lipid complex (ABLC), which is supplied for parenteral administration as a sterile suspension which must be filter sterilized and diluted before use with 5% dextrose solution, and amphotericin B colloidal dispersion (ABCD), which is supplied for parenteral administration in lyophilized form and is first reconstituted in sterile water before being diluted with 5% dextrose solution.

Pharmacokinetics

Amphotericin B is not absorbed following mucosal or cutaneous application. Minimal absorption occurs from

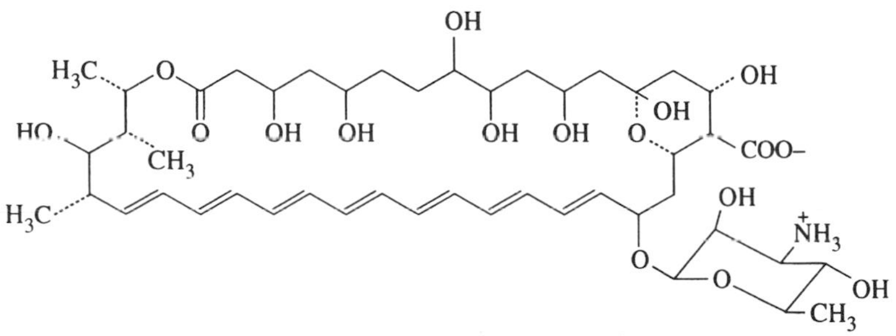

Figure 17.1 Structure of amphotericin B.

221

the gastrointestinal tract. Parenteral administration of a 1 mg/kg dose of the conventional formulation of the drug will produce serum concentrations of 1.0–2.0 mg/L (Bindschadler & Bennett, 1969). Less than 10% of the dose remains in the blood 12 h after administration and >90% of this is protein-bound. The remainder is thought to bind to tissue cell membranes, the highest concentrations being found in the liver and spleen (Christiansen *et al.*, 1985). Levels in cerebrospinal fluid (CSF) are <5% of the simultaneous blood concentration. Amphotericin B binds to tissues for long periods of time, re-entering the circulation slowly from these sites.

The pharmacokinetics of the different lipid-based formulations of amphotericin B are quite diverse. Large structures, such as ABLC, are rapidly removed from the blood, but smaller liposomes remain in the circulation for much longer periods.

The maximum serum concentrations obtained after parenteral administration of liposomal amphotericin B have ranged from 10 to 35 mg/L for a 3 mg/kg dose and from 25 to 60 mg/L for a 5 mg/kg dose (Janknegt *et al.*, 1992). Levels of 5–10 mg/L have been detected 24 h after a 5 mg/kg dose. In animals, administration of liposomal amphotericin B results in much higher drug concentrations in the liver and spleen than are achieved with conventional amphotericin B. Levels in renal tissue are much lower than those obtained with equivalent amounts of the conventional formulation. Human tissue distribution has not been studied in detail, but the highest drug levels occur in the liver and spleen.

The maximum serum concentrations obtained after parenteral administration of ABLC are lower than after administration of equivalent amounts of conventional amphotericin B because of the more rapid distribution of the drug to tissue. Maximum concentrations are 1–2 mg/L for a 5 mg/kg dose. In animals, administration of ABLC results in much higher drug concentrations in the liver, spleen and lungs than are achieved with conventional amphotericin B. Concentrations in renal tissue are much lower than those obtained with equivalent amounts of the conventional formulation. Human tissue distribution has not been studied in detail.

Maximum serum concentrations of about 2 mg/L have been obtained following a 1 mg/kg dose of ABCD, but levels in the blood decline soon after the end of the infusion as a result of the rapid distribution of the drug to tissue. In animals, administration of ABCD results in much higher drug concentrations in the liver and spleen than are achieved with conventional amphotericin B. Levels in renal tissue are much lower than those obtained with equivalent amounts of the conventional formulation. Human tissue distribution has not been studied in detail.

No metabolites have been identified, but it is thought that amphotericin B is metabolized by the liver. The conventional formulation of the drug has a β-phase elimination half-life of about 24–48 h and an third (γ-

phase) half-life of about 2 weeks (Atkinson & Bennett, 1978). Less than 5% of a given dose is excreted as unchanged drug in the urine.

Monitoring of blood concentrations of amphotericin B during treatment is seldom indicated. The optimum serum concentrations of the drug for particular fungal infections have not been determined. Although amphotericin B is nephrotoxic, high blood levels do not lead to greater impairment of renal function, nor does renal failure result in higher blood concentrations (Working Party of the BSAC, 1991). Haemodialysis does not reduce blood levels unless the patient is hyperlipaemic, in which case there is some drug loss resulting from adherence to the dialysis membrane.

Assay methods

Microbiological methods have been described for determination of amphotericin B concentrations in serum, CSF and other biological fluids. These methods need 24–36 h to perform and interference from other antifungal drugs is a potential problem. Several reversed-phase HPLC methods have been described (Table I). These methods offer much greater precision and interference from other drugs is not a problem.

Microbiological assay method

Preparation of drug solutions. To prepare a stock drug solution with a concentration of 10 g/L, weigh 50 mg of amphotericin B and make up to 5 mL with DMSO at room temperature. Allow to stand for 30 min to permit self-sterilization. This stock solution can be stored, protected from light, in small aliquots at –70°C for up to 1 month or at 4°C for 1 week. Further ten-fold dilutions should be prepared with sterile pooled human serum. Horse serum may be used instead, as it gives identical results. The following calibrator solutions should be prepared fresh on each occasion: 0.25, 0.5, 1.0, 2 and 4 mg/L.

Preparation of media. Suspend 62.5 g of Antibiotic Medium 12 (Nystatin Agar, code 0669; Difco Laboratories Ltd, West Molesey, UK) in 1 L of cold distilled water. Heat to boiling point to dissolve the agar and then sterilize for 15 min at 121°C. Dispense in 100 mL amounts.

Preparation of indicator strain inoculum. Inoculate a Sabouraud agar plate with a suitable sensitive assay strain, such as *Paecilomyces variotii* NCPF 2169, and incubate at 30°C for 2–3 days. Harvest the spores in sterile water and adjust the concentration of the suspension to about 1×10^6 spores/mL using a haemocytometer.

Test method. Melt 100 mL of agar, cool to 50°C and add 1 mL of spore suspension. Pour into a levelled 25 cm square assay plate. When set, place the plate in a 37°C incubator, with the lid removed, and leave for about 10 min. Cut 30 wells of 4 mm diameter in the agar and place 15 μL of each calibrator in each of three wells. Each

Table I. High-performance liquid chromatography assay method for amphotericin B

Authors	Extraction or deproteinization	Mobile phase	Stationary phase	Mode of separation	Flow rate (mL/min)	Detection	Wavelength of detection (nm)	Retention time (min)
Nilsson-Ehle et al., 1977	methanol	methanol:EDTA	C_{18}	partition RP	2.5	absorption	405	4.2
Mayhew et al., 1983	methanol	methanol:EDTA acetonitrile:EDTA (gradient elution)	C_{18}	partition RP	1	absorption (internal standard)	388	c.15
Golas et al., 1983	acetonitrile	sodium acetate: acetonitrile	C_{18}	partition RP	1.5	absorption (internal standard)	405	c. 5
Bach, 1984	pH Bond Elut column acetonitrile:EDTA	acetonitrile: methanol:water	C_{18}	partition RP	1.5	absorption (internal standard)	386	c. 4.8
Brassinne et al., 1987	methanol	EDTA:acetonitrile	C_{18}	partition RP	1.5	absorption (internal standard)	405	NS
Granich et al., 1986	Supelclean solid phase column	acetonitrile:EDTA	C_{18}	partition RP	1.5	absorption (internal standard)	382	4.9
Hosotsubo et al., 1988	acetonitrile	acetonitrile: acetate buffer	trimethyl silica	partition RP	1	absorption	405	NS
Kan et al., 1991	methanol	methanol: acetonitrile	C_{18}	partition RP	1.6	absorption	405	NS
Christiansen et al., 1985	ethanol:water	acetate buffer: methanol	C_8	partition RP	2	absorption	407	8.8

Abbreviations: C_8, octasilane; C_{18}, octadecylsilane; EDTA, ethylenediaminetetraacetetic acid; NS, not specified; RP, reversed phase.

sample should also be tested in triplicate. Incubate the plate at 30°C for 24 h. Measure the diameters of the zones of inhibition and construct a calibration curve from which the drug concentrations in the patient samples can be interpolated (see *Chapter 3*).

HPLC assay method

Preparation of drug solutions. To prepare stock solutions with a concentration of 100 mg/L, add 5 mL of DMSO to 5 mg of amphotericin B or *N*-acetylamphotericin B (internal standard). Then make each up to 50 mL with 1% aqueous sodium dodecylsulphate. These stock solutions can be stored at –70°C for at least 1 month.

The following amphotericin B calibrators should be prepared fresh in sterile horse serum on each occasion: 0.5, 1, 2, 4 and 8 mg/L. Prepare a 5 mg/L solution of *N*-acetylamphotericin B in sterile distilled water fresh on each occasion.

Extraction procedure. Add 100 µL of internal standard to 100 µL of each sample, calibrator and internal control. Then add 200 µL of acetonitrile, mix for 30 s and let stand for 30 min. Centrifuge at 10,000**g** for 5 min and add 100 µL of the supernatant to 100 µL of mobile phase. Re-centrifuge at 10,000**g** for 5 min and assay the supernatant.

HPLC. The stationary phase is Ultrasphere C_{18}, 5 µm (4.6 mm × 250 mm) (Beckman Instruments Inc., San Ramon, CA, USA). The mobile phase is composed of methanol: acetonitrile:2.5 mM disodium EDTA (250:175:100 by volume). A 30 µL aliquot of extracted sample is injected into the chromatograph and the solvent pumped through at a rate of 1.5 mL/min. The absorbance of the eluate is monitored at 382 nm.

Calibration procedure. The concentration of amphotericin B is determined by comparing the peak area ratios of the drug with those of the internal standard. A calibration curve is constructed with the five calibrators. If higher concentrations of amphotericin B are anticipated (e.g. in patients receiving liposomal or lipid formulations), an additional 16 mg/L calibrator should be included.

Materials for laboratory use

Amphotericin B can be obtained from the British Pharmacopoeia Commission (Stanmore, Middlesex, UK). *N*-Acetylamphotericin B can be obtained from Xechem (New Brunswick, NJ, USA). HPLC-grade acetonitrile, methanol and DMSO are used. All other solvents and inorganic compounds are of analytical grade.

Pitfalls and problems

High concentrations of conjugated bilirubin (3 mg/dL) can interfere with the HPLC assay of amphotericin B. Several methods have been described to overcome this problem; for example, see Bach (1984) and Granich *et al.* (1986). The simplest solution appears to be that of Hosotsubo *et al.*

(1988) who used a reversed-phase CLC-trimethylsilyl analytical column (5 µm, 150 × 6 mm) (Shimadzu, Kyoto, Japan).

If a microbiological method is to be used to measure amphotericin B concentrations in the presence of flucytosine, the latter must first be inactivated by incorporating cytosine into the medium. Prepare a stock solution (10 g/L) of cytosine in 0.1 M saline. Filter sterilize and add 1 mL of the stock solution to 9 mL of sterile distilled water. Add 1 mL of this solution to 100 mL of molten cooled base agar to give a final concentration of 10 mg/L cytosine in the medium. Perform the assay as described above but use *Chrysosporium pruinosum* ATCC 36374 (which is resistant to flucytosine) as the assay organism and include flucytosine-containing controls.

Flucytosine

Physicochemical properties

Flucytosine (old name 5-fluorocytosine) is a synthetic fluorinated pyrimidine with a molecular weight of 129.1 and empirical formula of $C_4H_4N_3OF$ (Figure 17.2). It has pK_a values of 2.9 and 10.7. It is soluble in water up to concentrations of 10 g/L and stock solutions have an almost indefinite storage life if held at –20°C.

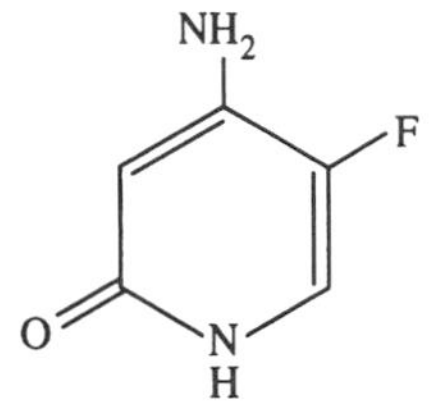

Figure 17.2 Structure of flucytosine.

Dosage forms

Flucytosine is available as tablets containing 500 mg of flucytosine and inactive additional ingredients and as an iv infusion containing 2.5 g of flucytosine in 250 mL of saline.

Pharmacokinetics

Oral administration of flucytosine leads to rapid and almost complete absorption of the drug. Peak serum concentrations are reached more quickly, in 1–2 h, in persons who have received several doses of the drug. Absorption is slower in persons with impaired renal function, but peak serum concentrations are higher. In adults with normal renal function a dose of 25 mg/kg given at 6 h intervals will produce peak serum concentrations of 70–80 mg/L and trough concentrations of 30–40 mg/L. There is slight accumulation of the drug during the first few days of treatment, but thereafter peak serum concentrations remain almost constant (Block & Bennett, 1972).

Flucytosine concentrations in the CSF are about 75% of the simultaneous blood concentration (Block & Bennett, 1972), and similar concentrations are found in pleural, peritoneal and joint fluids. The good penetration of flucytosine has been attributed to its low molecular weight and low protein binding (about 4%).

More than 90% of an oral dose of flucytosine is eliminated in the urine in unchanged form. The serum half-life of the drug is 3–6 h in persons with normal renal function, but much longer in patients with renal failure, necessitating modification of the dosage regimen.

Blood concentrations of flucytosine should be measured in all patients. This is especially important when there is renal impairment or when the drug is given together with amphotericin B, to ensure adequate concentrations and avoid excessive concentrations that may cause adverse effects. Levels should be determined twice weekly, or more frequently if renal function is changing. Blood should be taken just before a dose of flucytosine is due, and 2 h after an oral dose or 30 min after an iv dose since the distribution phase is completed within 30 min. The dose and interval should be adjusted so as to produce peak serum concentrations of about 70–80 mg/L and trough concentrations of 30–40 mg/L.

Assay methods

Several microbiological methods have been described for determining flucytosine concentrations in serum, CSF and other biological fluids. Agar diffusion methods are convenient to perform, but long incubation periods (at least 18 h) are required. If amphotericin B is also present in the specimen, this may need to be removed or inactivated. Fluorimetric (Wade & Sudlow, 1973; Richardson, 1975) and GLC methods have been reported, but HPLC has become the most popular non-biological method for measurement of the drug (Table II). Interference from other antifungal drugs is not a problem.

Microbiological assay method

Preparation of drug solutions. To prepare a stock solution with a concentration of 10 g/L, weigh 50 mg of flucytosine and make up to 5 mL with distilled water. Filter-sterilize and prepare a ten-fold dilution in sterile water. Then prepare the following calibrator solutions: 6.25, 12.5, 25, 50 and 100 mg/L. All solutions should be stored in small aliquots at −20°C until required. They are stable for up to 12 months provided that repeated freezing and thawing are avoided. Working calibrators made up in horse serum, pooled normal human serum and water have been found to give identical results (Working Group Report of the British Society for Mycopathology, 1984).

Preparation of media. Dissolve 6.7 g of Yeast Nitrogen Base (Difco) and 10 g of glucose in 100 mL of distilled water. Filter-sterilize this concentrated YNBG solution and store in 10 mL amounts at 4°C.

Table II. High-performance liquid chromatography assay methods for flucytosine

Authors	Extraction or deproteinization	Mobile phase	Stationary phase	Mode of separation	Flow rate (mL/min)	Detection	Wavelength of detection (nm)	Retention time (min)
Blair et al., 1975	none	ammonium phosphate	SO_3^-	cation exchange	0.24	absorption	280	13.5
Diasio et al., 1978	ultrafiltration	ammonium phosphate	SO_3^-	cation exchange	2	absorption	254	3.2
Bury et al., 1979	TCA	methanol:potassium dihydrogen phosphate	C_{18}	partition RP	1	absorption	280	c.5
Miners et al., 1980	TCA	potassium dihydrogen phosphate	C_{18}	partition RP	1.5	absorption (internal standard)	276	4
Warnock & Turner, 1981	acetonitrile	methanol:HSA	C_{18}	ion-pair	1	absorption	254	4.5
St-Germain et al., 1989	TCA	sodium acetate buffer	C_{18}	partition RP	1	absorption	276	NS
Schwertschlag et al., 1984	TCA:ammonium phosphate	methanol:water acetonitrile: phosphoric acid: ammonium phosphate	SCX	cation exchange	1	absorption (internal standard)	254	4

Abbreviations: C_8, octasilane; C_{18}, octadecylsilane; HSA, heptane sulphonic acid; NS, not specified; RP, reversed phase; SCX, strong cation exchanger; TCA, trichloroacetic acid.

Suspend 20 g of Purified Agar (Oxoid Ltd, Basingstoke, UK) in 900 mL of cold distilled water. Heat to boiling to dissolve the agar and then sterilize for 15 min at 121°C. Dispense in 90 mL amounts.

Preparation of indicator strain inoculum. Add 0.5 mL concentrated YNBG solution to 4.5 mL sterile water. Inoculate with a loopful of a sensitive indicator strain, such as *Candida kefyr* (*Candida pseudotropicalis*) NCPF 3234, harvested from a Yeast Morphology Agar (YMA; Difco) plate to obtain a concentration of about 1×10^7 cells/mL. Add the cell suspension to 10 mL of concentrated YNBG solution. The indicator strain should be maintained on YMA and subcultured at 1 week intervals.

Test method. This is similar to that described for amphotericin B except that the agar volume is 90 mL and 15 mL of cell suspension is added. Incubation is overnight at 37°C.

Urine specimens should be diluted with sterile 0.1 M saline or water to bring their drug levels within the calibrator range. The choice of water or saline does not influence the results.

HPLC assay method
Preparation of drug solutions. To prepare a stock solution with a concentration of 10 g/L, weigh 50 mg of flucytosine and make up to 5 mL with sterile distilled water. Prepare a ten-fold dilution with sterile distilled water. Both solutions should be stored frozen at –20°C.

From the stock solution prepare calibrators of 35, 50, 70 and 100 mg/L in horse serum or pooled normal human serum. These should be dispensed in small aliquots and stored at –20°C until required.

Extraction procedure. Add 100 μL of acetonitrile to 100 μL of the patient's sample or calibrator solution. Mix for 30 s and then stand for 10 min. Centrifuge at 10,000**g** for 5 min then transfer 100 μL of supernatant to a 10 mL glass tube and (in a fume cupboard) evaporate to dryness under nitrogen in a water bath at 37°C. Dissolve the residue in 100 μL of mobile phase and vortex-mix for 10 s. Re-centrifuge at 10,000**g** for 5 min, transfer the supernatant to another tube and centrifuge again for 5 min. This supernatant is injected.

HPLC. An ion-pair reversed-phase HPLC procedure is used. The column is a radial-PAK cartridge C_{18} (Waters), 8×100 mm. The mobile phase is methanol:water (15:85 v/v) containing 0.1% heptane sulphonic acid and 0.1% glacial acetic acid pumped at 2 mL/min. A 5 μL aliquot of the extracted sample is injected. Detection is by UV absorbance at 280 nm.

Calibration procedure. The flucytosine concentration is determined on the basis of peak area. The assay may be calibrated with the four calibrators.

Materials for laboratory use
Flucytosine can be obtained from Sigma and reference substance can be obtained from the European Pharmacopoeia Commission. HPLC-grade methanol, acetonitrile and heptane sulphonic acid are used. Other solvents are of analytical grade. The mobile phase is prepared with double-distilled water.

Pitfalls and problems
Therapeutic concentrations (1–2 mg/L) of traditional parenteral formulations of amphotericin B should not interfere with the microbiological assay of flucytosine. If the patient is also receiving liposomal amphotericin B or an azole such as fluconazole or itraconazole, flucytosine concentrations should be determined by HPLC.

Fluconazole

Fluconazole is a synthetic bis-triazole derivative with a molecular weight of 306 and the empirical formula $C_{13}H_{12}ON_6F_2$ (Figure 17.3). It is soluble in water up to concentrations of 6 g/L; the pK_a is 2.03 at 37°C.

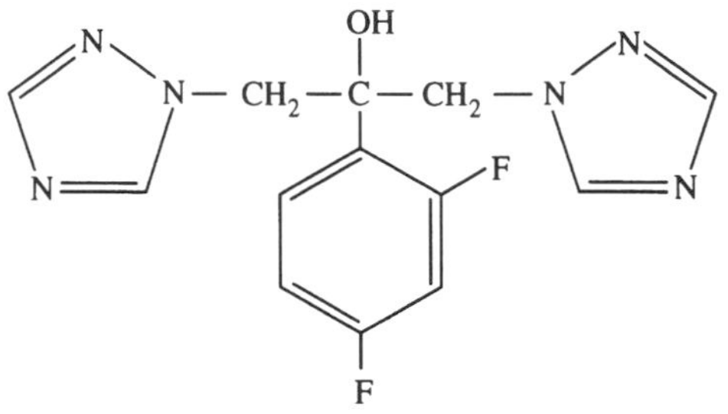

Figure 17.3 Structure of fluconazole.

Dosage forms

Fluconazole is available as oral capsules containing 50, 150 or 200 mg fluconazole together with inactive ingredients such as colloidal silicon dioxide, magnesium stearate, starch, lactose and sodium lauryl sulphate. It is also available as an iv infusion containing 2 mg/mL fluconazole in saline and an oral suspension which contains, in addition to fluconazole, citric acid, colloidal silicon dioxide, sodium benzoate, sodium citrate, sucrose, titanium dioxide, gum and flavouring.

Pharmacokinetics

Oral administration of fluconazole leads to rapid and almost complete absorption of the drug (Humphrey *et al.*, 1985). Identical serum concentrations are obtained after oral and parenteral administration indicating that first-pass metabolism does not occur. Blood concentrations increase in proportion to the dosage over a wide range. Two hours after a single 50 mg oral dose, serum concentrations in the region of 1 mg/L can be anticipated, but after repeated dosing this increases to approximately 2–3 mg/L (Brammer *et al.*, 1990). Administration with food does not affect absorption.

Oral or parenteral administration of fluconazole results in rapid and widespread distribution of the drug. Unlike other azole antifungals, the protein binding is low (about 12%), resulting in high levels of circulating unbound fluconazole (Humphrey *et al.*, 1985). Levels in the CSF are between 50 and 90% of the simultaneous blood concentrations (Tucker *et al.*, 1988). Levels in other host fluids are similar to the blood concentrations.

Unlike other azole antifungals, fluconazole is not metabolized in humans, but is excreted unchanged in the urine (Brammer *et al.*, 1990). The drug is cleared through glomerular filtration, but significant tubular reabsorption occurs. Fluconazole has a serum half-life of 25–30 h, but this is prolonged in renal failure, necessitating adjustment of the dosage regimen.

Blood concentrations of fluconazole are much more predictable than those of other azole drugs and there is no requirement for routine assays of blood concentrations. Excessive concentrations have not so far been associated with unwanted side-effects. Concentrations are unchanged in patients with AIDS and in bone marrow transplant recipients, and the reduction in concentration following concomitant administration with rifampicin is smaller than that seen with itraconazole or ketoconazole.

Assay methods

Fluconazole concentrations in serum, CSF and other biological fluids can be determined using microbiological methods, GLC (Wood & Tarbit, 1986) or HPLC (Table III).

Microbiological assay method
Preparation of drug solutions. To prepare a stock solution with a concentration of 1 g/L, weigh 10 mg of fluconazole and make up to 10 mL with sterile distilled water. Prepare a further ten-fold dilution with sterile water, horse serum or pooled normal human serum. The following calibrator solutions should be prepared in the same diluent: 3.12, 6.25, 12.5 and 25 mg/L. The calibrators should be dispensed in 0.5 or 1 mL amounts and stored at −20°C until required.

Preparation of media. Suspend 20 g of Bacto Agar (Difco) in 1 L of cold 0.2 M phosphate buffer, pH 7.5. Heat to boiling point to dissolve the agar and then sterilize for 15 min at 121°C. Dispense in 50 mL amounts.

Dissolve 29.34 g of HR medium (Oxoid) in 900 mL of distilled water. Then add 2 g of sodium bicarbonate, stir to dissolve and make up to 1000 mL. Filter-sterilize and store in 50 mL amounts at 4°C.

Preparation of indicator strain inoculum. Inoculate a YMA plate with a suitable sensitive indicator strain, such as

Table III. High-performance liquid chromatography assay methods for fluconazole

Authors	Extraction or deproteinization	Mobile phase	Stationary phase	Mode of separation	Flow rate (mL/min)	Detection	Wavelength of detection (nm)	Retention time (min)
Hosotsubo *et al.*, 1990	acetonitrile	acetonitrile: water	–	partition RP	1.5	absorption (internal standard)	210	9.3
Foulds *et al.*, 1988	ethyl acetate: hydrochloric acid ethyl acetate	methanol: phosphate buffer	–	partition RP	1	absorption (internal standard)	260	5

Abbreviation: RP, reversed phase.

C. kefyr NCPF 3234, and incubate at 37°C overnight. Prepare a suspension of 1×10^8 cells/mL in 5 mL of sterile distilled water using a haemocytometer.

Test method. Melt 50 mL of buffered agar and equilibrate to 56°C in a water bath. Add 1 mL of cell suspension to 50 mL of HR broth, pre-warmed to 37°C. Combine the inoculated broth with the molten agar and pour into a 25 cm square assay plates. The method is similar to that described for amphotericin B except that 6 mm wells are cut and 20 μL samples are placed in the wells. Incubation is overnight at 37°C.

As with flucytosine, urine specimens should be diluted with 0.1 M saline or water to bring their concentrations within the calibrator range.

HPLC assay method

Preparation of drug solutions. To prepare stock solutions with a concentration of 1 g/L, weigh 10 mg of fluconazole or UK-54,373 (internal standard) and make up to 10 mL in methanol. Both solutions can be stored at –40°C.

Fluconazole stock solutions of 5, 10, 20, 40 and 80 mg/L should be prepared fresh in methanol on each occasion. Prepare a 500 mg/L solution of internal standard in methanol fresh on each occasion.

Extraction procedure. Add 100 μL of each stock solution to 1 mL aliquots of horse serum and then add 100 μL of internal standard (500 mg/L concentration) to each tube. This makes calibrators equivalent to 0.5, 1, 2, 4 and 8 mg/L. Add 100 μL of methanol and 100 μL of internal standard to 1 mL of each patient sample. Then add 1 mL of 1 M sodium hydroxide and 4 mL of ethyl acetate. Mix for 5 min at 12 rpm on a rotary mixer, then centrifuge at 2500**g** for 5 min. Transfer the organic top layer to a 10 mL capped test tube. Add 2 mL of 1 M hydrochloric acid, mix on a rotary mixer for 5 min and centrifuge for 5 min. Discard the organic top layer and add 1 mL of 5 M sodium hydroxide to the remaining layer followed by 4 mL of ethyl acetate. Mix and centrifuge at 2500**g** for 5 min. Transfer the organic top layer to a 10 mL glass test tube and evaporate to dryness under nitrogen in a water bath at 37°C. This step should be performed in a fume cabinet.

Dissolve the residue in 100 μL of mobile phase and mix for 10 s.

HPLC. The stationary phase is a Spherisorb S5C$_8$ (25 cm × 0.5 cm) column (Anachem Ltd, Luton, UK). The mobile phase consists of 0.2 M tetramethyl-ethylenediamine (TEMED):acetonitrile (75:25 v/v), adjusted to pH 7.0 with concentrated orthophosphoric acid. Extracted sample (80 μL) is injected and the solvent pumped at a rate of 1 mL/min. The absorbance of the eluate is monitored at 260 nm.

Calibration procedure. The concentration of fluconazole is determined by comparison of the peak area ratios of the drug and the internal standard. A calibration curve is constructed using the five calibrators.

Materials for laboratory use

Fluconazole and the internal standard (UK-54,373) can be obtained from Pfizer Ltd, Sandwich, UK. HPLC-grade acetonitrile and methanol are used. The other solvents are of analytical grade. The inorganic reagents are prepared in double-distilled water.

Itraconazole

Physicochemical properties

Itraconazole is a synthetic dioxolane triazole derivative with a molecular weight of 705.6 and empirical formula of $C_{35}H_{38}Cl_2N_8O_4$ (Figure 17.4). Its pK_a is between 3 and 4. It is insoluble in water, but is soluble in dimethylformamide up to concentrations of 30 g/L and chloroform up to 300 mg/L.

Dosage forms

Itraconazole is available as an oral capsule formulation containing 100 mg of itraconazole as coated beads.

Pharmacokinetics

Absorption of itraconazole from the gastrointestinal tract is incomplete (about 55%) (Heykants *et al.*, 1989), but is

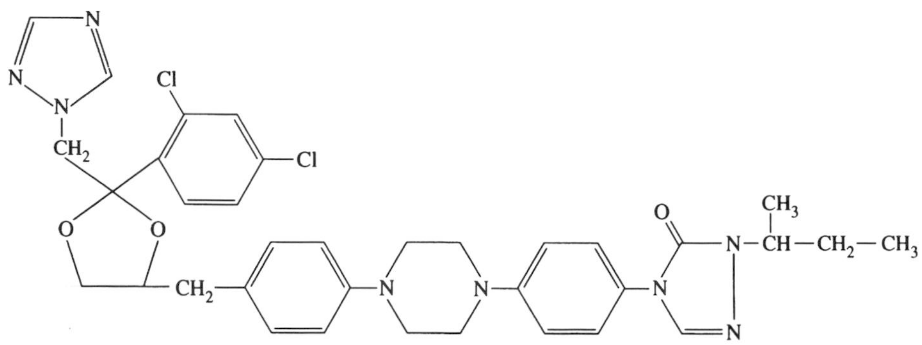

Figure 17.4 Structure of itraconazole.

improved if the drug is taken with food (Van Peer *et al.*, 1989). Administration of a single 100 mg dose will produce a peak serum concentration of 0.1–0.2 mg/L within 2–4 h. Much higher concentrations are obtained after repeated dosing, but there is much variation between individuals (Hardin *et al.*, 1988). Blood concentrations are reduced when gastric acid production is impaired.

Itraconazole is >99% protein-bound in human serum. Levels of itraconazole in host fluids, such as CSF, which are in equilibrium with unbound drug in serum, are low, but levels in tissue are much higher than in serum. Less than 1% of a given dose of itraconazole is excreted as unchanged drug in the urine.

Itraconazole is degraded in the liver to a large number of metabolites, most of which are inactive and these are excreted in the bile and urine (Heykants *et al.*, 1987). However, the major metabolite, hydroxyitraconazole, is biologically active. The serum half-life is approximately 20–30 h, increasing to 40 h after prolonged dosing (Hardin *et al.*, 1988).

Blood concentrations of itraconazole should be measured in patients with life-threatening fungal infections and in patients in whom poor absorption or drug interactions are anticipated. Low blood concentrations (<0.25 mg/L at 4 h post-dose) often predict failure of treatment or relapse. High serum concentrations have not so far been associated with unwanted side-effects. Levels should not be measured until the patient has reached steady state, typically after 1–3 weeks. The blood sample should be taken 4 h after an oral dose.

Assay methods

Itraconazole concentrations in serum and other biological fluids can be measured using microbiological methods or HPLC, but the latter gives up to ten-fold lower results (Warnock *et al.*, 1988). The higher concentrations detected with microbiological methods result from interference by the metabolite, hydroxyitraconazole (Heykants *et al.*, 1989). The difference in concentrations between the methods is not constant, nor is it proportional to the amount of drug present. For these reasons, HPLC is the preferred method for measuring itraconazole levels. It can detect concentrations as low as 0.01 mg/L and a number of procedures have been described (Table IV).

HPLC assay method

Preparation of drug solutions. To prepare stock solutions with a concentration of 1 g/L, weigh 10 mg of itraconazole or R51012 (internal standard), add 1 mL of dimethylformamide (DMF) and then make up to 10 mL with methanol. Prepare a further ten-fold dilution (100 mg/L) in methanol. Store at –20°C in 1 mL amounts.

Itraconazole stock solutions of 0.5, 1, 2, 5, 10 and 20 mg/L should be prepared in methanol. These can be used for up to 1 week if kept at 4°C.

Table IV. High-performance liquid chromatography assay method for itraconazole

Authors	Extraction or deproteinization	Mobile phase	Stationary phase	Mode of separation	Flow rate (mL/min)	Detection	Wavelength of detection (nm)	Retention time (min)
Woestenborghs *et al.*, 1987	heptane: isoamyl alcohol: sulphuric acid: heptane/isoamyl alcohol	acetonitrile: water: diethylamine	C$_{18}$	partition RP	0.5	absorption (internal standard)	263	5

Abbreviations: C$_{18}$, octadecylsilane; RP, reversed phase.

Freshly prepare a 10 mg/L solution of the internal standard R51012 in methanol on each occasion.

Extraction procedure. Add 100 μL of each methanolic stock solution to 1 mL aliquots of pooled human serum. Add 100 μL of internal standard (10 mg/L concentration) to each. This makes calibrators equivalent to 0.05, 0.1 , 0.2, 0.5, 1 and 2 mg/L. Add 100 μL of methanol and 100 μL of internal standard to 1 mL of each patient sample. Then add to all tubes 4 mL of 0.05 M phosphate buffer (pH 7.8) and 4 mL of heptane:isoamyl alcohol (98.5:1.5) (HIA). Mix for 10 min on a rotary mixer at 12 rpm and then centrifuge at 2500**g** for 10 min. Recover the top organic layer and add 4 mL HIA to the remaining layer. Repeat the mixing and centrifugation steps. Recover the top organic layer as before and combine it with the previous one. Add 3 mL of 0.05 M sulphuric acid to the combined layers. Repeat the mixing and centrifugation steps. Discard the top organic layer, then add 100 μL of 5 M potassium hydroxide solution followed by 2 mL of HIA. Repeat the mixing and centrifugation steps. Transfer the top organic layer to a new 10 mL glass tube. Add 2 mL of HIA to the remaining layer and repeat the mixing and centrifugation steps. Recover the top organic layer and combine it with the previous one. Then evaporate the combined layers to dryness under nitrogen at 55°C in a fume cabinet. Dissolve the concentrated samples in 50 μL of acetonitrile and 50 μL of double-distilled water.

HPLC. The stationary phase is Chrompak Hypersil 5 ODS (100 mm × 3 mm) (Chrompack UK Ltd, London, UK). The mobile phase consists of acetonitrile:water (60:40 v/v) with 0.03% diethylamine with the pH adjusted to 7.8 with concentrated orthophosphoric acid. A 5 μL aliquot of the extracted sample is injected and the mobile phase is pumped at a rate of 0.5 mL/min. The absorbance of the eluate is monitored at 263 nm.

Calibration procedure. The concentration of itraconazole is determined by comparison of the peak area ratios of the drug and the internal standard. A calibration curve is constructed using the six calibrators.

Materials for laboratory use

Itraconazole and the internal standard (R51012) can be obtained from Janssen Pharmaceutica (Beerse, Belgium). Glass-distilled water is used throughout the procedure. Acetonitrile, heptane, isoamyl alcohol and methanol are of HPLC grade. Sulphuric acid, diethylamine and orthophosphoric acid are of analytical grade.

Pitfalls and problems

It is possible to measure itraconazole and the active metabolite, hydroxyitraconazole, simultaneously in serum by altering the HPLC procedure described above. The solvent used in the extraction procedure is heptane: isoamyl alcohol (95:5) and the mobile phase is modified to acetonitrile:water (56:44 v/v) with 0.02% diethylamine (R. Woestenborghs, personal communication).

Ketoconazole

Physicochemical properties

Ketoconazole is a synthetic dioxolane imidazole derivative. The molecular weight is 531.4 and its empirical formula is $C_{26}H_{28}Cl_2N_4O_4$ (Figure 17.5). It is a weak dibasic compound with pK_a values of 6.5 and 2.9. It is almost insoluble in water (40 mg/L maximum). It is soluble in DMF up to a concentration of 10 g/L.

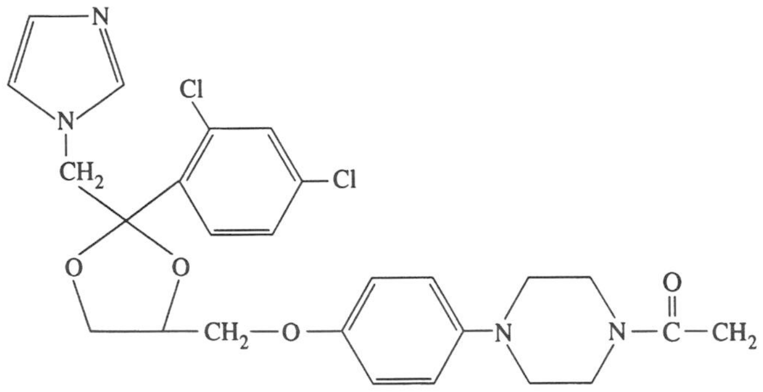

Figure 17.5 Structure of ketoconazole.

Dosage forms

Ketoconazole is available as tablets containing 200 mg ketoconazole and lactose, and as a suspension containing 20 mg/mL ketoconazole, sucrose, saccharin and sodium benzoate.

Pharmacokinetics

Ketoconazole is well absorbed after oral administration, peak blood concentrations being reached 2–4 h later. Absorption of the drug is delayed by food, but the peak concentration is unchanged (Daneshmend *et al.*, 1984). Two hours after a 400 mg dose, blood levels of about 5–6 mg/L can be anticipated, but there is much variation among individuals. Much higher concentrations are obtained with doses of 600–1000 mg.

Ketoconazole is 99% protein-bound in serum. Penetration of the drug into the CSF is poor, concentrations of around 5% of the simultaneous blood level being achieved in the lumbar CSF and 2% in the ventricular CSF (Craven *et al.*, 1983). Less than 1% of an oral dose is excreted unchanged in the urine.

Ketoconazole is metabolized in the liver and the metabolites are excreted in the bile. None of the metabolites is microbiologically active. The serum half-life appears to be dose-dependent. There is an initial distribution half-life of 1–4 h and an elimination half-life of 6–10 h.

Low blood concentrations of ketoconazole can be

anticipated in patients with AIDS and bone marrow transplant recipients. Drug levels are reduced by concomitant administration of antacids, H_2-antagonists and rifampicin. Low concentrations have been associated with treatment failure and serum concentrations may usefully be determined when this is suspected.

Assay methods

Microbiological methods have been described for the determination of ketoconazole in serum and other biological fluids. These methods are convenient to perform and have a lower limit of detection of about 0.1–0.3 mg/L. A number of HPLC methods with UV absorbance or fluorescence detection have also been described (Table V). These methods offer greater precision, but most are no more sensitive than plate methods. However, Pascucci *et al.* (1983) and Turner *et al.* (1986) have described reversed-phase HPLC methods in which the lower limit of detection is 0.04–0.05 mg/L.

Microbiological assay method

Preparation of drug solutions. To prepare a stock solution with a concentration of 10,000 mg/L, weigh 50 mg ketoconazole base and make up to 5 mL with DMF, and allow to stand for 30 min to permit self-sterilization. Ten-fold dilutions should be prepared in sterile horse serum or pooled normal human serum.

The following calibrator solutions should be prepared in horse serum or pooled human serum: 0.25, 0.5, 1, 2, 4, 8 and 16 mg/L. These give two possible ranges; five calibrators are used in each test either in the range 0.25–4 mg/L or 1–16 mg/L. The calibrators should be dispensed in 0.5 or 1 mL amounts and stored at –20°C until required. They are stable for up to 6 months provided repeated freezing and thawing are avoided.

Preparation of media. Dissolve 6.7 g of YNB (Difco), 10 g of glucose and 5.9 g of trisodium citrate in 100 mL of distilled water. Adjust the pH to 7.0 and filter-sterilize this concentrated YNBGC solution. Store in 10 mL amounts at 4°C.

Suspend 20 g of Purified Agar (Oxoid) in 900 mL of cold distilled water. Heat to boiling point to dissolve the agar and then sterilize for 15 min at 121°C. Dispense in 90 mL amounts.

Preparation of indicator strain inoculum. Inoculate a Sabouraud agar plate with a suitable sensitive indicator strain, such as *Candida albicans* ATCC 28516, and incubate at 37°C overnight. Harvest cells from this plate and prepare a suspension of 1×10^6 cells/mL in 20 mL of sterile water.

Test method. This is similar to that described for amphotericin B except that the agar volume is 90 mL and 10 mL of cell suspension is added. The volume applied to each well is 10 μL and incubation is overnight at 37°C.

Table V. High-performance liquid chromatography methods for ketoconazole

Authors	Extraction or deproteinization	Stationary phase	Mobile phase	Mode of separation	Flow rate (mL/min)	Detection	Wavelength of detection (nm)	Retention time (min)
Alton, 1980	ethyl acetate hydrochloric acid: ethyl acetate adsorption	CN	potassium dihydrogen phosphate buffer: acetonitrile	partition RP	2	absorption	205	c.6
Andrews *et al.*, 1981		C_{18}	methanol: phosphate buffer	partition RP	2	absorption	231	4.9
Mannisto *et al.*, 1982	heptane/isoamyl alcohol sulphuric acid heptane/isoamyl alcohol	C_{18}	acetonitrile:water: diethylamine	partition RP	1.5	absorption (internal standard)	254	NS
Pascucci *et al.*, 1983	adsorption	C_{18}	methanol:sodium phosphate	partition RP	1	absorption fluorescence (Ex.: 206 nm, Em.: 370 nm) (internal standard)	231	9.6
Swezey *et al.*, 1982	diethyl ether	C_{18}	acetonitrile: phosphate buffer	partition RP	1.5	fluorescence (internal standard)	206	5.2
Turner *et al.*, 1986	Bond-Elut C_{18} solid-phase column	C_{18}	acetonitrile: water:diethylamine	partition RP	0.6	absorption (internal standard)	254	5

Abbreviations: C_{18}, octadecylsilane; CN, nitrile; Em., emission; Ex., excitation; NS, not specified; RP, reversed phase.

HPLC assay method

Preparation of drug solutions. To prepare a stock keto-conazole solution with a concentration of 1 g/L, weigh 10 mg of ketoconazole and make up to 10 mL with methanol. Prepare a further ten-fold dilution in methanol: water (50:50) and store at –40°C.

Ketoconazole calibrators of 0.5, 1, 2, 5 and 10 mg/L should be prepared fresh in pooled normal human serum on each occasion. The internal standard is terconazole prepared as 50 and 200 mg/L aqueous solutions and stored at 4°C.

Extraction procedure for samples containing >1 mg/L of ketoconazole. Add 110 μL of terconazole (200 mg/L concentration) to 1 mL of patient sample or ketoconazole calibrator solution. Add 250 μL of 0.1 M sodium hydroxide to alkalinize the sample. Deproteinization and elimination of polar serum constituents is accomplished with Bond-Elut C_{18} sample preparation columns (3 mL) (Anachem). Precondition the columns with 6 mL of methanol followed by 6 mL of water. Add the sample to the column and then wash with 9 mL of water followed by 200 μL of methanol. Elute the drug from the column with 1 mL of methanol.

Extraction procedure for samples containing <1 mg/L of ketoconazole. Add 110 μL of terconazole (50 mg/L concentration) to 1 mL of patient sample or ketoconazole calibrator solution. Add 250 μL of 0.1 M sodium hydroxide to alkalinize the sample, then dilute with 3 mL of water. Extract the samples as described above. Then evaporate the eluate under a stream of nitrogen at 60°C and re-suspend the sample in 200 μL of mobile phase. Centrifuge the samples for 2 min at 10,000**g**.

HPLC. The stationary phase is a Chrompack glass cartridge containing Hypersil ODS (100 mm × 3 mm) (Chrompack UK). The mobile phase consists of water: acetonitrile (55:45 v/v) plus 0.05% diethylamine adjusted to pH 8 with orthophosphoric acid. For samples with an anticipated concentration of >5 mg/L, inject 10 μL. For samples with an anticipated concentration of 0.5–5 mg/L, inject 20 μL, and for samples with an anticipated concentration of 0.05–0.5 mg/L, inject 40 μL. The solvent is pumped at a rate of 0.6 mL/min and the eluent is monitored at 254 nm.

Calibration procedure. Ketoconazole is quantified by comparison of the peak area ratio of the drug with that of the internal standard. A calibration curve is constructed using the five calibrators.

Materials for laboratory use

Ketoconazole and terconazole (internal standard) can be obtained from Janssen Pharmaceutica. Acetonitrile and methanol are of HPLC grade. Sodium hydroxide, diethylamine and orthophosphoric acid are of analytical grade.

References

Alton, K. B. (1980). Determination of the antifungal agent, keto-conazole, in human plasma by high-performance liquid chromatography. *Journal of Chromatography* **221**, 337–44.

Andrews, F. A., Peterson, L. R., Beggs, W. H., Crankshaw, D. & Sarosi, G. A. (1981). Liquid chromatographic assay of keto-conazole. *Antimicrobial Agents and Chemotherapy* **19**, 110–3.

Atkinson, A. J. & Bennett, J. E. (1978). Amphotericin B pharmaco-kinetics in humans. *Antimicrobial Agents and Chemotherapy* **13**, 271–6.

Bach, P. R. (1984). Quantitative extraction of amphotericin B from serum and its determination by high-pressure liquid chromato-graphy. *Antimicrobial Agents and Chemotherapy* **26**, 314–7.

Bindschadler, D. D. & Bennett, J. E. (1969). A pharmacologic guide to the clinical use of amphotericin B. *Journal of Infectious Diseases* **120**, 427–36.

Blair, A. D., Forrey, A. W., Meijsen, B. T. & Cutler, R. E. (1975). Assay of flucytosine and furosemide by high-pressure liquid chromatography. *Journal of Pharmaceutical Sciences* **64**, 1334–9.

Block, E. R. & Bennett, J. E. (1972). Pharmacological studies with 5-fluorocytosine. *Antimicrobial Agents and Chemotherapy* **1**, 476–82.

Brammer, K. W., Farrow, P. R. & Faulkner, J. K. (1990). Pharmaco-kinetics and tissue penetration of fluconazole in humans. *Reviews of Infectious Diseases* **12**, *Suppl. 3*, S318–26.

Brassinne, C., Laduron, C., Coune, A., Sculier, J. P., Hollaert, C., Collette, N. *et al.* (1987). High-performance liquid chromato-graphic determination of amphotericin B in human serum. *Journal of Chromatography* **419**, 401–7.

Bury, R. W., Mashford, M. L. & Miles, H. M. (1979). Assay of flucytosine (5-fluorocytosine) in human plasma by high-pressure liquid chromatography. *Antimicrobial Agents and Chemotherapy* **16**, 529–32.

Christiansen, K. J., Bernard, E. M., Gold, J. W. M. & Armstrong, D. (1985). Distribution and activity of amphotericin B in humans. *Journal of Infectious Diseases* **152**, 1037–43.

Craven, P. C., Graybill, J. R., Jorgensen, J. H., Dismukes, W. E. & Levine, B. E. (1983). High-dose ketoconazole for treatment of fungal infections of the central nervous system. *Annals of Internal Medicine* **98**, 160–7.

Daneshmend, T. K., Warnock, D. W., Ene, M. D., Johnson, E. M., Potten, M. R., Richardson, M. D. *et al.* (1984). Influence of food on the pharmacokinetics of ketoconazole. *Antimicrobial Agents and Chemotherapy* **25**, 1–3.

Diasio, R. B., Wilburn, M. E., Shadomy, S. & Espinel-Ingroff, A. (1978). Rapid determination of serum 5-fluorocytosine levels by high-performance liquid chromatography. *Antimicrobial Agents and Chemotherapy* **13**, 500–4.

Foulds, G., Brennan, D. R., Wajszczuk, C., Catanzaro, A., Garg, D. C., Knopf, W. *et al.* (1988). Fluconazole penetration into cerebrospinal fluid in humans. *Journal of Clinical Pharmacology* **28**, 363–6.

Golas, C. L., Prober, C. G., Macleod, S. M. & Soldin, S. J. (1983). Measurement of amphotericin B in serum or plasma by high-performance liquid chromatography. *Journal of Chromatography* **278**, 387–95.

Granich, G. G., Kobayashi, G. S. & Krogstad, D. J. (1986). Sensitive high-pressure liquid chromatographic assay for amphotericin B which incorporates an internal standard. *Antimicrobial Agents and Chemotherapy* **29**, 584–8.

Hardin, T. C., Graybill, J. R., Fetchick, R., Woestenborghs, R., Rinaldi, M. G. & Kuhn, J. G. (1988). Pharmacokinetics of itraconazole following oral administration to normal volunteers. *Antimicrobial Agents and Chemotherapy* **32**, 1310–3.

Heykants, J., Michiels, M., Meuldermans, W., Monbaliu, J., Lavrijsen, K., Van Peer, A. *et al.* (1987). The pharmacokinetics of itraconazole in animals and man: an overview. In *Recent Trends in the Discovery, Development and Evaluation of Antifungal Agents* (Fromtling, R. A., Ed.), pp. 223–49. Prous Science Publishers, Barcelona.

Heykants, J., Van Peer, A., Van de Velde, V., Van Rooy, P., Meuldermans, W., Lavrijsen, K. *et al.* (1989). The clinical pharmacokinetics of itraconazole: an overview. *Mycoses* **32**, Suppl. 1, 67–87.

Hosotsuto, H., Takezawa, J., Taenaka, N., Hosotsubo, K. & Yoshiya, I. (1988). Rapid determination of amphotericin B levels in serum by high-performance liquid chromatography without interference by bilirubin. *Antimicrobial Agents and Chemotherapy* **32**, 1103–5.

Hosotsubo, K. K., Hosotsubo, H., Nishijima, M. K., Okado, T., Taenaka, N. & Yoshiya, I. (1990). Rapid determination of serum levels of a new antifungal agent, fluconazole, by high-performance liquid chromatography. *Journal of Chromatography* **529**, 223–8.

Humphrey, M. J., Jevons, S. & Tarbit, M. H. (1985). Pharmacokinetic evaluation of UK-49,858, a metabolically stable triazole antifungal drug, in animals and humans. *Antimicrobial Agents and Chemotherapy* **28**, 648–53.

Janknegt, R., de Marie, S., Bakker-Woudenberg, I. A. J. M. & Crommelin, D. J. A. (1992). Liposomal and lipid formulations of amphotericin B: clinical pharmacokinetics. *Clinical Pharmacokinetics* **23**, 279–91.

Kan, V. L., Bennett, J. E., Amantea, M. A., Smolskis, M. C., McManus, E., Grasela, D. M. *et al.* (1991). Comparative safety, tolerance, and pharmacokinetics of amphotericin B lipid complex and amphotericin B desoxycholate in healthy male volunteers. *Journal of Infectious Diseases* **164**, 418–21.

Mannisto, P. T., Mantyla, R., Nykanen, S., Lamminsivu, U. & Ottoila, P. (1982). Impairing effect of food on ketoconazole absorption. *Antimicrobial Agents and Chemotherapy* **21**, 730–3.

Mayhew, J. W., Fiore, C., Murray, T. & Barza, M. (1983). An internally standardized assay for amphotericin B in tissues and plasma. *Journal of Chromatography* **274**, 271–9.

Miners, J. O., Foenander, T. & Birkitt, D. J. (1980). Liquid-chromatographic determination of 5-fluorocytosine. *Clinical Chemistry* **26**, 117–9.

Nilsson-Ehle, I., Yoshikawa, T. T., Edwards, J. E., Schotz, M. C. & Guze, L. B. (1977). Quantitation of amphotericin B with use of high-pressure liquid chromatography. *Journal of Infectious Diseases* **135**, 414–22.

Pascucci, V. L., Bennett, J., Narang, P. K. & Chatterji, D. C. (1983). Quantitation of ketoconazole in biological fluids using high-performance liquid chromatography. *Journal of Pharmaceutical Sciences* **72**, 1467–9.

Richardson, R. A. (1975). Rapid fluorimetric determination of 5-fluorocytosine in serum. *Clinica Chimica Acta* **63**, 109–14.

St-Germain, G., Lapierre, S. & Tessier, D. (1989). Performance characteristics of two bioassays and high-performance liquid chromatography for determination of flucytosine in serum. *Antimicrobial Agents and Chemotherapy* **33**, 1403–5.

Schwertschlag, U., Nakata, L. M. & Gal, J. (1984). Improved procedure for determination of flucytosine in human blood plasma by high-pressure liquid chromatography. *Antimicrobial Agents and Chemotherapy* **26**, 303–5.

Swezey, S. F., Giacomini, K. M., Abang, A., Brass, C., Stevens, D. A. & Blaschke, T. F. (1982). Measurement of ketoconazole, a new antifungal agent, by high-performance liquid chromatography. *Journal of Chromatography* **227**, 510–5.

Tucker, R. M., Williams, P. L., Arathoon, E. G., Levine, B. E., Hartstein, A. I., Hanson, L. H. *et al.* (1988). Pharmacokinetics of fluconazole in cerebrospinal fluid and serum in human coccidioidal meningitis. *Antimicrobial Agents and Chemotherapy* **32**, 369–73.

Turner, C. A., Turner, A. & Warnock, D. W. (1986). High performance liquid chromatographic determination of ketoconazole in human serum. *Journal of Antimicrobial Chemotherapy* **18**, 757–63.

Van Peer, A., Woestenborghs, R., Heykants, J., Gasparini, R. & Cauwenbergh, G. (1989). The effects of food and dose on the oral systemic availability of itraconazole in healthy subjects. *European Journal of Clinical Pharmacology* **36**, 423–6.

Wade, D. & Sudlow, G. (1973). Fluorometric measurement of 5-fluorocytosine in biological fluids. *Journal of Pharmaceutical Sciences* **62**, 828–9.

Warnock, D. W. & Turner, A. (1981). High-performance liquid chromatographic determination of 5-fluorocytosine in human serum. *Journal of Antimicrobial Chemotherapy* **7**, 363–9.

Warnock, D. W., Turner, A. & Burke, J. (1988). Comparison of high performance liquid chromatographic and microbiological methods for determination of itraconazole. *Journal of Antimicrobial Chemotherapy* **21**, 93–100.

Woestenborghs, R., Lorreyne, W. & Heykants, J. (1987). Determination of itraconazole in plasma and animal tissues by high-performance liquid chromatography. *Journal of Chromatography—Biomedical Applications* **413**, 332–7.

Wood, P. R. & Tarbit, M. H. (1986). Gas chromatographic method for the determination of fluconazole, a novel antifungal agent, in human plasma and urine. *Journal of Chromatography* **383**, 179–86.

Working Group Report of the British Society for Mycopathology. (1984). Laboratory methods for flucytosine (5-fluorocytosine). *Journal of Antimicrobial Chemotherapy* **14**, 1–8.

Working Party Report of the British Society for Antimicrobial Chemotherapy. (1991). Laboratory monitoring of antifungal chemotherapy. *Lancet* **337**, 1577–80.

Chapter 18

Nitrofurans: nitrofurantoin, nitrofurazone and furazolidone

Richard Wise[a], Jenny M. Andrews[a] and Les O. White[b]

[a]*Department of Medical Microbiology, City Hospital NHS Trust, Birmingham B18 7QH;* [b]*Department of Medical Microbiology, Southmead Hospital, Bristol BS10 5NB, UK*

Introduction

The nitrofurans have been in clinical use for close to half a century (Stillman *et al.*, 1943), and compose a group of synthetic antimicrobials of which three (nitrofurantoin, nitrofurazone and furazolidone), are clinically useful. Nitrofurantoin is still used for treating urinary tract infections, nitrofurazone has been employed as a topical agent and a solution for bladder lavage, and furazolidone has been used in the therapy of gastrointestinal infections. Nitrofurantoin is currently available as 50 or 100 mg tablets, a suspension of 25 mg/mL and capsules containing 50 or 100 mg nitrofurantoin macrocrystals. There is also a modified-release preparation which comprises 25% macrocrystalline nitrofurantoin and 75% nitrofurantoin monohydrate blended in a patented dual delivery system (Spencer *et al.*, 1994).

The clinical laboratory will very rarely be required to undertake assays in this group of compounds but might be required to perform assays for research or development purposes. The advice below is largely based on that given in an earlier publication (Reeves *et al.*, 1978), together with more recent developments on HPLC analysis.

Nitrofurantoin

Nitrofurantoin (*N*-(5-nitro-2-furfurylidene)-1-aminohydantoin) (Figure 18.1) was marketed in 1953 and remains the most important of this group of compounds. Although it has been available for many years, nitrofurantoin remains a useful agent for the treatment of urinary tract infections. Resistance rates are low (for example 2% in *Escherichia coli* and 8% in *Klebsiella* spp. in the City

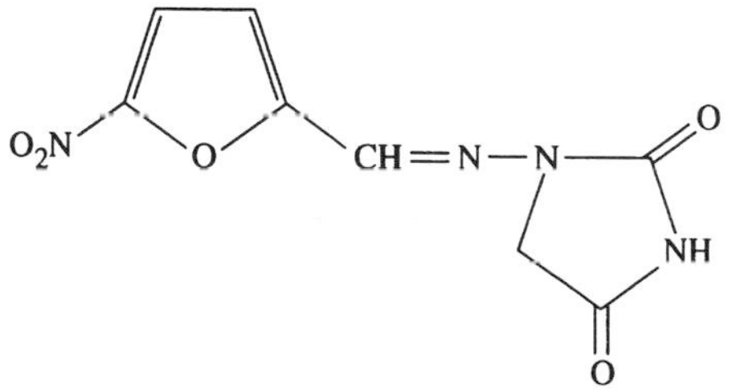

Figure 18.1 Nitrofurantoin.

Hospital, Birmingham, UK) and have not changed significantly since 1972 (Grüneberg, 1994). Nitrofurantoin also remains one of the relatively few agents which is commonly used in pregnancy.

The solubility of the crystalline compound in water is poor but increases significantly—although not to a useful degree—with increasing pH. The molecular weight is 238.2 and the pK_a 7.0. To solublize it, it is suggested that the initial solution should be in *N,N*-dimethylformamide (Chamberlain, 1976), but polyethylene glycol 300 has been used by some workers (Norfleet *et al.*, 1973).

Nitrofurantoin is well absorbed from the gastrointestinal tract, the absorption approaching 100%; however, therapeutically useful serum levels are not obtained due to metabolism and rapid renal elimination. Protein binding is 60–77% and the volume of distribution is around 0.6 L/kg. In patients with normal renal function the plasma elimination half-life is approximately 20 min (Reckendorf *et al.*, 1962), therapeutic concentrations appearing in the urine by 30 min post-dose (Paul & Paul, 1966).

Approximately 30–60% of a dose is eventually recoverable as active drug from the urine. The metabolism of nitrofurantoin to hydroxylamino compounds and aminofuraldehyde nitrofuroic acid takes place in most body tissues, especially the liver, and these microbiologically inactive metabolites may give the urine a brown colour. The degree of metabolism may be increased during therapy with hepatic enzyme-inducing agents such as barbiturates.

Although an iv preparation is not commercially available, following iv infusion serum concentrations are up to five times greater than those found with oral therapy; even so, these concentrations are not clinically useful. Nitrofurantoin penetrates into most tissues and body fluids (Paul & Paul, 1966) but therapeutic concentrations are only present in bile, urine and renal lymph. The renal elimination of nitrofurantoin is primarily by glomerular filtration; decreased renal clearance in uraemic patients and consequent higher serum concentrations may be associated with severe toxicity, presumably related to the accumulation of metabolites. A small part of the glomerular filtrate is reabsorbed by the tubules, this being enhanced by acid urine. The macrocrystalline form of nitrofurantoin is absorbed more slowly, so excretion is prolonged. Overall urinary recovery, however, is only slightly less than for the microcrystalline form.

Nitrofurazone and furazolidone

Nitrofurazone (Figure 18.2), molecular weight 198.1, is only available as topical preparations. It may be absorbed to a small degree through intact and broken skin and probably also from the bladder in view of the evidence for nitrofurantoin being absorbed from this site (Conklin & Hollifield, 1967).

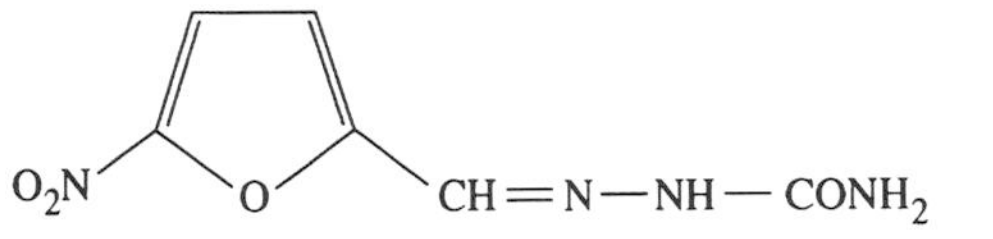

Figure 18.2 Nitrofurazone.

There is less information on furazolidone (Figure 18.3), molecular weight 225.2, than on nitrofurantoin, reflecting its clinical usage. It would appear that only metabolites are present in the urine and that early evidence of absorption and excretion resulted from a lack of specificity in the chemical assay procedure.

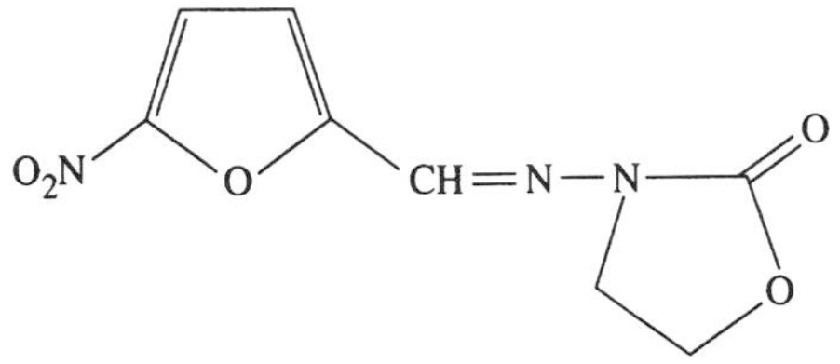

Figure 18.3 Furazolidone.

Determination of nitrofurantoin concentrations in body fluids

There are no indications for the routine assay of nitrofurans in patients. Such assays might, however, form part of a clinical or pharmacological research project. The standard assay is the spectrophotometric method described by Conklin & Hollifield (1965), which is highly sensitive and specific. Microbiological methods using *Bacillus subtilis* (Jones *et al.*, 1965) and *Streptococcus faecalis* (Gang & Shaikh, 1972) as assay organisms have been described. However, such methods are not as specific as the spectrophotometric method and, because pH has a marked effect on the antimicrobial activity of nitrofurantoin (Brumfitt & Percival, 1962), the pH adjustment of samples is crucial to assay accuracy.

The spectrophotometric method for the assay of urine concentrations is described below, together with an HPLC method.

Spectrophotometric assay (Conklin & Hollifield, 1965)

Reagents. The following reagents are required: crystalline nitrofurantoin (Eaton Laboratories, Norwich Pharma-

ceutical Company, Norwich, NY, USA), 0.1 N hydrochloric acid, *N,N*-dimethylformamide (ACS reagent grade, Eastman Laboratory Chemicals, Deeside, UK; see warnings below), benzethonium hydroxide, 1 M in methanol (Sigma Chemicals, Poole, UK; see warnings below), nitromethane (spectroscopy grade, Eastman; see warnings below), methanol (Spectrosol grade, BDH).

Nitrofurantoin calibrators. Prepare a 1 g/L solution by dissolving 50 mg of nitrofurantoin in 50 mL of *N,N*-dimethylformamide. From this make working calibrators of 60, 30, 10 and 5 mg/L by dilution into pooled human urine or distilled water. Both matrices give acceptable results. Nitrofurantoin calibrators must be stored away from direct sunlight or fluorescent lighting since either may cause photodegradation of the compound.

Benzethonium (hyamine) hydroxide reagent. Mix 1 mL of the 1 M benzethonium hydroxide solution with 25 mL of methanol to give 0.04 M solution.

Warnings. Benzethonium hydroxide is inflammable and can be explosive when mixed with air. It is toxic on inhalation and ingestion and causes burns to skin. *N,N*-Dimethylformamide is harmful if inhaled or if it comes into contact with skin or eyes. If inhaled remove subject to fresh air and if breathing becomes difficult obtain immediate medical advice. It also dissolves some plastics. Nitromethane forms a shock-sensitive explosive mixture when mixed with amines. It is combustible and harmful if inhaled. Also avoid skin contact.

It is therefore recommended that the greatest of care is taken, paying particular attention to the following precautions: prepare dilutions of reagents in a fume cupboard; do not pipette by mouth; wear protective gloves.

Equipment. The following equipment is required: volumetric glassware; volumetric pipettes; stoppered extraction tubes (20 mL) (Quickfit, Gallenkamp); glass test tubes (105 mm × 15 mm); a centrifuge (with head to take Quickfit tubes); a spectrophotometer.

Assay procedure. To 1 mL of each calibrator, test urine, internal controls and water blank dispensed into Quickfit extraction tubes, add 4 mL of 0.1 N HCl and 10 mL of nitromethane from a burette. Mix vigorously for 2 min, then centrifuge for 10 min. Remove 4 mL of the lower layer (nitromethane) and transfer to a 105 mm × 15 mm tube. If any samples appear cloudy, place them under warm tap water for about 1 min. Add 0.5 mL of the 0.04 M benzethonium hydroxide solution, mix and allow to stand for 1 min.

Read the absorbance of each sample at 400 nm using the water blank to set the machine and pure nitromethane to set zero absorbance.

Note: Samples should be read within 30 min of adding the benzethonium hydroxide. Plot absorbance against calibrator concentration to construct a linear calibration curve. Results for the test samples and internal controls are interpolated from this curve.

Comments. The method shows good accuracy between 5 and 100 mg/L; as the top calibrator is 60 mg/L and the the calibration curve may not be linear over 100 mg/L it is useful to include an internal control of 200 mg/L. If this internal control is assayed accurately it may not be necessary to repeat tests that give results between 60 and 200 mg/L. Urine samples can be stored in the dark at –20°C for up to 1 month. This method is specific for nitrofurantoin. Metabolites of nitrofurantoin or other antimicrobials do not interfere. The above method is used to assay nitrofurantoin in urine, modifications of this method have been described for assay in whole blood and plasma (Conklin & Hollifield, 1966) and bile (Conklin *et al.*, 1973).

HPLC methods for the assay of nitrofurans
Below are described three HPLC methods for nitrofurantoin, all of which could be adapted for the assay of other nitrofurans. HPLC assays are considerably more sensitive than the spectrophotometric method, and the method of Vree *et al.* (1979) requires only 10 μL of serum.

The HPLC method of Aufrere et al. (1977)
Application. Assays in plasma or urine.

Column. Microbondapak C_{18} (300 × 3.9 mm)

Mobile phase. methanol–water–glacial acetic acid (800:200:0.6 by volume) adjusted to pH 5.0 with 4 M sodium hydroxide. Flow rate: 2 mL/min.

Detection. UV absorption at 365 nm.

Internal standard. Furazolidone dissolved in DMSO, at 5 g/L for urine, and 5 mg/L for plasma.

Sample preparation. For plasma, mix 10 μL of internal standard with 200 μL of plasma sample and 300 μL methanol, centrifuge, and inject 15–60 μL of the clear supernatant. For urine, mix 3 μL of internal standard with 200 μL of urine, mix and inject 4–10 μL.

Comments. The quoted assay range was 0.02–200 mg/L. CVs were around 2%. Urine and serum samples stored at –10°C were stable for at least 20 days. Results compared well with the benzethonium method described above. The method could also be used for nitrofurazone. It could presumably be used to assay furazolidone using an alternative internal standard such as nitrofurantoin.

The HPLC method of Roseboom & Koster (1978)
Application. Plasma or urine.

Column. Lichrosorb 10 μm RP_{18} (300 mm long).

Mobile phase. Water containing 20–40% methanol and 0.5% glacial acetic acid pumped at 2.5 mL/min (20% methanol was optimum for nitrofurantoin).

Detection. UV absorbance at 365 nm.

Internal standard. 2,6-Dinitrophenol.

Sample preparation. Mix 1 mL of plasma with 20 μL of internal standard (125 mg/L aqueous solution), then add two drops of 4 M hydrochloric acid and 3 mL of dimethylformamide in dichloromethane (10% v/v). After mixing, centrifuge for 10 min and separate the organic phase through phase-separation paper. Reduce the volume to approximately 100 μL under nitrogen at 4°C and inject 25 μL. Dilute urines 1 in 10 and use 200 μL of internal standard.

Comments. The limit of sensitivity was around 0.01 mg/L in plasma. Other nitrofurans such as nitrofurazone and furazolidine could be assayed under very similar conditions. Metabolites did not interfere since cleavage of the furan ring results in loss of absorbance at 365 nm.

The HPLC method of Vree et al. (1979)
Application. Plasma or urine.

Column. Lichrosorb 5 μm RP_8 (150 mm × 4.6 mm).

Mobile phase. 5% Ethanol in water pumped at 1.6 mL/min.

Detection. UV absorption at 370 nm.

Sample preparation. No internal standard is used. Mix 10 μL of plasma with 500 μL 0.33 N perchloric acid. After mixing centrifuge for 10 min and inject 100 μL of the supernatant. Urine is treated similarly but centrifugation is unnecessary.

Comments. The lower limit of sensitivity was around 0.02 mg/L in plasma. Other nitrofurans, such as nitrofurazone and furazolidine, could be separated. Nitrofurantoin was unstable on the bench at room temperature, possibly as a result of photochemical degradation.

References

Aufrere, M. B., Hoener, B.-A. & Vore, M. E. (1977). High performance liquid-chromatographic assay for nitrofurantoin in plasma and urine. *Clinical Chemistry* **23**, 2207–12.

Brumfitt, W. & Percival, A. (1962). Adjustment of urine pH in the chemotherapy of urinary tract infections. A laboratory and clinical assessment. *Lancet i*, 186–90.

Chamberlain, R. E. (1976). Chemotherapeutic properties of prominent nitrofurans. *Journal of Antimicrobial Chemotherapy* **2**, 325–36.

Conklin, J. D. & Hollifield, R. D. (1965). A new method for the determination of nitrofurantoin in urine. *Clinical Chemistry* **12**, 925–31.

Conklin, J. D. & Hollifield, R. D. (1966). A quantitative procedure for the determination of nitrofurantoin in whole blood and plasma. *Clinical Chemistry* **12**, 690–6.

Conklin, J. D. & Hollifield, R. D. (1967). Studies on the movement of nitrofurantoin across the dog urinary bladder. *Investigative Urology* **5**, 244–9.

Conklin, J. D., Sobers, R. J. & Wagner, D. L. (1973). Further studies on nitrofurantoin excretion in dog hepatic bile. *British Journal of Pharmacology* **48**, 273–7.

Gang, D. M. & Shaikh, K. Z. (1972). Turbidimetric method for assay of nitrofuran compounds. *Journal of Pharmaceutical Science* **61**, 462–5.

Grüneberg, R. N. (1994). Changes in urinary pathogens and their antibiotic sensitivities, 1972–1992. *Journal of Antimicrobial Chemotherapy* **33**, *Suppl. A*, 1–8.

Jones, B. M., Ratcliffe, R. J. M. & Stevens, S. G. E. (1965). Comparative assays of some nitrofurans in urine. *Journal of Pharmacy and Pharmacology* **17**, *Suppl.*, 52S–55S.

Norfleet, C. M., Beamer, P. R. & Carpenter, H. M. (1953). Furadantin in infections of genito-urinary tract: first report. *Journal of Urology* **70**, 113–8.

Paul, H. E. & Paul, M. F. (1966). The nitrofurans—chemotherapeutic properties. In *Experimental Chemotherapy* (Schnitzer, R. J. & Hawking, F., Eds), Vol II, pp. 307–70, and Vol. IV, pp. 521–36. Academic Press, New York.

Reckendorf, H. K., Castringius, R. G. & Springler, H. K. (1962). Comparative pharmacodynamics, urinary excretion and half life determinations of nitrofurantoin sodium. In *Antimicrobial Agents and Chemotherapy* (Sylvester, J. C., Ed.), pp. 531–7. American Society for Microbiology, Washington, DC.

Reeves, D. S., Phillips, I., Williams, J. D. & Wise, R. (1978). *Laboratory Methods in Antimicrobial Chemotherapy.* Churchill Livingstone, Edinburgh.

Roseboom, H. & Koster, H. A. (1978). The determination of nitrofurantoin and some structurally related drugs in biological fluids by high-pressure liquid chromatography. *Analytica Chimica Acta* **101**, 359–65.

Spencer, R. C., Moseley, D. J. & Greensmith, M. J. (1994). Nitrofurantoin modified release versus trimethoprim or co-trimoxazole in the treatment of uncomplicated urinary tract infection in general practice. *Journal of Antimicrobial Chemotherapy* **33**, *Suppl. A*, 121–9.

Stillman, W. B., Scott, A. B. & Clampit, J. M. (1943) United States Patent no. 2,319,481, 18th May 1943.

Vree, T. B., Hekster, Y. A., Baars, A. M., Damsma, J. H. E., van der Kleijn, E. & Bron, J. (1979). Determination of nitrofurantoin (Furadantine) and hydroxymethyl-nitrofurantoin (Urfadyn) in plasma and urine of man by means of high-performance liquid chromatography. *Journal of Chromatography* **162**, 110–6.

Inhibitors of folate metabolism: trimethoprim, antibacterial sulphonamides and co-trimoxazole

Grace Sweeney[a] and Les O. White[b]

[a]*Department of Bacteriology, Southern General Hospital NHS Trust, Glasgow G51 4TF;* [b]*Department of Medical Microbiology, Southmead Hospital, Bristol BS10 5NB, UK*

Introduction

The 2,4-diaminopyrimidines (notably trimethoprim), the antibacterial sulphonamides and co-trimoxazole are discussed together in this chapter for the following three reasons: (i) All inhibit folate metabolism (Figure 19.1). Diaminopyrimidines inhibit dihydrofolate reductase (DHFR) enzymes and one of these used clinically, the antimicrobial trimethoprim, owes its selective toxicity to its higher affinity for bacterial rather than human DHFR. Antibacterial sulphonamides inhibit bacterial dihydropteroate synthetase (DHPS). They are selectively toxic because this enzyme is not found in humans. (ii) Trimethoprim and sulphonamides are said to exhibit antimicrobial synergy. (iii) Co-trimoxazole comprises trimethoprim plus sulphamethoxazole in the ratio 1:5 and it may be necessary to assay both drugs simultaneously in the same biological samples.

2,4-Diaminopyrimidines: trimethoprim

Introduction

Research on the synthetic 2,4-diaminopyrimidines in the 1960s confirmed these as inhibitors of the DHFR enzymes, their targets in the folate pathway (Burchall & Hitchings,

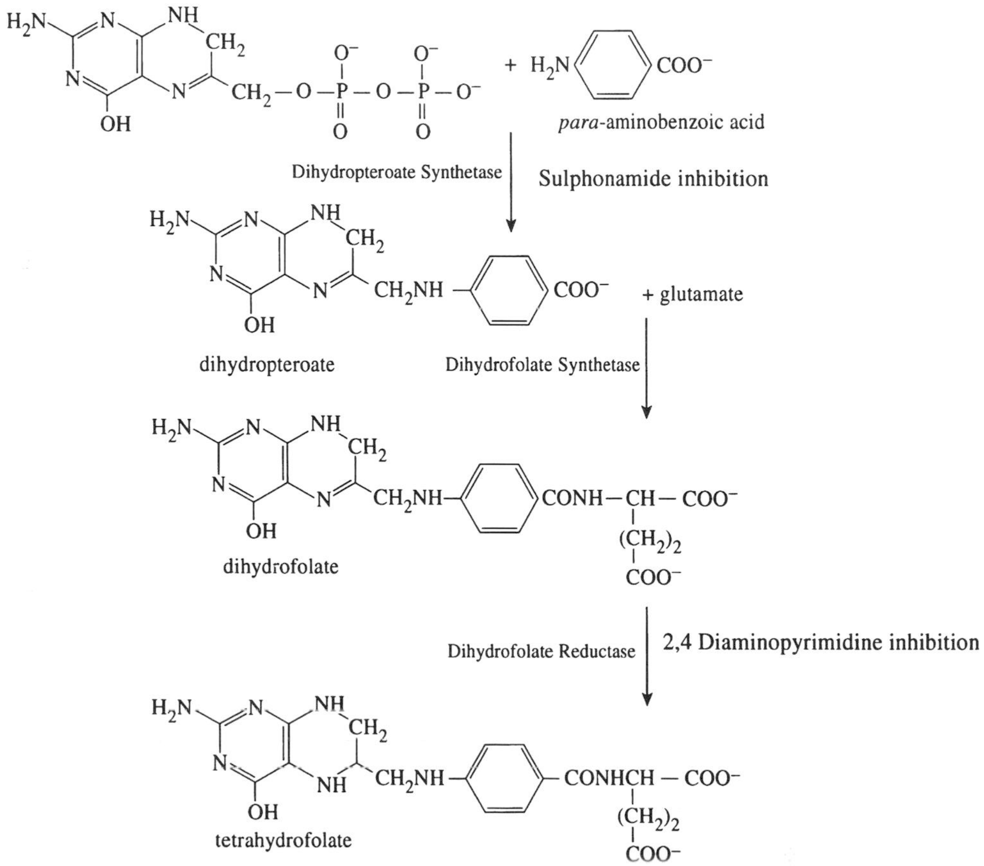

Figure 19.1 Pathway of folate metabolism.

1965). These folate antagonists are the structural analogues of part of dihydrofolate (Figure 19.1) and include the antibacterial agents trimethoprim, brodimoprim, metioprim, tetroxaprim and, more recently, epiroprim (Ro 11-8958; Locher *et al.*, 1996), the antimalarial drugs, pyrimethamine and proguanil, and anticancer agents such as methotrexate. Trimethoprim (Figure 19.2), which is used clinically as an antimicrobial, has a 50,000-fold higher affinity for the bacterial DHFR than the human enzyme, accounting for its selective toxicity (Burchall, 1979). However, it retains some affinity for the human DHFR and this may contribute to an adverse reaction (megaloblastic anaemia) seen occasionally following its use.

Properties of trimethoprim

Trimethoprim base, molecular weight 290.3, pK_a 7.3, should be stored below 25°C and protected from light.

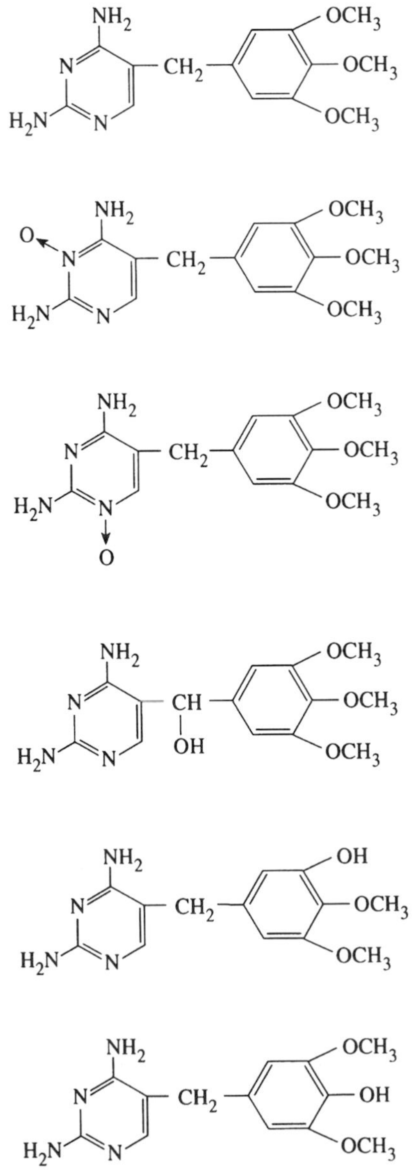

trimethoprim

trimethoprim-3-oxide

trimethoprim-1-oxide

alpha-hydroxytrimethoprim

3-hydroxytrimethoprim

4-hydroxytrimethoprim

Figure 19.2 Trimethoprim and its metabolites.

Reference substance is available from both The British Pharmacopoeia Commission Laboratory (Stanmore, UK) and the European Pharmacopoeia Commission (Strasbourg, France). The free base is insoluble in water at pH 7.2 but is soluble in lactic acid or dilute hydrochloric acid. The salt trimethoprim lactate is water-soluble. Trimethoprim is administered alone or in combination with a sulphonamide or sulphone for therapy of microbial infections.

Trimethoprim is lipophilic and is concentrated in many tissues. It potentiates the antimicrobial action of sulphonamides with demonstrable synergy *in vitro* and optimum activity noted at a ratio of 1:20 (Bushby, 1969). However, the poor distribution of the sulphonamides compared with trimethoprim into some tissues produces ratios below that required for optimal synergic bactericidal effect. The in-vivo role of this synergy remains controversial.

Trimethoprim is active against many Gram-negative bacilli including most Enterobacteriaceae, *Haemophilus* spp., *Pasteurella* spp., *Vibrio* spp. and *Bordetella pertussis* and some Gram-positive organisms such as staphylococci and streptococci. *Enterococcus faecalis* shows in-vitro sensitivity but can metabolize exogenous folate and may thus be considered potentially insusceptible. *Corynebacterium diphtheriae* and *Burkholderia cepacia* show sensitivity although other pseudomonads, *Mycobacterium tuberculosis* and *Nocardia* spp. do not (Hughes, 1997a). Although trimethoprim is bacteriostatic, in the presence of low concentrations of thymidine it is bactericidal against some strains.

Acquired resistance is common in Enterobacteriaceae and is due to genes coding for resistant DHFRs transmitted by transposons Tn*7*, Tn*402* and Tn*4132*. These confer high-level resistance with MICs >1 g/L (Amyes & Towner, 1990).

Indications include respiratory, urinary tract or prostatic infections and susceptible isolates from shigellosis or invasive salmonellosis.

Trimethoprim formulations

Trimethoprim is available as 100–200 mg tablets and a sugar-free suspension (10 mg/mL). Trimethoprim lactate solution for iv use (pH 4–4.5) equivalent to 20 mg/mL base is available in 5 mL ampoules.

Trimethoprim in combination with polymyxin B is available for topical ophthalmic use as either drops or ointment. Though reportedly as effective as chloramphenicol in therapy of eye infections (Study Group, 1989), this preparation contains thiomersal (0.05 mg/L) which may cause hypersensitivity particularly in contact lens users.

Co-trimoxazole (see below) comprises trimethoprim in combination with sulphamethoxazole at a ratio of 1:5.

Pharmacokinetics

Trimethoprim shows almost 100% oral absorption with peak plasma concentrations after 160 mg of 2.0 mg/L attained within 1–3 h. Following 160 mg iv, peak serum concentrations of 3–5 g/L are recorded 1.5 h post-dose. The mean half-life of elimination is around 10 h but there is considerable inter-patient variation. As a result of its high lipid solubility, the volume of distribution (V_d) is approximately 10-fold higher than that of sulphamethoxazole (69–133 L vs 10–16 L). In children aged <10 years, trimethoprim has a shortened $t_{1/2}$ of 5.6 h and this can increase with age to 16 h in adults >50 years old (Siber et $al.$, 1982). Protein binding is 42–46% and, as with other basic antimicrobials which bind to α-glycoprotein, serum albumin may not be the primary site of protein binding (Paap & Nahata, 1989).

The lipophilicity of trimethoprim facilitates penetration into body tissues including kidney, liver, lung, sputum, vaginal secretions, breast milk, prostatic fluid, bile, CSF and urine. With trimethoprim sequestered inside cells, higher concentrations are found within the tissues than in serum.

Almost 80% of orally administered trimethoprim is excreted unmetabolized in the urine, with the remainder being metabolized to 1-oxide-, 3-oxide-, 3-hydroxy-, 4-hydroxy- and α-hydroxy-trimethoprim (Figure 19.2; Reeves & Wilkinson, 1979). The metabolites account for only 10–15% of the drug in serum. The existence of a further, potentially toxic, metabolite, resulting from oxidation of the para-amino group to a hydroxylamine intermediate was postulated (van der Ven et $al.$, 1991).

Renal elimination of trimethoprim and its metabolites occurs partly by glomerular filtration and partly by tubular secretion and is promoted by acidification of urine.

Toxicity

High doses or prolonged therapy may cause suppression of haemopoiesis, megaloblastic anaemia or neutropenia in the folate deficiency state, for example, the elderly. Interference of trimethoprim with haematopoeisis is greatest where the daily dose exceeds 500 mg. Fewer adverse reactions are associated with trimethoprim alone than with co-trimoxazole although the former may cause rashes, nausea, vomiting, thrombocytopenia or, rarely, Stevens–Johnson syndrome. Trimethoprim triggers folate deficiency which can be counteracted by administration of folinic acid supplement. It is responsible for the hyperkalaemia seen in AIDS patients. Since trimethoprim hinders the tubular secretion of creatinine, high creatinine levels in the absence of uraemia are not necessarily indicative of renal dysfunction (Vree & Heckster, 1987).

Trimethoprim is contra-indicated in neonates, in those with a history of hypersensitivity, in pregnancy, and in those with porphyria and severe renal impairment except where monitoring of plasma trimethoprim concentrations is available. Trimethoprim should be used with care in the elderly if there is no alternative, as $t_{1/2}$ increases markedly in this group. Similarly, it should be used cautiously in those with severe hepatic dysfunction.

Concomitant use of rifampicin may cause increased elimination of pyrimidines including trimethoprim. Trimethoprim potentiates the anticoagulant activity of warfarin, prolongs the half-life of phenytoin, and may trigger nephrotoxicity in patients taking cyclosporin.

Assays

Chemical, microbiological and chromatographic assays are available. Chemical methods have used the natural fluorescence of trimethoprim or have involved derivatization. Although microbiological assays are easy to perform, problems of specificity are encountered. Sulphonamides and other agents may be present and the inactivation or removal of potential antagonists present in the medium is essential. In the presence of a sulphonamide, $para$-aminobenzoic acid must be added to inhibit its antimicrobial activity. HPLC assays promise high specificity and the ability to assay sulphonamide levels simultaneously. Trimethoprim can be easily detected by UV absorption at 254–280 nm. In alkaline solution trimethoprim is fluorescent and this property has been used for detection after HPLC.

Chemical assay methods

These were reviewed by Allen & Nimmo-Smith (1978) but have now been superseded by HPLC assays. The method of Schwartz et $al.$ (1969) is suitable for measuring serum concentrations and is outlined below. Trimethoprim is extracted and oxidized to 3,4,5-trimethoxybenzoic acid which is then determined fluorimetrically.

Assay of trimethoprim in body fluids by derived fluorescence (Schwartz et $al.$, 1969)

$Reagents.$ Prepare or purchase the following reagents (but see 'Expectations, problems and pitfalls' below): 0.1 N sodium carbonate; chloroform; 0.01 N and 1 N sulphuric acid; 0.1 M potassium permanganate in 0.1 N sodium hydroxide; 3 g/L ascorbic acid in 0.1 N sodium hydroxide (this must be freshly prepared for each assay; it is used as a less toxic alternative to 35% formaldehyde); a stock aqueous solution of trimethoprim 100 mg/L.

$Calibrators.$ As an alternative to the original calibration procedure, use the modified method of Allen & Nimmo-Smith (1978). Prepare working calibrators by mixing stock solution with drug-free plasma to give concentrations of 0.25, 0.5, 1, 1.5 and 2.5 mg/L. Also prepare a drug-free serum blank.

Procedure

1. Add 2 mL of samples, blank, calibrators and controls to 30 mL tubes.

2. Add 8 mL sodium carbonate solution and 10 mL

chloroform and mix by rotating the closed tubes end-over-end for 10 min.

3. Centrifuge (at 900**g** for 10 min)and transfer 7 mL of the lower chloroform layer to a new tube, add 4 mL 0.01 N sulphuric acid and re-extract and centrifuge as above.

4. Transfer 3 mL of the upper acid layer to a new tube and add 2 mL permanganate solution.

5. Mix and heat at 60°C for 10 min.

6. Add 0.2 mL ascorbic acid solution and mix.

7. Add 1 mL 1 N sulphuric acid, mix, and heat at 60°C until a clear solution is obtained (*c.*10 min) and cool.

8. Add 2 mL chloroform, shake for 10 min, centrifuge and transfer the chloroform layer to a quartz cuvette.

9. Measure the fluorescence (excitation 275 nm, emission 350 nm) in a suitable fluorimeter.

10. Deduct the blank reading from the other readings and construct a calibration curve. Read unknowns from this curve by interpolation. Dilute and re-assay any samples showing greater fluorescence than the top calibrator.

Expectations, problems and pitfalls. The lower limit of detection should be <0.05 mg/L and assay CVs should be 6% or better. Some metabolites are co-extracted but these give lower yields of 3,4,5-trimethoxybenzoic acid and therefore do not interfere significantly. Clean glassware should be used and plastic containers should be avoided to minimize the risk of high background fluorescence. Water should be double-distilled to avoid the risk of fluorescent impurities. Some batches of chloroform may also produce high background fluorescence. Urine samples should be diluted 1/20 before assay.

Bioassay

Use the method of Bushby & Hitchings (1968) modified for the assay of biological samples (Allen & Nimmo-Smith, 1978).

Indicator organism. Prepare a spore suspension of *Bacillus pumilis* CN 607 (Wellcome Research Laboratories) by growth in a liquid medium for several days. Determine the spore concentration by plating dilutions of a heat-shocked (50°C for 30 min) sample and dilute the remainder to a concentration of 10^6 spores/mL. Dispense into 500 μL portions and store at –20°C.

Medium. Oxoid sensitivity test agar containing 0.1% *para*-aminobenzoic acid (1 mL of a 1% solution to 100 mL of agar) incorporated into the molten agar. Addition of 5% lysed horse blood may improve assay sensitivity.

Calibrators. Stock calibrator solution (100 mg/L) is prepared by dissolving 1.3 g of trimethoprim lactate in water and making the solution up to 100 mL in a volumetric flask.

Final calibrators are prepared in a suitable matrix such as serum or (for urine samples) phosphate buffer pH 7.3. The claimed lower limit of detection is 0.02 mg/L for most biological fluids. Serum calibrators of 0.05, 0.1, 0.2, 0.4 and 0.8 mg/L are recommended for initial method work-up. A trimethoprim-free, sulphonamide-containing control sample should also be run to ensure the *para*-aminobenzoic inactivation is complete

Procedure

1. Pour three 100 mL layers of agar on to a 25 × 25 cm assay plate incorporating the spore suspension (500 μL), and lysed blood if required, into the central layer.

2. Cut 7 mm wells using a no. 4 cork borer.

3. Add the calibrators, controls and samples in triplicate as outlined in *Chapter 4.*

4. Incubate for 18 h at 28°C.

5. Accurately measure the zone diameters and construct the calibration curve. Where results of the controls are within 10% of their true antibiotic concentration, test sample results are accepted. If sulphonamide is present in the samples accept the results only if there is no zone of inhibition around the sulphonamide control.

Comments. Traces of thymidine in the medium will greatly reduce the sensitivity of the assay. Double zones may reflect the presence of antagonists in medium. Peak samples from patients may require two- to four-fold dilution in serum blank before assay. This method is useful for measuring levels in tissues or body fluids including urine where phosphate buffer (pH 7.3) is used as diluent to minimize pH effects.

HPLC assay of trimethoprim and other diaminopyrimidines

Some published methods for the assay of diamino-pyrimidines are given in Table I. Several permit assay in the presence of sulphonamides. Almost all use extraction procedures and internal standards. Weinfield & Macasieb (1979) described a normal phase method for trimethoprim metabolites. The 1-oxide metabolite eluted very close to the internal standard but the authors reported that this metabolite was not detected in any volunteers even when the internal standard was omitted. Two methods for trimethoprim assay are described below in detail. The latter method is suitable for therapeutic monitoring of co-trimoxazole as both trimethoprim and sulphamethoxazole are measured simultaneously.

HPLC assay of trimethoprim (or tetroxaprim) in serum, urine or milk (White & Reeves, unpublished)

Stationary phase. μBondapak C$_{18}$ (Waters, 300 × 4.6 mm).

Mobile phase. Methanol:water:heptane sulphonic acid solution (Fisons): acetic acid (310:592:47.5:1.9).

Detection. UV absorption at 254 nm.

Table I. Some HPLC assay methods for sulphonamides and diaminopyrimidines

Drug(s)	Sample	Stationary phase	Mobile phase	Detection	Sample preparation	Reference
Sulphadiazine, sulphamerazine, sulphamethoxazole, sulphanilamide, trimethoprim, brodimoprim	leucocytes	μBondapak C$_{18}$ 10 μm	water:ACN 80:20 (sulphonamides) water:MeOH:ACN + phosphate buffer 60:30:10 (others)	UV 254 or 280 nm	not stated	Climax et al. (1986)
Sulphonamides (various)	tissues	Chromopak CP C$_8$ 8 μm	ACN: ammonium acetate buffer pH 4.6 (30:70)	UV 254 nm	extraction procedure	Haagsma & van de Water (1985)
Sulphamethoxazole, N-acetyl metabolite; trimethoprim	serum, urine	Lichrosorb RP18 10 μm	EtOH:phosphate buffer pH 4.5 (26:74)	UV 260 nm	extraction procedure	Ascalone (1980)
Sulphadiazine, N-acetyl metabolite; trimethoprim	serum, urine	Lichrosorb RP18 Si60 10 μm	DCM:MeOH: 25% AmOH (80:19:1)	UV 289 nm	extraction procedure (as above)	Ascalone (1981)
Trimethoprim	sputum, saliva	Hi-Chrom ODS 5 μm	50 mM phosphoric acid + 0.1% HXS + 0.2% sodium nitrate:MeOH (30:70)	UV 235 nm	sputolysin then extraction	McIntosh et al. (1983)
Sulphamethoxazole, trimethoprim	serum, CSF	5μm C$_{18}$ column	ACN: Sorenson buffer pH 3 (25:75)	UV 235 nm	ACN protein precipitation extraction	Dudley et al. (1984)
Trimethoprim	plasma, urine	Brownlee RP8 10 μm	MeOH: 15 mM sodium borate pH 9 (35:65)	fluorescence[a] ex 279 nm em 370 nm	extraction procedure	Gautam et al. (1978)
Metioprim	body fluids	Nucleosil C$_{18}$ 5 μm	0.1 M phosphate buffer + 1% HAc, 1% EtAc:ACN (80:20)	UV 270 nm	chloroform extraction	Vergin et al. (1983)
Trimethoprim and metabolites	body fluids	μPorasil 10 μm	chloroform:(MeOH:water: AmOH 150:9:1) (500:25)	UV 280 nm fluorescence ex 290 nm em 340 nm	extraction procedure	Weinfield & Macasieb (1979)[b]
Sulphametrole N-acetyl metabolite	serum, urine	Lichrosorb RP8 5 μm[c]	phosphate buffer: MeOH (400:80)	UV 260 nm	perchloric acid protein precipitation	Hekster et al. (1981)
Sulphamethoxazole N-acetyl metabolite	serum, urine	Lichrosorb RP8 5 μm	0.067 M phosphate buffer pH 7: MeOH (400:80)	UV 260 nm	perchloric acid protein precipitation	Vree et al. (1978b)
Sulphadiazine, sulphamerazine, sulphamethazine	serum	μBondapak C$_{18}$ 10 μm[d]	ACN:1% HAc (13:87)	UV 254 nm	14% TCA protein precipitation	Goehl et al. (1978)
Sulphamethoxazole, sulphapyridine, N-acetyl metabolites	serum	Spherisorb ODS 10 μm	MeOH: 0.05 M phosphate buffer pH 7.4 (25:75)	UV 270 nm	extract into dichloromethane	Astbury & Dixon (1987)
Trimethoprim, sulphamethoxazole, N-acetyl metabolite	serum	Lichrosorb RP8 5 μm	0.67 M phosphate buffer pH 6.7: EtOH (400:80)	UV 260 nm	perchloric acid protein precipitation	Vree et al. (1978a)
Sulphonamides (various), N-acetyl and N-hydroxy metabolites	body fluids	various, mostly RP8 5 μm	various	UV 260–272 nm	perchloric acid protein precipitation	Vree et al. (1986)[e]

Abbreviations: ACN, acetonitrile; AmOH, ammonium hydroxide; DCM, dichloromethane; EtAc, ethyl acetate; EtOH, ethanol; HAc, acetic acid; HXS, hexane sulphonic acid; MeOH, methanol; TCA, trichloracetic acid; ex, excitation wavelength; em, emission wavelength.
[a] Fluorescent only at alkali pH. [b] Metabolites separated. [c] 40°C. [d] 29.6°C. [e] Unusual study of N-hydroxy metabolites.

[a] Column is at 29.6°C.
[e] Unusual study of N-hydroxy metabolites.

243

Calibrators. Calibrator solutions should be prepared in the same matrix as the samples. Up to four calibrators may be used covering the range 0.1–5 mg/L. A blank should also be run with every batch.

Procedure (serum). Mix serum with acetonitrile containing internal standard (0.5 mg/L tetroxaprim). Centrifuge and inject up to 100 μL of the clear supernatant.

Procedure (milk). Mix 2 mL of milk with 500 μL of 1 M sodium carbonate and 10 mL of chloroform and shake vigorously. Allow to settle, remove 8 mL of the chloroform and back-extract with 800 μL of 0.01 M sulphuric acid. Inject 100 μL of the acid layer.

Performance expectations. This extraction procedure eliminates interference found in normal milk and the limit of sensitivity is around 0.05 mg/L. Assay CVs should be better than 10%. Tetroxaprim may be assayed by this procedure using trimethoprim as the internal standard.

HPLC method for assay of trimethoprim and sulphamethoxazole in serum (Lovering & Shi, 1990)

Stationary phase. Spherisorb ODS or Hypersil 5 ODS (HPLC Technology, 100 × 4 mm).

Mobile phase. Methanol:water:orthophosphoric acid solution (28:71:1) containing 1 g/L heptane sulphonic acid and with the pH adjusted to 5.

Detection. UV absorption at 214 nm.

Calibrators. Aqueous calibrator solutions should be prepared and single-point calibration is suitable for therapeutic monitoring. The trimethoprim calibrator should be 2–10 mg/L and the sulphamethoxazole calibrator around 100 mg/L.

Procedure. Mix serum (100 μL) with an equal volume of methanol. Allow to stand for 10–15 min, centrifuge and inject up to 20 μL of the clear supernatant. The peak heights of the two drugs are very different and if an integrator is not being used a change in the sensitivity setting will be needed after the sulphamethoxazole has eluted. Suggested settings are 0.5 AUFS for sulphamethoxazole and 0.05 AUFS for trimethoprim. Sulphamethoxazole elutes first followed by an endogenous serum peak. The trimethoprim elutes last.

Performance expectations. The authors reported limits of sensitivity and reproducibility of 0.25 mg/L, CV 5.8% (trimethoprim) and 2.5 mg/L, CV 4% (sulphamethoxazole). Neither *N*-acetylsulphamethoxazole nor 18 other commonly prescribed antibiotics interfered with this assay.

Sulphonamides

Introduction

The activity of the dye sulphachrysoidine (prontosil rubrum), documented over 60 years go, heralded the development of the sulphonamides as chemotherapeutic agents (Domagk, 1935). The first of these, *para*-amino-benzene-sulphonamide (sulphanilamide), produced *in vivo* following the degradation of prontosil rubrum, was responsible for the antibacterial properties of the dye.

Since the discovery of the activity of sulphanilamide, numerous sulphonamide derivatives have been synthesized. All are derived chemically from sulphanilamide, a structural analogue of *para*-aminobenzoic acid (Figure 19.1). Because most protozoa and the majority of bacterial species cannot utilize exogenous folate, they require folic acid for nucleic acid synthesis and DNA replication. The sulphonamides act by competitively inhibiting the uptake of *para*-aminobenzoic acid by the microbial DHPS (Figure 19.1), thereby interrupting protein synthesis.

Modification to, or disruption of, the benzene ring results in loss of antimicrobial activity. The free amino group at N^4 is also essential for antimicrobial activity. Differences in substituents at the NH_2SO_2 moiety attached to C^1 (Figure 19.3) determine their pharmacological and pharmacokinetic properties.

Apart from being slightly more active than sulphanilamide, these compounds show similar antimicrobial spectra. The main differences between the compounds are their water solubility and elimination half-lives. Sulphonamides are often classified as being short-acting ($t_{1/2} < 7$ h), medium-acting ($t_{1/2} = 8$–20 h) or long-acting ($t_{1/2} > 30$ h).

Sulphonamides show activity against *Streptococcus pneumoniae*, *Streptococcus pyogenes*, staphylococci, *Corynebacterium diphtheriae* and Gram-negative bacteria such as the enterobacteria, *Neisseria* spp., *Haemophilus* spp. and bordetellae with sensitive strains inhibited by <32 mg/L. Actinomyces, nocardiae, brucellae, *Mycobacterium leprae* and the protozoans, *Pneumocystis carinii*, *Toxoplasma gondii* and *Plasmodium* spp. are also sensitive. *Pseudomonas aeruginosa*, *Mycobacterium tuberculosis* and *Enterococcus faecalis* are naturally resistant (Hughes, 1997b). Although their action is bacteriostatic it may be bactericidal where the thymidine content is low. Since their activity is inhibited by pus, their efficacy is reduced in suppurative lesions.

Within the Enterobacteriaceae, >40% of strains of *Escherichia coli*, *Salmonella* spp. and *Shigella* spp. are now resistant. This resistance is plasmid-mediated with the genes coding for synthesis of a new DHPS with reduced affinity (10,000-fold less) for sulphonamides though not for *para*-aminobenzoic acid. Excess production of DHPS or of *para*-aminobenzoic acid also confers resistance.

Former indications included meningococcal meningitis where the isolate was known to be sensitive. Less toxic agents are now preferred for uncomplicated urinary tract infections, acute exacerbation of chronic bronchitis, and meningococcal prophylaxis as >20% of isolates are resistant. Sulphonamides may have potential in the treatment of infections caused by *Nocardia asteroides* or *Haemophilus*

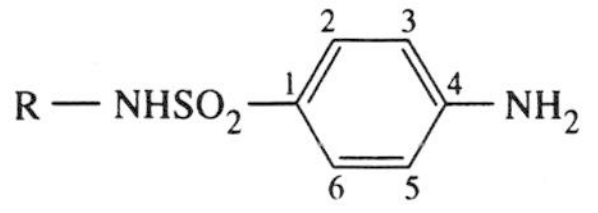

	R	Classification
Sulphanilamide	H	
Sulphacetamide	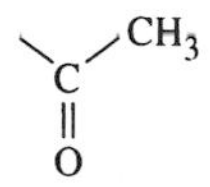	Medium-acting
Sulphathiazole	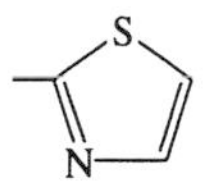	Short-acting
Sulphadiazine	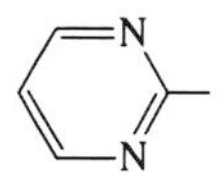	Medium-acting
Sulphadimidine	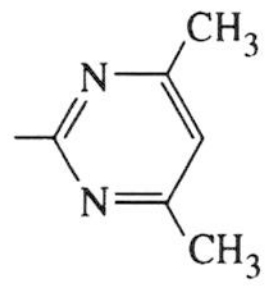	Long-acting
Sulphacarbamide	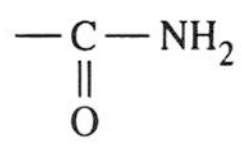	Short-acting
Sulphamethisole	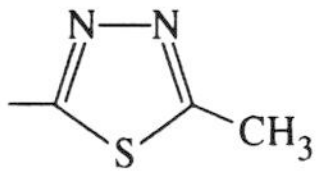	Short-acting
Sulphamethoxypyridazine	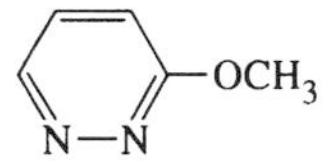	Long-acting
Sulphadimethoxine	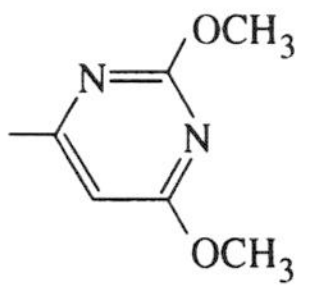	Long-acting
Sulphamethoxazole	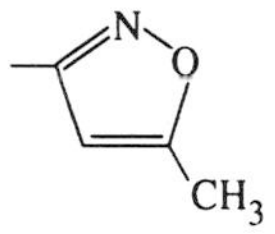	Medium-acting
Sulphametopyrazine	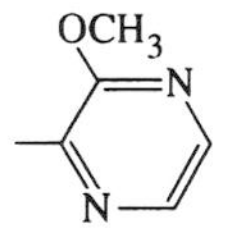	Long-acting
Sulphadoxine	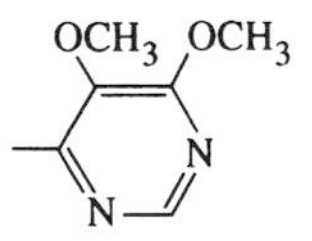	Long-acting

Figure 19.3 Antibacterial sulphonamides.

ducreyi where treatment options are limited and for treatment of *S. pneumoniae* in patients with β-lactam hypersensitivity.

With increasing reports of toxicity associated with their use and the emergence of sulphonamide resistance in bacteria, their usefulness diminished, precipitating their replacement by less toxic drugs. Except for a few specific indications, sulphonamides are no longer agents of first choice. However, the high mortality recorded within the last 20 years in immunosuppressed individuals following opportunistic protozoan infections has rekindled interest in the folate inhibitors as therapeutic agents.

Some sulphonamides are shown in Figure 19.3 in the order in which they were developed. Most are no longer used and the only antimicrobial sulphonamides currently available in the UK for systemic use are sulphadiazine, sulphadimidine, sulphamethoxazole (only available in combination with trimethoprim as co-trimoxazole), sulphametopyrazine and sulphadoxine (available only in combination with pyrimethamine for treatment of malaria).

Properties of sulphonamides

Sulphonamides are predominantly white, odourless, lipid-soluble, crystals or powders which are poorly water-soluble, but soluble in dilute mineral acid and alkali hydroxides. They are stable over a wide pH range in acid conditions and withstand heating but not UV irradiation. Storage is optimal at 25°C. As they contain ionizable groups, they are weak acids with pK_as of 4.6–7.4 for the systemic agents. The properties of some individual agents are given in Table II.

Sulphonamide formulations

Only a few sulphonamides remain in clinical usage. Sulphadiazine and sulphadimidine are available as tablets containing 500 mg of active sulphonamide and other inactive ingredients. Sulphadiazine sodium (1 g of active sulphonamide in 4 mL of fluid) is available for parenteral use. This highly alkaline solution (pH 10–11) can be given im if venous access is limited although iv administration is advised. Since it contains sodium thiosulphate it may trigger adverse effects and subcutaneous/intrathecal injections are contra-indicated because of potential toxicity and necrosis. Sulphadiazine is marketed in some countries in combination with other drugs. Sulphametopyrazine is available as 2 g tablets. Sulphadoxine in tablet form is available only in combination with pyrimethamine (500/25 mg). Similarly, sulphamethoxazole is available only in combination with trimethoprim as co-trimoxazole (see below).

Pharmacokinetics

When given orally, the drugs which have a free $-NH_2$ group at N^4 are readily absorbed from the intestine, achieving peak serum concentrations of 80–100 mg/L 2–4 h after a 1–2 g dose. In critically ill patients where absorption is variable, sulphonamides are given parenterally to rapidly

Table II. Physical, chemical and pharmacokinetic properties of antibacterial sulphonamides

Name	Mol. wt	pK_a	Water solubility at pH 7 (mg/L 25°C)	Elimination half-life (h)	Protein binding (%)	V_d (L/kg)	Metabolism (%)
Sulphadiazine	250.3	6.4–6.5	950	7–12	20–55	0.4	N-acetyl (slow), 13; N-acetyl (fast), 38; hydroxy, 30
Sulphadimidine	278.3	7.4	7000	1–11 (slow); 1–5 (fast)	90	0.6	N-acetyl (slow), 80; N-acetyl (fast), 90; hydroxy, slight
Sulphamethoxazole	253.3	5.7–6.1	1900	7–12	59–65	0.2	glucuronide, 15; N-acetyl, 65; hydroxy, 11
Sulphametapyrazine	280.3	6.4–7.2	2510	60–70	76–99	0.04	N-acetyl, 36; hydroxy, slight
Sulphadoxine	310.3	6.3	2387	150–200	92–95	0.1	glucuronide, 3; N-acetyl, 5; sulphate, 0.5

'Slow', slow acetylator phenotype; 'fast', fast acetylator phenotype.
From Reeves et al. (1978);Vree et al. (1986, 1987)

achieve therapeutic levels. The serum elimination half-lives ($t_{1/2}$) of individual compounds vary considerably (Table II) depending on the rate of metabolism and reabsorption.

The sulphonamides are conjugated by hepatocyte enzymes. Metabolism occurs by hydroxylation, deamination, acetylation and glucuronidation. The mixed-function oxidases and cytochrome P450 in the liver facilitate the slower processes of hydroxylation and deamination. Following passive reabsorption by renal tubules, the acetyl metabolites are excreted by active tubular secretion. Alkalinization of urine promotes excretion of sulphonamides though not that of their acetyl metabolites.

The N-aryl-, N-acetyl-transferases (NATs) found in leucocytes though predominantly in hepatocytes, acetylate at the N^4 position, producing conjugated derivatives with a higher polarity and protein binding than the parent compound. Individuals vary in their ability to conjugate these drugs and may be classed as 'fast' or 'slow' acetylators, the latter being an identifiable risk factor for hypersensitivity and crystalluria. In contrast to sulphamethoxazole, administration of sulphadimidine is useful for determining acetylator status (Bozkurt et al., 1990). The degree of acetylation varies with the particular sulphonamide. The rate of the competitive processes of hydroxylation and acetylation is determined by the structure of the sulphonamide, the acetylator phenotype and the activity of the mixed-function oxidases in the liver. The rate of this reaction is influenced by methyl groups on the N^1 substituent of the pyrimidine ring increasing the electronegativity of the sulphonamide. The acetylation–desacetylation equilibrium is dependent on acetylator phenotype. Hydroxylation of the moiety at N^1 confers antibiotic activity. Under acidic conditions reversion of metabolites to the parent compound may occur and is of importance in patients where creatinine clearance is <5 mL/min. For further information on the metabolism of sulphonamides, see Vree & Hekster (1987) and Cribb et al. (1996).

After absorption, sulphonamides distribute into ocular, pleural, synovial and peritoneal fluids, and into bile, saliva and urine. They diffuse across the placenta and are excreted into breast milk. CSF concentrations may rise to approximately 20–80% of serum concentrations.

Though dependent on urinary pH, >90% of the total dose is voided in the urine. The parent compound and its metabolites are excreted, mainly by glomerular filtration and by tubular reabsorption with the glucuronides detected in urine.

Since the sulphonamides are highly metabolized to mainly inactive derivatives, avoidance of these drugs in renal impairment is prudent. However, they can be used in the treatment of patients with renal dysfunction if monitoring facilities are available. There is a risk of tubular reabsorption and accumulation of the less polar, conjugated metabolites which are responsible for renal toxicity i.e. crystalluria and interstitial nephritis. These derivatives have a lower activity than the parent compound but an increased $t_{1/2}$ following oxidation. With a creatinine clearance of <30 mL/min, accumulation of these polar derivatives is marked and this may contribute to the hypersensitivity noted. Acetyl metabolite accumulation triggers desacetylation and this will cause an apparent rise in parent drug concentrations in the serum of anuric patients (Vree & Hekster, 1987).

Toxicity

Fever, headache, nausea and vomiting are frequent, with the commonest side-effects being skin rash and agranulocytosis, the latter being reversible on cessation of sulphonamides. Other blood dyscrasias—including haemolytic anaemia, marrow depression, thrombocytopenia and, rarely, Stevens–Johnson syndrome—are indications for withdrawal of drug. Crystalluria was noted with the use of long-acting sulphonamides; it could be reversed by adequate fluid intake; this side-effect is less frequent with the newer, more soluble compounds although it is reported increasingly following sulphadiazine therapy in AIDS patients. Phenylbutazone and ethyldicoumarol displace sulphonamides from binding sites. As a result of high protein binding, bilirubin, warfarin, tolbutamide and thiopentone may be displaced by sulphonamides from albumin binding sites. The complication of idiosyncratic haemolytic anaemia associated with sulphonamide oxidation of haemoglobin to methaemoglobin is seen in 1% of patients with glucose-6-phosphate dehydrogenase deficiency.

Contra-indications include a history of sulphonamide sensitivity, liver or renal impairment where monitoring facilities are lacking or use in the elderly where folate deficiency is common. Sulphonamides are contra-indicated in infants (<6 weeks old) and in late pregnancy because elevated levels of foetal bilirubin may cause kernicterus as a result of displacement from plasma proteins. Sulphonamides are physically incompatible with chloramphenicol, aminoglycosides, tetracyclines and synthetic penicillins.

Sulphonamide assays

An extensive range of methods is available for monitoring these agents, including colorimetry, paper and absorption chromatography, TLC, spectrofluorimetry, differential pulse polarography, GLC with flame ionization detection and various forms of HPLC, i.e. anion and cation exchange, gradient, normal phase, reversed phase and reversed phase-ion pairing. Bratton & Marshall (1939) first assayed active sulphonamide in deproteinized plasma by diazotizing the *para*-amino group, coupling this to a chromogen and measuring the resultant red dye colorimetrically. The N^4-acetyl derivative, being detectable only after acid hydrolysis, was assayed indirectly. However, acid hydrolysis may be incomplete and underestimate total

drug (Bury & Mashford, 1979). Unclean glassware can give spurious results in colorimetric assays (Reeves *et al.*, 1978). Although this protocol, reproduced recently (Stokes *et al.*, 1993), is still in use, it has been superseded by HPLC methods which are specific, selective and rapid.

The original HPLC analyses were hindered by complex sample extraction procedures, i.e. liquid/liquid or solid/liquid, with evaporation under nitrogen and resolvation before assay. Internal standardization was essential to monitor the reproducibility of drug recovery. Other disadvantages included co-extraction of endogenous substances, long analysis times, lack of metabolite detection and assay of individual antimicrobials. Simpler sample preparation with acid or methanol precipitation of proteins has replaced these laborious pre-treatments. Some HPLC procedures for the assay of sulphonamides and their metabolites are outlined in Table I. An HPLC method for the assay of sulphamethoxazole in the presence of trimethoprim is described in the first part of this chapter dealing with trimethoprim. A chemical method and a bioassay applicable to most sulphonamides are outlined below.

Chemical assay for sulphonamides

In this method (from Bratton & Marshall (1939) as modified by Reeves *et al.* (1978)) sulphonamides are diazotized in acid solution and converted to a coloured compound which is measured by absorption at 545 nm. *N*-Acetyl metabolites do not react but can be estimated after desacetylation in strong acid. The assay without acid hydrolysis is said to measure 'free sulphonamide' whereas the assay result after acid hydrolysis is said to measure 'total sulphonamide'.

Reagents. Prepare the following aqueous solutions using high quality distilled or deionized water and analytical grade chemicals: 20% trichloracetic acid (TCA); 0.5% ammonium sulphamate; 3 M hydrochloric acid; 10% sodium nitrite. Prepare the following aqueous reagents freshly for each assay: 0.1% sodium nitrite (from the 10% solution); 0.05% *N*-(1-naphthyl)ethylenediamine dihydrochloride (N-NED).

Calibrators. Prepare calibrator solutions of 10, 25, 50, 100 200 and 400 mg/L in water.

Procedure

Initial sample preparation

1. Dilute any samples expected to have a high concentration 1/5 in water.

2. Mix duplicate 200 μL of samples, calibrators, controls and water blank with 200 μL of distilled water.

3. To each tube add 180 μL of 20% TCA, mix well, allow to stand for 5 min and centrifuge. (This step disrupts any protein binding and precipitates serum proteins.)

4. Transfer 2 mL of each clear supernatant to a clean tube. Separate the duplicates into two sets. One set is for free sulphonamide and the other set for total sulphonamide (see above).

Assay of total sulphonamide

1. Add 200 μL of hydrochloric acid and mark the meniscus level on each tube.

2. Mix well, cover each tube, and place either in a boiling water bath for 60 min or in a steamer for 90 min.

3. Remove the tubes, cool, and add distilled water to make up any volume loss.

4. Add 200 μL of 0.1% sodium nitrite, mix well and allow to stand for 10 min.

5. Add 200 μL of ammonium sulphamate, mix well and allow to stand for 5 min.

6. Add 1 mL of N-NED, mix well and allow to stand for 5 min.

7. Read absorbance at 540–545 nm against the water blank.

8. Plot the calibration curve (drug concentration against optical density) which should be linear. Determine the unknown concentrations by interpolation. Dilute and re-assay any samples giving a colour density above the top calibrator.

Assay of free sulphonamide

1. Add 200 μL of 0.1% sodium nitrite, mix well and allow to stand for 10 min.

2. Add 200 μL of ammonium sulphamate, mix well and allow to stand for 5 min.

3. Add 1 mL of N-NED, mix well and allow to stand for 5 min.

4. Read absorbance at 540–545 nm against the water blank.

5. Plot the calibration curve (drug concentration against optical density) which should be linear. Determine the unknown concentrations by interpolation. Dilute and re-assay any samples giving a colour density above the top calibrator.

Comments. Very clean glassware is essential. New soda-glass tubes should be acid-washed before use. Aqueous calibrators can be used because recovery in the TCA precipitation step is virtually 100%.

Trimethoprim does not interfere and CVs of around 1% are possible. Hydroxy metabolites cannot be differentiated. It is possible to automate this assay.

Bioassay for sulphonamides (Reeves *et al.*, 1978)
Indicator organism. Spores of *Bacillus pumilis*.

Medium. Oxoid Diagnostic Sensitivity Test (DST) agar (Unipath) containing lysed horse blood (5 mL/100 mL of agar). Pour the plate (25 cm × 25 cm) incorporating enough spores to give almost-confluent growth and store overnight at 4°C.

Calibrators. Prepare serum calibrator solutions of 2.5, 5, 10, 20 and 40 mg/L. If urine samples are being assayed prepare similar calibrators in phosphate buffer pH 7.3. A sulphonamide-free, trimethoprim-containing control sample should also be run if appropriate.

Removal of trimethoprim (if appropriate). Mix the serum sample with 200 g/L Dowex 50-X8 ion-exchange resin. Repeat this procedure. Inoculate the plate with the absorbed serum. Perform the same procedure on a trimethoprim-containing control.

Procedure. Cut 9 mm wells using a no. 5 cork borer. Add the calibrators, controls and samples in triplicate as outlined in *Chapter 4*. Incubate for 18 h at 37°C. After incubation, measure accurately the zone diameters and construct the calibration curve. Where results of the controls are within 10% of their true antibiotic concentration, test sample results are accepted.

Comments. Other antimicrobials will interfere unless removed or inactivated. Because some batches of media contain sulphonamide antagonists, which will reduce zone size, antagonist-free medium should be used. Microbiologically active metabolites (for example, hydroxy metabolites) will be measured as well as parent drug. This may produce a large error in urine assays with some sulphonamides.

Co-trimoxazole

Introduction

Sulphamethoxazole is available only in combination with trimethoprim as co-trimoxazole. Where the individual components may be bacteriostatic this combination is bactericidal. Synergy occurs *in vitro* over a range of concentration ratios with the optimum being 1:20 trimethoprim:sulphamethoxazole. This property was successfully exploited when the combination of trimethoprim:sulphamethoxazole in a fixed ratio of 1:5 was released in 1968 (UK) as co-trimoxazole for oral use. Although this formulation was calculated to produce a 1:20 concentration ratio in serum, trimethoprim:sulphamethoxazole ratios ranging from 1:0.04 to 1:40 are found in healthy tissues (Reeves & Wilkinson, 1979).

Current indications are limited to prophylaxis (low dosage) or treatment (high dosage) of *P. carinii* pneumonitis, nocardiosis and toxoplasmosis in immunosuppressed, organ transplant, oncology and AIDS patients. Co-trimoxazole may also be of use for therapy of protozoan infections such as those caused by *Blastocystis hominis*, *Isospora belli* or cyclospora, as adjunct therapy for *B. cepacia* or *Stenotrophomonas maltophilia*, histoplamosis and for treatment of actinomycosis in patients allergic to β-lactams or for brucellosis. Co-trimoxazole should only be used to treat urinary tract infections, acute otitis media in childhood, or exacerbations of chronic bronchitis where the use of single-agent therapy has been precluded.

Formulations

Co-trimoxazole is available as a generic preparation or (in the UK) as Bactrim (Roche) or Septrin (Wellcome). Oral formulations include tablets (480 mg), dispersible tablets (480 mg), forte-tablets (960 mg) and oral suspension (480 mg/5 mL) and paediatric suspension (240 mg/5 mL). All contain ingredients in addition to the active drugs. A concentrated parenteral preparation (5 mL of a 96 g/L solution) can be used to prepare an iv infusion.

Pharmacokinetics

Following 1 g sulphamethoxazole po, a peak serum sulphonamide concentration of >50 mg/L is detected within 2–4 h and the $t_{1/2}$ of elimination is 8–12 h. After infusion, the peak generally occurs at 1.5 h. Although peak plasma concentrations are higher and attained more quickly after iv rather than oral administration, there is no significant difference in $t_{1/2}$ of elimination. Consequently, there may be no therapeutic advantage of parenteral over oral administration except where infection is acute and the patient is requiring treatment urgently, receiving other iv fluids, unable to accept oral therapy, or where absorption is dubious.

Sulphamethoxazole is extensively metabolized by acetylcoenzyme A and the N-acetyltransferases to the N^4-acetyl metabolite, conjugated with glucuronic acid to form the N^1-glucuronide, and metabolized by cytochrome P450 to the 5-hydroxy derivative (Figure 19.3). Acetylsulphamethoxazole, 5-hydroxysulphamethoxazole glucuronide, acetyl-5-hydroxysulphamethoxazole (and glucuronide) and 5-hydroxymethyl metabolites have all been detected (Vree *et al.*, 1986) . The sulphamethoxazole in plasma will be 15–25% acetyl metabolite and 75–85% unchanged parent drug. These percentages vary depending on liver enzymes and renal clearance which is pH-dependent. Protein binding of sulphamethoxazole and acetyl sulphamethoxazole is *c.* 65% and 75% respectively though the volumes of distribution are the same. The $t_{1/2}$ of the acetyl metabolite is only 2–3.5 h in normal subjects and it has no antimicrobial activity. Although responsible for tubular necrosis the hydroxy metabolite constitutes <5% of passively excreted drug, and has around 33% of the antimicrobial activity of the parent compound (Nouws *et al.*, 1985). A further metabolite, an N^4-hydroxylamine derivative, whose conversion is mediated by N-acetyltransferase 2 (Cribb *et al.*, 1996) is cytotoxic for T-cells and may confer hepatoxicity (Rieder *et al.*, 1992).

Trimethoprim metabolites were discussed earlier in this chapter.

Distribution of sulphamethoxazole is similar to that of other sulphonamides (see above). The rapid penetration of trimethoprim into the CSF in uninflamed meninges followed by slower diffusion of sulphamethoxazole was confirmed by Dudley *et al.* (1984). Following trimethoprim–sulphamethoxazole 5 + 25 mg/kg iv, the peak serum trimethoprim concentration was noted <1 h after infusion whereas that of sulphamethoxazole was delayed 4 h from the end of the 2 h infusion. The CSF ratios of these components were 1:3 to 1:34 with peak concentrations of 12–50% and 20–60% respectively of the simultaneous serum concentrations. Their $t_{1/2}$ values in CSF were 3.4 and 5.7 h respectively.

Though pH-dependent, approximately 10–20% of sulphamethoxazole is excreted in urine as active drug with 45–70% as the acetyl metabolite and <5% 5-hydroxymethylsulphamethoxazole (Cribb *et al.*, 1996). Hydroxy metabolites are found in renal calculi (Woolley *et al.*, 1980). Although the risk of accumulation is lowered by giving fluid and alkalinizing urine, these steps do not promote clearance of acetyl sulphamethoxazole. Reduced clearance of sulphonamide occurs in elderly patients and those with a creatinine clearance of <30 mL/min.

Effect of renal impairment

The use of co-trimoxazole in individuals in whom it is important to preserve residual renal function is unwise unless therapeutic monitoring is performed. In renal impairment the protein binding of sulphamethoxazole is decreased, and the V_d and $t_{1/2}$ of the parent drug (14 h) and acetyl metabolite (40 h) increase (Vree & Hekster, 1987), eventually triggering desacetylation of the metabolite back to the parent compound. The dosing interval should be increased and the maintenance dose adjusted accordingly when the creatinine clearance is <30 mL/min to prevent accumulation of drug and metabolites (Siber *et al.*, 1982; Sattler & Jelliffe, 1994).

Use of co-trimoxazole in dialysis patients is not recommended unless assays are performed. During haemodialysis almost half (44%) of the trimethoprim and 57% of the sulphamethoxazole are removed with $t_{1/2}$ values of 6 and 3.1 h respectively (Nissenson *et al.*, 1987). The N^4-acetyl metabolite accumulates, having a $t_{1/2}$ of up to 40 h. In CAPD, <6% of the drugs are removed in the dialysate and the $t_{1/2}$ of trimethoprim is up to 28 h (Walker *et al.*, 1989).

Adverse reactions and toxicity

There have been reports of aseptic meningitis in adults and intracranial hypertension in children in association with co-trimoxazole (or its individual components). Although reversible on cessation of drug, re-exposure to either component may cause recurrence. Fatalities have also been recorded (Committee on the Safety of Medicines, 1985)

Monitoring of clinical parameters, including blood counts and liver and renal function, is vital in early detection of adverse effects. Renal toxicity (hypersensitivity, crystalluria or interstitial nephritis) may develop during treatment. Toxicity with rash is the commonest manifestation in the immunosuppressed and often develops during co-trimoxazole therapy. The highest incidence, 40–80%, occurs in AIDS patients, frequently prompting a change of therapy or, controversially, desensitization (Gordin et al., 1984; van der Ven et al., 1994, Cribb et al., 1996). This is in marked contrast to <8% toxicity in the HIV-seronegative population and may be partly attributable to sustained high levels in the HIV-positive patients. Although the severity of the reactions noted is significantly lower with lower dosages, the reduction may not prevent toxicity, only delay it, further supporting a dosage-related problem (Siber et al., 1982; Schneider et al., 1992).

Although toxicity is commonly attributed to the sulphonamide component, the specific roles of the rarely quantified acetyl metabolite or of trimethoprim in toxicity are not fully elucidated. Increased production of the reactive N^4-hydroxylamine, which is converted into a nitroso derivative that binds to proteins, induces haemolytic anaemia and hepatotoxicity (Rieder et al., 1992). This derivative may mediate the hypersensitivity reactions seen in immunodeficient patients either by its immunogenicity or because lack of the antioxidant glutathione prevents detoxification and causes accumulation (van der Ven et al., 1994). Factors implicated in the toxicity of co-trimoxazole include cumulative dosing, slow acetylation, the activity of metabolites and glutathione deficiency, a condition common in the elderly and HIV-infected individuals. For further discussion see Cribb et al. (1996).

Therapeutic monitoring

A serum trimethoprim concentration of 2 mg/L is adequate for urinary tract infections as a result of the very high (c.100-fold) concentration seen in urine. Monitoring serum concentrations in patients treated with trimethoprim alone for urinary tract infection is not necessary.

For most infections, a serum sulphonamide concentration of 50–150 mg/L is advocated and assay is unnecessary. However, monitoring of serum co-trimoxazole is advised if prescribed for infants <6 weeks old, those receiving the high-dosage regimens and individuals with renal dysfunction (Siber et al., 1982; Paap & Nahata, 1989). Both drugs should be assayed although some clinicians may request only trimethoprim levels. The target plasma ratio of trimethoprim:sulphamethoxazole (1:20), which is variable in vivo, should not be used to estimate the concentration of one drug from measured serum levels of the other. Some methods described earlier in this chapter

are suitable for therapeutic monitoring. The HPLC methods that assay both drugs simultaneously are the most convenient and give a same-day result.

High dosage regimens

Although co-trimoxazole treatment of pneumocystis pneumonia is advised only where assay facilities are available, routine monitoring is a contentious issue and rarely practised. A consensus on optimal treatment regimen has not been reached and conventional high dosages (20 + 100 mg/kg/day), extrapolated from paediatric regime, or maintenance dosages (15 + 75 mg/kg/day), of trimethoprim plus sulphamethoxazole, frequently result in excessive serum levels. Despite extensive data confirming that the elimination $t_{1/2}$ may be >9 h, frequent qds prescribing occurs with rapid accumulation and subsequent intolerance, commonly resulting in drug withdrawal or change to alternative, possibly less effective, therapy. Although the relationship between serum concentrations and efficacy or toxicity is not conclusively resolved, pre- and post-dose monitoring during treatment of P. carinii pneumonia is advised to confirm adequate absorption, avoid the build-up of excessive serum concentrations and minimize risk of adverse effects.

The optimal time for monitoring is not universally agreed, with samples collected from 5 min to 6 h after dosing (Vohringer & Arasteh, 1993).

Since serum or plasma trimethoprim levels of <5 mg/L 1 h post-dose were associated with P. carinii pneumonia treatment failure in childhood, therapeutic targets of 3–5 mg/L and 100–150 mg/L, 2 h post-dose were proposed by Hughes et al. (1975, 1978). Subsequently the clinical benefit of therapeutic monitoring with dosage adjustment of a tds dosing regime to give peak serum concentrations of 5–8 mg/L was proven and deemed tolerable in 31 of 36 adult AIDS patients (Sattler et al., 1988). The side-effects associated recently with trimethoprim, decreased with lower dosages of 12 mg/kg/day (Sattler et al, 1988; Hughes & Killmar, 1996). Myelosuppression seen with serum trimethoprim levels >20 mg/L may, however, still occur in some patients with levels within the range 5–10 mg/L or with sulphamethoxazole levels of >200 mg/L (Siber et al., 1982; Fong, 1988; Sattler & Jelliffe, 1994).

The contribution made by the acetyl metabolite to the total sulphonamide content is often overlooked. This metabolite, a predictor of toxicity, contributes significantly to the total sulphonamide by an additional 5–23% or 53–77% of serum content in those with normal or impaired renal function respectively. Where total sulphonamide was held below 100 mg/L by adaptive control, efficacy was maintained with tolerable sequelae without need for therapy change (Joos et al., 1995).

A pharmacokinetic model was devised recently to predict loading and maintenance doses of co-trimoxazole for individual AIDS patients to achieve a narrower target

Table III. High dosage co-trimoxazole regime: therapeutic range for serum concentrations

	Trimethoprim (mg/L)	Sulphamethoxazole (mg/L)
Pre-dose	5^a or $5–7^b$	<100 or $<120^a$
Post-dose	$5–8^a$ or $(8–10)^b$	$50–150$ or $120–150^a$
Potentially toxic	>20	>200

The pre-dose sample should be taken no more than 30 min before the next dose. The post-dose sample should be taken 1.5–2 h after an oral or iv dose.
[a]In serious infection such as *P. carinii* pneumonia (Sattler *et al.*, 1988).
[b]Steady-state levels following infusion in renal failure (Sattler & Jelliffe, 1994).

range with a trimethoprim peak after the completion of a 1.5 h infusion, of 10 mg/L, with trough of 5–7 mg/L at 8 h post-dose (Sattler & Jelliffe, 1994)

Renal impairment

When creatinine clearance is <30 mL/min in patients receiving co-trimoxazole, protein binding decreases with increase in concentration and $t_{1/2}$ of the less polar N^4-acetyl metabolite and the parent drug. Monitoring of 12 h post-dose serum samples of sulphamethoxazole is advised every 2–3 days during treatment. Where total sulphamethoxazole is >150 mg/L, no further co-trimoxazole should be given until this is <120 mg/L. Where clearance is <15 mL/min, co-trimoxazole use is not supported unless dialysis facilities are available. To achieve a narrower target range for trimethoprim in renally impaired patients, an algorithm has been produced which may be applicable in those with stable chronic renal failure (Sattler & Jelliffe, 1994).

References

Allen, J. G. & Nimmo-Smith, R. H. (1978). Trimethoprim. In *Laboratory Methods in Antimicrobial Chemotherapy* (Reeves, D. S., Phillips, I., Williams, J. D. & Wise, R. Eds), pp. 227–31. Churchill Livingstone, Edinburgh.

Amyes, S. G. B. & Towner, K. J. (1990). Trimethoprim resistance; epidemiology and molecular aspects. *Journal of Medical Microbiology* 31, 1–19.

Ascalone, V. (1980). Assay of trimethoprim, sulfamethoxazole and its N^4-acetyl metabolite in biological fluids by high-pressure liquid chromatography. *Journal of High Resolution Chromatography and Chromatography Communications* 3, 261–4.

Ascalone, V. (1981). Assay of trimethoprim, sulfadiazine and its N^4-acetyl metabolite in biological fluids by normal-phase high-performance liquid chromatography. *Journal of Chromatography* 224, 59–66.

Astbury, C. & Dixon, J. S. (1987). Rapid method for the determination of either plasma sulphapyridine or sulphamethoxazole

and their acetyl metabolites using high-performance liquid chromatography. *Journal of Chromatography* 414, 223–7.

Bozkurt, A., Basci, N. E., Isimer, A., Tuncer, M., Erdal, R. & Kayaalp, S. O. (1990). Sulphamethoxazole acetylation in fast and slow acetylators. *International Journal of Clinical Pharmacology, Therapy and Toxicology* 28, 164–6.

Bratton, A. C. & Marshall, E. K. (1939). New coupling component for sulphanilamide determination. *Journal of Biological Chemistry* 128, 537–50.

Burchall, J. J. (1979). The development of the diaminopyrimidines. *Journal of Antimicrobial Chemotherapy* 5, *Suppl. B*, 3–14.

Burchall, J. J. & Hitchings, G. H. (1965). Inhibitor binding analysis of dihydrofolate reductases from various species. *Molecular Pharmacology* 1, 126–36.

Bury, R. W. & Mashford, M. L. (1979). Analysis of trimethoprim and sulfamethoxazole in human plasma by high-pressure liquid chromatography. *Journal of Chromatography* 63, 114–7.

Bushby, S. R. M. (1969). Combined antibacterial action *in vitro* of trimethoprim and sulphonamides. The *in vitro* nature of synergy. *Postgraduate Medical Journal* 45, *Suppl.*, 10–8.

Bushby, S. R. M. & Hitchings, G. H. (1968). Trimethoprim, a sulphonamide potentiator. *British Journal of Pharmacology and Chemotherapy* 33, 72–90.

Climax, J., Lenehan, T. J., Lambe, R., Kenny, M., Caffrey, E. & Darragh, A. (1986). Interaction of antimicrobial agents with human peripheral blood leucocytes: uptake and intracellular localization of certain sulphonamides and trimethoprims. *Journal of Antimicrobial Chemotherapy* 17, 489–98.

Committee on Safety of Medicines. (1985). Deaths associated with co-trimoxazole, ampicillin and trimethoprim. *Current Problems* 15, 1–2.

Cribb, A. E., Lee, B. L., Trepanier, L. A. & Spielberg, S. P. (1996). Adverse reactions to sulphonamide and sulphonamide–trimethoprim antimicrobials: clinical syndromes and pathogenesis. *Adverse Drug Reactions and Toxicological Reviews* 15, 9–50.

Domagk, G. (1935). Ein beitrag zur chemotherapie der bakteriellen Infektionen. *Deutsche Medizinische Wochenschrift* 61, 250–3.

Dudley, M. N., Levitz, R. E., Quintiliani, R., Hickingbotham, J. M. & Nightingale, C. H. (1984). Pharmacokinetics of trimethoprim and sulfamethoxazole in serum and cerebrospinal fluid of adult patients with normal meninges. *Antimicrobial Agents and Chemotherapy* 26, 811–4.

Fong, I. W. (1988). Correlation of side effects of trimethoprim/sulfamethoxazole with blood levels in AIDS patients treated for *Pneumocystis carinii* pneumonia. In *Program and Abstracts of the Twenty-eighth Interscience Conference on Antimicrobial Agents and Chemotherapy, Los Angeles, California, 1988*. Abstract 1226, p. 328. American Society for Microbiology, Washington, DC.

Gautam, S. R., Chungi, V. S., Bourne, D. W. A. & Munson, J. W. (1978). HPLC assay for trimethoprim in plasma and urine. *Analytical Letters* B11, 967–73.

Goehl, T. J., Mathur, L. K., Strum, J. D., Jaffe, J. M., Pitlick, W. H., Shah, V. P. *et al.* (1978). Simple high-pressure liquid chromatographic determination of trisulfapyrimidines in human serum. *Journal of Pharmaceutical Sciences* 67, 404–6.

Gordin, F. M., Simon, G. L., Wofsy, C. B. & Mills, J. (1984). Adverse reactions to trimethoprim–sulphamethoxazole in patients with the acquired immunodeficiency syndrome. *Annals of Internal Medicine* 100, 495–9.

Haagsma, N. & van de Water, C. (1985). Rapid determination of five sulphonamides in swine tissue by high-performance liquid chromatography. *Journal of Chromatography* 333, 256–61.

Hekster, Y. A., Vree, T. B., Damsma, J. E. & Friesen, W. T. (1981). Pharmacokinetics of sulphametrole and its metabolite N^4-acetylsulphametrole in man. *Journal of Antimicrobial Chemotherapy* **8**, 133–44.

Hughes, D. T. D. (1997a). Diaminopyrimidines. In *Antibiotic and Chemotherapy*, 7th edn (O'Grady, F., Lambert, H. P., Finch, R. & Greenwood, R. G., Eds), pp. 346–56. Churchill Livingstone, Edinburgh.

Hughes, D. T. D. (1997b). Sulphonamides. In *Antibiotic and Chemotherapy*, 7th edn (O'Grady, F., Lambert, H. P., Finch, R. & Greenwood, R. G., Eds), pp. 460–8. Churchill Livingstone, Edinburgh.

Hughes, W. T. & Killmar, J. (1996). Monodrug efficacies of sulfonamides in prophylaxis for *Pneumocystis carinii* pneumonia. *Antimicrobial Agents and Chemotherapy* **40**, 962–5.

Hughes, W. T., Feldman, S. & Sanyal, S. K. (1975). Treatment of *Pneumocystis carinii* pneumonitis with trimethoprim–sulfamethoxazole. *Canadian Medical Association Journal* **112**, Suppl., 47–50.

Hughes, W. T., Feldman, S., Chaudhary, S. C., Ossi, M. J., Cox, F. & Sanyal, S. K. (1978). Comparison of pentamidine isoethionate and trimethoprim–sulphamethoxazole in the treatment of *Pneumocystis carinii* pneumonia. *Journal of Pediatrics* **92**, 285–91.

Joos, B., Blaser, J., Opravil, M., Chave, J.-P. & Luthy, R. (1995). Monitoring of co-trimoxazole concentrations in serum during treatment of *Pneumocystis carinii* pneumonia. *Antimicrobial Agents and Chemotherapy* **39**, 2661–6.

Locher, H. H., Schlunegger, H., Hartman, P. G., Angehrn, P. & Then, R. L. (1996). Antibacterial activities of epiroprim, a new dihydrofolate reductase inhibitor, alone and in combination with dapsone. *Antimicrobial Agents and Chemotherapy* **40**, 1376–81.

Lovering, A. M. & Shi, Y. G. (1990). A simple, straightforward HPLC method for the simultaneous assay of trimethoprim and sulphamethoxazole (co-trimoxazole). In *Infection 1990 Programme, Abstracts and Participants,* Abstract 28, Association of Medical Microbiologists, London.

McIntosh, S. J., Platt, D. J., Watson, I. D., Guthrie, A. J. & Stewart, M. J. (1983). Liquid chromatographic assay for trimethoprim in sputum and saliva. *Journal of Antimicrobial Chemotherapy* **11**, 195–6.

Nissenson, A. R., Wilson, C. & Holazo, A. (1987). Pharmacokinetics of intravenous trimethoprim–sulphamethoxazole during hemodialysis. *American Journal of Nephrology* **7**, 270–4.

Nouws, J. F. M., Vree, T. B. & Hekster, Y. A. (1985). In vitro antimicrobial activity of hydroxy- and N^4-acetyl sulphonamide metabolites. *Veterinary Quarterly* **7**, 70–2.

Paap, C. M. & Nahata, M. C. (1989). Clinical use of trimethoprim/sulfamethoxazole during renal dysfunction. *DICP* **23**, 646–54.

Reeves, D. S. & Wilkinson, P. J. (1979). The pharmacokinetics of trimethoprim and trimethoprim/sulphonamide combinations, including penetration into body tissues. *Infection* **7**, Suppl. 4, S330–41.

Reeves, D. S., Bywater, M. J. & Holt, H. A. (1978). Sulphonamides. In *Laboratory Methods in Antimicrobial Chemotherapy* (Reeves, D. S., Phillips, I., Williams, J. D. & Wise, R., Eds), pp. 222–6. Churchill Livingstone, Edinburgh.

Rieder, M. J., Sisson, E., Bird, I. A. & Almawi, W. Y. (1992). Suppression of T-lymphocyte proliferation by sulphonamide hydroxylamines. *International Journal of Immunopharmacology* **14**, 1175–80.

Sattler, F. R. & Jelliffe, R. W. (1994). Pharmacokinetic and pharmacodynamic considerations for drug dosing in the treatment of *Pneumocystis carinii* pneumonia. In *Pneumocystis carinii Pneumonia*, 2nd Edn (Walzer, P. D., Ed.), pp. 467–85. Marcel Dekker, New York.

Sattler, F. R., Cowan, R., Nielsen, D. M. & Ruskin, J. (1988). Trimethoprim–sulphamethoxazole compared with pentamidine for treatment of *Pneumocystis carinii* pneumonia in the acquired immunodeficiency syndrome. A prospective, non crossover study. *Annals of Internal Medicine* **109**, 280–7.

Schneider, M. M. E., Hoepelman, A. I. M., Eeftinck Schattenkerk, J. K. M., Nielsen, T. L., van der Graaf, Y., Frisson, J. P.. *et al.* (1992). A controlled trial of aerosolized pentamidine or trimethoprim–sulphamethoxazole as primary prophylaxis against *Pneumocystis carinii* pneumonia in patients with human immunodeficiency virus infection. *New England Journal of Medicine* **327**, 1836–41.

Schwartz, D. E., Koechlin, B. A. & Weinfeld, R. E. (1969). Spectrofluorimetric method for the determination of trimethoprim in body fluids. *Chemotherapy* **14**, Suppl., 22–9.

Siber, G. R., Gorham, C. C., Ericson, J. F. & Smith, A. L. (1982). Pharmacokinetics of intravenous trimethoprim–sulphamethoxazole in children and adults with normal and impaired renal function. *Reviews of Infectious Diseases* **4**, 566–78.

Stokes, E. J., Ridgway, G. L. & Wren, M. W. D. (Eds) (1993). *Clinical Microbiology*, 7th edn, pp. 276–8. Edward Arnold, London.

Study Group. (1989). Trimethoprim–polymyxin B sulphate ophthalmic ointment versus chloramphenicol ophthalmic ointment in the treatment of bacterial conjunctivitis—a review of four clinical studies. *Journal of Antimicrobial Chemotherapy* **23**, 261–6.

van der Ven, A. J. A. M., Koopmans, P. P., Vree, T. B. & van der Meer, J. W. M. (1991). Adverse reactions to co-trimoxazole in HIV infection. *Lancet* **338**, 431–3.

van der Ven, A. J. A. M., Koopmans, P. P., Vree, T. B. & van der Meer, J. W. M., (1994). Drug intolerance in HIV disease. *Journal of Antimicrobial Chemotherapy* **34**, 1–5.

Vergin, H., Bishop-Freudling, G. B., Kaiser, W., Köhler, M., Strobel, K. & Reutter, F. W. (1983). Pharmacokinetics of metioprim in normal subjects and patients with impaired renal function. *Antimicrobial Agents and Chemotherapy* **24**, 190–3.

Vohringer, H. F. & Arasteh, K. (1993). Pharmacokinetic optimisation in the treatment of *Pneumocystis carinii* pneumonia. *Clinical Pharmacokinetics* **24**, 388–412.

Vree, T. B. & Hekster, Y. A. (1987). Clinical pharmacokinetics of sulfonamides and their metabolites: an encyclopedia. *Antibiotics and Chemotherapy* **37**, 1–214.

Vree, T. B., Hekster, Y. A., Baars, A. M., Damsma, J. E. & van der Kleijn, E. (1978a). Determination of trimethoprim and sulfamethoxazole (co-trimoxazole) in body fluids of man by means of high-performance liquid chromatography. *Journal of Chromatography* **146**, 103–12.

Vree, T. B., Hekster, Y. A., Baars, A. M., Damsma, J. E. & van der Kleijn, E. (1978b). Pharmacokinetics of sulphamethoxazole in man: effects of urinary pH and urine flow on metabolism and renal excretion of sulphamethoxazole and its metabolite N_4-acetylsulphamethoxazole. *Clinical Pharmacokinetics* **3**, 319–29.

Vree, T. B., Tijhuis, M. W., Hekster, Y. A. & Nouws, J. F. M. (1986). HPLC analysis, antimicrobial activity and pharmacokinetics of sulfonamides and their N_4-acetyl and N_1-hydroxysulphonamide derivatives in man. In *HPLC in Medical Microbiology* (Reeves, D. S. & Ullman, U., Eds), pp. 165–78. Gustav Fisher Verlag, New York.

Walker, S. E., Paton, T. W., Churchill, D. N., Ojo, B., Manuel, M. A. & Wright, N. (1989). Trimethoprim–sulfamethoxazole pharmacokinetics during continuous ambulatory peritoneal dialysis (CAPD). *Peritoneal Dialysis International* **9**, 51–5.

Weinfeld, R. E. & Macasieb, T. C. (1979). Determination of trimethoprim in biological fluids by high-performance liquid chromatography. *Journal of Chromatography* **164**, 73–84.

Woolley, J. L., Ragouzeous, A., Brent, D. A. & Sigel, C. W. (1980). Sulfonamide crystalluria: isolation and identification of sulfa-methoxazole and four metabolites in urinary calculi. In *Current Chemotherapy and Infectious Disease* (Nelson, J. D. & Grassi, C., Eds), pp. 552–4. American Society for Microbiology, Washington, DC.

Index

Index